0123
4567
8910

Mathematics
Teaching

WITHDRAWN

0123 4567 8910

Mathematics Teaching

KENNETH J. TRAVERS
University of Illinois at Urbana-Champaign

LEN PIKAART
Ohio University

MARILYN N. SUYDAM
The Ohio State University

GARTH E. RUNION
Illinois State University

HARPER & ROW, PUBLISHERS
New York   Hagerstown   San Francisco   London

Sponsoring Editor: George A. Middendorf
Project Editor: Robert Ginsberg
Designer: T. R. Funderburk
Production Supervisor: Kewal K. Sharma
Compositor: Monotype Composition Company, Inc.
Printer and Binder: Halliday Lithograph Corporation
Art Studio: Vantage Art, Inc.

Mathematics Teaching

Library of Congress Cataloging in Publication Data
Main entry under title:

Mathematics teaching.

    Bibliography: p.
    Includes index.
    1.   Mathematics —Study and teaching.   I.   Travers, Kenneth J.
QA11.M396   510′.7′1   76-54749
ISBN 0-06-045233-1

To our parents

Those with us . . .
And those who are not.

Contents

Preface

Is it possible to become an excellent teacher merely by reading a book? Emphatically, we say no! The most valuable learning about how to teach effectively seems to come from watching skilled teachers, practicing under the guidance of excellent supervisors, and learning from one's own experiences. But all teachers, and particularly beginning teachers, need to devote time to planning and evaluating as well as to teaching. We have written this book with the purpose of helping you develop competence in all three of these fundamental aspects of instruction: planning, teaching, and evaluating. We have included many other topics which we believe to be essential for effective teaching, such as classroom management, individualizing, and continuing professional development. All topics are illustrated by means of examples designed to assist you both as a student teacher and as an in-service teacher.

In Chapter 1 we present a view of mathematics teaching which will be helpful as you develop your own response to the question "Why teach mathematics?" In Chapter 2 we present a model for mathematics teaching based upon Bloom's Taxonomy of Educational Objectives. We have found that this conceptual framework provides a useful guide to the decision-making which goes on in day-to-day teaching. This model is developed throughout the book.

This book has several chapters which we believe to be distinctive and especially relevant to mathematics teaching today. Chapter 9 is devoted to the use of computers in the mathematics classroom; four major instructional uses of computers are discussed and sample programs are given for each. Chapters 3–4 and Appendix C provide an introduction to the BASIC programming language, programming exercises, and several longer programs of use to the classroom teacher.

In Chapter 10 we discuss the nature of research in mathematics education and the use of research in your own teaching. Appendix A is an annotated bibliography of research that will be of particular significance to secondary school mathematics teaching. Throughout the book, "Research Highlights" provide brief summaries of research relevant to the emphases of each chapter. Through the Research Highlights alone you will become acquainted with a substantial body of research in secondary school mathematics education; by combining the resources of the Research Highlights, Appendix A, and Chapter 10, you will have the essential ingredients of a sound background for a professional user of research—and perhaps a producer of research as well.

Another novel feature of the book is the use of Critical Incidents taken from our own experience to create a setting for study and discussion in each chapter. We strongly believe that much of the important learning about how to teach results from making decisions, proposing solutions, and analyzing the results of our actions in actual or simulated problem situations. We expect that you will have access to actual teaching situations as you study and practice methods of teaching. But we are also convinced that there is much to be gained from realistic simulated problem situations which demonstrate the specific goals of a given chapter. Therefore, each chapter begins with a list of its major goals. You can use these advance organizers as cues for what to look for in the chapter and as a self-check, when you have completed the chapter, to ensure that you have studied all the important points.

Appendix B provides a variety of problem-solving activities and Appendix D lists resources which can be used to enrich your mathematics teaching.

A book is the product of the efforts and the influence of many persons, of whom we can name only a few here. Our own teachers and professors have had a profound effect on us, and to them we express our appreciation. The comments by Jeremy Kilpatrick and James W. Wilson, of the Department of Mathematics Education, University of Georgia, and Kenneth Retzer, of the Department of Mathematics, Illinois State University, on early versions of this manuscript were of great assistance. Michael Covington at the University of Georgia and Jeffry S. Gordon at the University of Illinois, Urbana, provided suggestions on the sections dealing with computers, and Valerie Keighan at Illinois State University assisted in the compilation of Appendix D. Special thanks are due those who typed the manuscript, in particular Leonora P. Bradner, Beverly Brooks, Barbara J. Bullock, and Marilyn Parmantie.

Finally, we express our appreciation to our families: Janny, Stacey, Wendy, Kelly, Connie, Len Jr., Bill, Lori, Cindy, Laura, Amy, and Mindy. They helped us maintain the faith that this book would eventually be completed.

0123
4567
8910

Mathematics
Teaching

Why Teach Mathematics?

AFTER STUDYING CHAPTER 1, YOU WILL HAVE:

★ reviewed some of the practical and professional aspects of mathematics.
★ considered certain disciplinary values of mathematics.
★ surveyed some of the cultural aspects of mathematics.
★ received suggestions for developing your personal philosophy of mathematics teaching.

CRITICAL INCIDENTS

1. Willy struggles with the quadratic formula for most of algebra class, but can't make it "work." Willy angrily slams his book and mutters, "What good is all this stuff, anyway?"
2. John has decided to be a mathematics teacher. He has always liked mathematics and done well in his mathematics courses. But he knows many people who do not like mathematics and find it difficult. He wonders how he will communicate his enthusiasm to his future students.
3. Margo, who has been teaching mathematics for three months, finds herself in the middle of a heated discussion in her freshman algebra class. Several of her best students suddenly announce that they are not going to take any more mathematics, since only one year is required. Margo is at a loss as to

what to say; she has always signed up for mathematics courses without even asking herself why.

4. Bill is having his first interview for a teaching position. After a few preliminaries, the assistant superintendent looks over his glasses at Bill and says, "Tell me, young man, what is *your* philosophy of mathematics teaching?"

WELCOME TO MATHEMATICS TEACHING!

Since you are reading this book, we assume you have made the important decision. You want to be a mathematics teacher. Welcome to this exciting and challenging profession! If you are like most people, you have mixed feelings about what teaching holds for you. We hope that this book will help you sort out your thinking about teaching and prepare you to be successful in the classroom.

You probably have a lot of questions about teaching. Why be a teacher at all? In particular, why be a mathematics teacher? Questions which your students might ask may bother you even more. "Why do I have to study mathematics?" Or, as Willy asks, "What good is all this stuff?" Most of you probably have asked these questions on more than one occasion. Let's consider some of the answers that you may have heard.

ANSWER 1: "MATHEMATICS IS USEFUL"

Ever since the schools of the ancient Greeks, over 2000 years ago, mathematics has been a key subject in the curriculum. The four liberal arts—the *quadrivium*, consisting of arithmetic, geometry, astronomy, and music—were basically mathematical studies. Schooling was not intended for the masses of people who did the work of the world, but for the select few who were to rule. The origins of the term *liberal education* lie in Aristotle's belief that education should be liberal (that is, for free men) rather than suited for practical or vocational purposes (Butts, 1955, p. 58). Hence our conception today views a liberal education as the opposite of an education specifically designed to prepare one to earn a living.

Education in the early American colonies was influenced by this elitist view. Schools were primarily college preparatory, and the curriculum of the first American college, Harvard, reflected the elements of the medieval liberal arts: grammar, rhetoric, logic, arithmetic, geometry, and astronomy (Butts, 1955, p. 264). But as commerce and industry became important to the livelihood of North Americans, there were demands for schools to provide an education responsive to society's practical needs: Teach navigation! Give training in accounting! We need suveyors! To do this, Benjamin Franklin established academies, the best known of which later became the University of Pennsylvania. Franklin proposed three departments in his schools, English, Latin, and mathematics. Students were permitted to select studies on the basis of the occupations for which they were preparing from a wide variety of subjects: rhetoric, logic, history, sciences, agriculture, gardening, and mechanics. Mathematics remained central to this curriculum with geometry, arithmetic, and trigonometry taught as they applied to the trades (see Figure 1–1).

The criterion of usefulness raises problems, however, for times change more quickly than do curriculums. Benjamin, in *The Saber-tooth Curriculum* (1939), tells a parable of a primitive society which develops a curriculum for its school. Only the truly useful and essential subjects are to be taught, including "woolly-horse-clubbing" to provide hides for clothing and "fish-grabbing-by-the-bare-hands" to permit the catching of fish from the village stream. Then time passes. Woolly horses become extinct. The village stream becomes muddy, then dries up. But the schools continue to teach woolly-horse-clubbing and fish-grabbing-by-the-bare-hands since, as Benjamin observes, "You must know that there are some eternal verities, and the saber-tooth curriculum is one of them!" (p. 44).

RESEARCH HIGHLIGHT | WHAT ARITHMETIC SHALL WE TEACH?

In the 1930s, Guy Wilson and his associates (Wilson and Dalrymple, 1937) conducted a survey of the business and social usage of fractions in order to help determine the content of the mathematics curriculum for schools. The occurrence of various fractions in the work of persons in department stores, hotels, and other businesses was tabulated. Wilson concluded his investigations by noting:

> Needed mastery in fractions for common usage is limited to halves, thirds, fourths, eighths, and twelfths. Crossing of denominators seldom goes further than halves with fourths. Subtraction of fractions seldom occurs. . . . Division of a fraction by a fraction almost never occurs. Indications are that this simple program of mastery in fractions can be best accomplished through an objective, non-manipulative procedure. Any further program in fractions should be left to learning on the job, when and if needed; they are no part of the grade task of the schools. (p. 347)

A difficulty with the matter of practicality, therefore, is that what is useful today may not be useful tomorrow. Price (1961) has observed that the teaching of trigonometry and of logarithms provides examples of this. College graduates 300 years ago in the American colonies were likely to be sea captains, surveyors, or ministers. The sea captain needed trigonometry for navigation, the surveyor used it for establishing property lines for new communities, and the minister needed the subject for his study of astronomy and for calculating the date for Easter (p. 9). Logarithms were essential for performing complex calculations until the past decade. But with the advent of electronic calculators and computers, their importance has diminished (see Figure 1–2). Both topics do have importance, however, when studied as functions, and they will undoubtedly remain in the high

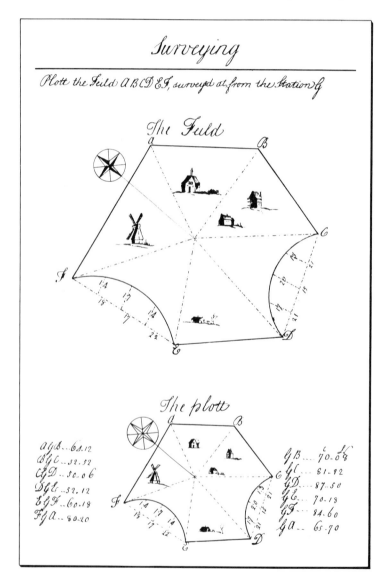

Figure 1–1 Uses of mathematics in surveying and astronomy are featured in these pages from a manuscript written about the time that Franklin's academies were established. The manuscript, in the rare books collection of the University of Illinois, Urbana, gives no indication of authorship, but bears the title inscription, "Mathematics and Its Applications, Date: About 1756." The book was acquired by the University in 1919.

Astronomy

Problem: 1ˢᵗ

To find the latitude of a place by the Circumpolar Stars.

Rule: Half the difference between the greatest and least Altitudes, (if they are both on the same side the Zenith) added to the least, or subtract from the greatest Altitudes, will be the Latitudes of the place

But if the said Altitudes are on different Sides of the Zenith, than half the difference between the least Altitudes & the Supplement of the greatest, added to the least will be the latitudes of the place

Ex: 1ˢᵗ

An Observer takes the lowest Alt: of a star 8.30 & the highest 46.40 both on the same side of the Zenith. Quare what lat: he is in

Ex: 2:

An Observer takes the lowest Altitude of a star 8.30, and its highest 46.40 both on diff: Sides of the Zenith, what Lat: is he in

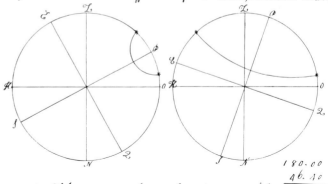

			180.00
			46.40
greatest Alt:	46.40	Suppᵗ: greatest Alt:	133.20
Least D:̊	8.30	Least Alt:	8.30
Difference	38.10	Difference	124.50
½ diff:	19.05	½ difference	62.25
Least Alt:	8.30	Least Altit:	8.30
Latitude of the observer	27.35 N	Latitude of the observer	70.55 N

school curriculum, taught with this new emphasis. Similarly, although relatively little use may be made of common fractions in daily living (see Research Highlight, "What Arithmetic Shall We Teach?"), particularly as conversion to the metric system of measurement becomes universal, they are still an important

Figure 1–2 The computer has made possible new approaches to mathematics teaching. (Photo courtesy Illinois State University Photographic Services.)

component of the curriculum, since the study of rational numbers provides an essential background for the study of algebra.

It is important to distinguish between immediate and eventual usage. Some topics in mathematics can be applied by the student here and now: percentage, for example, for determining purchase value at a sale, and series, when computing costs of borrowing money. Other topics—such as the general solution for the quadratic equation—may not have immediate application, but form part of the foundation upon which more powerful mathematics can be built. Many mathematical concepts themselves have been formulated without immediate application, later to become essential conceptual tools in other areas of study. Complex numbers arose through work in algebra but remained mysterious and obscure to most persons, as the original term *imaginary* implies. However, functions of complex variables are now used in the analysis of electronic circuitry, of wing lift in aerodynamics, and of water seepage in dams (Aleksandrov et al., 1963, p. 5).

From geometry we have the example of the work of Lobachevsky, whose contributions were so profound and revolutionary that he has been called "the Copernicus of geometry" (Bell, 1937, p. 294). Lobachevsky challenged the truth of Euclid's Fifth Postulate,* assumed to be reasonable or necessary by scholars for 2000 years or more. Lobachevsky was careful to label his new geometry *imaginary* because he could not see any meaning for it in the actual world, even though he was confident that such meaning would be found. Eventually, theories

---

* Euclid's Fifth Postulate may be stated: "If a straight line falling on two other lines makes the interior angles on the same side of the first line sum to less than 180 degrees, then the two lines if extended will intersect on the side of the transversal containing the angles."

of non-Euclidean spaces became the basis for the theory of relativity (Aleksandrov et al., 1963, p. 6).

We have seen, therefore, that mathematics is a useful, indeed an essential component of society. However, this utility may not be apparent to our students without appropriate emphases made by the teacher (see Research Highlight, "Of What Use Is It?").

RESEARCH HIGHLIGHT | OF WHAT USE IS IT?

Bell (1970) conducted an investigation which focused on the uses of mathematics. As one phase of his study, he analyzed verbal problem material in pre- and post-1960 algebra textbooks. Data about such factors as the context used in stating a problem, the mathematical model used, and the reality of the problem were tabulated. He came to the same "gloomy conclusion" that Thorndike reached in a similar study in the 1920s: "In general, teachers and youngsters who seek understanding of or information about the usefulness of mathematics will get little help from school algebra books."

Bell proceeded to carry out a limited study during a full year with an algebra class and found that "a substantial emphasis on the uses of mathematics can be made to harmonize with a variety of objectives within that course." Thus, with some careful work, the uses of mathematics can be made apparent to students.

ANSWER 2: "MATHEMATICS DISCIPLINES THE MIND"

One of the oldest theories of education is that of mental discipline. According to this belief, the mind is a muscle which is strengthened by mental exercise. Jeremiah Day, while president of Yale College, spoke for the mental discipline tradition of teaching mathematics when he declared in the preface of his introductory algebra book, published in 1823:

> If the design of studying the mathematics were merely to obtain such a knowledge of the practical parts, as is required for transacting business; it might be sufficient to commit to memory some of the principal rules, and to make the operations familiar, by attending to the examples. . . . But a higher object is proposed, in the case of those who are acquiring a liberal education. The main design should be to call into exercise, to discipline, to invigorate the powers of the mind. It is the *logic* of the mathematics which constitutes their principal

value, as a part of a course of collegiate instruction. The time and attention
devoted to them, is for the purpose of forming *sound reasoners*, rather than expert
mathematicians. (pp. iii–iv)

The style and flavor of Day's textbook, as suggested by the sample page on the
nature of division, bears out his convictions of the true worth of mathematical
study (Figure 1–3). Mathematics is done because of the great amount of practice
it affords the mind; it is a sort of mental jogging to build up the mind and keep it
fit. Accordingly, mathematics textbooks—the barbells and skipping ropes for the
mind—were designed to provide this needed exercise.

Let's talk about mental discipline. There is a measure of truth in it, to be sure.
When students are learning mathematical skills, they need repetition and practice
as they do in learning any other skills. The design and use of strategies for the
teaching of skills will be developed fully in Chapter 3. The point we wish to
make here is that skill learning does not transfer automatically, that when students
practice a skill they are learning only that one, not another.

Early in this century, when mental discipline was a prevailing point of view in
teaching, researchers took a close look at the exercise-the-mind or mental discipline view of learning. Thorndike (1924) looked for evidence that studying one
subject would tend to help a student achieve more in another subject (see Research Highlight, "Mental Discipline in High School Studies"). No such evidence
was found. Educational psychologists today point out that if you want learning to
transfer from one subject to another, you must teach that way. If you want
students to learn about proof in both geometry and algebra, for example, you
should do proofs in both areas, discuss common characteristics, and point out
differences. On the other hand, if you teach mathematics only because you
believe that somehow it "sharpens the mind," the main result could well be that
students learn only that the subject is dull, repetitious, and difficult.

Now that we have disposed of mental discipline as a viable theory of education, we must hasten to say that there is disciplinary value in the study of mathematics—in the development of sound work habits, the capacity to work independently, and the acquiring of problem-solving skills and strategies. There is
indeed a wealth of self-discipline which attends the analysis of a problem, the
identification of what is given and what is to be solved, the selection of a strategy
to solve the problem, and the interpretation of the obtained results. And the
resulting sense of accomplishment can be enormously satisfying to student and
teacher alike. Perhaps it is the disciplinary value of mathematics which has led to
its use in some instances as a "proving ground." Personnel departments, for
example, have been known to use success in trigonometry or college algebra as
a criterion for screening candidates for management positions in their companies.

There is too the discipline imposed by mathematics itself—such as the study
of mathematical structures and of deductive proof. One instance of the power
of mathematical deduction was the discovery of the distant planet Neptune
(Aleksandrov et al., 1963, p. 4). This planet is so far away that it had escaped

RESEARCH HIGHLIGHT | MENTAL DISCIPLINE IN HIGH SCHOOL STUDIES

In the early 1920s, Thorndike (1924) studied nearly 9000 students in an attempt to isolate the effects of studying one particular school subject on general gain in achievement. The research was conducted in order to examine the validity of the commonly held belief that certain subjects, notably Latin, should be studied because of their value in improving the "thinking ability" of pupils. Thorndike concluded his rather extensive investigation by stating:

> By any reasonable interpretation of the results, the intellectual values of studies should be determined largely by the special information, habits, interests, attitudes, and ideals which they demonstrably produce. The expectation of any large differences in general improvement of the mind from one study rather than another seems doomed to disappointment. The chief reason why good thinkers seem superficially to have been made such by having taken certain school studies, is that good thinkers have taken such studies, becoming better by the inherent tendency of the good to gain more than the poor from any study. When the good thinkers studied Greek and Latin, these studies seemed to make good thinking. Now that the good thinkers study Physics and Trigonometry, these seem to make good thinkers. If the abler pupils should all study Physical Education and Dramatic Art, these subjects would seem to make good thinkers. These were indeed a large fraction of the program of studies for the best thinkers the world has produced, the Athenian Greeks. After positive correlation of gain with initial ability is allowed for, the balance in favor of any study is certainly not large. Disciplinary values may be real and deserve weight in the curriculum, but the weights should be reasonable. (p. 98)

The results of this important study were later corroborated by Wesman (1945).

detection by the usual telescopic techniques. In 1846 the astronomers Adams and Leverrier noticed irregularities in the motion of the planet Uranus; they made mathematical calculations and concluded that these deviations were caused by the gravitational attraction of another planet. Leverrier then proceeded to calculate the location of such a planet and communicated his conclusions to a colleague who was doing observations with a telescope. Sure enough, the unknown planet was discovered exactly where Leverrier, on the basis of his calculations, said it would be.

One further aspect of discipline must be mentioned. Mathematics, with its

DIVISION.   **47**

PROMISCUOUS EXAMPLES.

1. Divide $12aby + 6abx - 18bbm + 24b$, **by** $6b$.
2. Divide $16a - 12 + 8y + 4 - 20adx + m$, by 4.
3. Divide $(a - 2h) \times (3m + y) \times x$, by $(a - 2h) \times (3m + y)$.
4. Divide $ahd - 4ad + 3ay - a$, by $hd - 4d + 3y - 1$.
5. Divide $ax - ry + ad - 4my - 6 + a$, by $-a$.
6. Divide $amy + 3my - mxy + am - d$, by $-dmy$.
7. Divide $ard - 6a + 2r - hd + 6$, by $2ard$.
8. Divide $6ax - 8 + 2xy + 4 - 6hy$, by $4aaxy$.

129. From the nature of division it is evident, that the value of the quotient depends both on the divisor and the dividend. With a given divisor, the greater the dividend, the greater the quotient. And with a given dividend, the greater the divisor, the less the quotient. In several of the succeeding parts of algebra, particularly the subjects of fractions, ratios, and proportion, it will be important to be able to determine what change will be produced in the quotient, by increasing or diminishing either the divisor or the dividend.

If the given dividend be 24, and the divisor 6; the quotient will be 4. But this same dividend may be supposed to be multiplied or divided by some *other* number, before it is divided by 6. Or the *divisor* may be multiplied or divided by some other number, before it is used in dividing 24. In In each of these cases, the quotient will be altered.

130. In the first place, if the given divisor is contained in the given dividend a certain number of times, it is obvious that the same divisor is contained,

In *double* that dividend, *twice* as many times;
In *triple* the dividend, *thrice* as many times, &c.

That is, if the divisor remains the same, *multiplying the dividend* by any quantity, is, in effect, *multiplying the quotient* by that quantity.
Thus, if the constant divisor is 6, then $24 \div 6 = 4$ the quotient.

Multiplying the dividend by 2, $2 \times 24 \div 6 = 2 \times 4$
Multiplying by any number n $n \times 24 \div 6 = n \times 4$

Figure 1–3 A page on division from Day's *Introduction to Algebra* (1823). (Courtesy Rare Book Room, University of Illinois, Urbana.)

endless possibilities for providing long, difficult, and tedious exercises, has found itself to be a favorite source of "disciplinary" assignments for misbehaving students: "Johnny, because you were talking in class today you are to do ten extra problems from the C exercises." Such a classroom management technique has clear implications for Johnny, of course: He is being told that mathematics is a form of punishment. Apparently his teacher feels that way about the subject, too.

In defense of our teaching specialty, we would do well to mount a crusade against using mathematics as a source of punitive exercises. But it is sobering to reflect upon the many reasons why mathematics has been so often thus chosen.

ANSWER 3: "MATHEMATICS IS BEAUTIFUL"

The beauty of mathematics has long been an important justification for studying the subject (see Figure 1–4). Perhaps the most ardent artist-mathematician was G. A. Hardy, who wrote the captivating little book, A *Mathematician's Apology* (1940), in which he tries to identify the main reasons why anyone should study mathematics. Hardy asserts: "The mathematicians' patterns, like the painters' or

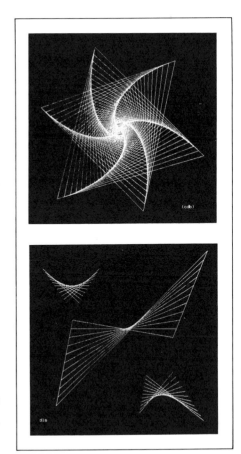

Figure 1–4 Graphic displays on PLATO, the teaching computer system of the University of Illinois, Urbana. The PLATO system is described in more detail in Chapter 9. (Courtesy Computer-Based Education Research Laboratory, University of Illinois, Urbana.)

the poets', must be *beautiful*; the ideas, like the colours or the words, must fit together in a harmonious way. Beauty is the first test: there is no permanent place in the world for ugly mathematics" (p. 24).

While Hardy is aware of the beauty of art, or of poetry, it is the beauty of ideas which attracts him to mathematics. To illustrate his conviction, he refers to two famous theorems of Greek mathematics, calling them "simple" but "of the highest class" (p. 32). They are Euclid's proof that there are infinitely many prime numbers and Pythagoras's proof that $\sqrt{2}$ is irrational (pp. 32–36).

It is essential that when we teach mathematics, we make every effort to ensure that the power of the ideas we are dealing with does not escape our students. As Hardy has shown, the examples need not be complex to be persuasive. One powerful idea, for instance, comes from that of chance outcomes such as the tossing of a balanced coin. No matter how many times the coin is tossed, the probability that the next toss is heads remains .5. But toss 10,000 such times and one can be confident that the number of heads and the number of tails will be divided about evenly. The problems presented in this book, notably in Chapter 5 and in Appendix B, provide a rich source of important, beautiful ideas in mathematics which can be grasped and appreciated by high school students.

One of the enormous rewards of teaching is that students become fascinated by ideas. Abbott's *Flatland* (1963), an account of living in a world of only two dimensions, and Gamow's *One, Two, Three . . . Infinity* (1961), which contains a wealth of material that includes a chapter about the world of four dimensions, are two references you will find useful. Perhaps one of the best collections of mathematical ideas presented in a form suitable for many secondary school students is Jacobs's *Mathematics, A Human Endeavor; A Textbook for Those Who Think They Don't Like the Subject* (1970). You will also want to refer to our list of valuable resource books for beginning mathematics teachers (pp. 419–421).

The attractiveness of ideas is not fixed, but varies from student to student and is dependent upon such factors as the way in which the topic is taught. For example, Price (1961) observed that the study of heat flow and the distribution of temperatures in a solid body, a problem of great importance in modern technology, was too advanced to be studied in high school. However, now that computers are available in schools, a solution to the problem is accessible using random walk procedures (see, for example, Nievergelt et al., 1974, pp. 158–160). A program written by a high school student to solve one version of the heat flow problem appears in Appendix C (pp. 537–538).

Another aspect of the beauty of mathematics is found in the world around us. The hexagons of snowflakes, the hexagonal prisms of the honeycomb, and the crystal structures of minerals are all excellent subjects for geometric study (see Figure 1–5). The occurrence of patterns in nature has held the fascination of mathematicians and nonmathematicians as well (see Figure 1–6). And there is the Golden Section, known by the ancient artists, architects, and sculptors

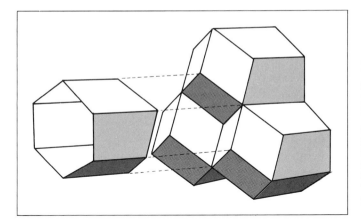

Figure 1–5 Much mathematics can be gleaned from a study of the cells of a honeycomb. Each cell is a hexagonal prism with a trihedral base formed by three congruent rhombi. It can be shown using methods of the calculus that for a given volume the minimum surface of the prism is achieved when the acute angles of the rhombus are 70° 32′, which is the angle the bees use in constructing their cells. (For more on the mathematics of the honeycomb, see Siemens, 1965, 1967.)

and still of interest today (Figure 1–7). The Golden Section is formed in the segment AB when the point C in $\overline{AB}$ is such that

$$\frac{AB}{AC} = \frac{AC}{CB}$$

The value of each of the ratios in the proportion is frequently called τ (tau) and has an approximate value of 1.618. A Golden Rectangle is a rectangle the ratio of whose longer side to the shorter side is τ (see Figure 1–8). One useful reference on the Golden Section, suitable for both high school students and teachers, is Runion (1972).

DEVELOPING THE "BIG PICTURE"

Mathematics teachers can be expected to be able to make statements about why they teach mathematics, why mathematics is important, and how students learn mathematics. Much of this book is devoted to helping you formulate such statements. We have just considered some reasons for the importance of mathematics in the secondary school curriculum. We suggest that these reasons might form a basis upon which Bill, in our fourth critical incident, could develop a statement concerning his own philosophy of mathematics teaching.

TEACHING MATHEMATICS

All of the persons in the critical incidents for this chapter have one problem in common: They need to know how mathematics fits into the "big picture" of daily living. Alfred North Whitehead, a famous mathematician and educator who spent a great deal of time talking to mathematics teachers about sound education

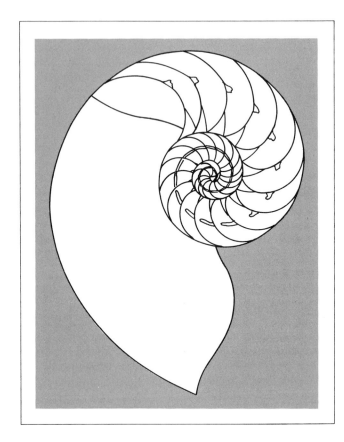

Figure 1–6   The spiral formed by a Chambered Nautilus very closely approximates that of a logarithmic or equiangular spiral which has a constant curvature. This spiral also manifests itself in the shape of certain animal horns as well as in the shape of galaxies. (See Lockwood, 1967, and Maor, 1974, for more on the logarithmic spiral.)

principles, once warned about this primary danger facing the teacher. He observed (1956) that the teacher must avoid presenting mathematics as isolated "scraps of gibberish." He protested the fragmentation which often occurs in teaching, the presentation of bits and pieces of information, rules, and formulas without any clear picture of where they fit into the overall scheme of things. Students may cram their minds full of this "inert knowledge" and reproduce the material on a test, but for many of them the entire procedure is without apparent meaning or purpose.

During the curriculum reform movement of the 1960s (treated in Chapter 13 of this book), significant steps were taken to remedy the fragmentation problem.

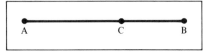

Figure 1–7   Point C forms the Golden Section on segment AB.

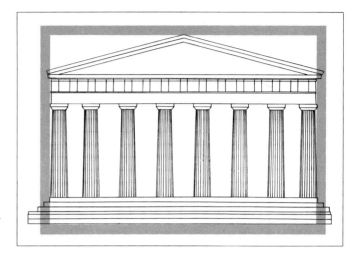

Figure 1–8 The Parthenon in Athens, Greece, comes very close to fitting into a Golden Rectangle.

Many contemporary textbooks employ a "spiral" approach that begins early in the child's education. The student meets major mathematical ideas (number, measure, variable, function, etc.) and returns to them repeatedly at successively more complex levels. But the teacher needs to continue the task of integrating mathematics with other elements of the school experience. Helping the student to realize the relationships of the various components of mathematics to each other and to other fields of learning is challenging—and vital.

Teaching consists of more than merely telling students what you think they should know. Students must be learners—and they need the opportunity to learn in many ways. The teacher can become a guide, a resource person, one who helps to create proper conditions for learning, and one who provides inspiration and leadership. The growing popularity of mathematics laboratories has been one response to a need to provide richer and more meaningful learning environments for all students, not just those who have a talent for or interest in mathematics (mathematics laboratories are discussed in Chapter 8 of this book).

In the second critical incident, John is concerned about sharing his feelings toward mathematics. Whitehead may be helpful here; he talks about a love and enthusiasm for mathematics, together with a respect, virtually a reverence, for the student:

> The environment within which the mind is working must be carefully selected. It must, of course, be chosen to suit the child's stage of growth, and must be adapted to individual needs. In a sense it is an imposition from without: but in a deeper sense it answers to the call of life within the child. In the teacher's consciousness the child has been sent to his telescope to look at the stars, in the child's consciousness he has been given free access to the glory of the heavens. (Whitehead, 1956, pp. 43–44)

So whoever would teach must surely have some commitment both to the subject and to those who are to be taught. Small likelihood that the teacher will ever have the child use the telescope if the teacher has never gazed in wonder at the stars!

We might summarize the chapter thus far by attempting to respond to Willy, who wants to know "What good is all this stuff, anyway?" You may not be surprised to find there is no single right answer to Willy's question. One important approach for the teacher is to try to determine what Willy is *actually* saying. He may be saying simply that he is frustrated, or that he wants the teacher's attention. He has to say something to somebody, so he asks a question for which he may not be expecting an answer.

The fact that Willy has asked this sort of question is an important cue, even if you cannot provide an answer which will satisfy Willy (or you). Willy is telling you how he feels, and it is important that you know how each of your students feels about what he or she is doing. A good response from you might be to help reduce Willy's anxiety and frustration. Give him something to do in mathematics

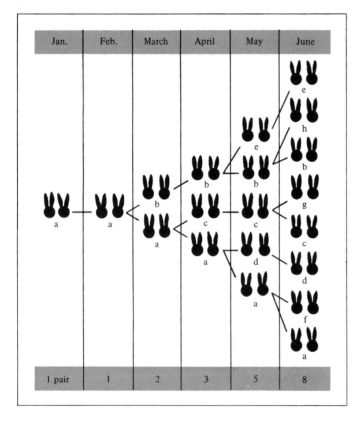

Figure 1–9 The Rabbit Problem. Another kind of beauty found in mathematics is that derived from numerical and geometric patterns which arise in various settings. This problem was posed by Leonardo of Pisa (Fibonacci): Suppose we have two rabbits (a male and a female) such that (1) the rabbits begin to produce young two months after their own birth, and (2) after reaching the age of two months each pair of rabbits produces another mixed pair every month thereafter. How many pairs of rabbits will there be at the end of 12 months? The pattern of resulting pairs (1,1,2,3,5,8,13, . . .) forms the famous Fibonacci sequence. This same sequence can be found in another familiar pattern of numbers known as Pascal's Triangle by adding entries along the diagonals as indicated. Pascal's Triangle is also the source for quite a number of other patterns (see Huntley, 1970, for some of these).

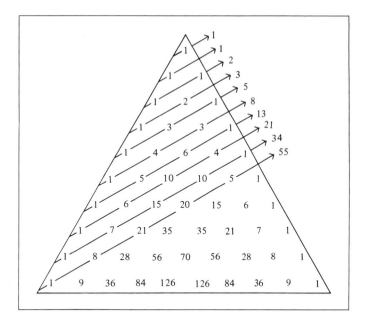

which will give him a successful experience. Or have him tell you (maybe in the form of a story problem or an activity) what part of mathematics is important to him. You might try to find out what Willy's interests are and then try to gear some class activities along those lines. But whatever you do, it is essential that you recognize Willy's existence as an individual who has feelings, needs, and goals.

TEACHING MATHEMATICS TO PEOPLE

Yes, you want to teach mathematics, and we have explored justifications for the traditionally important role which mathematics has had in the curriculum. But there is another aspect of teaching mathematics—teaching mathematics *to people*. Not only have you chosen a career in mathematics, but you have chosen teaching from among the many mathematics-oriented careers which are available today. You have said "I want to work with people." There is something within you that places a priority on having a part in helping others grow—intellectually, socially, and emotionally.

In this book we have tried to integrate research knowledge, educational principles, theories, and our own experience-based hunches in order to provide the groundwork for those who want to teach mathematics. Someday a classroom door will close behind you and you will be alone with and responsible for two or three dozen young people—you will be their teacher. Your decisions, your actions, your attitudes will affect each person in that room. The importance of teachers in the lives of students is portrayed dramatically, and tragically, in the following true story, written by a former high school teacher and guidance counselor.

Cipher in the Snow

BY JEAN MIZER TODHUNTER

It started with tragedy on a biting cold February morning. I was driving behind the Milford Corners bus as I did most snowy mornings on my way to school. It veered and stopped short at the hotel, which it had no business doing, and I was annoyed as I had to come to an unexpected stop. A boy lurched out of the bus, reeled, stumbled, and collapsed on the snowbank at the curb. The bus driver and I reached him at the same moment. His thin, hollow face was white even against the snow.

"He's dead," the driver whispered.

I didn't register for a minute. I glanced quickly at the scared young faces staring down at us from the school bus. "A doctor! Quick! I'll phone from the hotel. . . ."

"No use. I tell you he's dead." The driver looked down at the boy's still form. "He never even said he felt bad," he muttered, "just tapped me on the shoulder and said, real quiet, 'I'm sorry. I have to get off at the hotel.' That's all. Polite and apologizing like."

At school, the giggling, shuffling morning noise quieted as the news went down the halls. I passed a huddle of girls. "Who was it? Who dropped dead on the way to school?" I heard one of them half whisper.

"Don't know his name; some kid from Milford Corners," was the reply.

It was like that in the faculty room and the principal's office. "I'd appreciate your going out to tell the parents," the principal told me. "They haven't a phone, and anyway, somebody from the school should go there in person. I'll cover your classes."

"Why me?" I asked. "Wouldn't it be better if you did it?"

"I didn't know the boy," the principal admitted levelly. "And in last year's sophomore personalities column I noted that you were listed as his favorite teacher."

I drove through the snow and cold down the bad canyon road to the Evans place and thought about the boy, Cliff Evans. His favorite teacher! Why, he hasn't spoken two words to me in two years! I could see him in my mind's eye all right, sitting back there in the last seat in my afternoon literature class. He came in the room by himself and left by himself. "Cliff Evans," I muttered to myself, "a boy who never talked." I thought for a minute. "A boy who never smiled. I never saw him smile once."

The big ranch kitchen was clean and warm. I blurted out my news somehow. Mrs. Evans reached blindly toward a chair. "He never said anything about bein' ailing."

His stepfather snorted. "He ain't said nothin' about anything since I moved in here."

Mrs. Evans got up, pushed a pan to the back of the stove, and began to untie her apron. "Now hold on," her husband snapped. "I got to have breakfast before I go to town. Nothin' we can do now anyhow. If Cliff hadn't been so dumb, he'd have told us he didn't feel good."

After school I sat in the office and stared bleakly at the records spread out before me. I was to close the boy's file and write his obituary for the school paper. The almost bare sheets mocked the effort. "Cliff Evans, white, never legally adopted by stepfather, five half brothers and sisters." These meager strands of information and the list of D grades were about all the records had to offer.

Cliff Evans had silently come in the school door in the mornings and gone out the school door in the evenings, and that was all. He had never belonged to a club. He had never played on a team. He had never held an office. As far as I could tell, he had never done one happy, noisy kid thing. He had never been anybody at all.

How do you go about making a boy into a zero? The grade school records showed me much of the answer. The first and second grade teachers' annotations read "sweet, shy child"; "timid but eager." Then the third grade note had opened the attack. Some teacher had written in a good, firm hand, "Cliff won't talk. Uncooperative. Slow learner." The other academic sheep had followed with "dull"; "slow-witted"; "low IQ." They became correct. The boy's IQ score in the ninth grade was listed at 83. But his IQ in the third grade had been 106. The score didn't go under 100 until the seventh grade. Even timid, sweet children have resilience. It takes time to break them.

I stomped to the typewriter and wrote a savage report pointing out what education had done to Cliff Evans. I slapped a copy on the principal's desk and another in the sad, dog-eared file; slammed the file; and crashed the office door shut as I left for home. But I didn't feel much better. A little boy kept walking after me, a boy with a peaked face, a skinny body in faded jeans, and big eyes that had searched for a long time and then had become veiled.

I could guess how many times he'd been chosen last to be on a team, how many whispered child conversations had excluded him. I could see the faces and hear the voices that said over and over, "You're dumb. You're dumb. You're just a nothing, Cliff Evans."

A child is a believing creature. Cliff undoubtedly believed them. Suddenly it seemed clear to me: When finally there was nothing left at all for Cliff Evans, he collapsed on a snowbank and went away. The doctor might list "heart failure" as the cause of death, but that wouldn't change my mind.

We couldn't find 10 students in the school who had known Cliff well enough to attend the funeral as his friends. So the student body officers and a committee from the junior class went as a group to the church, looking politely sad. I attended the service with them and sat through it with a lump of cold lead in my chest and a big resolve growing in me.

I've never forgotten Cliff Evans or that resolve. He has been my challenge year after year, class after class. Each September, I look up and down the rows carefully at the unfamiliar faces. I look for veiled eyes or bodies scrouged into a seat in an alien world. "Look kids," I say silently, "I may not do anything else for you this year, but not one of you is going out of here a nobody. I'll work or fight to the bitter end doing battle with society and the school board, but I won't have one of you leaving here thinking yourself into a zero."

Most of the time—not always, but most of the time—I've succeeded.

WHAT DO *YOU* TEACH?

There is an old conundrum which asks "Do you teach mathematics or do you teach children?" The answer is that you teach mathematics to children. Mathematics is useful, has disciplinary values, and is beautiful. But you teach mathematics to students so that they will find it useful and be able to apply it in day-to-day situations. You teach mathematics so that young people will come to realize the disciplinary aspects of mathematics, learn skills of mathematical problem-solving, and attain good work habits. You teach mathematics so that others can see its beauty and have their lives made richer and more meaningful. You help to see that students do not lose career opportunities because they lack the proper background, that students are more adequately prepared for living in our complex technological society—in short, that students become better educated. All this, and more, can be your contribution to humanness through your chosen profession of teaching.

Activities

1. Suppose you were to develop a mathematics curriculum based solely on the criterion of usefulness. How would you define usefulness? What

general topics might be included, say, in such a junior high school course?

2. Recall the discussion of mental discipline and consider your own experiences in mathematics classes both in school and college. To what extent is mental discipline still a commonly held theory for justifying mathematics teaching?

3. Mathematics is an important subject in many vocations. Try to list those vocations that definitely will *not* use mathematics 20 years from now. What doors will be closed to the student who doesn't study mathematics now in preparation for future work?

4. Identify examples from your own experience (think of your home town or your college campus) which illustrate the beauty of mathematics. What mathematical principles are used, for example, in architecture or in city planning or in the fine arts?

5. Assume that you are minister of mathematics education in a totalitarian state. Devise a mathematics curriculum which will provide the necessary knowledge for the following vocations:
   a. social worker
   b. sales manager
   c. artist
   d. carpenter
   To what extent would a general curriculum for all vocations be appropriate? What sort of specialization would be needed?

6. Imagine that you are a member of a mathematics department and wish to introduce a new topic, such as probability and statistics, to the curriculum.
   a. Identify a topic of your choosing. Justify your choice: why is the topic an important one to include in the curriculum?
   b. What *unique* contribution would the topic make to the student's education?
   c. How would you decide what to remove from the curriculum to make room for your new topic?

7. As an exercise in broadening your background in the role of mathematics in the sciences, make a survey of magazines, journals, and books to identify areas of applications. Find examples which would be of interest to secondary school students. You may want to design a bulletin-board display on Mathematics: Queen and Servant of the Sciences. Perhaps you can interview various persons—technicians, scientists, physicians— to obtain firsthand information on mathematical applications in various fields.

8. List three reasons which Margo, in the third critical incident, could give her freshmen for continuing the study of mathematics throughout high school. Compare your reasons with those prepared by fellow students

in your class. Identify those reasons which would be valued most highly by high school students.

9. Bill, in the fourth critical incident, had a common experience in his interview for his first teaching position. Put yourself in his shoes and prepare a statement of about 200 words which summarizes your "philosophy of mathematics teaching." Tell why you believe mathematics teaching is important; the statement might include the reasons for your decision to become a mathematics teacher. Keep this statement for future reference—at the end of the present semester, or after your first year of teaching, you will be able to see how your perceptions of teaching change with experience.

References

Abbott, Edwin A. *Flatland: A Romance of Many Dimensions*. New York: Barnes and Noble, 1963.

Aleksandrov, A. D. et al., eds. *Mathematics, Its Content, Methods and Meaning*, Vols. 1–3. Cambridge, Massachusetts: M.I.T. Press, 1963.

Bell, Eric T. *Men of Mathematics*. New York: Simon and Schuster, 1937.

Bell, Max S. Studies with Respect to the Uses of Mathematics in Secondary School Curricula. (The University of Michigan, 1969.) *Dissertation Abstracts International* 30A: 3813–3814; March 1970.

Benjamin, Harold. *The Saber-tooth Curriculum*. New York: McGraw-Hill, 1939.

Bergamini, David et al., eds. *Mathematics*. Life Science Library. New York: Simon and Schuster, 1956.

Butts, R. Freeman. *A Cultural History of Western Education*. New York: McGraw-Hill, 1955.

Day, Jeremiah. *An Introduction to Algebra: Being the First Part of a Course of Mathematics. Adopted to the Methods of Instruction in the American Colleges*. 3rd ed. New Haven, Connecticut: Howe and Spalding, 1823.

Gamow, George. *One, Two, Three . . . Infinity*. Rev. ed. New York: Viking, 1961.

Hardy, G. H. *A Mathematician's Apology*. Cambridge, England: The University Press, 1940.

Huntley, H. E. *The Divine Proportion. A Study in Mathematical Beauty*. New York: Dover Publications, 1970.

Jacobs, Harold R. *Mathematics, A Human Endeavor; A Textbook for Those Who Think They Don't Like the Subject*. San Francisco: Freeman, 1970.

Lockwood, E. H. *A Book of Curves*. New York: Cambridge University Press, 1967.

Maor, Eli. The Logarithmic Spiral. *Mathematics Teacher* 67: 321–327; April 1974.

Nievergelt, Jurg; Farrar, J. Craig; and Reingold, Edward M. *Computer Approaches to Mathematical Problems*. Englewood Cliffs, New Jersey: Prentice-Hall, 1974.

Price, G. Baley. Progress in Mathematics and Its Implications for the Schools. In *The Revolution in School Mathematics*. Washington, D.C.: National Council of Teachers of Mathematics, 1961, pp. 1–14.

Runion, Garth E. *The Golden Section and Related Curiosa*. Glenview, Illinois: Scott, Foresman, 1972.

Siemens, David F., Jr. The Mathematics of the Honeycomb. *Mathematics Teacher* 58: 334–337; April 1965.

Siemens, David F., Jr. Of Bees and Mathematicians. *Mathematics Teacher* 60: 758–761; November 1967.

Thorndike, E. L. Mental Discipline in High School Studies. *Journal of Educational Psychology* 15: 1–22; January 1924; 15: 83–98; February 1924.

Todhunter, Jean Mizer. Cipher in the Snow. *Today's Education* 64: 66–67; March–April 1975.

Wesman, A. G. A Study of Transfer of Training from High School Subjects to Intelligence. *Journal of Educational Research* 39: 254–264; December 1945.

Whitehead, Alfred N. *Aims of Education.* New York: New American Library, 1956.

Wilson, Guy and Dalrymple, Charles O. Useful Fractions. *Journal of Educational Research* 30: 341–347; January 1937.

A Model of
Mathematics Teaching

AFTER STUDYING CHAPTER 2, YOU WILL HAVE:

★ learned a way to conceptualize and categorize the goals in teaching mathematics.
★ examined a model which is based on the interrelationships of the goals, processes, and content of mathematics teaching.
★ gained a perspective on mathematics teaching which includes a consideration of what students believe about their capabilities and mathematics learning.

CRITICAL INCIDENTS

1. Neal's homework papers are peculiar. He refuses to attempt any of the word problems assigned but completes almost all of the computation exercises. When 12 factoring exercises were assigned from page 115, Neal factored all of the 60 trinomials on the page and missed only two! On another day he turned in a page with only his name and the problem numbers for the two word problems assigned.

This chapter is adapted from an article which appeared in *Arithmetic Teacher* (Pikaart and Travers, 1973). © 1973 National Council of Teachers of Mathematics. Used by permission.

2. An assignment early in the year in a geometry course calls for justifying each statement in a proof. At the beginning of the next class period, Ann says she thinks the assignment was silly: "Anyone can look at the figure and see that the theorem is always true."
3. At a PTA meeting Bob Brown's father says, "I don't know why Bob is doing poorly. He can't learn the quadratic formula. I know it, but I can't seem to help him. What do you think I should do?"

TOWARD A THEORY

Does every mathematics teacher have a theory of teaching mathematics? If asked, all teachers can provide some description of what they teach, how they teach, and how well their students learn. Such responses would be reflections of the individual teacher's theory. They would indicate what mathematics the teacher views as important and, therefore, what mathematics he or she teaches; they would reveal the teacher's beliefs about how students should be taught and about how well students can learn the subject. But does the teacher have a theory? What is a *theory* of teaching mathematics?

If by *theory* we mean a complete characterization of teaching mathematics, then such a theory does not exist. If it did, it would contain sets of hypotheses, principles, and other generalizations which would serve as guides for mathematics teachers in all situations. Two characteristics of a useful theory are completeness and consistency, and these are characteristics which hold important roles in mathematics. A complete theory would provide a basis for deciding what to do in every situation—that is, a situation could not occur in teaching mathematics such that the theory would not be applicable. A consistent theory would provide a single course of action in every situation covered by the theory—that is, in a given situation, the theory would not specify both an action and a denial or contradiction of the action. Goedel's Proof, now considered a classic in mathematics, demonstrated the futility of attempting to develop a complete and consistent mathematical system (Nagel and Newman, 1956, pp. 1668–1695), and the analogy to developing a complete and consistent theory of mathematics teaching is obvious. However, just as mathematicians continue to develop useful, consistent mathematical systems, albeit incomplete, so it is reasonable for mathematics educators to continue to develop a useful, consistent theory of mathematics teaching. Such an incomplete theory might be called a *model*.

WHAT IS A MODEL?

A model represents a system and enables testing of the system in its components. A system is sometimes difficult to communicate, so models are used to help in the communication. Models can be physical constructions, like wind tunnels and space capsules, or they can be symbolic representations.

Models are used in many different aspects of daily life. When a salesman explains the cost of financing a new car, he might use a model as simple as the familiar equation, Interest = Principal × Rate × Time. Or he might refer to a chart that gives the total amount paid and the cost of the loan for any particular number of payments. Charts, tables, and diagrams all may be used as models for depicting the relationships among interest rates, purchase price, and the number and amount of payments.

In psychology, a familiar model is one used for depicting stimulus-response learning. For example, recall the experiment in which Pavlov found that dogs, which salivated at the sight of food, began salivating at the sound of a bell that preceded the appearance of the food. Diagramatically, this phenomenon can be expressed as in Figure 2–1 (Gage, 1963, p. 95). The stimulus S_1 (food) brings about the response R (the production of saliva). Eventually the dog learns that the sound of a bell (S_2) precedes food, and S_1 is replaced by S_2.

These examples are simple. For a phenomenon such as learning the long division algorithm, the models become more complex; but they also become more essential for a clear analysis and discussion of what is entailed in such phenomena.

MODELS OF EDUCATIONAL OBJECTIVES

A model that has become a classic in education is one devised by Bloom and his associates for classifying goals of learning (Bloom, 1956; Krathwohl et al., 1964; Bloom, Hastings, and Madaus, 1971). This model involves a two-dimensional framework, with content or subject matter as one dimension and student performance as the other dimension. There are six behavioral levels in the cognitive domain: knowledge, comprehension, application, analysis, synthesis, and evaluation. In the affective domain, there are five levels: receiving, responding, valuing, organization, and characterization by a value or value complex.

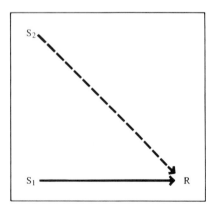

Figure 2–1 Artificial stimulus.

A model similar to the cognitive portion of Bloom's scheme was developed for NLSMA, the National Longitudinal Study of Mathematical Abilities (Begle and Wilson, 1970). Where the Bloom model was designed for use in all content areas and in different phases of the instructional process, the NLSMA model was developed only for evaluating mathematics achievement. The levels of behavior in the NLSMA scheme are similar to those of the cognitive dimension in the Bloom model, except that the classification has been reduced to four levels: computation, comprehension, application, and analysis. Another adaptation of the cognitive aspects of the Bloom model has been proposed by Avital and Shettleworth (1968), who reduced the classification to five levels by incorporating evaluation in the categories of analysis and synthesis. Excellent discussions and comparisons of these models and others are provided by Weaver (1970, pp. 355–366) and Begle and Wilson (1970, pp. 367–404). For a comprehensive description of an application of the Bloom model in evaluating secondary school mathematics learning, the reader is referred to Wilson (1971, pp. 643–696).

With the notable exception of the Bloom model and its application by Wilson, little attention has been paid to that vastly important array of educational objectives which are affective in nature. We believe that any analysis of mathematics teaching must take into account attitudes, emotions, and beliefs, as well as such cognitive components as comprehension and application (see Research Highlight, "Your Attitude Affects!").

RESEARCH HIGHLIGHT | YOUR ATTITUDE AFFECTS!

Many studies have been designed to investigate the relationship between the attitudes of teachers and the attitudes and achievement of students. Among these studies is one by Phillips (1970), who hypothesized that attitudes toward mathematics are developed over a period of years. He therefore looked at data comparing the attitudes and achievement of students with the attitudes of their teachers for the previous three years. Most-recent-teacher attitude toward mathematics was found to be significantly related to student attitude, but not to achievement in mathematics.

The type of teacher attitude encountered by the student for either two or three of the previous three years, however, was found to be significantly related to the student's present attitude toward mathematics—and to student achievement as well. The students with the highest attitudes were in the classes of teachers with highly positive attitudes toward mathematics.

A SIMPLIFIED MODEL

As used by Begle and Wilson (1970), the term *model* denotes "an organizational framework; it represents a categorization system with some stated rules and relations for using the system" (p. 372). Thus, even if there is at present little hope of constructing a complete theory, there exists the prospect of developing useful models of mathematics teaching. The mathematics teachers who respond to questions about their teaching probably base their replies on some model or mental construct of mathematics teaching. This model may be implicit or explicit, but it is likely that each teacher has somehow organized his or her teaching tasks in his or her mind. Such a model may be expected to be incomplete; hopefully, it will not be inconsistent.

In developing the proposed model, we were motivated to find a scheme that would help teachers describe specific learning goals that would be comprehensive, flexible, simple, and functional. The model should be comprehensive in that it should take into account every worthwhile goal of secondary school mathematics. A flexible model is one that can be adapted to different points of view concerning what mathematics should be taught, how it should be taught, and how students learn. The model should be simple in the sense that it should include only a small number of categories. However, it should also be functional in the sense that the categories should be natural classifications that can be readily applied in the teaching process—that is, the model should be useful to teachers.

As in the models just described, ours takes into account two dimensions of education: content and learner behavior or goals. However, we add a third dimension: teacher behavior or process—what the teacher does to try to bring about the desired changes in the learner. These three components are represented in the model in Figure 2–2.

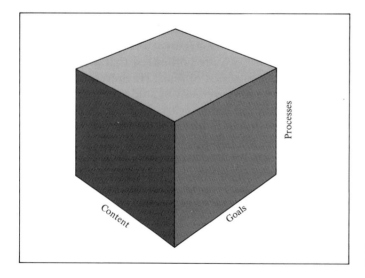

Figure 2–2 Overview of the model of mathematics teaching.

This chapter will develop and describe a particular model for mathematics teaching in order to provide a framework which will be useful throughout the remainder of the text, as the teaching and the learning of mathematics are analyzed and discussed. You have the options of accepting, rejecting, or modifying our model. Ultimately, we hope, each teacher will develop his or her own model; we present this one as a starting point. All of the components will be fully described later. In many ways, this model is similar to those proposed by others and contains characteristics of many of them (see Bloom, Hastings, and Madaus, 1971, pp. 5–41).

Acquiring a sound model of mathematics teaching does not guarantee that a prospective teacher will be successful in a classroom—but it helps. The model provides a frame of reference within which the teacher can contemplate and discuss his subject with students, colleagues, and parents. The model also provides a framework for the teacher in decision-making situations. Teachers must make decisions as they plan for instruction, they must make new decisions as they teach, and they must make still more decisions as they evaluate the effectiveness of their teaching. Figure 2–3 depicts some of the considerations, decisions, and activities of teaching; it highlights many of the topics we will consider in this book.

GOALS IN SECONDARY SCHOOL MATHEMATICS

Here we will elaborate the goals dimension of our model of mathematics teaching. These goals will be classified by characteristic similarities, and these classifications will serve as organizers for examining common teaching strategies in Chapters 3–6. The three major classifications of goals (see Figure 2–4) are:
- Knowledge
- Understanding
- Problem Solving

These classifications will serve as headings under which we may examine the goals of mathematics teaching—both cognitive (subject-matter oriented) goals and affective (emotional or belief) goals. There are several classification schemes of goals in mathematical teaching, but ours is particularly useful because each classification is directly associated with a set of planning activities, with a set of instructional strategies, and with a set of evaluation techniques. In the following subsections, each classification will be defined and described. (These same goal classifications will be discussed in much greater detail in Chapters 3–6, where emphasis will be placed on selecting appropriate goals, strategies of teaching, and techniques for evaluating attainment of each set of goals.)

In the description of the goal classifications, the basic referent is an instructional unit. A unit is considered to be a sequence of lessons with a central theme. The length of time needed to teach a unit may vary from a single class period to several weeks. Authors of secondary school mathematics textbooks organize

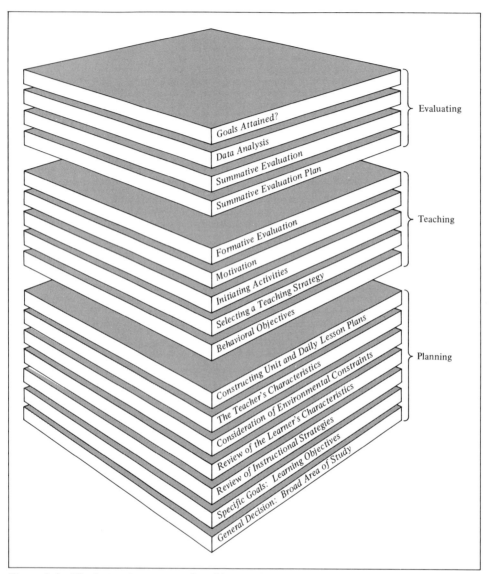

Figure 2–3   The instructive process of teaching a unit.

courses into chapters and sections of chapters. Either these sections or the chapters themselves may be considered as instructional units. For example, units in Algebra I might deal with such topics as: number systems, open sentences, graphing, quadratics, or simultaneous equations and inequalities. Each of these could be divided into smaller instructional units and into daily lessons. The choice

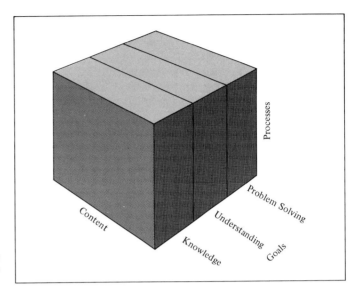

Figure 2–4 The model of mathematics teaching with the three major classifications of the goals dimension.

of an instructional unit as a basis for classifying goals is somewhat arbitrary. Clearly, goals can and should be specified for entire courses, for instructional units, and for daily lessons.

KNOWLEDGE GOALS

The first classification of goals is that of *knowledge*. Goals at this level are most often the easiest to think about, to relate, to describe, to teach, and to evaluate. (Note that we make no mention of the ease of learning these goals.) The general category of knowledge may be naturally partitioned into two subclasses, *facts* and *skills*. Facts are the specific statements the students are to recognize or recall in the unit. In a sense, these are a "know that" part of knowledge. Skills, on the other hand, are the procedures, often computational algorithms, in which the student is to become proficient during the unit. These are the "know how" part of knowledge. Both facts and skills have been categorized as knowledge because achievement in both types of goals depends on recognition or recall of facts or basic skills. For example, a student may know a statement of the distributive property, of the Pythagorean theorem, or of the quadratic formula; that is, he can write or recite the statement when given an appropriate stimulus. In particular, a student could make a statement equivalent to one of the following:

▶ *Distributive Property*: A system S with operations $\oplus$ and $\otimes$ is said to have a distributive property of $\otimes$ over $\oplus$ provided that for any x, y, and z in S,

$$x \otimes (y \oplus z) = (x \otimes y) \oplus (x \otimes z)$$

▶ *Pythagorean Theorem:* In any right triangle the square of the measure of the hypotenuse is equal to the sum of the squares of the measures of the two legs.

▶ *Quadratic Formula:* If $ax^2 + bx + c = 0$ is any quadratic equation such that a, b, and c are real numbers and $a \neq 0$, then the elements of the solution set are

$$\frac{-b + \sqrt{b^2 - 4ac}}{2a}$$

and

$$\frac{-b - \sqrt{b^2 - 4ac}}{2a}$$

In each of these examples, student learning could simply be the ability to answer correctly a request to state the property or the theorem—such as "Tell me what it means for an operation $\otimes$ to distribute over an operation $\oplus$ in a number system" or "What is a statement of the Pythagorean Theorem?" Later it will be seen that *understanding* statements such as these imply that the student is able to apply the statement; for now, our focus is on knowledge of the statement alone.

It is also important to note that the assertions we call "facts" are facts in a slightly restricted sense—to the extent that they are universally true in the system considered. For example, the Pythagorean Theorem is not true when applied to spherical triangles. The statement of the theorem given above could be modified to include a quantifier such as "in any plane right triangle," but this is usually understood by the student and the teacher. The need for such quantifiers is best determined by the teacher.

Skills, used here as a category, is intended to imply basic skills. (Higher order skills such as proving theorems or solving problems are reserved for the higher level category of *Problem Solving.*) Basic skills may be characterized as the ability to solve routine exercises. However, it is important to note that "basic" is a relative term—relative to the level and ability of the student. Finding the derivative of a polynomial function might be a basic skill for a college student but an understanding or problem solving goal for an advanced student in secondary school. Similarly, at one stage in an Algebra I course, factoring a polynomial would be a basic skill; but earlier in the course the same problem might have been an application (understanding) of the distributive property. Skills are usually based on understanding; students learn certain generalizations which can then be applied in a routine way to develop skills. Thus the classification scheme presented here does *not* indicate the sequence in which learning takes place.

Another characteristic of basic skills is that they are often reduced to simple computational algorithms. Many examples are to be found in arithmetic, such as the addition, subtraction, multiplication, and division of whole numbers, integers, or rational numbers. You are certainly familiar with the algorithms indicated in these exercises:

$$
\begin{array}{r}
317 \\
\times 24 \\
\hline
1268 \\
634 \\
\hline
7608
\end{array}
$$

$$
\begin{array}{r}
348 \;\; R6 \\
23)\overline{8010} \\
-69 \\
\hline
111 \\
-92 \\
\hline
190 \\
-184 \\
\hline
6
\end{array}
$$

$(-7) + 5 = -2$

$(-3) \times (-9) = 27$

$\frac{1}{2} \div \frac{2}{3} = \frac{1}{2} \times \frac{3}{2}$
$\qquad = \frac{3}{4}$

$\frac{1}{2} + \frac{2}{3} = \frac{3}{6} + \frac{4}{6}$
$\qquad = \frac{7}{6}$
$\qquad = 1\frac{1}{6}$

Solving these exercises clearly requires skills. It is simply not efficient to solve them in other than a routine way. Similarly, secondary school mathematics courses involve many basic skills such as factoring, graphing, drawing sketches, making estimates of answers, performing synthetic division, completing the square of a quadratic, or finding the determinant of a matrix. With each of these, if the teacher wants his students to be able to use readily a particular procedure, then the teacher's task is to teach a skill.

UNDERSTANDING GOALS

To demonstrate that he has acquired an understanding, a student would not simply state it; he would show that he could apply the understanding in a particular situation. Thus the term *understanding* is defined as a level of learning a general statement which enables a student to apply it in problem situations. Here are two examples:

▶ The students are to understand that $x!$ means $x(x - 1)(x - 2) \cdots \cdots 2 \cdot 1$.
▶ The students are to understand that if two parallel lines are cut by a transversal, the alternate interior angles are congruent.

There are many kinds of understandings. Some of the common types in mathematics are theorems, generalizations, and principles. All these are placed under the general classification of "understandings" and are defined to be *statements* which can be applied. In this section, different types of understandings will be distinguished; however, it will be useful to consider first how understandings are stated.

To state an understanding, a teacher should be able to write a clear, precise statement of that understanding. Too often teachers say that they teach their students to understand some topic in mathematics. What could it mean "to understand number" or "to understand rational numbers" or "to understand functions?" To appreciate the vagueness of such phrases, consider several different ways students could, for example, "apply rational numbers" or "apply understandings of rational numbers." The possibilities are many. The problem is that phrases like "understand rational numbers" do not specify what the student is to understand. Other examples of vague specification of understandings are statements which imply the understanding of a theorem, principle, or generalization. These statements may be precise if the speaker and the listener are in agreement about the meaning of the theorem, principle, or generalization. Most mathematics teachers would comprehend statements such as "My students understand the Pythagorean Theorem" or "Bill understands the Fundamental Theorem of Arithmetic." It would probably be clear in the first example that the students understand (are able to apply in a problem situation) the statement "that if ABC is a triangle in a plane and the angle at C is a right angle, then

$$AB^2 = BC^2 + AC^2$$

where the notation AB means the measure of line segment AB."

A help in making precise statements of an understanding is to say ". . . understand *that* . . . ." It is difficult to follow the word *that* with anything other than a precise statement of the understanding. In considering the following examples, think how one might determine if a student had learned each understanding and also think of the various ways one might interpret the vague statement of understanding.

UNDERSTANDING STATEMENTS

VAGUE	PRECISE
The student should understand the term *triangle*.	The student should understand *that* a triangle is a three-sided closed polygon. (Or: *that* a triangle is a closed polygon with three vertices.)
The student should understand *rational numbers*.	The student should understand *that* a rational number is an ordered pair (a, b) of integers such that $b \neq 0$. (Or: *that* a rational number may be represented as a repeating decimal.)
The student should understand what is meant by *relation*.	The student should understand *that* a relation is a set of ordered pairs of numbers.

It might be argued that the "vague" statements above do, in fact, imply the more precise statements. For example, wouldn't a teacher who says his or her students are to understand the term *triangle* really mean that the students are to understand that "a triangle is a three-sided closed polygon *and* that it is a closed polygon with three vertices?" If this is the case and the speaker, the listener, and, in particular, the student understand, fine! However, it is important and useful for beginning teachers to state explicitly what their students are to learn so that they will then be able to make better judgments of how to teach and how to evaluate learning—and thus will better promote learning in their students.

What are the distinctions among the different kinds of understandings one finds in mathematics? Henderson (1969) has done more than any other author in clarifying these distinctions and in specifying the instructional moves (particular teaching acts) and strategies (sequences of moves) associated with each type of understanding.

In discussing *concepts*, Henderson points out:

> It is difficult to distinguish between teaching a concept and teaching the meaning of a term or expression which designates the concept. A teacher does about the same thing when he teaches the concept of an ellipse, for example, as when he teaches what the term *ellipse* means. Hence it is not profitable pedagogically to distinguish between these two activities. A concept will be regarded as the meaning of a team. (p. 7)

Some educators envision a concept as a well-defined set of ordered pairs which forms a one-to-many relation. For example, the concept *triangle* may be thought of as a set of ordered pairs (triangle, x) where x is an instance of a triangle. With such an understanding, a student would be expected to be able to identify within a universe (of all geometric figures, say) all elements which are triangles and those which are not.

Another point of view, one that we hold, is to accept the notion that a concept is the meaning of a term and to observe that there are many levels of concepts. Some concepts, like that of *triangle*, are relatively simple and can be taught readily. In this case the definition, which is a generalization and an understanding, may be sufficiently clear to make teaching the concept easy. Other concepts, like *number*, are very complex and are taught over a period of many years. With these more complex concepts, it makes pedagogical sense to specify *levels of understandings*.

To explain further this point of view, we consider the concept *number*. A first grader's concept might be the cardinal number of a set; that is, the student understands that a (whole) number is a label attached to a (finite) set, and that the same label is attached to any sets which can be matched in one-to-one correspondence. (Note that a student can understand—that is, apply—that statement without being able to verbalize it. In other words, he or she can acquire the understanding before learning the same statement as knowledge!) Later the elementary school student's concept is broadened. The student learns that rational numbers are numbers of a special kind. Still later he or she will acquire the concepts of real numbers and complex numbers. At any given time, the student has attained a particular level of understanding of numbers. Everyone, even a mathematics teacher, continues to broaden a concept of numbers throughout his or her lifetime. One can study new number systems, like quaternions, and one can learn new understandings about more familiar systems. Space does not permit listing here all the understandings associated with the concept *number*, but a few examples will indicate how statements of understandings can be used to indicate the level of attainment of a concept.

The student should understand:

- that whole numbers indicate the numerosity of finite sets (e.g., 5 is associated with $\{^*, 1, 0, m, -3\}$).
- that the integers are needed to solve equations of the form $x + a = 0$ where a is a whole number.
- that π is irrational.
- that every real number can be associated with a point on the number line.
- that the integers under addition and multiplication form a ring.
- that matrix multiplication is not commutative.

Henderson observes, "The term *principle* is used to denote a generalization . . . or a prescription . . . " (Henderson, 1969, p. 13). He points out that true generalizations can be rephrased into prescriptions.

Consider the following generalization: For all real numbers, a, b, c, d, where c, d are not zero,

$$\frac{a}{c} \times \frac{b}{d} = \frac{a \times b}{c \times d}$$

A familiar formulation of this generalization is the prescription, "To multiply two fractions, multiply the respective numerators and denominators." In mathematics, principles appear formally as axioms and (true) theorems and less formally as properties of systems. In all cases, they appear as general statements in that they are true within some mathematical system—that is, they are universally quantified.

There is a common ground between principles as prescriptions and skills. Principles, as understandings, are tested by requiring the student to apply them. The process of applying the principle is not the same as the process of using the associated skill. The distinction, which is subtle, is a matter of degree and pedagogy. Henderson notes that prescriptions are easier to teach than generalizations, particularly to slow learners, but he warns, "excessive use of prescriptions, while producing students who are good manipulators, will not produce students who have much depth of understanding . . . " (p. 14). Thus a principle which is classified as an understanding can be reduced to a skill if it is taught in a certain way (such as by repeated application to similar problems).

PROBLEM-SOLVING GOALS

The highest level of mathematical learning is *problem solving*. Intuitively, these goals should be thought of as problem solving in a very broad sense. This category is defined to encompass all mathematical learning goals not included in *knowledge* or *understanding*. Problem solving is the ability to use previous mathematical learnings (1) to formulate hypotheses and test them, or (2) to prove theorems, or (3) to solve nonroutine problems.

Clearly, goals in this classification are the most difficult to evaluate. Test items on speeded examinations are particularly inappropriate. It is hard to estimate how much time is reasonably needed for a student to solve a mathematical problem, to construct a proof, or to formulate a hypothesis.

The goals in this classification strongly emphasize the *processes* of mathematics. The student must do something or think something. Goals in the other classifications are not process oriented in the sense that the student selects the process. To demonstrate that he has acquired knowledge, the student recalls a fact and states it. To demonstrate that he understands a statement, he applies it. In such an instance, he may use one or more mathematical processes; but the nature of the exercise is such that a particular generalization, principle, theorem, definition, or axiom is applicable. The student would typically select the appropriate understanding and apply it. In this way, the problem or exercise is an instance of, or closely associated with, the particular understanding. To demonstrate that he had learned a basic skill or algorithm, the student would employ a mathematical process; but, again, the process is determined by the exercise. The problem would be routine.

Items designed for students to demonstrate that they have achieved goals in the problem-solving classification will be solved by mathematical processes. Typically, two major characteristics of these items are that (1) usually more than one process will be needed, and (2) there will be more than one avenue for the student to follow to solve the problems. How many methods can you find for solving this problem:

> What is the greatest number of line segments one can draw to connect 10 points arranged on a circle? (Several ways to attack this problem will be presented in Chapter 5.)

Here are examples of other nonroutine problems which could be used to teach problem solving or to evaluate attainment of problem-solving goals:

1. A bee and a lump of sugar are located at different points inside a triangle (Figure 2–5). The bee wishes to reach the lump of sugar, while traveling a minimum distance, under the requirement that it must touch all three sides of the triangle before coming to the sugar. What is the path (Pollak, 1970, p. 316; credited to A. N. Kolmogorov)?
   A similar problem based on the same idea is: Find a path of minimum distance between two points on the same side of a given line, which touches the line (Figure 2–6).
2. Truman Botts has discussed a problem which has an easy solution, but which can be modified to lead to some very interesting consequences:

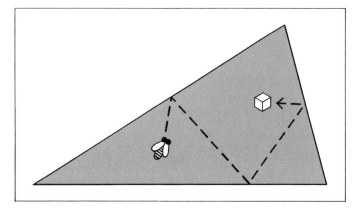

Figure 2–5 A bee and a lump of sugar.

Find three distinct natural numbers such that the sum of their reciprocals is an integer. That is, find positive integers A, B, C, and D such that

$$\frac{1}{A} + \frac{1}{B} + \frac{1}{C} = D$$

(Rapaport, 1963, p. 15; Botts, 1965a and 1965b).

3. Problems based on "tricks" are not recommended. For example, the next number in the sequence 42, 20, 4, 16, 37, 58, 89, 145, . . . , is found by squaring the digits in the last term and adding these squares. The next term is $1^2 + 4^2 + 5^2 = 42$.

COMPLEXITY AND HIERARCHY

The three classifications of goals may be considered as levels of goals proceeding from level (1) Knowledge to level (2) Understanding and to level (3) Problem Solving. These are levels of cognitive complexity—that is, level (3) is cognitively more complex than level (1). Achievement in these levels may or may not be hierarchical. That is, it is not necessary to achieve level (1) goals before level (2) goals. Often the reverse sequence is used in teaching. Students acquire understandings and then, through practice in applying the understandings, acquire basic skills. However, as noted earlier, students can acquire skills through pre-

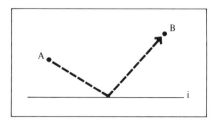

Figure 2–6 The problem is to find a path of minimum distance between points A and B that touches line i.

scriptions without achieving understanding. In general, it is assumed that level (2) goals contribute to success in level (3) goals, but the relationship is not always prerequisite.*

GOALS IN THE AFFECTIVE DOMAIN

Goals in the affective (emotional and attitudinal) domain are extremely important in learning and teaching mathematics; unfortunately, research in this domain is difficult to conduct and much remains to be explored. Affective goals can be identified and described within the existing classification of goals. Indeed, there is an important interaction of goals in the cognitive and affective domains.

Goals in the affective domain are beliefs and attitudes. In a sense, these beliefs and attitudes about mathematics, as about other subjects, are transitory—they may change. Surely they are influenced by every mathematics teacher encountered by the student (see Research Highlight, "Your Attitude Affects!" p. 27). And some negative attitudes and beliefs may deter students from learning mathematics. Consider, for example, the implications of the statements "I never could learn mathematics" and (with pride) "I was a mathematics dropout."

What are the affective goals in mathematics teaching? The cliché "Nothing breeds success like success" comes to mind. Successful students can use their knowledge, understanding, and problem-solving abilities to form very positive beliefs about mathematics. Although it does not necessarily follow that they will like mathematics, there appears to be a high probability that their ability will contribute to an attitude that mathematics is worthwhile, useful, and important. In this way, cognitive learning in mathematics provides a frame of reference for positive beliefs and attitudes, which in turn stimulate cognitive learning. Regrettably, the reverse cycle is often observed. A lack of success may contribute to negative beliefs and attitudes, which may cause even poorer performance (see Research Highlight, "Communicating with Students").

The distinction between cognitive and affective goals may be artificial. Cognitive problem-solving goals, for example, may be dependent on affective attitudes and beliefs. A student's belief that he can find a solution to a problem may be the impetus he needs to succeed. The interrelationships between cognitive and affective goals lead us to use the same categories for both. We wish to keep the classification scheme simple, but we are also impressed with the importance of emphasizing goals in both domains. Therefore, we believe that by using the same classification scheme, teachers will be more likely to capitalize on these interrelationships.

* Avital and Shettleworth propose a model similar to the one being constructed here. They also conceive a dimension of cognitive complexity which has five levels. However, they assume that "performance on an item in one category implies mastery of related materials in the lower categories." It is left to the reader to decide which model is more useful. (See Avital and Shettleworth, 1968, p. 4.)

RESEARCH HIGHLIGHT | COMMUNICATING WITH STUDENTS

Kester (1969) explored the way in which teachers communicate their expectations about student performance to their students. During a nine-week period, four one-hour observations of teacher-pupil interaction were made in junior high school classes. Teachers communicated with their allegedly bright pupils in a more friendly, encouraging, accepting manner. As the pupil's positive communication to the teacher increased, the teacher's communication to the pupil tended to be positive. And teachers communicated both more frequently with these pupils and in longer exchanges.

Meanwhile, the less bright student received less attention. It was perhaps to be expected that approval would generate more responses—but teachers have a responsibility toward everyone in their classes. Attention to each student could have a payoff in terms of increased response from each one.

Let us consider how such a classification may be employed. Suppose an Algebra I class is to begin a unit on graphing continuous functions in a plane. The teacher might identify the following affective goals for the unit.

The student should believe:
1. That graphs can be drawn by specifying only a finite number of points.
2. That continuous graphs can be used to determine points of the function not in the original set of data points.
3. That a graph enables a reader to "see" a function or to "see" how a function behaves at different neighborhoods of its domain.
4. That graphs are useful "pictures" of real-life situations.

Within our proposed classification, at least two of these goals can be taught and learned as Knowledge. That is, there is information about the world and about the role of mathematics in the world that is essentially factual but has a strong affective component. (Consider the central role of mathematics in the space program, for example.) Teachers can do much to help their classes build up a rich storehouse of knowledge about mathematics that will assist in later learning situations.

Beliefs can also be understandings. But, using our interpretation of understanding, how does one apply a belief? Refer again to the goals for the Algebra I class just listed. Teachers can provide opportunities for fostering, testing, and developing belief systems concerning mathematics. In the hypothesized situation, the teacher could encourage students to acquire the beliefs by including real data samples, such as age-weight data, time and temperature measures of an automobile engine, or time and height measures of a projectile. The teacher might also assign experiments which can be graphed, such as temperature-pressure studies of a gas. You may think of specific class activities which would be likely to contribute to each of the four beliefs. (Note that simply telling them the appropriate statement is permissible, but you cannot be certain the statement will be accepted as a belief.)

To what extent can affective goals be considered as belonging to *Problem Solving?* Clear-cut decisions are not easily made here, or anywhere else in the classification scheme. But recall the famous anecdote about the mathematician Karl F. Gauss (Bell, 1956). At the age of ten, Gauss discovered that the sum of the first n natural numbers is

$$\frac{n(n + 1)}{2}$$

A teacher might tell the story to his class as purely factual information, but with the expectation that the students would be encouraged to believe that they, too, might discover some mathematics, or at least realize that mathematics is man-made.

To evaluate beliefs is more difficult. True-false items one would naturally associate with each belief are unlikely to indicate accurately the students' beliefs. Such items are useful in determining a student's knowledge, but it is clear that what the student says to a teacher may not agree with what the student believes. The teacher could ask students if they believed statements such as "Graphs are useful pictures." But again the replies may be better indicators of how much the students want to please the teacher than of what the students actually believe. Alternatively, over a long period, the teacher might observe how often students elect to draw graphs when another method, such as listing several more data points, is available.

This example indicates the problem of evaluating goals in the affective domain. It is simply harder than evaluating goals in the cognitive domain. Most teachers, reasonably, rely on their observation and intuition about students to evaluate affective domain goals. They decide if students want to use their mathematical learning. They observe the students to see if they react enthusiastically, for example, to the information they can determine from a graph or to the generalizations they can make from looking at graphs. Thus teachers who are trying to evaluate achievement of affective domain goals are collecting information on how often and in what ways students apply cognitive learnings. These teachers are also sensitive to and interested in the ways students reveal their attitudes and beliefs. (Other ideas about evaluating affective domain goals are presented in Chapter 6.)

A THREE-DIMENSIONAL MODEL

The major emphasis in describing the model this far has been in the classification of goals for mathematics learning. However, there are two other major dimensions of the model: content and teaching processes.

CONTENT DIMENSION

Selection of a particular content scale will depend upon the particular use for the model. In this book, the overall scale will be secondary school mathematics, grades 7–12. If a school mathematics department were to use the model, the faculty members might choose the same scale, and major subsections of the scale might be the courses offered by the department. Such a list might include: General Mathematics I, General Mathematics II, Algebra I, Geometry, Algebra II, Probability, Computer Mathematics, Senior Mathematics (Function Theory), and Calculus. In contributing to the department model, a particular teacher might focus only on those courses he or she teaches. For example, one teacher might teach two sections of Algebra I, one section of Geometry, one section of Probability, and one section of General Mathematics I. Such a teacher would list those course titles as major headings in the content dimension of the model (see Figure 2–7).

For the model to be useful, however, the content dimension must be refined to specify particular content elements. The teacher in the example above should focus his or her attention, at some time, on Algebra I, specifying the content of the course by listing instructional unit topics. Such a list might include:

- Introduction to Open Sentences
- Introduction to the Ordered Field Axioms
- Graphing
- Linear Equations and Inequalities
- Polynomial Functions and Relations

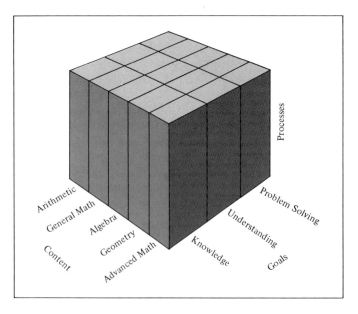

Figure 2–7 The model of mathematics teaching with course titles as the major divisions of the content dimension.

- The System of Complex Numbers
- Logarithmic, Exponential, and Circular Functions
- Systems of Equations

But even these unit titles are not sufficiently specific to be elements of the content dimension; so the teacher should list the mathematical topics to be included in each unit. Thus, by successive refinements, from teaching assignments to course titles to unit titles to topics, the teacher might finally come to list the following topics for the unit on "Introduction to the Axioms for an Ordered Field":

- The Set of Real Numbers
- Commutativity of Addition
- Commutativity of Multiplication
- Closure of Addition
- Closure of Multiplication
- Associativity of Addition
- Associativity of Multiplication
- The Existence of an Identity (0) for Addition
- The Existence of an Identity (1) for Multiplication
- Distribution of Multiplication over Addition
- The Existence of Additive Inverses
- The Existence of Multiplicative Inverses
- The Three Order Axioms
- The System of Real Numbers
- Subsystems of the Real Numbers

PROCESS DIMENSION

Now that the content and goal dimensions of the model have been described, the teacher might focus on the teaching processes—planning, teaching, and evaluating. These three major activities constitute the major headings of the process dimension of the model (see Figure 2–8). During the planning stage the teacher should consider the content by goals matrix and list the specific goals to be attained by students in the unit. Within the unit, there may be knowledge goals, understanding goals, and problem-solving goals for each topic as well as affective domain goals for the entire unit.

Such a list of goals serves as a basis for the teaching and evaluating phases. Before interactive teaching begins, the teacher should develop a plan for teaching each lesson. This plan would serve as a guide for the teaching phase, but during the teaching process it may be necessary or wise for the teacher to modify the plan. Part of the lesson plan will include some in-class evaluation activities. This feedback, or formative evaluation, helps the teacher decide how to make on-the-spot changes in the lesson plan or to modify future plans.

The list of goals developed in planning also serves as a basis for developing

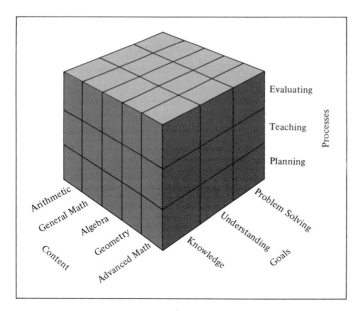

Figure 2–8 The model of mathematics teaching with the three activities that constitute the major headings of the processes dimension.

the evaluation plan. This plan will include unit tests, daily quizzes, and less formal evaluation plans (e.g., observations). The distinction between the total unit or summative evaluation and the formative evaluation conducted during classes is admittedly ambiguous. Overall, the teacher's purpose is to help students acquire the goals of the unit. In doing this, the teacher collects evidence daily on how well the students are achieving. Ultimately he or she will want to determine whether the students have achieved all the goals.

Details of implementing the proposed model will be presented in Chapters 3 to 6, and elsewhere, in this book; this chapter has been devoted to introducing and describing the model, which may be depicted by the three-dimensional object in Figure 2–9. The model may be entered along any dimension; that is, the teaching of mathematics may be thought of initially in terms of content, goals, or processes. Ultimately, all dimensions should be considered. Some authors organize books along a content dimension, others along the process dimension; we have elected to organize this book along the goal dimension.

But the model is not complete. It does not provide a basis for answering all the questions teachers of mathematics must answer, and so it is not presented as a *theory*. Mention has not been made of the environment for teaching, the use of research results, or the process of constructing lesson plans. These and other important considerations will be presented in succeeding chapters.

Critical Incidents Revisited

Consider the critical incidents at the beginning of this chapter. Can you guess what problem is indicated in each incident?

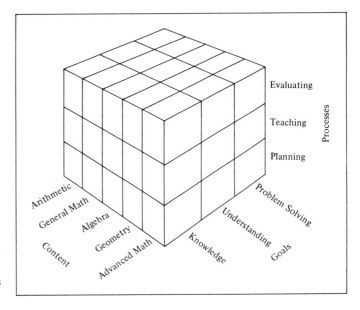

Figure 2–9 The model of mathematics teaching.

Neal, in the first critical incident, can demonstrate skills. Perhaps he devotes his efforts to identifying and remembering the procedures to use in solving routine exercises. However, he avoids working on new problems. Part of Neal's difficulty may be solved by setting goals in the affective domain. Neal may well have the ability to achieve goals in the problem-solving area, but he may lack the self-confidence to try.

In the second critical incident, Ann may well be verbalizing her beliefs about studying geometry. Here, even more than in the first critical incident, the problem appears to be affective. Ann may be frustrated because she cannot complete the assignment or she may not understand the role of deduction in mathematics. Perhaps it would be worthwhile to work with Ann by (1) providing her with theorems that are very easy to prove to see if the assignment was too difficult, (2) demonstrating uses and abuses of logical reasoning in nonmathematical settings to show her the wide application of deduction, (3) discussing some of the theorems and the proofs with her to determine her degree of understanding of the assignment, or (4) providing less obvious theorems for her to prove to see if she simply was not challenged sufficiently.

Bob, in the third critical incident, is described as not *knowing* the quadratic formula. The problem here may well be the distinction between *knowing* and *understanding*. We might guess that Bob can learn to remember the formula and the problem is that he does not understand it—that is, he cannot apply it. Although the affective domain may be involved, we suggest the problem is cognitive. The teacher could work with Bob to help him understand that the formula provides a means for solving quadratics and to show him how to use the formula.

The intent of these critical incidents is to show how the categorization of goals in mathematics teaching is useful and to demonstrate the interrelationship between the cognitive and affective domains. The critical incidents present problems that can be dealt with effectively by utilization of the model of mathematics teaching described in this chapter.

Activities

1. Almost all prospective teachers of mathematics have been greatly influenced by one or two outstanding teachers during their careers as students. Select one teacher who contributed to your decision to teach mathematics and describe this person's teaching.

2. Consider the list of mathematical topics suggested for the unit on "Introduction to the Axioms for an Ordered Field." Prepare a list of the knowledge, understanding, and problem-solving goals you would want students to achieve in this unit. Select one goal from each level and propose an evaluation item for it.

3. Theory was discussed in this chapter. Consider the following theories taken from science, mathematics, and education:
   Science:
      Molecular Theory
      Wave Theory of Light
      Theory of Relativity
   Mathematics:
      Theory of Equations
      Probability Theory
   Education:
      Stimulus-response Theory
      Gestalt Theory
      Piaget's Developmental Theory
   Select one of these theories or another of your choosing. Briefly describe the theory and discuss how well it qualifies as a theory. In particular, consider its completeness and consistency.

4. a. Select one of the following topics:
   (1) Factoring of polynomials in Algebra I
   (2) Introduction to coordinate graphing in General Mathematics
   (3) Finite mathematical induction in Algebra II
   (4) A theorem from Geometry
   (5) Continuity in Calculus
   b. State some reasonable assumptions you might make about a class to whom you would teach the topic you selected.
   c. Prepare a list of cognitive goals for a unit on your topic and classify each as knowledge, understanding, or problem solving.
   d. Prepare a list of affective goals and classify each as in part c.

    e. Write a paragraph about how you would propose teaching a lesson for the goals listed in parts c and d.

    f. Prepare a lesson plan for the goals you stated in parts c and d.

    g. Describe your evaluation plans to determine if the goals of parts c and d are acquired.

    h. If you have access to one or two students at the appropriate grade level, teach the lesson you planned in part f and conduct the evaluation from part g.

    i. Report your results.

5. Examine the work of the following authors referenced at the end of the chapter:

    a. Benjamin S. Bloom, J. Thomas Hastings, and George F. Madaus (1971)

    b. J. Fred Weaver (1970)

    c. Shmuel M. Avital and Sara J. Shettleworth (1968)

    d. Edward G. Begle and James W. Wilson (1970)

Compare the models proposed by these authors to the one proposed in this chapter. Consider the advantages and disadvantages of each for teachers of mathematics. Include any suggestions for the models which you believe would improve their usefulness for you.

6. For each of the following sample test items, specify a goal level and a goal which could be associated with the item. That is, given the test item, specify a goal and the level of the goal which the item could be used to evaluate.

    a. If $|x| = |-3|$ then $x =$ _____.

    b. Factor $x^3 - 27$.

    c. $(96 \times 57) + (43 \times 96) + (57 \times 57) + (43 \times 57) =$ _____.

    d. State the definition of the absolute value of a number.

    e. What is the greatest possible distance between a point in the plane and a nearest point which has integer coordinates?

7. Describe what you consider to be your strengths as a teacher, using the model proposed in this chapter. Consider the content, goal, and process dimensions in describing your background, education, and experience for effective teaching.

8. You are enrolled in a course for teachers. List your own goals in this course using the model presented in this chapter as a guide. What elements of the content, goal, and process dimensions do you hope to acquire in this course? List any other elements you hope to learn. (You might find the table of contents for this book helpful in organizing your thoughts.)

References

Avital, Shmuel M. and Shettleworth, Sara J. *Objectives for Mathematics Learning.* Bulletin 3. Toronto: Ontario Institute for Studies in Education, 1968.

Begle, Edward G. and Wilson, James W. Evaluation of Mathematics Programs. In *Mathematics Education* (edited by Edward G. Begle). Sixty-ninth Yearbook of the National Society for the Study of Education, Part I. Chicago: University of Chicago Press, 1970.

Bell, Eric T. The Prince of Mathematics. In *The World of Mathematics*, Vol. 1 (edited by James R. Newman). New York: Simon and Schuster, 1956.

Bloom, Benjamin S., ed., *Taxonomy of Educational Objectives: The Classification of Educational Goals*. Handbook I: *Cognitive Domain*. New York: David McKay, 1956.

Bloom, Benjamin S.; Hastings, J. Thomas; and Madaus, George F. *Handbook on Formative and Summative Evaluation of Student Learning*. New York: McGraw-Hill, 1971.

Botts, Truman. Problem Solving in Mathematics, I. *Mathematics Teacher* 58: 496–501; October 1965.

Botts, Truman. Problem Solving in Mathematics, II. *Mathematics Teacher* 58: 596–600; November 1965.

Gage, N. L. Paradigms for Research on Teaching. In *Handbook of Research on Teaching* (edited by N. L. Gage). Chicago: Rand McNally, 1963.

Gibb, E. Glenadine; Jones, Phillip S.; and Junge, Charlotte W. Number and Operation. In *The Growth of Mathematical Ideas Grades K–12*. Twenty-fourth Yearbook of the National Council of Teachers of Mathematics. Washington: The Council, 1959.

Henderson, Kenneth B. *Teaching Secondary School Mathematics: What Research Says to the Teacher*, Number 9. Washington: National Education Association and Association of Classroom Teachers, 1969.

Henderson, Kenneth B. Concepts. In *The Teaching of Secondary School Mathematics*. Thirty-third Yearbook of the National Council of Teachers of Mathematics. Washington: The Council, 1970.

Kester, Scott Woodrow. The Communication of Teacher Expectations and Their Effects on the Achievement and Attitudes of Secondary School Pupils. (The University of Oklahoma, 1969.) *Dissertation Abstracts International* 30A: 1434–1435; October 1969.

Krathwohl, D. R. et al. *Taxonomy of Educational Objectives: The Classification of Educational Goals*. Handbook 2: *Affective Domain*. New York: David McKay, 1964.

Nagel, Ernest and Newman, James R. Goedel's Proof. In *The World of Mathematics*, Vol. III (edited by James R. Newman). New York: Simon & Schuster, 1956, pp. 1668–1695.

Phillips, Robert Bass, Jr. Teacher Attitude as Related to Student Attitude and Achievement in Elementary School Mathematics. (University of Virginia, 1969.) *Dissertation Abstracts International* 30A: 4316–4317; April 1970.

Pikaart, Len. A Simplified Taxonomic Model for Teachers of Mathematics in Elementary School. In *Report of the TTT Project Science and Mathematics: Tri-University Project in Elementary Education*. New York: New York University, 1971.

Pikaart, Len and Travers, Kenneth J. Teaching Elementary School Mathematics: A Simplified Model. *Arithmetic Teacher* 20: 332–342; May 1973.

Pollak, Henry O. Applications of Mathematics. In *Mathematics Education* (edited by Edward G. Begle). Sixty-ninth Yearbook of the National Society for the Study of Education, Part I. Chicago: University of Chicago Press, 1970.

Rapaport, Elvira, translator. *Hungarian Problem Book*. SMSG New Mathematical Library, Vol. 12. New York: Random House, 1963.

Romberg, Thomas A. and Wilson, James W. The Development of Mathematics Achievement Tests for the National Longitudinal Study of Mathematical Abilities. *Mathematics Teacher* 61: 489–495; May 1968.

Sawyer, W. W. *A Path to Modern Mathematics*. Baltimore: Penguin Books, 1966.

Swafford, Jane Oliver. A Study of the Relationship Between Personality and Achievement in Mathematics. (The University of Georgia, 1969.) *Dissertation Abstracts International* 30A: 5353; June 1970.

Weaver, J. Fred. Evaluation and the Classroom Teacher. In *Mathematics Education* (edited by Edward G. Begle). Sixty-ninth Yearbook of the National Society for the Study of Education, Part I. Chicago: University of Chicago Press, 1970.

Wilson, James W. Evaluation of Learning in Secondary School Mathematics. In Benjamin S. Bloom, J. Thomas Hastings, and George F. Madaus. *Handbook on Formative and Summative Evaluation of Student Learning*. New York: McGraw-Hill, 1971.

Knowledge and Skill

AFTER STUDYING CHAPTER 3, YOU WILL HAVE:
- ★ seen that knowledge and skill goals are important parts of mathematical learning.
- ★ examined several statements of knowledge and skill goals and learned how to specify such goals.
- ★ found out what strategies are useful in teaching these goals.
- ★ been given an opportunity to learn about the computer programming language BASIC.

CRITICAL INCIDENTS
1. Bill says to his teacher, "I could have solved Problem 3, but I couldn't remember the formula for sin $2x$."
2. The teacher asks Lori for the decimal expansion of 1/7. Lori says, "I remember, it's $.\overline{142857}$." When asked for the decimal expansion of 3/7, Lori replies, "That's easy, it's $.\overline{428571}$."
3. Mrs. Keene, the algebra teacher, announces: "Class, for the quiz tomorrow make sure you know the quadratic formula and how to solve quadratics by factoring."

4. Jack, now a college student, tells his friend, "The only way I could get through mathematics courses in high school was to memorize everything."
5. Sally, a senior in high school, tells her parents: "All of mathematics was invented a long time ago. It's old stuff and all our teachers do is tell us the facts and how to use them."

THE ROLE OF SKILLS AND KNOWLEDGE

Basic skills and knowledge comprise a large part of mathematical learning. Students are often required to memorize mathematical facts or to become proficient in using algorithms. Many teachers require students to know the quadratic formula, to memorize the square roots of the natural numbers 2, 3, and 5 to three decimal places, or to be able to factor appropriate binomial expressions. On a more elementary level, most students know a great many mathematical facts such as:

- $2 \times 3 = 6$
- The area of a rectangle equals the base times the height
- 22/7 is an approximation to π

And most students are able to demonstrate such skills as:

- Multiplying two-digit whole numbers
- Changing a number to scientific notation
- Reducing a fraction to lowest terms

This chapter explores the teaching and learning of mathematics knowledge and skills. Teachers must decide how much and what parts of the content should be learned as knowledge and skills. They need to know effective methods to teach these kinds of mathematical learnings and ways to evaluate student acquisition of them. Thus this chapter deals with planning and teaching, two of the *processes* of teaching knowledge-level goals depicted in the model (Figure 3–1). (*Evaluating*, the third process, will be the focus of Chapter 6.) Throughout this chapter we will emphasize the teacher's decisions and the effects of these decisions on the students' mathematical learning and on their affective beliefs about themselves and about mathematics.

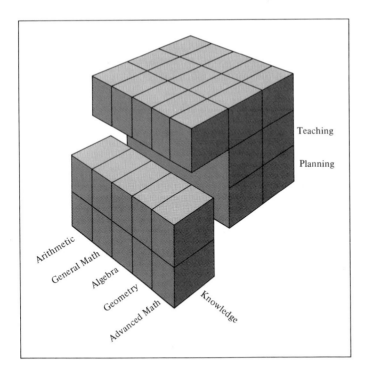

Teaching

Planning

Arithmetic

General Math

Algebra

Geometry

Advanced Math

Knowledge

Figure 3–1　Teaching and planning, as two of the processes of teaching knowledge-level goals in the model of mathematics teaching, are the subject of Chapter 3.

PLANNING

SELECTING ITEMS OF KNOWLEDGE AND SKILLS

One of the more important decisions a teacher makes is to select that content which is to be learned as knowledge or skill. This decision is essentially a process of selecting mathematical facts and procedures which are to become automatic responses for the students. When the student is asked for a fact or presented with an exercise requiring a skill, he or she is able to respond with very little reflection, recalling and stating the fact or performing the skill immediately.

In order to point up the importance of a teacher's decision in selecting content for this automatic response category, let us consider two hypothetical extremes. Mrs. Alpha views mathematical learning as composed almost entirely of facts and skills. She can take any problem and determine the basic facts and skills needed for its solution. She may even view her students as "human computers" and see her task as one to load their memories with information and to program them to perform operations. However, she is probably aware that her students are unlike computers in that information cannot simply be read into their memories —nor is one set of clear instructions sufficient to teach them a skill. If Mrs. Alpha avidly wants her students to be successful, she probably selects a small amount of content and uses a great amount of repetition. She drills students on the facts, and she gives them exercise after exercise in practicing the skills. Alternatively, Mrs. Alpha may believe that "even good teachers cannot be successful with all students"—that is, she expects some students will fail to learn the facts or skills she teaches. In this case, she may "cover" more material, but again she tells students the facts they are to learn and shows them the skills. Those students who are willing and able to memorize the material are successful; the others fail.

Another hypothetical teacher is Mr. Omega. He believes that the number of facts and skills in mathematics should be kept to a minimum. However, even Mr. Omega admits that all students must learn a few basic facts and skills. He expects his students to know the addition and multiplication facts and the basic computation algorithms.* His classes are very different from Mrs. Alpha's; Mr. Omega and his students derive almost everything in mathematics. His lessons include a great variety of problems, each different from the other. Many of his tests are "open-book" because "students in his classes do not have to memorize." Like Mrs. Alpha, Mr. Omega's views of how students learn mathematics determines how much his students learn. If he puts a high priority on success, his students will learn less content but learn it well. If he puts a high priority on the amount of material taught, more material might be covered, but his students may be less successful. In either case, Mr. Omega probably does not cover as much material as Mrs. Alpha simply because it takes longer to derive everything.

Now consider the student reactions in classes taught by Mrs. Alpha and Mr. Omega. Mrs. Alpha's students are probably much more inclined to exemplify the attitudes depicted in the fourth and fifth critical incidents at the beginning of this chapter. On the other hand, if Bill (in the first critical incident) were in Mrs. Alpha's class, she might say, "I told the class they would need to know the double angle formulas for the quiz." And if Bill were in Mr. Omega's class, he might hear, "But you could have derived it easily." In whose class would you expect to find Lori, the student of the second critical incident?

Students in either classroom could be unhappy about their learning of mathematics. Mrs. Alpha's students might see little justification in memorizing vast quantities of information which they may regard as being of very little use outside the classroom. They probably find little excitement in learning mathematics, and they may believe that all mathematics exists as a body of knowledge to be memorized. Students in Mr. Omega's class may also find mathematics disagreeable; they may feel frustrated that they have to work so hard

---

* Mr. Omega could have taken an even more extreme position. He might require only that his students be able to count. Thus, if they fail to recall an addition fact, they could rely on their understanding of the addition process to determine the sum of two numbers by counting. A forgotten multiplication fact might lead to a repeated addition problem. But such an extreme position seems unreasonable, even for the hypothetical Mr. Omega.

PI IS NO MYSTERY BUT HOMEWORK IS

Some people will go to any lengths to achieve glory. In fact, 16-year-old Mark Peel is willing to go to the 5,000th decimal place.

Mark, a junior in Thornton Township High School, wants to earn a place in The Guinness Book of World Records by reciting the infinite computation of the mathematical symbol Pi to its 5,000th decimal place.

Pi is the mathematical expression of the ratio of the circumference of a circle to its diameter, approximately equal to 3.1416. But Pi is not a finite number—there is no limit to the number of decimal places to which it can be computed.

Mark told Richard Toland, a City News Bureau of Chicago reporter, that his determination to assault the Pi record began last Dec. 17, when he began writing the decimal places of Pi on Thornton's mathematics department blackboard at 7:40 a.m. and by 9:04 a.m. had reached the 5,000th decimal place without error.

His math teacher, Miss Shirley Vallort, verified the results, altho she looked upon the feat with a jaundiced eye.

"Obviously, I've never seen anyone do anything like that before," she said. "I hope he gets in the book [of records]. He spent a lot of time studying those numbers which he could have been spending on homework."

Despite Mark's wizardry with Pi, Miss Vallort gave him a "C" in her college algebra course last year because he let homework assignments slide.

Tribune Photo by Walter Kale

Mark Peel of Dolton . . . a memory for figures.

Altho he did not earn high marks in math, Mark did win a $150 bet with his father that he could carry Pi to the 5,000th place.

Mark, who lives at 14924 Edbrooke Av., Dolton, will be vying for a record currently held by Timothy Pearson of England, who was 13 when he recited 1,210 digits in 4 minutes and 20 seconds on Dec. 21, 1973.

Mark hopes to recite 1,600 decimals in four minutes and finish the entire 5,000 "to make sure [the record] lasts."

Reprinted courtesy of *Chicago Tribune*, Friday, April 25, 1975

Figure 3–2   An extreme example of one high school student's mastery of one skill.

to obtain even simple results. Perhaps some have learned that they can memorize particular facts to save time in solving problems. They may believe Mr. Omega is hypocritical in telling them not to memorize material for they have learned that memorizing enables them to obtain high scores on their tests.

In practice, teacher beliefs generally fall between those of Mrs. Alpha and Mr. Omega. Usually teachers develop their beliefs about the role of knowledge and basic skills subconsciously. Their beliefs may be discerned by observing how and what they teach or by examining the tests they use. However, what a teacher believes about the role of knowledge and skills in mathematics will influence his or her goals and teaching strategies—and, consequently, what his or her students learn.

How can mathematics teachers determine what emphasis to place on the learning of facts and basic skills? There is no easy answer. Mathematics teachers have chosen a career in mathematics because they like it, they have studied it for many years, and they have accumulated a vast background in the subject. Thus they may not be sensitive to the achievement level and needs of their students, most of whom will not major in mathematics. For example, many secondary school mathematics teachers will know that 1.414214 is a close approximation to $\sqrt{2}$ or that 3.1415927 is a good approximation to π. The teachers must decide if either of these facts is important enough or useful enough to require students to learn them. Again, it may be helpful for the *teachers* to know many things which are not particularly useful for their students. Almost all mathematics teachers know the standard algorithm for calculating the square root of a whole number, but it is difficult to justify requiring this skill of all secondary school students.

There are, however, items of knowledge and skills which are useful for students, even if the items are retained only for a school year. For example, in Algebra I, it is probably worthwhile for students to know that 1.41 is a good approximation to $\sqrt{2}$ and 3.14 is a good approximation to π; in Algebra II, students will probably memorize the definitions of the six basic trigonometric functions; junior high school students usually know the area formulas for triangles and other common figures (see Figure 3–2).

Ultimately, you must decide which content elements are to be learned as items of knowledge and as basic skills by your students. It is important to realize that (1) items of knowledge and skills are appropriate parts of the mathematics curriculum, (2) it is no easy task to decide what parts of the content are to be learned as knowledge or skills, (3) a balance must be maintained between knowledge and skill learning on the one hand and other kinds of learning, and (4) the particular balance may influence the students' attitudes and feelings about mathematics. To decide what items of knowledge and skills are to be learned, teachers might ask themselves how often the item will be used and how difficult it will be for students to derive or look it up when they need it. After these decisions have been made for a unit of instruction, a reasonable next step is to specify the goals.

SPECIFYING STUDENT GOALS

You may find it difficult to find time, once you are a full-time teacher, to specify goals for all instructional units. However, pre-service experience in specifying objectives can help develop the pedagogical skill. When you are student teaching, you will have more preparation time than that typically available to regular teachers. Some of this time can be used profitably to specify goals and further develop this teaching skill. Later, when you are teaching, you may be able to plan units of instruction effectively without taking the time to specify each goal. Alternatively, groups of teachers may work together to prepare lists of goals for a particular course, and such lists may serve as course guides. The more fortunate teachers are provided released time for such activities. Often state or regional curriculum guides and teacher's editions of textbooks are available as aids in specifying objectives. However, because you as a teacher will hold the final responsibility for the content of your courses, it is important that you be able to write clearly stated goals for student learning.

In general, a useful format for knowledge goals is: *"The student should know that . . . ."* For example:

1. *The student should know that* 3.14 is the value of π to the nearest hundredth.
2. *The student should know that* for any real number x and any natural number n, x^n is defined to be a product of x taken as a factor n times and n is called the exponent of x.
3. *The student should know that* an isosceles triangle has two congruent sides.
4. *The student should know that* if a line is perpendicular to one of two parallel lines in the same plane, the line is perpendicular to the other parallel line.

In each example, the word *that* is followed by a specific statement of the item of knowledge to be learned. Use of *that* is highly recommended to avoid vague statements such as "The student will know the definition of x^n." It is difficult to complete a sentence in the recommended format without being explicit (see Research Highlight, "Do Objectives Help?").

Specifications of skills require a different format. These should be clear statements of what the student should be able to do after instruction in the unit. For example:

1. *The student will be able to* plot ordered pairs of real numbers in the Cartesian plane.
2. *The student will be able to* factor (over the ring of integers) all factorable quadratic expressions.
3. *The student will be able to* specify the product of two binomial expressions.
4. Given the measure of two sides of any right triangle, *the student will be able to* find the measure of the third side.
5. *The student will be able to* rationalize the denominator of any algebraic expression.

RESEARCH HIGHLIGHT | DO OBJECTIVES HELP?

Does explicitly stating objectives result in students having higher achievement? Piatt (1970) conducted a study in which the experimental group scored significantly higher than the control group on tests of computation and concepts. No significant differences were found on an application test. The experimental group consisted of 300 randomly selected students taught by 11 seventh-grade mathematics teachers who were trained in defining, writing, and implementing educational objectives stated in behavioral terms. The control group consisted of 300 randomly selected students whose teachers did not have this specific training.

This is a type of study that might be carried out, on a smaller scale, with your own classes and those of several other teachers. Try writing explicit objectives for some classes and not writing objectives for others. It might help you to realize the value of knowing just what you're trying to do!

A detailed discussion of the distinction between basic skills and understandings will be presented in the next chapter.

A SAMPLE OF GOALS

In this section we present examples of typical knowledge and skill goals for several secondary school mathematics courses. It is to be emphasized that ultimately the choice of content and classification of goals is the responsibility of each teacher, whether explicitly specified as such by a list of goals or implicitly determined by the choice of teaching and evaluation strategies.

Arithmetic (seventh and eighth grades)

KNOWLEDGE GOALS:

1. The student will know that the additive inverse of any integer a is the integer denoted by ^-a such that $a + {}^-a = 0$. For example, the additive inverse of 7 is $^-7$, of $^-4$ is $^-(^-4) = 4$.
2. The student will know that the formula for finding the volume of a right prism is $V = B \cdot h$, where B is the area of the base and h is the altitude of the prism (Figure 3–3).

SKILL GOALS:

1. The student will be able to inscribe a regular hexagon, pentagon, square, or triangle in a circle using a protractor.

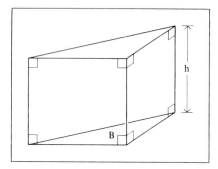

Figure 3–3 The formula for finding the volume of a right prism is $V = B \cdot h$, where B is the area of the base and h is the altitude of the prism.

2. The student will be able to classify a given set of triangles by the measure of their sides as isosceles, scalene, or equilateral.

General mathematics
KNOWLEDGE GOALS:

1. The student will know that $<$ is the symbol for "is less than."
2. The student will know that each object in a set is called a *member* or *element* of the set.

SKILL GOALS:

1. The student will be able to match a finite set of rational numbers with their associated points on a number line.
2. The student will be able to add any two rational numbers whose numerators and denominators are less than 20.

Algebra
KNOWLEDGE GOALS:

1. The student will know that if two or more numbers are multiplied, each of the numbers is called a factor of the product.
2. The student will know that $\in$ is the symbol for "is an element of" and $\notin$ is the symbol for "is not an element of."

SKILL GOALS:

1. The student will be able to find, using a prime factorization technique, the greatest common divisor of two numbers, neither of which contains more than four digits.
2. The student will be able to solve any first-degree equation in two unknowns with integral coefficients.

Geometry
KNOWLEDGE GOALS:

1. The student will know that a set B is convex if and only if the line segment determined by any two points in B is entirely in B.

2. The student will know that two coplanar lines m and n are parallel if and only if $m \cap n = \emptyset$, that is, m and n have no points in common.

SKILL GOALS:
1. The student will be able, given any triangle, to locate its centroid using only a compass and straightedge.
2. The student will be able to find the measure of any exterior angle of a triangle knowing the measures of its two remote interior angles.

Advanced Mathematics
KNOWLEDGE GOALS:
1. The student will know that the period of the sine function denoted by $y = \sin x$ is 2π.
2. The student will know that $\log 10 = 1$.

SKILL GOALS:
1. The student will be able to find the determinant of any matrix by expanding by minors.
2. The student will be able, given the graph of any trigonometric function, to determine the amplitude, period, and phase-shift of the function (Figure 3–4).

TEACHING ITEMS OF KNOWLEDGE AND SKILL

Once a teacher has decided upon a set of knowledge and skill goals, he or she must decide what to do in order that students may achieve them. The major characteristic of knowledge and skill goals is that they require automatic responses from the student. When asked to state the commutative property, the student recites it. When presented with a quadratic expression and asked to factor it, the

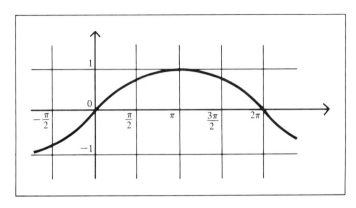

Figure 3–4 The graph of a trigonometric function.

student initiates a procedure to determine the factors. In both of these examples the teacher would typically expect students to understand what they are saying or doing, but, because the goals have been classified as knowledge and skills, the aim of instruction would be automatic (quick, efficient) responses.

BASIC STRATEGIES

How are automatic responses learned? The key is *repetition*. Consider some elemental learnings such as learning your own name, learning to count, or learning to type. Each of these examples may be described as a goal:

1. The student knows that his name is John Smith.
2. When presented a set with less than 20 elements, the student can state the number of elements in the set.
3. The student should be able to type at least 50 words per minute with less than three errors in five minutes.

Goal 1 is an item of knowledge while Goals 2 and 3 are skills. How could these goals be taught?

Almost every person learns his name at a very early age. Family members and others around a child speak to him, even before he has learned to talk at all. People use the child's name. At first they may call him "John" and use sentences like "John is a big boy." After John has passed through a babbling stage, he begins to use words and receives reinforcement in the form of smiles, reciprocal communication, pats, kisses, and so on for using words like "Dada," "Mommy," and "John." The process continues. After he learns to say "John" he learns to say "John Smith." Finally,through a series of repetitions and reinforcement, he learns to respond with "John Smith" when asked "What is your name?"

In a similar way, a student may be taught to count or to type. At early stages he may be taught to count only to three or to type the letters "a s d f j k l ; (the "home" keys on a typewriter). As he develops these simpler skills through practice, he is taught to count higher or to use more keys on the typewriter. More practice is necessary for him to become proficient.

Thus, with both knowledge and skills, practice is a fundamental part of learning. Other useful characteristics which can be abstracted from the examples cited are partial learning and motivation. Often it is easier initially to teach part of an item of knowledge or of a skill. We teach the child to say "John" first and later to say "John Smith." Motivation is a difficult topic but an extremely important one. Clearly the student will learn little if he or she has little motivation to learn. To achieve Goals 1 and 2 above, the student might be motivated by adult praise and attention. Motivation for an older child, such as one who is learning to type, is more complex. Although he or she may be motivated by praise and attention from his family, peers, and/or teachers, other factors may influence him. The child may want to learn to type because of a desire to use the family typewriter, because his teachers require typewritten papers, because several of his friends can type, because he realizes it is a useful skill, or because he wants to earn a high grade (or avoid a low grade) in Typing I. Success in early typing experiences may be an important motivational factor. And some people learn for their own satisfaction—learning itself is motivating.

REPETITION

To achieve the automatic characteristic of knowledge and skill learning, a student must employ some form of repetition or practice in the learning process. To clarify the role of repetition, consider the following experiment:

> Select a friend to help you. When you have your friend's attention, repeat this sentence *at least* five times: "A gloob is a partice." To provide variety, you might emphasize a different word each time. After this treatment, ask your "student" this question: "What is a gloob?"

Did your "student" learn the item of knowledge? Chances are that he or she did, even though it meant learning a nonsense "fact." The subject cannot understand this item of knowledge; it has no meaning because "gloob" and "partice" are meaningless words. However, the subject can correctly respond to the question "What is a gloob?" Thus it appears that repetition helped in teaching this knowledge.

The experiment can be easily refined to obtain other insights into the learning of items of knowledge. As an investigator you might construct other "facts" to be learned and vary the number of repetitions from zero to ten. Replicating the study by testing additional subjects would provide more data which would be influenced less by individual behavior. You might ask the subject to repeat the sentence after you in order to involve him more effectively in the learning act. And you might vary the length of time between the "treatment" or conditioning phase of the study and the evaluation phase. This time period might vary from a few seconds to a day or two. Typically, you could expect to find that retention decreases when the time period lengthens.

In general, in this small study, you have been investigating the phenomena of the learning and forgetting curves. It is well known that during an instructional treatment a student's achievement level in the area of the instruction may be expected to increase. Even while avoiding any precise definitions of achievement and time, we may generally expect to be able to establish some plot or graph of achievement over a time scale, as illustrated in Figure 3-5. This curve, called a learning curve, characterizes the learning process. At point A on the curve, the subject begins a period of instruction at some initial achievement level. After a period of instruction, the student is at point C, which corresponds to a higher level of achievement. Often in arriving at the higher level indicated by point C

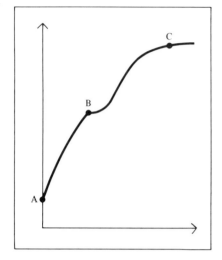

Figure 3–5 A learning curve.

the student will experience a plateau, indicated near point B. The characteristic of a plateau is that achievement increases little, or perhaps not at all for a period of time—that is, the slope of the learning curve decreases.

After a relatively high level of achievement has been attained, the phenomenon of the forgetting curve occurs, as illustrated in Figure 3–6. At point D the instruction has been terminated and immediately the subject begins to forget—his achievement decreases. The general drop in achievement is very rapid at first, but gradually the rate decreases to a new, relatively stable, achievement level, as indicated near point E. Usually the new achievement level is higher than the original achievement level (point A, Figure 3–5). It is often surprising, even to teachers, that students forget such a large percentage of what they learn.

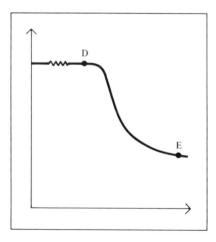

Figure 3–6 A forgetting curve.

Now, happily, a new phenomenon can occur if a new period of instruction is provided. A new learning curve is established which typically has a higher rate of increase in achievement than the original—that is, students can acquire a high achievement level in less time with subsequent instruction; the forgetting curve decreases less rapidly and typically the new lower level is associated with a higher achievement than the preceding level. Thus, with several instructional periods, a composite learning and forgetting curve for a student might appear as in Figure 3-7 (the phenomenon of plateaus has been omitted to simplify the sketch).

Over a period of time, repetition can be expected to increase achievement in learning items of knowledge and skills. With each repetition the item of knowledge or the skill becomes more assuredly fixed in the student and more easily available to him or her. If the student uses the knowledge or skill, he or she may be expected to retain this high level of achievement (see Research Highlight, "Reviewing and Retaining").

PARTIAL LEARNING

In learning items of knowledge and skills, few, if any, students are able to acquire the teacher's goal at once. The typical pattern is that part of the knowledge or skill is learned and then more is added. At the beginning of a school year, for example, teachers often memorize their students' names, sometimes even before they see the students in class. Some teachers proceed by developing a seating chart of the class and memorizing the names in sequence from the chart. During this process the teacher may first memorize the names in the first row, then those in the first and second rows, then those in the first three rows, and so on. This is a typical sequence for memorizing a large amount of information. The procedure has built into it a substantial review, particularly of the names of students in the first row.

After the names in an entire class have been memorized, the teacher may

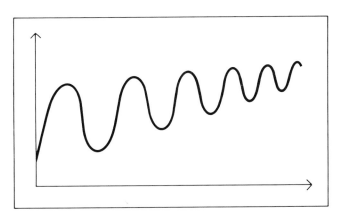

Figure 3-7 A learning and forgetting curve.

RESEARCH HIGHLIGHT | REVIEWING AND RETAINING

In a study conducted by Gay (1973), review of four rules involving principles from algebra and geometry was given to seventh and eighth graders either one day, one week, or two weeks after original learning. Three weeks from the day of original learning, the students were administered a delayed retention test on the four rules. All review groups gained considerably more than a no-review group, but the timing of the reviews was not significant.

In a second experiment, two reviews were given: after one and two days, one and seven days, or six and seven days. While all review groups retained significantly more than a no-review group, the group with both an early and a delayed review (after one and seven days) retained significantly more than the group with two early reviews.

It was also found that, in general, students in all groups needed the same number of examples to reach criterion at the time of the first review as at the time of original learning. At the time of the first review, however, the work took approximately 50 percent less time. At the time of the second review, students required half as many examples to reach criterion as at the time of learning, with time being 75 percent shorter.

Gay concluded that with respect to the retention of mathematical rules, one review is more effective than no review, regardless of temporal position. Optimal retention over a three-week interval is obtained with two reviews, one early and one delayed.

test himself or herself by starting with any row and thinking of the names in that row. The teacher might also begin at the back end of a row and review the student names. Often at this stage in the process the teacher engages in overlearning, memorizing the list so well that he or she will be unlikely to forget it and continuing to study even after he or she can list all the names in sequence. You may be able to recall the similarity of this sequence of events to the way you learned the multiplication tables in elementary school. Other mathematical knowledge and skills can be learned in much the same way. Thus, to learn that a good approximation to π is 3.1415927, the student might first concentrate on 3, then 3.14, then 3.1415, and so on until he can easily repeat the whole sequence. Similarly, learning a skill can be effectively improved by building on previous skills (Sobel, 1970, p. 302).

To teach students to simplify arithmetic expressions containing parentheses, for example, the teacher might first explain (that is, teach the *understanding*) that parentheses are used to specify the order of simplifying the expression. Then, when students are armed with the properties of integers (which are understandings), progressively more difficult exercises could be presented, as in this sequence:

$$3(2 - 1)$$
$$-3(5 + 2)$$
$$-3(6 - 12)$$
$$(5 - 3)(4 - 1)$$
$$. \ . \ . \ .$$
$$\{[(3 - 2)(5 + 2) + 5] - 3\}[(4 + 1)(2 + 3) + (7 - 4)(6 + 4)]$$

MNEMONICS

When we recognize that an important part of mathematics learning is memorizing, then we accept the role of mnemonics to promote this aspect of learning. Mnemonic devices are aids to memory and are used by everyone. Most mnemonics are created by individuals for a specific memory task.

Mnemonics may range from simple phrases to diagrams. Most people have heard the phrase "My Dear Aunt Sally," which is used to remember that *multi*plication and *d*ivision precede *a*ddition and *s*ubtraction when parentheses are not employed to specify an order of operations. And everyone who has taken a course in algebra has probably seen the diagram (Figure 3–8) which indicates all partial products in the product of two binomials. Some people recall the word *FOIL* to remember to calculate the products of the *f*irst terms, *o*utside terms, *i*nside terms, and *l*ast terms when multiplying binomials.

The learning of algorithms may be helped by the use of mnemonics. The mental pictures we all retain of the long division algorithm or multiplication algorithm serve as mnemonics for the procedures to follow. Diagrams in geometry might be considered mnemonics to help students recall the essential characteristics of a statement, theorem, or axiom. Figure 3–9 shows a mnemonic device for how the trigonometric functions vary from quadrant to quadrant. Recalling this picture can help greatly in sketching any of the functions. And some people remember "All Skiers Take Chances" as an aid in recalling the signs of the sine, cosine, and tangent functions in each quadrant: in I all are positive, in II only the sine is positive, in III only the tangent is positive, and in IV only the cosine is positive.

Figure 3–8 Mnemonic for the product of two binomials.

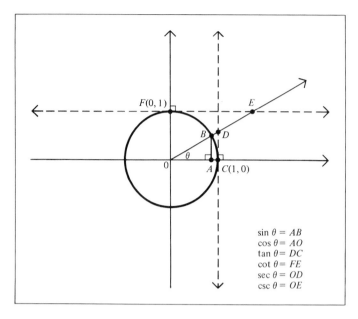

$\sin \theta = AB$
$\cos \theta = AO$
$\tan \theta = DC$
$\cot \theta = FE$
$\sec \theta = OD$
$\csc \theta = OE$

Figure 3–9 A mnemonic device for how the trigonometric functions vary from quadrant to quadrant.

Recognizing the role and use of mnemonics in mathematics is helpful. Because learning items of knowledge and skills are memory related, mnemonics have special significance in this type of learning. Mnemonics are less appropriate for learning items of understanding and problem solving.

VARIETY

Although variety might well be considered as part of the motivation process, we consider it separately here to emphasize its importance in teaching items of knowledge and skills. To practice a skill or to learn an item of knowledge well, repetition is highly important. Some students may have acquired enough internal motivation to be willing to expend long periods of time and individual effort in drill exercises. However, experienced teachers know that few students generally exhibit such motivation. These teachers attempt to provide a wide variety of activities for their students. They know that the problem of practice cannot be solved by a deceptively simple solution such as assigning lengthy homework drills or providing pages of similar exercises to be completed in class. Instead, the effective teacher plans for repetition and practice while being careful to keep the student from becoming discouraged or bored. Games may be devised to practice skills. Novel problems are located and employed that require the knowledge or skill to be learned (for examples see Appendix B). The effective teacher is mindful of changing the pace in the learning environment by varying individual work, group activities, and games.

The context and format of the practice exercises may be varied. Perhaps the students are to work on selected exercises from the textbook for ten minutes. Then the exercises are discussed by some students (a review). A little later the teacher may employ oral problems to continue to provide a practice session. As the items of knowledge or skills become fixed in the students' minds, a team game might be played (see Research Highlight, "A Gaming Note"). Here, for example, is a simple classroom game that provides practice in the skill of multiplying binomial expressions:

> The class is divided into two teams, the North Mathletes and the South Mathletes. Each member of a team constructs an exercise in multiplying binomials. Students are permitted to use exercises in their textbooks if they want, but they are advised to construct exercises they can solve. Opposing team members exchange exercises so that each student works on one problem. However, no one is to put his name on his solution because scores are to be tallied for teams, not individuals. The solutions are then returned to the originating teams and checked as a joint team effort. Each team's score is calculated by assigning one point for each correct solution and a bonus of one point for each incorrect solution of the opposing team which the team can solve on the chalkboard.

RESEARCH HIGHLIGHT | A GAMING NOTE

A series of studies at Johns Hopkins University focused on various ways in which the use of learning games (such as "Equations") affect student achievement and attitude in mathematics. Edwards and DeVries (1972) reported on one of the studies, in which seventh graders played games and were rewarded for winning by being given recognition in a classroom newsletter. Praise was given either to the team collectively or to individual students.

The attitudes of low- and average-ability students in team-reward classes were more positive toward mathematics class than were the attitudes of similar students in individual-reward classes. However, the reverse was true for high-ability students: they preferred to have *individual* recognition. You might consider these differing reactions as you plan games and select scoring procedures.

Finally, a small number of inherently interesting problems may be assigned as homework. This procedure intercepts the forgetting curve and may increase the achievement level the following day. Homework problems can serve also as a basis for review (repetition) on the succeeding day in order to reactivate the learning process.

The very nature of variety in teaching makes it impossible to specify the form of variety in a textbook; it depends on the teacher's own creative talents. However, it is clear that variety is a necessary element of effective teaching.

LESSON PLANS

The sequence of steps described thus far to develop a unit of instruction is:
- Determine the content area of the unit.
- Decide upon the student goals for the unit.
- Consider appropriate teaching strategies.

The next appropriate step is:
- Construct lesson plans for the unit.

Lesson plans are written outlines of lessons which are used by teachers as guides in teaching classes. We do not contend that all lesson plans should be written out in detail, or even that they necessarily be written. As teachers obtain more experience, they are able to determine accurately what materials are useful for them in their teaching. Beginning teachers usually find it helpful initially to prepare lesson plans. At first, the plans may be detailed. However, we recommend that the plans *not* be "scripted." Key questions or statements, such as definitions, may well be indicated in the plan, but it is usually a handicap to work with a written dialogue or a scripted talk to be presented in class. An outline or list of anticipated steps can be readily scanned as you teach; too many words, as in a dialogue, cannot, and they therefore prove less useful.

A more detailed discussion of the construction of lesson plans and sample plans is presented in Chapter 7. At this time we simply note that the preparation of lesson plans is an important part of the planning process (see Research Highlight, "Presenting Effective Lessons").

BASIC PROGRAMMING

You may be familiar with some computer programming language, such as FORTRAN, PL/I, ALGOL, or BASIC. We have selected BASIC because it is reasonably powerful and one of the easiest, most readily available computer languages that mathematics teachers may expect to encounter. An important feature of the language is that programs may be written after learning as few as two commands. Then new and more useful commands may be taught as they are needed. Thus, in the school setting, students can write programs after minimal instruction and then learn more and more about BASIC.

We introduce BASIC here for two reasons: (1) It may serve as a new learning experience in which you can identify the items of knowledge and skills to be

RESEARCH HIGHLIGHT | PRESENTING EFFECTIVE LESSONS

In a study by Smith (1974), 20 algebra teachers were given a list of lesson objectives. Each lesson was tape-recorded, and five variables were quantified: total lesson time, total time of teacher talk, total teacher words, frequency of "OKs" uttered by the teacher, and frequency of "uhs" uttered by the teacher. Coders analyzed the tapes for seven other variables including: the frequency of vagueness terms, the frequency of teacher-initiated student responses, the frequency of student-initiated student responses, the frequency of examples and applications, and the manner in which lesson objectives were met and dealt with.

Correlation coefficients indicated that the variable pertaining to lesson objectives was positively correlated with posttest achievement, as was the frequency of "OKs." The frequency of vagueness terms was negatively correlated with posttest achievement, as was the percentage of teacher talk.

Lesson effectiveness appears to depend on such factors as explicit objectives, reinforcement, and clarity of presentation.

learned. In Chapter 4 we extend this learning and emphasize understandings in using BASIC, and in Chapter 5 we will examine problem solving with computers. (2) This development of BASIC will provide a background for the presentation in Chapter 9 of instructional uses of computers in secondary schools. If you have already studied BASIC, you will be able to read through these sections rapidly. If not, you can become familiar with the language through careful study of these sections.

KNOWLEDGE OF BASIC

The BASIC programming language contains several items of knowledge to be learned. We have listed these items in a beginning section of Appendix C. There you will find definitions of program statement, numeric constant, string constant, numeric variable, string variable, arithmetic expression, and relational expression. Also you will find lists and definitions of the BASIC operators, functions, and commands. Together these items comprise an adequate and functional basis for using BASIC.

BASIC SKILLS: WRITING SIMPLE PROGRAMS

The first skill in learning BASIC is to write simple programs. BASIC commands, functions, and expressions may be combined to form statements and programs.

In this section we present some elementary programs to illustrate how they are constructed. Note that we use a box ⬛ to indicate a program and a scroll 〰 to indicate the output of a program. This notation is employed throughout the text.

```
10   LET X=100
20   PRINT X,X*2,X*3
30   END
```

This program will print

```
100          200          300
```

If each "," is replaced by a ";" in line 20, the output will appear closer together as:

```
100    200    300
```

```
100   READ A
200   LET X=3.14*A↑2
300   LET Y=3.14*2*A
400   PRINT A,X,Y
500   NØDATA 900
600   DATA 1,5,7,10
700   GØTØ 100
900   END
```

This program will calculate the areas and circumferences of circles with radii A = 1, 5, 7, and 10. The output would appear as:

```
1        3.14        6.28
5        78.50       31.40
7        153.86      43.96
10       314         62.80
```

Note that line 400 could be replaced by the following to obtain a more descriptive output:

```
400   PRINT "IF CIRCLE RADIUS=";A;",AREA=";X;"AND CIRCUMFERENCE=";Y
```

Alternatively, headings could be typed by adding the following lines

```
50  PRINT   "RADIUS    AREA   CIRCUMFERENCE"
60  PRINT   "_____  ____  _____"
```

The following program permits the user to input the number of scores he wishes to enter. Each score is requested in the loop from line 80 to line 120. After N scores are entered, the mean is calculated and reported. If you compare the program and the output for it, you may observe that a delimiter (, or ;) at the end of a PRINT statement will inhibit a carriage return, and that in this example the "?" for the INPUT statement will be printed. (Note that italics are used to indicate input provided by the computer user.)

```
40   REM TØ CALCULATE A MEAN
50   LET S=0
60   PRINT "HØW MANY SCØRES";
70   INPUT N
80   FØR I=1 TØ N
90   PRINT "SCØRE NUMBER ";I;"=";
100  INPUT X
110  LET S=S+X
120  NEXT I
130  LET M=S/N
135  PRINT
140  PRINT "THE MEAN ØF THE ";N;" SCØRES IS ";M
150  END
```

```
HØW MANY SCØRES?        6
SCØRE NUMBER 1=?        90
SCØRE NUMBER 2=?        80
SCØRE NUMBER 3=?        70
SCØRE NUMBER 4=?        85
SCØRE NUMBER 5=?        85
SCØRE NUMBER 6=?        90
THE MEAN ØF THE 6 SCØRES IS 83.3333333333
```

The next program uses IF-THEN and a clever way to determine if a number D is a divisor of a number M (see line 40) and N (see line 50). Note also that the PRINT statements in lines 5 and 75 are each equivalent to a carriage return. The technique in lines 80-100 is useful to permit the user to end or to execute the program again.

```
  1  REM GREATEST CØMMØN FACTØR ØF M<N
  5  PRINT
 10  PRINT "WHAT INTEGERS FØR GCF";
 20  INPUT M, N
 25  PRINT "GREATEST CØMMØN FACTØR ØF ";M;" AND";N;"IS";
 30  FØR D=M TØ 1 STEP −1
 40  IF M/D <> INT(M/D) THEN 70
 50  IF N/D <> INT(N/D) THEN 70
 55  PRINT D
 60  GØ TØ 75
 70  NEXT D
 75  PRINT
 80  PRINT "ANØTHER GCF (1=YES)";
 90  INPUT A
100  IF A=1 THEN 10
110  END
```

```
WHAT INTEGERS FØR GCF? 12, 18
GREATEST CØMMØN FACTØR ØF 12 AND 18 IS 6
ANØTHER GCF (1=YES)? 1
WHAT INTEGERS FØR GCF? 12, 19
GREATEST CØMMØN FACTØR ØF 12 AND 19 IS 1
ANØTHER GCF (1=YES)? 0
```

The final program demonstrates that string variables may be inputs and used in the program. Also the addition problem consists of two randomly generated whole numbers between 1 and 100, inclusive. Note the technique to obtain X and Y in lines 600 and 650.

```
400  PRINT "WHAT IS YØUR NAME";
450  INPUT A$
500  PRINT "HELLØ,"; A$
550  PRINT "HERE IS AN ADDITIØN PRØBLEM"
600  LET X=INT(100*RND(1)+1)
650  LET Y=INT(100*RND(1)+1)
700  PRINT "   ";X
750  PRINT "+";Y
800  PRINT "____"
```

```
 850   INPUT Z
 900   IF Z=X+Y THEN 1050
 950   PRINT "SØRRY, YØU GØØFED. ";X;"+";Y;"=";X+Y
1000   STØP
1050   PRINT "GREAT," A$; ",YØU GØT IT!"
1100   END
```

```
WHAT IS YØUR NAME? BILL
HELLØ, BILL
HERE IS AN ADDITIØN PRØBLEM
      39
    +52
    ----
?  91
GREAT, BILL, YØU GØT IT!
```

PROGRAMMING EXERCISES

Several simple exercises follow which may be used to practice programming in BASIC. To learn the language you will have to practice. With study you can acquire (1) knowledge of the meaning of the operations, functions, and commands and (2) basic skill in writing simple (for now) programs.

1. Write a program to print the multiples of five up to 100.
2. Write a program to present a two-digit subtraction example and check the answer.
3. Write a program to print the prime numbers less than 100.
4. Write a program to find the smallest natural number having at least 20 factors.

CRITICAL INCIDENTS REVISITED

The critical incidents presented at the beginning of this chapter were selected to raise questions about teaching mathematical knowledge and skills. After studying this chapter, you should be convinced that items of knowledge and skills are important parts of mathematical learning, that you are able to specify such goals, and that you have learned to associate specific teaching strategies with these goals. Finally, if you studied the section on BASIC programming, you should have acquired fundamental knowledge about the language and sufficient programming skills to write short, specific programs, such as those required in the exercises in that section. Study of the sections on BASIC in subsequent chapters will provide an opportunity for you to achieve understanding and problem-solving goals in the use of BASIC.

Now let's reconsider the critical incidents. Bill (in the first critical incident) could not solve a problem because he did not remember the double angle formula for sin 2x. Bill believes he could have solved the problem if he had acquired the following knowledge goal: the student will know that $\sin 2x = 2 \sin x \cos x$. Bill could have used the following formula if he knew and understood it:

$$\sin (A + B) = \sin A \cos B + \cos A \sin B$$

We cannot tell from reading the critical incident whether Bill's teacher had selected a goal on a knowledge level or on an understanding level. The teacher's classification of the goal is important. In this incident it might well have been the case that Problem 3 was designed to see if the students *knew* the double angle formula.

The second critical incident demonstrates that students may have knowledge of relatively obscure facts. Here Lori may have learned the interesting relationship among the decimal representations of fractions with denominators of 7:

$$\tfrac{1}{7} = .142857\overline{142857}$$
$$\tfrac{2}{7} = .285714\overline{285714}$$
$$\tfrac{3}{7} = .428571\overline{428571}$$
$$\tfrac{4}{7} = .571428\overline{571428}$$
$$\tfrac{5}{7} = .714285\overline{714285}$$
$$\tfrac{6}{7} = .857142\overline{857142}$$

We might speculate that Lori became interested in this surprising result at some time and memorized the sequence of digits in the decimal representation of 1/7. With this knowledge she is able to specify the other representations indicated by simply determining the first digit. For example, 3/7 is slightly less than 1/2, so she can guess that the decimal expansion of 3/7 starts with .4, and she knows the sequence of digits that follows.

Lori's motivation for acquiring the knowledge is relevant. When she first saw the relationship among the decimal representations, she was probably interested in the fact. However, she appears to have memorized the sequence of digits. Perhaps her teacher or one of her parents encouraged her to memorize the sequence; we don't know, but we do see evidence that she memorized. And we see that she enjoys having acquired the knowledge ("That's easy").

The third critical incident is an example of a teacher advising students to acquire knowledge and skill goals ("make sure you know the quadratic formula and how to solve quadratics by factoring"). We might feel that teachers should not *threaten* students with a quiz, and we did not select this incident as an example of a threat; rather, we present it as an example of a teacher indicating clearly to students that particular knowledge and skill goals have been selected. We hope the students understand that they are being advised to acquire the indicated goals.

The fourth and fifth critical incidents were selected to emphasize that students may perceive all mathematics as knowledge (the fourth) or as knowledge and skills (the fifth). Although the focus in this chapter has been that knowledge and skills are important, we want students to believe that the study of mathematics also includes attaining understanding and problem-solving goals. In these critical incidents we encounter again the interrelationship of goals in the affective domain and goals in the cognitive domain. Hopefully, your students will believe that items of knowledge and basic skills in mathematics are useful, but these learnings are not sufficient for an adequate foundation in any mathematics course. Subsequent chapters provide descriptions of strategies for teaching understanding and problem-solving goals.

The strategies presented in this chapter for teaching knowledge and skill goals are (1) repetition, (2) partial learning, (3) use of mnemonics, and (4) variety. The techniques described in these sections are useful in helping students to achieve knowledge and skill goals.

Activities

1. Select a standard textbook for any secondary school course, pick a chapter of interest to you, and prepare a list of knowledge and skill goals for a unit of instruction.
2. Make a list of the square roots, to three decimal places, of 2, 3, 5, 6, 7, 8, and 10. Memorize the list. Describe the strategy you used. In what order did you memorize the list? Some square roots are easy to memorize, others are not. Why? How long did it take you to learn all these square roots?
3. One day after completion of Activity 2, test yourself. How many of the square roots were you able to recall? Restudy the list, noting the time it takes you to master them.
4. Select a small number of related knowledge and skill goals—perhaps only two of each type. Develop *two* lesson plans for these goals. In the first lesson plan, assume that you are a teacher who believes mathematics is a body of facts and skills which students should learn. In the second lesson plan, assume that you are a teacher who believes this knowledge to be of little value but that students should learn processes and explore mathematical ideas (see Chapter 7 for two sample lesson plans). Which plan would you prefer to use? Why?
5. What is the importance of memorizing mathematical knowledge and learning skills? Recall some of the facts and skills you mastered when you were a calculus student. Look in a textbook to identify items you have forgotten, but which you would like to be able to recall.
6. How would you reply to the student in the first critical incident at the beginning of this chapter if you were his teacher?

7. Refer to the fourth and fifth critical incident. Suggest reasons why students might say such things. How could you as a teacher conduct your teaching so that students do not acquire such beliefs?
8. Write BASIC computer programs for the exercises listed in the chapter.
9. Refer to Appendix C. Suppose you are to teach a group of seventh-grade students an introductory unit on BASIC programming. List the items of knowledge and skills you would select so that they could begin writing programs in only a few days, perhaps even one or two.
10. If you have access to students, prepare a brief lesson (about 20 minutes long) to teach them items of knowledge and/or a basic skill. Teach the lesson to about three students in a group, being sure to include an evaluation, and write a reaction to your teaching. What parts were successful? What parts would you change?

References

Brown, John Kenneth, Jr. Textbook Use by Teachers and Students of Geometry and Second-Year Algebra. (University of Illinois at Urbana-Champaign, 1973.) *Dissertation Abstracts International* 34A: 5795–5796; March 1974.

Edwards, Keith J. and DeVries, David L. *Learning Games and Student Teams: Their Effects on Student Attitudes and Achievement.* Baltimore, Maryland: Johns Hopkins University, 1972. ERIC Document No. ED 072 391.

Gay, Lorraine R. Temporal Position of Reviews and Its Effect on the Retention of Mathematical Rules. *Journal of Educational Psychology* 64: 171–182; April 1973.

Kemeny, John G. and Kurtz, Thomas E. *BASIC Programming.* 2nd ed. New York: Wiley, 1971.

Piatt, Robert George. An Investigation of the Effect the Training of Teachers in Defining, Writing and Implementing Educational Behavioral Objectives Has on Learner Outcomes for Students Enrolled in a Seventh Grade Mathematics Program in the Public Schools. (Lehigh University, 1969.) *Dissertation Abstracts International* 30A: 3352; February 1970.

Smith, Lyle Ross. Aspects of Teacher Discourse and Student Achievement in Mathematics. (Texas A&M University, 1973.) *Dissertation Abstracts International* 34A: 3716; January 1974.

Sobel, Max A. Skills. In *The Teaching of Secondary School Mathematics.* Thirty-third Yearbook of the National Council of Teachers of Mathematics. Washington: The Council, 1970.

BASIC Language Reference Manual. Publication Number 60306200 Revision C. Sunnyvale, California: Control Data Corporation, 1972.

Introduction to an Algorithmic Language (BASIC). Reston Virginia: National Council of Teachers of Mathematics, 1969.

Programming Languages: PDP-8 Family Computer, Vol. 2. Maynard, Massachusetts: Digital Equipment Corporation, 1970.

Understanding

AFTER STUDYING CHAPTER 4, YOU WILL HAVE:

★ learned the distinction among axioms, definitions, theorems, principles, and concepts.
★ examined statements of understanding goals.
★ explored these strategies for teaching understandings: authority teaching (including telling, use of analogies, and demonstrations), interaction and discussions, discovery teaching, and laboratory experiences.
★ learned the key elements of a discovery lesson.
★ extended your understanding of BASIC programming.

CRITICAL INCIDENTS

1. Carol asks, "Miss Hemmerly, do we have to memorize the proof of the theorem for the quiz?"
2. Sid explains his solution: "We know that $f(x)$ is continuous for all positive x. Also we found that $f(2) < 0$ and $f(3) > 0$. So, if we apply the Intermediate Value Theorem, we know that there is an x, such that $2 < x < 3$ and $f(x) = 0$."

3. Barb describes her solution to the homework problem: "All we have to do to solve the inequality $|r + 3| < 5$ is to use the definition of absolute value:

$$|x| = \begin{cases} x, \text{ if } x \geq 0 \\ -x, \text{ if } x < 0 \end{cases}$$

These two conditions lead to either

$$-3 \leq r < 2$$

or

$$-8 < r < -3$$

Therefore, any r such that $-8 < r < 2$ will be in the truth set."
4. Bill argues that $x^2 + 4 = 0$ has no solution, but Jim points out that it does if x can be a complex number.

THE ROLE OF UNDERSTANDINGS

The basis of mathematical thought is the set of understandings that has been developed over the ages. As Thurstone said, "It is the faith of all sciences that an unlimited number of phenomena can be comprehended in terms of a limited number of concepts or ideal constructs" (Thurstone, 1953, p. 51). Mathematics, as a natural science, is a collection of specific statements and generalizations which describe the behaviors, operations, and interrelationships of elements within the several systems of mathematics.

There are a great many mathematical systems, among them whole numbers, integers, plane Euclidean geometry, quaternions, finite dimensional vector spaces, and real numbers. Suppose someone wanted to study one of these systems. One way to begin would be to attempt to list all the statements which are believed to be true within the system. For example, the following are only a few statements one might choose for the whole number system (which includes the set W of whole numbers; the operations of addition, subtraction, multiplication, and division; and the relations of "equal to" and "less than"):

Assume a, b, and c are arbitrary elements of W.

1. $a + b = b + a$.
2. $a - b = a + (-b)$.
3. There is an element 0 such that $a + 0 = a$.
4. If $a < b$ and $b < c$, then $a < c$.
5. For every element a in W, there is another element in W called the *successor of a*, denoted by $s(a)$.
6. 0 is not the successor of any element in W.

7. If $s(a) = s(b)$ then $a = b$.
8. 21 is in W.
9. $21 + 17 = 38$.
10. 7 and 1 are the only divisors of 7.

The list could become very long. However, some of the statements are logical consequences of others. Thus the student might reasonably attempt to find a finite set of statements which would be sufficient to derive all (or most) of the remaining statements.

Such a set of statements is fundamental within the system, although no particular set is unique. Another person might select another set of basic statements. It is in the nature of mathematics to study a small number of statements that can be used to derive other statements in the system. All the statements of the system may be classified as *understandings*; to understand many or all of these statements is to understand the system. But all of the statements have other characteristics. Some may be classified as generalizations and others as specific instances. Some are definitions, others may be axioms, and others may be theorems.

It is the purpose of this chapter to consider various types of understandings and the ways in which understandings may be effectively taught.

TYPES OF UNDERSTANDINGS

SPECIFIC INSTANCES

Statements which are true in one particular case will be called specific instances. In the context of a particular problem, discussion, or observation a specific instance may be noted. The following examples might be specific instances in particular problems or discussions:

▶ $3 < 6$
▶ Jack is 18 years old.
▶ 3.1416 is a close approximation to π.
▶ $3 + 4 = 7$

By definition, specific instance statements have little generality. They are classified as understandings because they are items to be understood, but the application of specific instances is limited to a particular case. To understand a specific instance is to comprehend the meaning of its statement—to be able to apply it.

GENERALIZATIONS

By far the most important understandings in mathematics are generalizations, that is, statements which are true for all instances in a given universe or system.

Generalizations are valuable to the student because he can apply them in many instances. However, the student must *understand* the generalization. It typically is not sufficient for the student to learn the generalization as an item of knowledge (which would imply only that he can repeat it). Rather, he must acquire understanding of the statement so that he is able to apply it or use it in appropriate situations.

GENERALIZATIONS IN MATHEMATICAL SYSTEMS

Generalizations are used so widely in mathematics that special classifications of them are common. Within mathematical systems three types of generalizations are identified: axioms, definitions, and theorems. This is a classification of generalizations according to the way we determine truth value: assumed, stipulated, and proved or provable, respectively.

Axioms may be defined as "statements accepted without proof." Like other general statements, axioms are statements which may be applied throughout a mathematical system. But axioms have another important function. In every mathematical system some terms are designated "primitive terms" (often called "undefined terms"). The axioms of the system provide information about the primitive terms; that is, the primitive terms are defined to the extent that the axioms of the system stipulate their properties. For example, in geometry the following are often primitive terms: "point," "line," and "between." A common axiom is: "Two points lie on exactly one line." Thus, whatever interpretation or model we have in mind for "point" and "line," the terms in the system must satisfy the constraint that every pair of points is coincident with only one line. Axioms are classified as generalizations because, by definition, they are true statements in the mathematical system and apply to all relevant instances. Euclid and other mathematicians have sometimes distinguished between axioms and postulates. However, in modern day mathematics it is common to use the terms interchangeably.

Definitions are another class of generalizations found in mathematical systems. A definition is a statement of the meaning of a term or a phrase. Thus, as universally true statements in a mathematical system, definitions are classified as generalizations. The following are examples of definitions:

1. Triangle ABC is the union of $\overline{AB}$, $\overline{BC}$, and $\overline{CA}$ where A, B, and C are noncollinear points.
2. $x - y$ means $x + (-y)$. (definition of subtraction)
3. $\dfrac{a}{b} + \dfrac{c}{d} = \dfrac{ad + bc}{bd}$ where a, b, c, and d are integers, $b \neq 0$, $d \neq 0$. (definition of addition of rationals)
4. Two lines that meet so as to form a right angle are *perpendicular*.

Definitions are statements of double implication. They can always be written in a form using "if and only if." In general, they state that a term or a phrase may be used whenever certain conditions are met and that whenever the term or phrase is used, the stated conditions hold. Definitions 1 and 4 above, for example, may be rewritten as follows:

1. A figure is a *triangle* if and only if there exist noncollinear points A, B, and C such that the figure is the union of $\overline{AB}$, $\overline{BC}$, and $\overline{CA}$.
4. Two lines are *perpendicular* if and only if the lines intersect and form a right angle.

Theorems are another class of generalizations in mathematical systems. Theorems are general statements which can be shown to be logical consequences of the axioms, definitions, and previously proven theorems in a mathematical system. Thus the theorems of a mathematical system are classified as generalizations in a mathematics course. Most theorems may be written in the form of a statement of implication: "if . . . , then . . . ." The following are two different statements of the same theorem:

1. Two points on the same side of a line *l* have no point of *l* between them.
2. If A and B are two points on the same side of a line *l*, then there is no point of *l* between A and B.

Some other theorems may be double implications or biconditional. Such theorems may be written as two theorems, each a statement of implication. For example:

3. $a \cdot b = 0$ if and only if $a = 0$ or $b = 0$
4. If $a \cdot b = 0$ then $a = 0$ or $b = 0$
    and
If $a = 0$ or $b = 0$ then $a \cdot b = 0$

The terms *lemma* and *corollary* merely refer to special theorems. A lemma is a theorem which typically is easier to prove than a theorem associated with it and is used in proving the associated theorem. A corollary is a consequence of a main theorem and it typically is easy to prove after the main theorem is established. Lemmas and corollaries are special theorems and thus are classified as generalizations.

PEDAGOGICAL GENERALIZATIONS

Primitive terms, axioms, definitions, and theorems (together with a calculus of logic) are the formal components of a mathematical system. Axioms, definitions, and theorems are all generalizations and must be taught as such. These classifications are especially useful in studying formal mathematical systems. However, much of the study of mathematics is not a formal study of mathematical systems but a study of interrelationships between many elements in mathematical systems and the use of these interrelationships in solving problems, developing mathematical skills, and acquiring mathematical knowledge. Thus two other classes of generalizations are of practical value in the *teaching* of mathematics.

Principles, as we indicated in Chapter 2, are either prescriptions or understandings. When principles are prescriptions, they are taught as skills. Thus teachers who decide to teach the use of the quadratic formula as a skill will teach their students the prescription:

▶ If $ax^2 + bx + c = 0$, where a, b, and c are real numbers, and $a \neq 0$; then

$$x = \frac{-b \pm \sqrt{b^2 - 4ac}}{2a}$$

That is, to find the value of x *such that* $ax^2 + bx + c = 0$, simply let

$$x_1 = \frac{-b + \sqrt{b^2 - 4ac}}{2a} \quad \text{and} \quad x_2 = \frac{-b - \sqrt{b^2 - 4ac}}{2a}$$

Students could learn to solve equations such as $3x^2 + 4x + 7 = 0$ as a skill and yet not understand the quadratic formula. A principle is not necessarily an understanding.

When principles are taught as understandings, they are not skills. *Implicit in the goals of teaching understandings are student abilities of applying, deriving, or deducing a consequence of a generalization.* Thus, referring to the example of the quadratic formula, the student who has learned the generalization as an understanding can (1) apply it, but not as a prescription; (2) derive it, but not as a result of merely memorizing the proof; or (3) use it to deduce a consequence such as:

▶ The sum of the roots of a quadratic equation in the form

$$ax^2 + bx + c = 0 \text{ is } \frac{-b}{a}$$

Axioms, definitions, and theorems may all be principles. The classification of a generalization in a mathematical system does not influence the roles of these generalizations in student applications. Thus students who are learning about a formal mathematical system will use the terms *axiom*, *definition*, and *theorem* for the appropriate types of generalizations in the system, while students who are solving problems or exercises will use all of these generalizations as principles which they may apply. A generalization may be taught as an item of knowledge, as a skill, or as an understanding. If it is taught as an item of knowledge, the goal for the student is to be able to recall and state it. If it is taught as a skill or prescription, the goal for the student is to use it to solve routine exercises. If a generalization is taught as an understanding, then the goal for the student is to apply it, derive it, or use it to deduce a consequence. Axioms, definitions, and theorems may be taught so that they are learned as knowledge, skills, or understandings.

Concepts are our final classification of generalizations in mathematics and are perhaps the most difficult to discuss because of the great variety and complexity of concepts which occur in mathematics classrooms. As we indicated in Chapter 2, concepts are associated with such mathematical terms as "triangle," "whole numbers," "least upper bound," "limit of a function," and "integration." Some concepts may be defined in a single statement, but teaching and learning the concepts may involve much more. Other concepts are acquired over a long period of time, perhaps even years. The concepts of number, measure, and proof are introduced in early grades and taught throughout secondary school and college. We emphasize the idea that concepts are associated with terms, or labels of the concepts, and we define the teaching of concepts to be the teaching of the meaning of terms.

Concepts help unify many mathematical ideas. Difficult though they may be to describe, concepts are a major part of mathematical learning; they are important because they help students understand and communicate mathematical ideas.

How does one describe a concept? Take, for example, the concept of right triangle. How can you tell if a student has acquired this concept? First we must agree on what the student is to learn. Some teachers might decide that their students should be able to use accurately the term "right triangle." Their students would merely have to demonstrate that they could distinguish in a set of figures between those which are right triangles and those which are not. But there is much more to understand about right triangles. Does the concept include understanding the Pythagorean Theorem, understanding that the sine of either acute angle in a right triangle is the length of opposite side divided by the length of the hypotenuse, and so forth? To describe adequately what is to be learned about a concept, the student may find a generalization helpful in being precise.

Thus, with a particular concept, you may have several associated generalizations that students are to learn. As a teacher you must decide whether each of these generalizations is to be learned as an item of knowledge, as a skill, or as an understanding. Consider for example the concept of parallelism. The following generalizations about parallelism have been assigned to categories of learning. Would you classify them differently?

- Parallel lines do not intersect. (Knowledge)
- The alternate interior angles formed by a transversal and two parallel lines are congruent. (Skill)
- If l_1 and l_2 are parallel lines and k is a line parallel to l_1, then k and l_2 are parallel. (Understanding)

CATEGORIES OF GENERALIZATIONS

We have explored several classifications of statements in mathematics teaching. Now we summarize these different classifications. First we noted that statements could be classified as specific instances or as generalizations, and we argued that generalizations are extremely important in learning mathematics. Next we noted that generalizations in the context of a mathematical system may be classified as axioms, definitions, and theorems. However, in teaching mathematics it is often useful to classify generalizations as principles or concepts. Specific instances are taught as facts. Finally, generalizations may be classified by the goals of teaching; the categories are knowledge, skills, and understandings. Figure 4–1 depicts all these ways of classifying statements in mathematics.

PLANNING TO TEACH UNDERSTANDINGS

STATING UNDERSTANDINGS

In stating understandings it is important to state precisely *what* the student is to understand. Avoid labels of the understanding or concept. Do not say that students are "to learn right triangles" or that they are "to understand the Pythagorean Theorem." Say rather that "The student is to understand that the area of a right triangle is equal to one half the product of the length of the triangle's legs" or "The student is to understand that in a right triangle with legs a and b and hypotenuse c, $a^2 + b^2 = c^2$." In this way you state the understanding precisely.

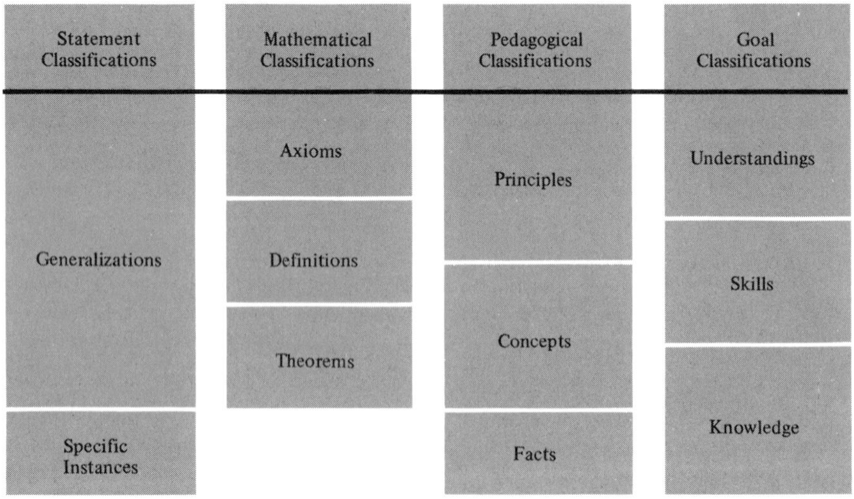

Statement Classifications	Mathematical Classifications	Pedagogical Classifications	Goal Classifications
Generalizations	Axioms	Principles	Understandings
	Definitions		Skills
	Theorems	Concepts	
Specific Instances		Facts	Knowledge

Figure 4–1   Classifications of (true) statements.

As shown in Chapter 2, the most effective way to state understanding goals is to use the word *that* to indicate specifically what the student is to understand. All understandings may be stated in this way. Axioms, definitions, theorems, and principles are themselves explicit; therefore, stating them is a relatively straightforward task. Concepts, typically identified simply as terms, require more effort. The teacher must decide which understandings related to the concept that the students are to learn. These understandings will define the level of attainment of the concept at that time.

A SET OF TYPICAL UNDERSTANDING GOALS

The following are examples of understanding goals appropriate for the mathematics courses indicated.

Arithmetic (seventh and eighth grades)

1. The student will understand that the relative error of a measurement is the greatest possible error divided by the recorded measurement, i.e.,

$$\text{relative error} = \frac{\text{greatest possible error}}{\text{recorded measurement}}$$

2. The student will understand that the set of integers is closed with respect to addition and subtraction.
3. The student will understand that any decimal numeral may be written in scientific notation.

General mathematics

1. The student will understand that the volume of a rectangular solid is the product of its length, width, and height (Figure 4–2).
2. The student will understand that the precision of a measurement varies directly with the size of the smallest unit with which the measurement is made.

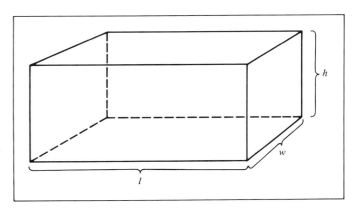

Figure 4–2 An understanding goal for general mathematics students: that the volume of a rectangular solid is the product of its length (l), width (w), and height (h).

3. The student will understand that to find the value of an expression that contains more than one operation, the multiplications and divisions in the order in which they occur (from left to right) are done first and then the additions and subtractions are performed—unless parentheses have been used to indicate another order to evaluate the expression.

Algebra

1. The student will understand that the product of the slopes of two perpendicular lines is -1, if neither line is vertical (Figure 4–3).
2. The student will understand that an irrational number has a nonrepeating, nonterminating decimal expansion.
3. The student will understand that one method of solving quadratic equations is by completing the square.

Geometry

1. The student will understand that corresponding parts of congruent triangles are congruent.
2. The student will understand that the distance between any two points in space is

$$\sqrt{(x_2 - x_1)^2 + (y_2 - y_1)^2 + (z_2 - z_1)^2}$$

where (x_1, y_1, z_1) and (x_2, y_2, z_2) are the coordinates of the two points (Figure 4–4).

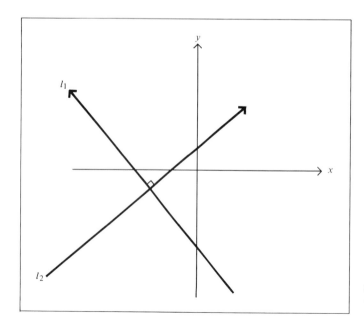

Figure 4–3   An understanding goal for algebra students: that the product of the slopes of two perpendicular lines is -1, if neither line is vertical.

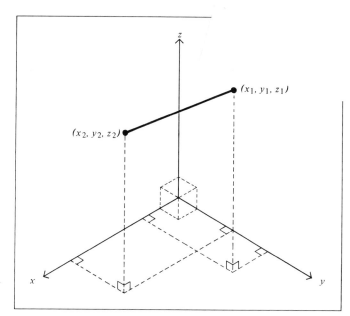

(x_1, y_1, z_1)

(x_2, y_2, z_2)

Figure 4–4 Diagram of an understanding goal for geometry students (Example 2).

3. The student will understand that the sum of the (degree) measures of the angles of a triangle is 180.

Advanced mathematics
1. The student will understand that over the set of complex numbers, $(x - r)$ is a factor of a polynomial $P(x)$ if and only if r is a root of $P(x) = 0$.
2. The student will understand that for all real numbers x and y, exactly one of the following statements is true:

$$x < y \qquad x = y \qquad x > y$$

3. The student will understand that the sum of the first n terms of a geometric progression whose first term is a and whose common ratio is r (where $r \neq 1$) is given by the expression

$$\frac{a - ar^n}{1 - r}$$

TEACHING UNDERSTANDINGS

The distinguishing characteristic of understandings is that they are to be applied, derived, or used to deduce a consequence. Given an exercise which can be solved with the aid of an understanding, the student should be able to use the acquired understanding and perhaps some items of knowledge and some skills to solve the exercise. Knowing the generalization is not in itself sufficient for using it. Therefore the techniques of repetition, partial learning, and mnemonics which are appropriate in teaching items of knowledge and skills are not appropriate in teaching understandings. However, the other method of teaching knowledge and skills, variety, is appropriate in all mathematical learning. Several other appropriate strategies for teaching understandings will be described in this section.

AUTHORITY TEACHING

One of the more commonly used strategies in teaching mathematics is authority. The teacher, as an authority, simply states the understanding to be learned. Often the authoritarian statement of the understanding is accompanied by a justification that explains it. For example, a teacher might say:

▶ The area of a right triangle is one-half the product of the lengths of the legs. Thus, in this "3, 4, 5" right triangle (Figure 4–5), the area is $6m^2$, that is

$$A = \frac{(3m)(4m)}{2}$$

Notice that if we put two such triangles together they would form a rectangle of sides $3m$ and $4m$, and the area would be $12m^2$. Each triangle is half this area, or $6m^2$ (Figure 4–6).

There are several techniques a teacher may use in authority teaching. Three of them are telling, analogy, and demonstration.

Telling is defined as stating an understanding without justification; it is the quickest way to present an idea, the most commonly used method of beginning teachers, and probably the least effective way to teach the understanding. It is quick because the teacher, a person with a substantial background of knowledge, understanding, and experience in mathematics, can efficiently express important understandings. The method appeals to beginning teachers because telling (or stating) understandings closely parallels the form in which they have retained

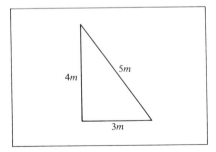

Figure 4–5.

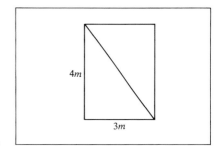

Figure 4–6.

understandings, regardless of how they learned them. Finally, telling is least effective in the sense that it is an insufficient technique by itself; that is, simply telling the student an understanding is seldom sufficient for student learning.

There are counterexamples to this last generalization, as in cases where a student has thought about ideas related to an understanding. At the right time, your pointing it out may be sufficient for him or her to learn the understanding. Suppose that a student has learned to plot quadratic functions on a coordinate plane and knows how to find the roots of quadratic equations. It might be the case that a teacher could say: "See, the real roots of the quadratic equation, if they exist, are the x-coordinates or abscissa of the points where the graph of the function crosses the x-axis—that is, they are the values of x for which $f(x)$ is zero." Thus it could happen that, for the first time, the student would understand that the zeros of a quadratic expression are related to points of the corresponding function at its intersection with the x-axis. The student might apply this generalization in solving the quadratic:

$$f(x) = x^2 - 4x - 5$$
$$= (x + 1)(x - 5)$$

thus,

$$f(x) = 0, \quad \text{when} \quad x = -1 \quad \text{or} \quad x = 5$$

Therefore, the graph of $f(x)$ will cross the x-axis at $(-1, 0)$ and $(5, 0)$. In this example, the understanding was probably learned by the student because he or she had already learned skills, items of knowledge, and other understandings about zeros of quadratic expressions and about graphs of quadratic functions.

Typically, however, teaching understandings requires more than simply telling. If telling were sufficient, mathematics teaching would be far easier! Something must happen within students for them to learn an understanding; therefore, although telling may be a part of the teaching act, teachers have learned to employ other techniques.

Analogy is one of the techniques used to help students learn understandings. As described by Polya (1973, p. 37), "Analogy is a sort of similarity. Similar

objects agree with each other in some respect, analogous objects *agree in certain relations* of their respective parts." Use of analogies is classified here as a technique under authority teaching because the validity of the analogy is based on the teacher's authority, the analogy is presented to the students by the teacher (or by a textbook or an activity card), and student participation in such a learning experience is essentially mental.

Analogies are employed to suggest to students that statements of understandings are true in one system because they are true in another, more familiar but similar, system. Here are examples of analogies that could be used in mathematics classrooms:

1. Plotting a point in the coordinate plane is like locating an address on a city map (Figure 4–7).
2. If we think of a function as a machine in which we enter a number and receive a single answer, then the inverse of a function may be thought of as a machine which works in reverse (Figure 4–8).
3. A point might be thought of as small dot on a piece of paper.
4. A ray is similar to a narrow beam of light from a flashlight.

In each analogy some relatively familiar system is described and an understanding is explained by indicating, noting, or implying that the (implicit) understanding will be true in a new, less familiar system. Analogies are useful because they provide students with mental images or constructs to acquire the understanding. Of course there is a danger that students might overgeneralize or assume that all analogies work. For example, the following surprise is based on assuming that a property of real numbers may be applied to complex numbers.

$$\sqrt{-1} = i$$

so

$$\sqrt{-1} \cdot \sqrt{-1} = i \cdot i$$

but

$$\sqrt{-1} \cdot \sqrt{-1} = \sqrt{(-1) \cdot (-1)}$$
$$= \sqrt{1}$$
$$= 1$$

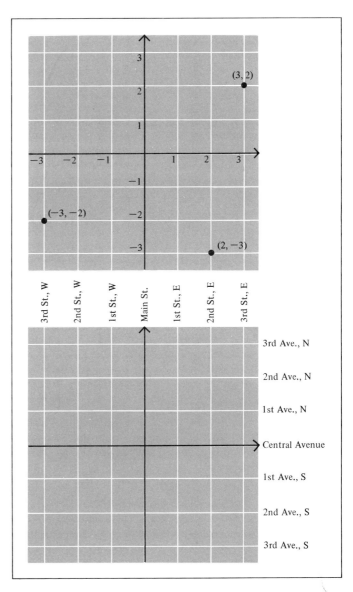

Figure 4–7 Coordinate plane and city street map.

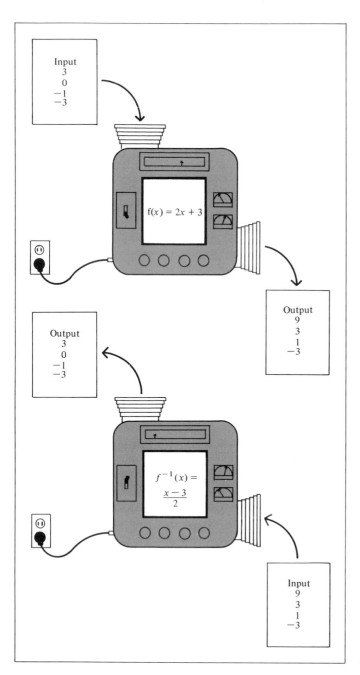

Figure 4–8 Function machines: $f(x)$ and $f^{-1}(x)$.

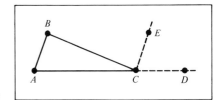

Figure 4–9

and

$$i \cdot i = i^2$$
$$= -1$$

thus

$$1 = -1$$

Analogies are often used to introduce understandings. The mathematical soundness of analogies is questionable, but their pedagogical use is well established by tradition. As all teachers use analogies, it is important that students become able to distinguish between arguments based on analogies and arguments based on standard rules of logic.

Demonstration is a third category of authority teaching. In using this method, the student carefully directs the sequence of information for the students to acquire the understanding. However, unlike telling and analogy methods, the teacher uses the properties of the mathematical system to justify or deduce the understanding.

Teacher (and textbook) proofs of theorems may be classified as demonstrations. Consider the following proof that the exterior angle of a triangle is equal in measure to the sum of the measures of the remote interior angles (Figure 4–9).

▶ Suppose that $\angle BCD$ is an exterior angle of $\triangle ABC$ such that $\angle CBA$ and $\angle CAB$ are the remote interior angles. Let E be a point in the B-side of $\overleftrightarrow{AD}$ such that $\overleftrightarrow{CE} \parallel \overleftrightarrow{AB}$. Then $\angle ECD$ and $\angle BAC$ are corresponding angles. Therefore,

$$\angle ECD \cong \angle BAC.$$
$$\angle ABC \text{ and } \angle ECB \text{ are alternate interior angles.}$$
$$\angle ABC \cong \angle ECB.$$

Thus,

$$m(\angle BCD) = m(\angle ECD) + m(\angle ECB)$$
$$= m(\angle BAC) + m(\angle ABC)$$

Such a demonstration proof might be employed to show students how to construct proofs, or it might be employed to teach the understanding (that is, the theorem) specified.

One way for students to demonstrate that they have acquired an understanding which is a theorem is to prove it. But remember that students can memorize proofs of theorems! Their ability to recall a proof does not necessarily demonstrate their understanding of the theorem; it may simply indicate *knowledge* of the proof. The role of memory causes difficulty in making clear distinctions between knowledge of a proof and understanding. The students might remember (in the example above) that using the line through the vertex of the exterior angle which is parallel to the opposite side of the triangle is the key to the proof. Most teachers would believe that the student had not memorized the proof in this case. Student memorization of proofs for examinations may be minimized by avoiding a request for proofs which have been shown in class, by presenting partial proofs, and by asking for proofs of other, usually simpler, theorems.

Algebra is a rich source of theorems, and many are proved by teacher demonstrations. For example, justification of the quadratic formula is most often shown by a teacher demonstration of completing the square of

$$ax^2 + bx + c = 0$$

Teacher demonstrations may be used with many other types of understandings besides proofs. Here are a few examples:

1. To justify the definition—that is, to show that a given definition is reasonable or consistent: For positive integers m and n,

$$x^{-m/n} = \frac{1}{x^{m/n}}$$

The teacher might show that the definition preserves and extends properties of the following other definitions:

$$x^m \cdot x^n = x^{(m+n)}$$
$$x^0 = 1$$
$$x^m \div x^n = x^{(m-n)}$$

2. Either numerical or geometrical explanations might be employed to justify any of the following:

$$(ax + b) \cdot (cx + d) = acx^2 + (ad + bc)x + bd$$
$$(ax + b)^2 = (ax)^2 + 2abx + b^2$$
$$(x + y)^2 = x^2 + 2xy + y^2$$
$$(x + y)(x - y) = x^2 - y^2$$

(See, for example, Fig. 4–10, p. 99.)

3. Following are two demonstrations to show that $1.000 = .999 \ldots$:

a.
$$\tfrac{1}{3} = .3333 \ldots$$
$$\tfrac{2}{3} = .6666 \ldots$$
$$\tfrac{1}{3} + \tfrac{2}{3} = 1$$

and by addition

$$\tfrac{1}{3} + \tfrac{2}{3} = .9999\ldots$$

Thus,

$$1 = .9999\ldots$$

b. Let

$$x = .9999\ldots$$
$$10x = 9.999\ldots$$
$$10x - x = 9x$$

But

$$(9.999\ldots) - (.9999\ldots) = 9.000$$

Thus,

$$9x = 9.000$$
$$x = 1.000$$

But recall,

$$x = .9999\ldots$$

Thus,

$$1 = .9999\ldots$$

Telling, using analogies, and demonstrating are categories of authority teaching. In a given lesson a teacher might use all or any combination of the three. A pure form of each strategy would involve only the teacher talking, writing, and showing. But few teachers would rely on a pure form of any one of these methods to teach understandings; it is unreasonable to expect students to simply sit, watch, and listen. Instead, there will be interaction between students and teacher.

INTERACTION AND DISCUSSION

Observation of almost any mathematical classroom during other than a testing period will quickly verify that mathematics teaching involves interaction between the teacher and the students. Most teachers create this interaction by asking questions. Thus, when any of the methods discussed in the previous section are employed, the teacher will frequently interject questions to provide a means for active student participation in the class—instead of mere passive (receiving) participation. Student replies to teacher questions also give the teacher information about the conduct of the lesson. Do the students understand? Are they listening? Are there problems I did not anticipate? Is Marilyn bored? Is Jim keeping up with the explanation? Such classes could be called discussions; the teacher presents ideas and asks questions to stimulate student participation. In some cases a student response leads to other questions which can be, and are, answered by the teacher or by other students. Teachers who are effective in leading discussions are most often skillful in asking questions.

Questioning   In recent years asking questions has almost become a science

© 1974 by United Feature Syndicate, Inc.

of its own. Hopefully, as more is learned about questioning techniques, teachers will be better able to use questions effectively (see Research Highlight, "What Will You Say?"). At present there are several "rules of thumb" teachers can rely on:

1. In general, ask a question, *then* call on a student to respond. This procedure stimulates thinking by all students. If you call on a particular student first, you run the risk of asking only that student to think about the question. Say "What theorem can we use? (pause) Sue?" Don't say "Sue, what theorem can we use?"

2. Do not ask questions unless you are prepared to *accept* the answer. The reply to "Do you want to learn to solve simultaneous linear equations?" might be no. If you ask for factors of 84, be prepared for any of several correct answers.

3. Try to *use* an answer to your question. Correct replies should certainly be acknowledged, but often a wrong answer can be used, particularly if the teacher is able to analyze it and guess at the cause of the error. For example:

Teacher: Can someone factor $x^2 + 4x - 5$?
  Bill: How about $(x - 5)(x + 1)$?
Teacher: Bill, that's close, but let's multiply $(x - 5)(x + 1)$. (pause) What would that be?
  Bill: $x^2 - 4x - 5$, Oh, I see. Let's try $(x + 5)(x - 1)$.
Teacher: Right!

4. Ask questions of students in different parts of the room.

5. Sometimes redirect questions to other students. This technique is useful to maintain student involvement.

6. If a student gives a wrong answer, try asking the question he answered. For example:

Teacher: What is $9 \cdot 7$, Sue?
  Sue: 16.
Teacher: What is $9 + 7$, Sue?
  Sue: Oops! $9 \cdot 7 = 63$.

RESEARCH HIGHLIGHT | WHAT WILL YOU SAY?

Much research in recent years has focused on the kinds of questions teachers ask and on patterns of teacher-student interaction. Stilwell (1968) found that teacher-talk consumed approximately three times as much time as student-talk. Less than three percent of all problem-solving time involved the method of solving a problem. Fey (1969) also found that teachers spoke more than students, leaving responding as the major student activity. Friedman (1973) reported on the median percentage of questions asked by teachers in teaching a geometry theorem. He coded 20 percent of the questions at the memory level, considerably less than typically asked by teachers of other subjects. Comprehension questions were asked most frequently; the median was 56 percent. Application questions accounted for 18 percent. Only four of the 1,841 questions he recorded were higher-level questions. Kysilka (1970) found that mathematics teachers asked more convergent and procedural questions, made more directing and describing statements, rejected fewer student responses, and talked more than social studies teachers. But students volunteered less frequently in mathematics classes.

Among the other studies in which teacher and student behaviors were analyzed are: Cooney and Henderson (1972), Gregory and Osborne (1975), Hernandez (1973), Hoffman (1973), Lockwood (1971), Strickmeier (1971), and Wolfe (1972). A look at these various descriptions of how teachers act may help you in analyzing your own behavior in a classroom.

Questions may be categorized by the types of responses to them. Some questions may be asked for information, such as "What are some whole number factor pairs of 84?" Other questions might ask for processes: "What might we do to factor $x^2 + 17x - 84$?" or "How could we find the bisector of $\overline{AB}$?" Finally, some questions may request conjectures or hypotheses, such as "Do you think this expression has a real solution?" or "How many line segments do you think we could draw to connect nine points, no three of which are collinear?" If only one or two types of questions are asked in a classroom, a teacher cannot expect his students to be thinking along the lines of another type.

Cybernetics The term *cybernetics* describes corrective or adjustive actions taken on the basis of feedback. The word is derived from the Greek *kybernetes*, which refers to the helmsman on a ship. Think of the helmsman. As he steers, he moves the helm one way, then another. However, he must have some point of reference, be it a compass direction, a point on the coast, or a star. If he lacks a point of reference, he cannot obtain the feedback necessary to decide his "steering strategy."

The thermostat on a home heating or air conditioning system is another cybernetic device. The current temperature of a room provides data for the operation of systems which results in a relatively constant temperature.

When we teach, we plan and present a lesson; we have our goals, our reference points on the horizon. But we search for feedback to keep us aiming for the goal. True, we could teach a lesson in a room equipped with a one-way mirror and a loudspeaker so that students in an adjoining room could see and hear us without our seeing or hearing them. However, if we cannot observe or hear our students, much of the effectiveness of the lesson is likely to be lost. An important feature of computer assisted instruction (CAI) lessons, which will be discussed in Chapter 9, is that feedback in terms of student responses is processed to decide which part of the program should be presented next.

Our point is that the effectiveness of teaching in general, and that of teaching understandings in particular, is dependent on cybernetics—on clear goals and provisions for feedback. Some of these provisions for feedback may be planned before the lesson; many will occur to the effective teacher during class. Questioning is an important technique for obtaining feedback, but observing students and their reactions to the lesson can also provide you with useful clues.

Interaction lessons may be used effectively in mathematics. In general, such lessons are efficient in terms of ideas being presented quickly by the teacher and in terms of understanding being determined by asking questions or observing students.

DISCOVERY

Teaching by discovery is discussed a great deal but commonly not well defined. Some teachers claim to be using discovery when they ask open-ended questions like the following:

1. What is a formula for the coefficient of the jth term $(1 \leqslant j \leqslant n + 1)$ of $(a + b)^n$?
2. What is the maximum number of line segments connecting n points?
3. What is the best way to mow a rectangular lawn?

Other teachers (and some textbook authors) classify the following as discovery:

4. To find the product $(x + y) \cdot (x + y)$, we could consider x and y as measures of lengths. Thus, a square plot with $(x + y)$ on each side would have an area

$$(x + y) \cdot (x + y)$$

It can be seen that the area (Figure 4–10) is also:

$$x^2 + xy + xy + y^2 = x^2 + 2xy + y^2$$

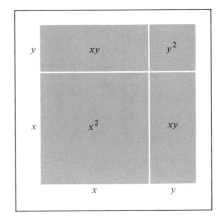

Figure 4–10.

Thus we see that:

$$(x + y) \cdot (x + y) = x^2 + 2xy + y^2$$

We argue that a useful definition of discovery teaching falls between these extremes. The first three examples might be classified as "open search" problems (to be considered in the next chapter); the last example is simply a demonstration or analogy (previously discussed). Discovery will be defined in this section to include these key elements: motivation, a primitive process, an environment for discovery, conjectures, verification, and application. Before discussing these elements, let's consider a simple example taken from a textbook (Beberman and Vaughan, 1964, p. 60). Please try the problem.

Here is a start on a multiplication table (Figure 4–11). The first entry, ⁻14, tells you that

$$^+21 \cdot {}^-(\tfrac{2}{3}) = {}^-14$$

Complete the table.

As students work on the example, what do you expect to happen? At first, students might diligently carry out the multiplication calculations called for. They may write

$$
\begin{array}{r}
{}^+120 \\
\times \quad {}^+21 \\
\hline
120 \\
240 \\
\hline
{}^+2520
\end{array}
$$

and fill in the appropriate box. Some may note that ⁺2520 appears elsewhere in the table and continue to find other products. Some students might observe that some classmates completed the table quickly. Why? It seems likely to expect that

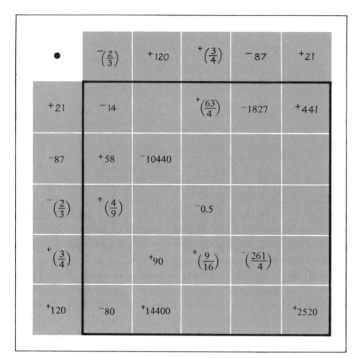

Figure 4–11 Multiplication table (from Max Beberman and Herbert Vaughan, *High School Mathematics*, © 1964 by D. C. Heath and Company. Reprinted by permission of the Publisher).

almost all of the students would discover before they carried out all the multiplications that only one answer ($^-87 \cdot \, ^-87$) does not already appear in the table. In this way, the students demonstrate their ability to apply the commutative property for multiplication of rational numbers.

Imagine what would happen if this exercise were used in a classroom. If the teacher presents the problem in a way which encourages pupils to start it, he has provided a *motivation*. The *primitive process* in this example is skill in multiplying signed rational numbers. Assuming that this skill has been taught (and learned), each student will have a way to proceed. *Conjecture* is provided for implicitly. We suspect that students will realize that there might be a "quicker way" to do this problem. As they use the primitive process, they will receive more and more clues that they can use the data in the table to complete it. If they conjecture that, say,

$$^+\tfrac{3}{4} \times \, ^+120 = \, ^+120 \times \, ^+\tfrac{3}{4}$$

they can *verify* the specific conjecture and extend it to other cases. That is, after they verify the conjecture, they can *apply* it to complete the table.

Thus the example could provide students with all the key elements of a discovery experience: motivation, a primitive process, an environment for discovery, an opportunity to make conjectures, a method to verify conjectures, and a provision for applying the generalizations. Now let us examine each element.

Motivation is a necessary element of all mathematics teaching. However, in discovery lessons, motivation takes on new significance. Students must have an internal wish at least to begin the task at hand if there is to be any hope that they can discover an associated generalization.

There are several techniques available to teachers to stimulate student motivation. In the example, the motivation element indicated was minimal—simply a request to try to complete the table. The teacher might have hinted at the prospect of discovering a generalization by saying "Before you complete this table I believe you will find (discover) a way to do it quickly" or "Can you find a simple way to complete this table?"

Wills discusses the use of a target task, "a criterion task introduced at the beginning of a lesson" (Wills, 1970, p. 282). Such tasks may serve to motivate students by clarifying a goal for them or to provide a specific accomplishment which the teacher may use as evidence of achieving the generalization. Two target tasks provided by Wills (p. 285) are:

1. How many vertices has each 1776-gon?
2. What is the [greatest] number of nonoverlapping regions formed in a circle by 1999 chords?

Such tasks present a problem which appears difficult, but which may be readily solved with the generalization to be discovered. Often, when target tasks are presented, there is an implicit method for solving the problem: This method is to change the problem to an easier one and look for a pattern. Thus, with a little experience, a student might begin the first problem by determining how many vertices are in a triangle, a quadrilateral, a pentagon, and a hexagon.

The teacher might encourage the students by suggesting that they make a table of the number of sides of the n-gon and the number of vertices. Such suggestions can serve to motivate students or to provide a primitive process.

Other methods of motivating students are to challenge them or to promise a "payoff." For example, the teacher might ask "Can we find a quick way to determine the number of nonoverlapping regions formed in a circle by 1999 chords?" In this way, the second target task above is presented as a challenge. An example of a promised benefit might be:

▶ We're going to learn how to sketch graphs of equations like the following by just looking at them:

$$y = 2(x + 1)^2 + 3$$
$$y = 3(x - 1)^2 + 2$$

We will follow the development of this brief unit in the next subsection.

A *primitive process* is the beginning *student* activity in a discovery lesson. It is a key ingredient because its existence provides an immediate focus for the students. In the discovery of the commutative property for multiplication of rational numbers (above), the primitive process was simply multiplication of rational numbers. Let's examine how a primitive process may be employed.

Suppose we decide upon the following goal:

▶ The students are to understand that the graph of $y = a(x - k)^2 + p$ has the properties:

1. If $a > 0$, the graph opens upward and the curve has a lowest point (k, p). If $a < 0$, the graph opens downward and has a highest point (k, p).

2. The graph has the line $x = k$ as its axis. (*Intermediate Mathematics*, 1965, p. 215)

Suppose also that we decide to teach this understanding by discovery. Assume that the students have acquired skill in plotting functions by selecting values for the independent variable, determining the corresponding values for the dependent variable, plotting several ordered pairs of points, and drawing a smooth curve through the points. This skill is the primitive process they will use in the lessons.

Here we outline two days of lessons.* Two sets of tasks are provided for each day.

Day 1: A class discussion of plotting the function $y = x^2$ is designed to provide a review of the primitive process. Students will then be asked to plot the following functions, with Exercises 1–5 on one graph and 6–10 on another:

1. $y = 5x^2$ 6. $y = -5x^2$
2. $y = 2x^2$ 7. $y = -2x^2$
3. $y = x^2$ 8. $y = -x^2$
4. $y = \frac{1}{2}x^2$ 9. $y = -\frac{1}{2}x^2$
5. $y = \frac{1}{10}x^2$ 10. $y = -\frac{1}{10}x^2$

Figures 4–12 and 4–13 show graphs of these sets of functions. Before any class discussion of these graphs, the teacher will ask students to sketch quickly graphs of some other functions, such as:

11. $y = 10x^2$ 13. $y = -\frac{1}{20}x^2$
12. $y = 4x^2$ 14. $y = -8x^2$

The class discussion will be designed (1) to insure that the first ten graphs were drawn fairly accurately, (2) to see if the students could "guess" how to plot the last four functions, and (3) to attempt to generalize how the shape and location of $y = ax^2$ will vary for different values of a.

---

* Adapted with permission from *Intermediate Mathematics*, Part 1, 1965, pp. 206–215.

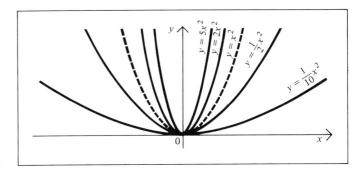

Figure 4–12.

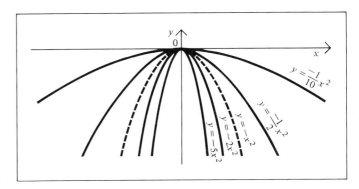

Figure 4–13.

Then students will be asked to plot the following functions:

1. $y = x^2 + 2$ 5. $y = -x^2 + 2$
2. $y = x^2 + 1$ 6. $y = -x^2 + 1$
3. $y = x^2 - 1$ 7. $y = -x^2 - 1$
4. $y = x^2 - 2$ 8. $y = -x^2 - 2$

The graphs of these functions are shown in Figures 4–14 and 4–15. Again students will be asked to sketch graphs such as:

9. $y = x^2 + \frac{1}{2}$ 10. $y = -x^2 - \frac{1}{2}$

and the teacher will lead a discussion analogous to the one following the first exercise set, but teaching generalizations about the effect on graphs of varying c in $y = ax^2 + c$.

Day 2: This lesson will follow the same pattern as the previous one. The exercises in the first group are selected to emphasize the role of k in $y = a(x - k)^2$ and those in the second set are selected to emphasize the role

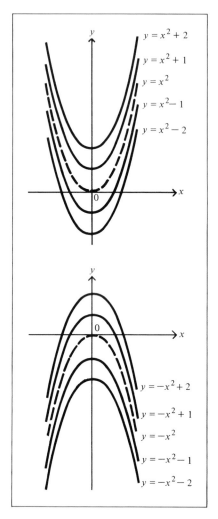

Figure 4–14.

Figure 4–15.

of p in $y = a(x - k)^2 + p$. For simplicity and brevity, we show in Figure 4–16 only the graphs of the following functions:

1. $y = 2(x + 3)^2 + 1$
2. $y = 2(x - 3)^2 + 1$

In this example the motivation provided was to plot seemingly complicated functions rapidly. As students apply the primitive process in plotting functions in these lessons, we hope—and expect—they will discover generalizations about the roles of the constants a, k, and p in $y = a(x - k)^2 + p$. Several exercises should be provided to develop, test, verify, and apply the generalizations.

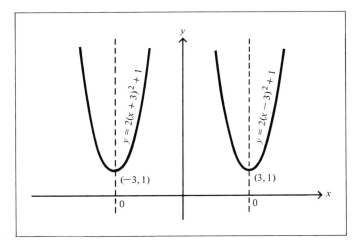

Figure 4–16.

It should be emphasized that the understanding goal stated before the lessons was presented for you. The teacher of these lessons would *not* tell the students the understanding; to do so would be to exclude the opportunity for discovery. However, at appropriate times, the teacher might help students focus on the roles of the constants by providing hints or asking questions such as "What is the difference between the graphs of $y = ax^2$ when a is greater than zero and when a is less than zero?" In each exercise set the students could employ the primitive process. If such a primitive process were not available, or developed by the teacher, we would not classify the lesson as discovery; rather, it would be *open search*.

In the earlier problem of finding the number of vertices of a 1776-gon, the primitive process could be to count the vertices in an n-gon, with n a small natural number. Each student might collect data such as shown in Table 4–1.

Table 4–1.

Number of Sides n	Number of Vertices in an n-gon
3	3
4	4
5	5
6	6
. . .	. . .
1776	1776

In the problem of finding the number of nonoverlapping regions formed in a circle by 1999 chords, students may collect the data shown in Table 4–2.

Table 4–2.

Number of Chords	Number of Nonover-lapping Regions
1	2
2	4
3	7
4	11
5	16
. . .	. . .
n	(?)
1,999	1,999,001

In each of these examples, the other key elements of discovery must be present. Students must have an *environment for discovery, an opportunity to make conjectures*, and *a provision for applying the discovered generalization*. The teacher can provide these elements by asking questions, by encouraging students to make guesses, by permitting them to try out their conjectures on easier problems, and by insuring that students apply the discovered generalization in a seemingly different task. Teachers can encourage students to look for generalizations, they can provide time for students to try their ideas, they can help students organize data—but they cannot make the discoveries for the students! If the teacher wants the student to discover, he or she must be patient and avoid telling (or letting another student tell) the discovery.

Verbalization of discovery has purposely not been included as a key element. Often such verbalization interferes with learning the generalization. Encourage students to apply the discovered generalization, not to state it!

We might say that the student is in a state of *unverbalized awareness*. That is, the student has discovered the generalization and is able to apply it, but he or she has not volunteered or been required to verbalize it. Thus the student might well be able to reply correctly to the following:

1. How will the graph of $y = 30x^2$ be different from the graph of $y = 5x^2$?
2. What is the difference between the graphs

$$y = 3(x - 1)^2 + 10$$

and

$$y = -3(x - 1)^2 + 10?$$

3. Without plotting the graph of $y = (x - 18)^2 + 3$ can you tell if it will have an axis of symmetry and, if so, what is the equation of this axis?

These questions are very different from the following:
- What is the generalization you are using?
- Can you tell me what you have done?
- How could we describe this process for all cases?

In a now classic study by Hendrix (1947), there was evidence that premature verbalization of a generalization might interfere with student ability to apply it at a later time (for a review of this and related studies, see Henderson, 1963). However, sometimes the student may well benefit from learning a statement of a generalization. The point here is that a distinction must be made between unverbalized awareness of a generalization and learning a statement of generalization. The teacher must decide when each learning should occur (see Research Highlight, "To Speak or Not to Speak").

RESEARCH HIGHLIGHT | TO SPEAK OR NOT TO SPEAK

The question of whether or not students should verbalize a generalization has been debated for years. The research by Hendrix (1947) cited in this chapter has formed a focal point for most of these discussions. Researchers have approached the question in several ways.

Retzer and Henderson (1967) taught selected logic concepts to a group of students in grades seven and eight. The students were given practice in applying these concepts in writing generalizations for a programmed unit on vectors. It was found that they were able to state correctly the relations they discovered significantly better than students who had not had the logic materials. The researchers suggest that, if logical components of universal generalizations are taught explicitly, students could then be asked for immediate verbalization of a newly discovered generalization.

Three methods of teaching a unit on exponents to eighth graders were compared by Neuhouser (1965). In one method, the rule was explained and students were given examples for practice. In a second method, the student was helped to state the rules after discovery. Students not required to verbalize achieved significantly higher scores.

Flaherty (1973) found no significant differences, in problem-solving score or time, between students required to verbalize and those who remained silent. The overt verbalization students made significantly more computational errors, however.

LABORATORIES

Similar to discovery is the laboratory method of teaching understandings. In Chapter 8 you will find several ways to individualize mathematics instruction; the use of laboratories is one of them. We introduce it here because it can be used

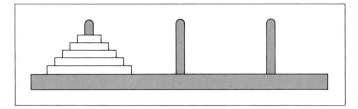

Figure 4–17 The "Tower of Hanoi" puzzle.

effectively in teaching understandings. Often in a laboratory approach the teacher prepares *activity cards* or *task cards* which provide instructions for an individual student or for a small group of students. Typically the physical embodiment of a mathematical idea, principle, concept, or other understanding is employed in an inductive sequence. For example, an activity card may call for a student to add a sequence of equal weights to a spring and to record its length. A plot of spring length as a function of load weight on the spring will provide data for determining the relation between these variables—a generalization. Here are descriptions of two other laboratory experiences:

A laboratory experience might be designed to determine the minimum number of moves to solve the "Tower of Hanoi" puzzle (Figure 4–17). The rules for this puzzle are: (1) move one disc at a time from one peg to another, (2) never place a larger disc over a smaller one, and (3) the puzzle is completed when all the discs have been moved to the center peg. In this laboratory experience, the students might be asked to solve the puzzle first for one disc, then two, then three, and so forth. Each time they could try to determine the minimum number of moves. Eventually, the students could determine the minimum number of moves to solve the puzzle for five discs or for 30 discs.

Another laboratory experience might be based on using simple "Artin Braids" to explore physical embodiments of a noncommutative group.* Figure 4–18 is an illustration of a device for constructing braids. Nine removable, equally spaced pegs are mounted on a board so that elastic cord or rubber bands may be stretched between the pegs on two rows. The six possible braids (labeled A, B, C, D, E, and I) are pictured in Figure 4–19 and an example of braid composition (*) is illustrated in Figure 4–20. After a description of the braids and explanation of braid composition, students are asked to complete the table in Figure 4–21. After completing the table, students may determine if the set of braids together with the operation of braid composition (*) satisfied the requirements of a mathematical group.

* The description of this laboratory experience is an adaptation from a set of activities developed by Larry L. Hatfield, Department of Mathematics Education, University of Georgia. (See also Fitzgerald et al., 1967, pp. 59–63.)

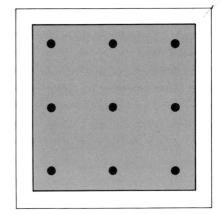

Figure 4–18   Peg board for braid construction.

Is the system closed, is it associative, is there an identity element, and, if so, does every element have an inverse? For example, students are asked to see if the following is true, as part of the verification of the associative property:

Is $(D * B) * E$ the same as $D * (B * E)$?

Finally, students are asked if the system is commutative.

There is substantial similarity between the discovery method and laboratory experiences. A laboratory experience can be designed so that the lesson is not discovery by telling the student initially what the generalization is, but often the laboratory experience will contain all the key elements of a discovery lesson. Almost all laboratory experiences involve physical embodiments and often include obtaining physical measures, while most discovery lessons focus on abstract situations (but such a distinction is probably artificial).

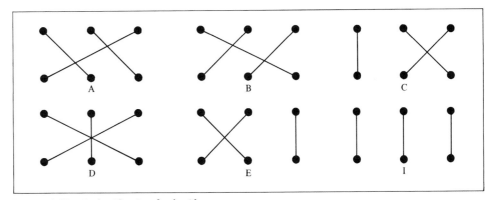

Figure 4–19   A classification for braids.

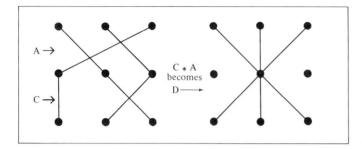

Figure 4–20 A sample braid composition.

MOVES AND STRATEGIES

In teaching understandings a teacher may select a particular sequence of moves or specific teaching acts. Henderson (1970, p. 192) defines such a sequence of moves as a strategy. We have described four general approaches to teaching understandings: authority, interaction, discovery, and laboratories. In preparing to teach a lesson a teacher may select any of these methods or a combination of them. The use of a method will, in general, incorporate particular characteristics. If we consider the methods as points on a continuum from authority to interaction and discussion to discovery and laboratory, we would note that *generally* (1) the *time* to teach an understanding increases, (2) the amount of *overt student involvement* and *student talk* increases, and (3) the prospect of student *retention* of the understanding increases along the continuum.

We have presented these approaches to teaching understandings to exemplify the range of appropriate strategies. However, there is also a considerable overlap

Top Braid (written second)

*	I	A	B	C	D	E
I						
A						
B						
C						
D						
E						

Bottom Braid (written first)

Figure 4–21 A table of braid compositions.

among approaches. We would expect to see interaction and discussion in almost every lesson. It might not, and need not, be clear to a student that a particular lesson is a discovery lesson or a laboratory lesson. We discussed each of the approaches separately, but in practice they are often used in combination.

Not only must a teacher select appropriate moves for teaching an understanding and sequence these moves into a strategy, but he or she must avoid strategies which could reduce the level of the goal. If students do not learn an understanding, it may be because an overanxious teacher has selected an inappropriate strategy. For example, suppose we are to teach students the generalization that the distance between point P at (x_1, y_1) and Q at (x_2, y_2) may be found by the following:

$$D(P, Q) = \sqrt{(x_2 - x_1)^2 + (y_2 - y_1)^2}$$

Suppose also that the teacher presented a developmental lesson such as this:

▶ We note that the point R at (x_2, y_1) could be thought of as a vertex for the right triangle PQR (Figure 4–22). The measure of the distance between P and R is $(x_2 - x_1)$. The distance between R and Q is $(y_2 - y_1)$. Thus the measure of the hypotenuse $\overline{PQ}$ is given by:

$$PQ = \sqrt{(x_2 - x_1)^2 + (y_2 - y_1)^2}$$

If the teacher gave a short quiz which required students to apply the generaliza-

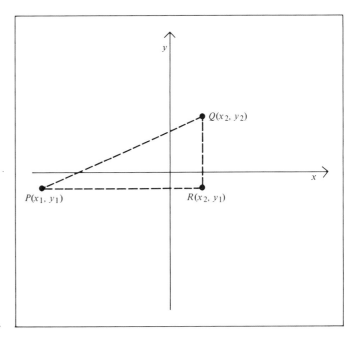

Figure 4–22.

tion, perhaps by finding the distance between several pairs of points, and if several students were not successful, the teacher might decide that reteaching the generalization was necessary. Now some teachers, anxious for their students to succeed, might be tempted to try some of the following procedures:

1. State the distance formula and tell the students to memorize it.
2. As a class, determine the distance between ten pairs of points in the co-ordinate plane. In finding each distance, emphasize the procedure to follow.
3. Assign a lengthy homework exercise set so that each student will have an opportunity to "see how the distance formula works."
4. Give the class a short quiz each day until almost all demonstrate mastery of applying the distance formula. The quiz will consist of stating the distance formula and finding the distance between two pairs of points.

Such activities actually change the learning from an understanding to an item of knowledge (Activities 1 and 4) and/or a skill (Activities 2, 3, and 4). These activities are not appropriate for teaching understandings!

In the situation described, the teacher realizes that the students have not learned the generalization and that reteaching is necessary. The teacher must select teaching strategies which are appropriate for teaching understandings if he or she wants students to acquire the understanding. Thus the teacher might try some of the following activities:

1. Design a set of laboratory experiences to present the generalization inductively. For example, the following activity cards might be prepared:
   a. With graph paper plot and find the distance between pairs of points on the same vertical line in the first quadrant, in the first and fourth quadrant, in the second quadrant, in the second and third quadrant, and in the third quadrant.
   b. Similarly, find the distance between pairs of points on the same horizontal line.
   c. Using graph paper, plot pairs of points, draw vertical and horizontal lines through them, and apply the Pythagorean Theorem. (The theorem would be written on the activity card and students would be asked to find the length of each side of the right triangle they construct.)
   d. Given the distance formula and the coordinates of two points, find the distance between the points.
2. Prepare another developmental lesson. Use the overhead projector and graph paper. Call on students who were successful on the quiz to explain how they would find the distance between two points. Try to emphasize the construction of a right triangle and use of the Pythagorean Theorem.
3. Prepare a great many rectangles of various sizes. Distribute them to the students. Ask each student to measure the length, width, and diagonal

of two rectangles. Record the data and lead students to see that the Pythagorean Theorem may be used to obtain the length of the diagonal. Later, distribute graph paper and have students plot the vertices. You might give them the coordinates for the lower left corner of their rectangle and ask them to plot the other three vertices, with the longest side on a horizontal line. Draw the sides of the rectangle and one of its diagonals. Use the points to determine the length of the diagonal.

Each of these activities is designed to present a slightly different approach to teaching the understanding of the distance formula. Using these or similar activities, the teacher can avoid changing the level of learning from understanding to that of knowledge or skill. If the level were changed, the students might well be able to perform on a test; that is, they could find the distance between two points on a coordinate graph. However, the knowledge or skill will probably not be a sufficient background for more advanced study such as approximating the length of a curve.

BASIC PROGRAMMING REVISITED

We presented an introduction to BASIC Programming in Chapter 3. In this section we provide an opportunity for you to extend your understanding of the BASIC language and your programming ability. We will discuss the construction of programming loops and debugging programs, and we will offer a number of exercises in program construction. Appendix C contains sample problems appropriate for secondary school students which you may wish to try.

FOR/NEXT LOOPS

The FOR/NEXT commands may be used to construct loops in a program. Loops permit a block of instructions to be used repeatedly during a run of the program. Here is a simple program to count in base five:

```
100  PRINT "COUNTING IN BASE FIVE:"
110  FOR X=0 TO 4
120  FOR Y=0 TO 4
130  FOR Z=0 TO 4
140  PRINT 100*X+10*Y+Z;
150  NEXT Z
160  NEXT Y
170  NEXT X
180  END
```

The output would appear as follows:

0	1	2	3	4	10	11	12	13	14	20	21
22	23	24	30	31	32	33	34	40	41	42	43
44	100	101	102	103	104	110	111	112	113	114	120
121	122	123	124	130	131	132	133	134	140	141	142
143	144	200	201	202	203	204	210	211	212	213	214
220	221	222	223	224	230	231	232	233	234	240	241
242	243	244	300	301	302	303	304	310	311	312	313
314	320	321	322	323	324	330	331	332	333	334	340
341	342	343	344	400	401	402	403	404	410	411	412
413	414	420	421	422	423	424	430	431	432	433	434
440	441	442	443	444							

Note that there are three *nested* loops—that is, some loops are contained within larger loops. In BASIC, loops must be nested. We could not interchange the line numbers for lines 150 and 160 because the loop for Z would not be nested within the loop for Y. Also note that there is a "trick" in line 140 which uses base ten arithmetic to print a number which will appear to be in base five. The same output would be produced by the following program which has been written without use of FOR/NEXT commands:

```
100   PRINT "COUNTING IN BASE FIVE:"
110   LET X=Y=Z=0
140   PRINT 100*X+10*Y+Z;
150   LET Z=Z+1
160   IF Z>4 THEN 180
170   GOTO 140
180   LET Z=0
190   LET Y=Y+1
200   IF Y>4 THEN 220
210   GOTO 140
220   LET Y=0
230   LET X=X+1
240   IF X>4 THEN 260
250   GOTO 140
260   END
```

DEBUGGING

One of the first clichés beginning computer programmers encounter is "Garbage in, garbage out." There is probably no programmer who has not submitted programs for execution which contained errors. Finding these errors can be very difficult. (Suppose the two letter "oh's" in line 210 were typed as zeros!) Happily, most BASIC compilers provide diagnostic checks to help you locate program errors. But there are ways for errors to appear in programs which will not be caught by diagnostic checks. (Suppose line 240 was entered as "IF X>4 THEN 180". We would have programmed an "infinite loop" which would not terminate until a time limit was exceeded—if the computer provided such a control!)

The point is that programming errors do occur. Several helpful procedures are available to locate errors (see Kemeny and Kurtz, 1971, pp. 39–43). One technique is to "play computer" yourself. Go through your program, recording current values of variables, and obey the instructions as if you were the computer.

Another useful technique, particularly if you obtain surprising output or no output at all, is to include "trace" statements at various places in the program. These can take the form, PRINT "TRACE n", where n is the line number of the trace statement. By including several such statements, you may be able to determine which sections of your programs are running correctly and which are candidates for close examination.

In some programs you may be able to try a run with simple data. If you know what answer should be calculated with simple data, you may be able to find where the error occurred.

EXERCISES

Here we present several exercises intended (1) to provide an opportunity for you to write simple programs and improve your programming ability and (2) to exhibit a list of problems relevant for use in secondary school mathematics classes.*

Write a program that will find:
1. the union and the intersection of two sets of whole numbers.
2. the least common multiple of any three numbers.
3. the circumference and area of a circle.
4. the sum of arithmetic progressions.
5. the slope of a line through two given points.
6. the product of two binomials.

Critical Incidents Revisited

All of the critical incidents at the beginning of this chapter are examples of student behavior.

Carol, in the first critical incident, asked Miss Hemmerly if the class had to memorize the proof of a theorem. This incident was selected to emphasize that seldom, if ever, would we want a student to memorize a proof. If Carol *understands* the proof, that is, if she understands how the theorem may be proven,

---

* These exercises were adapted from a mimeographed workshop handout prepared by Larry L. Hatfield, Department of Mathematics Education, University of Georgia, Athens, Georgia.

she should be able to construct the proof. One distinction to be made here is between *knowing* a proof (memorizing it) and *understanding* a proof. As indicated in the chapter, a student may remember some key ideas for a proof without memorizing the proof. Another important distinction is to be made between understanding a theorem and understanding a proof of the theorem. A student who understands a theorem (which is a generalization) is able to apply (that is, use) the theorem, either in exercises or in proving other theorems. A student who understands the proof of a theorem is able to use the key ideas (generalizations) to construct the proof. A student who *understands the proof of a theorem* is able to demonstrate a degree of *understanding the theorem*. That student can argue logically to show why the theorem holds. Another way to demonstrate *understanding the theorem* is to be able to use it in deducing another theorem or consequence.

The second and third critical incidents are examples of students demonstrating their understanding of generalizations. In the second incident Sid applies the intermediate value theorem to show that $f(x)$ has a real root between $x = 2$ and $x = 3$. In the third incident Barb applies the definition of the absolute value of a number to solve the inequality.

Bill, in the fourth critical incident, seems to have missed the understanding that all quadratic equations are solvable in the system of complex numbers.

There are a great many teaching techniques that are used to help students like Carol, Sid, Barb, and Bill understand generalizations. In this chapter, we have examined different classifications of generalizations. Within a mathematical system generalizations may be classified as axioms, definitions, or theorems. However, the same generalizations may be classified as principles or concepts. Finally, generalizations may be classified by instructional goals as knowledge, skills, or understandings. Meanings of all these terms have been discussed and some hints have been provided for stating understanding goals.

The major portion of the chapter has been devoted to an examination of ways to teach understanding goals. The approaches discussed include authority teaching, interaction and discussion teaching, discovery teaching, and uses of laboratory experiences.

Authority teaching includes telling, use of analogies, and demonstrations. Questioning techniques and cybernetics were emphasized as a part of interaction and discussion teaching. The key ingredients in discovery teaching (and in laboratory experiences) were motivation for the discovery, availability of a primitive process, use of conjectures, verification, and application. We stressed that caution should be exercised in forcing students to verbalize a generalization they have discovered.

Finally, the chapter included another step in the process of learning BASIC programming. The use of FOR/NEXT loops, methods of debugging programs, and sample problems from secondary school mathematics were presented.

Activities

1. Pick a secondary school mathematics course and list three understandings in that course which were not mentioned in this chapter.
2. Select an understanding appropriate for Algebra I and outline:
   a. an authoritative lesson, and
   b. a discovery lesson, to teach the understanding.
3. Select an understanding in secondary school geometry and outline:
   a. a discussion lesson, and
   b. a sequence of laboratory experiences, to teach the understanding.
4. Plan a demonstration lesson to teach that

$$(x + y) \cdot (x - y) = x^2 - y^2$$

   Include at least one analogy.
5. Plan a discovery lesson to teach the understanding in Exercise 4.
6. Plan a discussion lesson to teach the following:

$$\sin^2 x + \cos^2 x = 1$$

   Include several questions you would ask students. Select two of your questions and list five possible (correct or incorrect) responses.
7. Outline a discovery lesson to teach the following generalization:
   The square of a number x which is divisible by 5 is

$$(x - 5) \cdot (x + 5) + 25$$

   Thus,

$$25^2 = 20 \times 30 + 25 = 625$$
$$55^2 = 50 \times 60 + 25 = 3025$$

   Extend this idea to find a quick way to square any rational number ending in 5. For example:

$$(3.5)^2 = 3 \times 4 + .25 = 12.25$$

8. Select two exercises listed in the section on BASIC programming in this chapter and write programs for them.
9. Select one of the exercises in Appendix C and construct a BASIC program for it.
10. Use the following generalization: The measure of an angle which intercepts two arcs of a circle is one-half the difference of the measures of the intercepted arcs. Describe how the generalization could be taught as:
    a. an item of knowledge
    b. a skill
    c. an understanding

References

Beberman, Max and Vaughan, Herbert E. *High School Mathematics*. Boston: Heath, 1964, p. 60.

Cooney, Thomas J. and Henderson, Kenneth B. Ways Mathematics Teachers Help Students Organize Knowledge. *Journal for Research in Mathematics Education* 3: 21–23; January 1972.

Fey, James Taylor. Patterns of Verbal Communication in Mathematics Classes. (Columbia University, 1968). *Dissertation Abstracts* 29A: 3040; March 1969.

Fitzgerald, William M. et al. *Laboratory Manual for Elementary Mathematics*. Boston: Prindle, Weber, and Schmidt, 1967.

Flaherty, Eileen Gertrude. Cognitive Processes Used in Solving Mathematical Problems. (Boston University School of Education, 1973.) *Dissertation Abstracts International* 34A: 1767; October 1973.

Friedman, Morton Lawrence. The Development and Use of a System to Analyze Geometry Teachers' Questions. (Columbia University, 1972.) *Dissertation Abstracts International* 33A: 4215–4216; February 1973.

Gregory, John William. The Impact of the Verbal Environment in Mathematics Classrooms on Seventh Grade Students' Logical Abilities. (Ohio State University, 1972.) *Dissertation Abstracts International* 33A: 1585; October 1972.

Gregory, John W. and Osborne, Alan R. Logical Reasoning Ability and Teacher Verbal Behavior within the Mathematics Classroom. *Journal for Research in Mathematics Education* 6: 26–36; January 1975.

Henderson, Kenneth B. Research on Teaching Secondary School Mathematics. In *Handbook of Research on Teaching* (edited by N. L. Gage). Chicago: Rand McNally, 1963, pp. 1014–1019.

Henderson, Kenneth B. Concepts. In *The Teaching of Secondary School Mathematics*. Thirty-third Yearbook of the National Council of Teachers of Mathematics. Washington: The Council, 1970.

Hernandez, Norma G. A Model of Classroom Discourse for Use in Conducting Aptitude-Treatment Interaction Studies. *Journal for Research in Mathematics Education* 4: 161–169; May 1973.

Hendrix, Gertrude. A New Clue to Transfer of Training. *Elementary School Journal* 48: 197–208; December 1947.

Hoffman, Joseph Raymond, Jr. A Heuristic Study of Key Teaching Variables in Junior High School Mathematics Classrooms. (University of Illinois at Urbana-Champaign, 1972.) *Dissertation Abstracts International* 34A: 665; August 1973.

Kemeny, John G. and Kurtz, Thomas E. *BASIC Programming*. 2nd ed. New York: Wiley, 1971.

Kysilka, Marcella Louise. The Verbal Teaching Behaviors of Mathematics and Social Studies Teachers in Eighth and Eleventh Grades. (University of Texas at Austin, 1969) *Dissertation Abstracts International* 30A: 2725; January 1970.

Lockwood, James Riley. An Analysis of Teacher Questioning in Mathematics Classrooms. (University of Illinois at Urbana-Champaign, 1970.) *Dissertation Abstracts International* 31A: 6472–6473; June 1971.

Neuhouser, David Lee. A Comparison of Three Methods of Teaching a Programmed

Unit on Exponents to Eighth Grade Students. (Florida State University, 1964.) *Dissertation Abstracts* 25: 5027; March 1965.

Polya, George. *How to Solve It*. 2nd ed. Princeton: Princeton University Press, 1973.

Retzer, Kenneth A. and Henderson, Kenneth B. Effect of Teaching Concepts of Logic on Verbalization of Discovered Mathematical Generalizations. *Mathematics Teacher* 40: 707–710; November 1967.

Stilwell, Merle Eugene. The Development and Analysis of a Category System for Systematic Observation of Teacher-Pupil Interaction During Geometry Problem-Solving Activity. (Cornell University, 1967.) *Dissertation Abstracts* 28A: 3038; February 1968.

Strickmeier, Henry Bernard, Jr. An Analysis of Verbal Teaching Behaviors in Seventh Grade Mathematics Classes Grouped by Ability. (University of Texas at Austin, 1970.) *Dissertation Abstracts International* 31A: 3428: January 1971.

Thurstone, L. L. *Multiple Factor Analysis*. Chicago: University of Chicago Press, 1953, p. 51.

Wills, Herbert. Generalizations. In *The Teaching of Secondary School Mathematics*. Thirty-third Yearbook of the National Council of Teachers of Mathematics. Washington: The Council, 1970.

Wolfe, Richard Edgar. Strategies of Justification Used in the Classroom by Teachers of Secondary School Mathematics. *School Science and Mathematics* 72: 334–338; April 1972.

Intermediate Mathematics, Part 1. Stanford, California: School Mathematics Study Group, 1965, pp. 206–215.

Problem Solving

AFTER STUDYING CHAPTER 5, YOU WILL HAVE:
* ★ learned the characteristics of a mathematical problem for the student.
* ★ seen how one problem might be solved in many different ways.
* ★ explored a great many mathematical problems, including applications, abstract problems, open search problems, projects, and proofs.
* ★ examined several ways teachers can help students become problem solvers.

CRITICAL INCIDENTS
1. Miss Manning begins her class with the problem "Where should you sit in a movie theater?"
2. One night Cindy finds a problem in a book, thinks about it, and suddenly discovers a solution. She is so excited she telephones her mathematics teacher and describes her solution. The problem is: "If two prime numbers have a difference of two, they are called *twin primes*. Show that, except for 3 and 5, the sum of each pair of twin primes is divisible by 12."
3. Joe reads the problem "How should the green, yellow, and red lights at an intersection be timed?" Joe asks, "Is this mathematics?"
4. Barb looks at the question "What is the area of a triangle with a base of 4 *cm* and an altitude of 3 *cm*?" She says immediately, "It's 6 *cm²*."

CLASSIFICATION OF PROBLEMS

Problem solving qualifies as the ultimate justification for teaching mathematics, yet it is the hardest to teach and often the most neglected part of the mathematics curriculum. Although we might be tempted to argue that the beauty of mathematics, its aesthetic appeal, and its role in the history of man warrants its inclusion in the school's curriculum, we saw in Chapter 1 that such an argument is an inadequate justification for the amount of time devoted to mathematics, except perhaps for such relatively small populations as future mathematicians or future mathematics educators.

Knowledge, basic skills, and understandings are important components of mathematical learning, but ultimately a student learns mathematics in order to solve a great variety of problems. Most teachers agree that we cannot predict what problems our students of today will be called on to solve in the next 5, 10, 20, or 30 years. Thus the task is to prepare students in secondary school for solving the unknown, and often unpredictable, problems they will face the rest of their lives. Many students will continue to study mathematics beyond secondary school; those who do will rely on the background they acquire under your guidance. You will have to decide what problem-solving abilities your students should acquire and how to teach these abilities.

The result of the revolution in school mathematics which occurred during the 1950s and 1960s was a significant improvement in the teaching and learning of mathematical understanding (see Chapter 13). Outcomes of that revolution include a more precise use of language in mathematics textbooks and classrooms, an increased emphasis on unifying ideas in mathematics instruction, a general resequencing of mathematical objectives within the curriculum, and better teacher preparation. However, relatively little change occurred in the teaching of problem solving. We believe there will be another revolution in the teaching of mathematics, that it may have already begun, and that it will focus on improving student problem-solving ability.

This chapter discusses problem solving, presents ways to teach mathematical problem solving, and describes some techniques which you may use in teaching students how to solve problems.

WHAT IS A PROBLEM?

The word *problem* is derived from the Greek *problema*, which translated literally means "something thrown forward" (from *ballein*, "to throw"). Less inspiring but more common is the standard definition, "a question raised for inquiry, consideration or solution . . . a source of perplexity . . . ." (*Webster's New Collegiate Dictionary*, 1975.) For our use then, a problem is a perplexing question or situation. It is extremely important to note that a problem is not simply a question or situation—it must be perplexing. The question in the fourth critical incident at

the beginning of this chapter is most likely not a problem for Barb or for anyone reading this book. The answer, 6 cm², is immediate, probably a result of
understanding that the area of a triangle may be found by using $A = \frac{1}{2}bh$ (an item
of knowledge) and skill in multiplying. The question in this critical incident was
planned as a nonproblem, but the questions in the other three critical incidents
were selected because they may represent reasonable problems for you.

A question or a situation can be judged perplexing, and thus be a problem,
only in relation to a *person* and a *time*. At some time in the past, the fourth critical
incident was a problem to every reader (and every author) of this book. Now it is
not. Hence teachers must select questions and situations which will be problems
for their students.

Another implication of the definition is the idea that a question or situation
must be accepted by the student as a problem. "Perplexing" implies that the
question or situation is of some interest and that the student will accept it. If a
reader is not interested in any of the first three critical incidents, then each
critical incident, at least for now, is not a problem for that reader.

The characteristics of a problem for a student, then, are that:
1. It is a question or a situation.
2. It is accepted by the student.
3. At the time it is presented to the student, there is some blockage or challenge
   so that the solution is not immediate (see Figure 5–1).

Problem solving in mathematics, the topic of this chapter, may be described
as the process of arriving at solutions to problems which involve the use of mathematics. Such a definition excludes many questions and situations which are called

Figure 5–1 A problem.

problems by others. For example, we distinguish between *exercises* and *problems.* Exercises are employed in mathematics teaching to provide practice in learning skills or as applications of understandings which have been recently taught. Problems, unlike exercises, require the student to use *synthesis* or *insight.* To solve a problem the student must draw upon previously learned items of knowledge, skills, and understandings, but now he uses them in a new situation; he synthesizes previous learnings. This distinction between problem and exercise may sometimes be ambiguous, particularly because the definition of "problem" includes the sometimes unobservable requirements of acceptance by and challenge to the student. However, a teacher familiar with a group of students will usually be able to distinguish between questions and situations which are exercises for his students and those which are problems.

Not only does the definition of problem solving exclude student exercises, it also excludes solving trick questions. Consider the following question:

What is the next number in the sequence 14, 17, 50, _____?

Each term in the sequence was determined by taking the sum of the squares of the digits of the previous term. Thus the next term will be 25 ($5^2 + 0^2 = 25$). We consider such a question a trick. However, there is an interesting conjecture associated with the indicated sequence:

Start with any number. Form a sequence such that each term is the sum of the squares of the digits of the preceding term. At some point in the sequence thus formed either each term will be 1 or the repeating sequence 16, 37, 58, 89, 145, 42, 20, 4 will be encountered.

A final characteristic of problems, as we apply the term, is that often there will be several different ways for students to attack them. Let us consider the ways of solving one problem:

What is the greatest number of line segments one can draw to connect ten points arranged on a circle?

There are several ways to attack this problem; if you have not solved it before, we ask you to stop here, close the book, and try to solve it.

In solving such a problem students might first read it carefully and try to identify a simple generalization or process stored in their minds which they could apply. If none occurs, which we expect in this case, the student might next try to examine the constraints or determine what he or she can use. Different students might employ different attacks on the problem. Descriptions of some of these approaches follow. In each it is assumed the student understands the problem: Ten points are arranged on a circle (Figure 5–2). The student is to determine the maximum number of segments which can be drawn to connect these points, labeled $A, B, \ldots, J$.

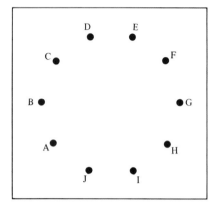

Figure 5–2   Ten points arranged on a circle.

Solution A

Starting with a diagram like Figure 5–2, the student might see that:
1. there are nine line segments between point A and each of the other nine points.
2. there are eight other line segments between point B and the remaining (excluding A) points.
3. each point, going clockwise, will be the endpoint of $x - 1$ new line segments where x is the number of new line segments at the preceding point. The generalization holds through point I, which is the endpoint of only the line segment IJ. Thus the solution is
$9 + 8 + \cdots + 2 + 1 = 45$.

Solution B

This solution is very similar to A, with the difference that the student draws, or imagines to be drawn, all segments. He notes that each point is an endpoint of nine segments, but as he goes clockwise from point to point, he finds some segments have already been accounted for. As a matter of fact, he will have accounted for each line segment twice by counting *both* endpoints. His solution might be

$$\frac{10 \cdot 9}{2} = 45$$

Solution C

Another student might ask herself "How many line segments can I draw with two, three, or four endpoints on a circle?" She might try to draw some simple cases (Figure 5–3). As she tries a specific number of points, say p, she first draws the p segments in the perimeter (if $p \geqslant 3$). Next, she picks two adjacent points and draws $(p - 3)$ new segments from each. When she considers the next point, say, in a clockwise direction, she finds three segments have been drawn, so she can draw only $(p - 4)$ new ones. So she might speculate that a general formula is:

$$p + (p - 3) + (p - 3) + (p - 4) + \cdots + 2 + 1$$

If $p = 10$, she has:

$$10 + 7 + 7 + 6 + 5 + 4 + 3 + 2 + 1 = 45$$

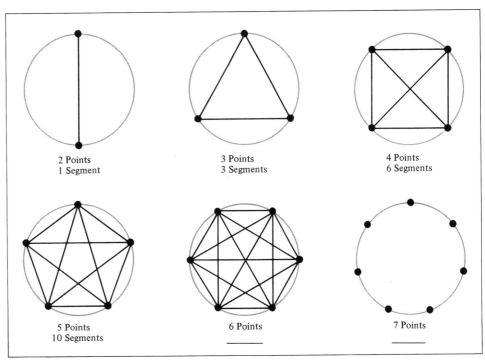

Figure 5–3 Segments for two through seven endpoints.

Solution D

Alternatively, the student, using the diagrams in Figure 5–3, might construct a table of values for points and segments and then look for a pattern.

number of points, p	2	3	4	5	6
number of segments, s	1	3	6	10	15

There are several procedures for determining the number of segments where $p = 10$. The basis for most lie in the theory of finite differences. These processes have been studied and described (Sobel and Maletsky, 1975, pp. 32–34). For example, the student might look at the differences in values for s. Then he might look at the differences of these differences.

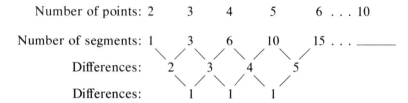

By observing the pattern and hypothesizing a generalization he could inductively complete the table and determine that $s = 45$ when $p = 10$.

Solution E

Again, by using the finite difference techniques indicated in Solution D, the student might hypothesize that there is a quadratic equation which relates p to s, that is,

$s = Ap^2 + Bp + C$

Knowing s when $p = 2$, $p = 3$, and $p = 4$, the student could observe that:

$$1 = 4A + 2B + C, \quad \text{when } p = 2$$
$$3 = 9A + 3B + C, \quad \text{when } p = 3$$
$$6 = 16A + 4B + C, \quad \text{when } p = 4$$

Using differences, and differences of differences, with the above equations, the student would find that:

$$2 = 5A + B$$
$$3 = 7A + B$$

so that

$$1 = 2A$$

and by substituting,

$$A = \tfrac{1}{2}, \; B = -\tfrac{1}{2}, \; \text{and} \; C = 0$$

Thus the student would find that

$$s = \tfrac{1}{2}p^2 - \tfrac{1}{2}p$$

Checking when $p = 10$, we find

$$s = \tfrac{1}{2}(10)^2 - \tfrac{1}{2}(10)$$
$$= 45$$

Solution F

An insightful student might observe that the problem is simply to determine how many pairs of points one can select from ten points. That is,

$$\binom{10}{2} = \frac{10!}{2! \, 8!} = 45$$

It is easy to imagine a situation in which a student has been drilled on the use of binomial coefficients and Solution F would be automatic. This problem would evaluate problem-solving goals only if it were nonroutine. If the student has solved the problem, or a very similar one, before, that student's solution might simply be a demonstration of knowledge, skill, or understanding.

WHAT KINDS OF PROBLEMS?

The classification of problems in mathematics seems almost limitless. There are mixture problems, distance-rate-and-time problems, "real-life" problems, number theory problems, proofs, insightful problems, open-search problems, and so on and on. Although there appears to be no particular advantage in trying to develop a partitioning of mathematical problems, a discussion of several classes of problems may be useful.

Problem types are found in many older textbooks. This approach to teaching problem solving consisted of classifying mathematics problems into types, providing students with a discussion which was seldom read or understood, presenting a small number of examples, and asking students to solve many problems which were similar to the examples. We can find *work* problems, *motion* problems, *age* problems, *mixture* problems, and others. Here are examples:

(1) Bill can mow the lawn in 4 hours and Fred can mow it in 5 hours. How long will it take the boys to mow it together? (A *work* problem.)
(2) How much water must be added to a quart of 10 percent solution of sulphuric acid to obtain a 6 percent solution? (A *mixture* problem.)

The fallacy of teaching problem types is that "problem solving" is then taught as a skill. It may seem that a student who can answer many questions of a "type" has acquired the ability to solve each problem; most often, however, students who are successful have acquired only the *skill* of solving that type of problem, and, with most skills, if practice is not continued, the skill recedes. And, because the skill has been acquired in a relatively homogeneous set of experiences (such as solving similar exercises) there is very little transfer to problems or situations which differ from the problem set.

In many textbooks problems still are organized by problem types. One author has presented a fairly extensive treatment of solving problem types; some examples from that study appear in Figures 5–4 and 5–5 (Ranieri, 1961, pp. 28–29).

Applications of mathematics can be employed to stimulate student interest in problem solving and to demonstrate the wide range of physical situations in which mathematics is used to find solutions or to reach decisions. Problems derived from believable real-world situations may be of special interest to students. For example, the question in the third critical incident ("How should the green, yellow, and red lights at an intersection be timed?") is considered from both theoretical and practical viewpoints in a modern textbook which unifies science, social science, and mathematics (*The Man-made World*, 1971, pp. 5–19, 40–43). Perhaps there is no perfect solution to this problem; but traffic lights are timed, and students can learn to consider some of the crucial variables associated with the physical, real-world situation. Using such problems helps convince students that mathematics is not simply a body of knowledge assembled for youngsters to learn, but a modern, useful tool.

The mere fact that a problem is taken from the real world, however, does not assure its pedagogical value. The teacher may have to modify the problem by reducing its scope, simplifying the context, or by restating it in terms which the students can understand. As will be discussed in Chapter 9, the availability of calculators and computers has lowered the "calculation barrier" of many interest-

UNIFORM MOTION – SAME DIRECTION

1. Key –   The distance that one travels equals the distance that the other travels.

$$\therefore D_1 = D_2$$

(Note: "D" with the subscript represents the distance.) When the starting and finishing points of both travelers are the same but one of the two journeys a greater distance because of a different route, then the "key" becomes $D_1 + C = D_2$, where "D_1" is the shorter distance and "C" is a constant—the additional distance traveled.

2. Problem –   Jane left her home for school at 8:25 A.M. walking at the rate of 4 miles per hour. At 8:30 A.M. her brother departed and running the same route, met Jane 10 minutes later. At what speed did the brother run?

3. "Step-by-Step" Procedure

A) Representations

Let x = Speed of the brother in m.p.h.

D = (Rate) (Time) = RT

B) Figure

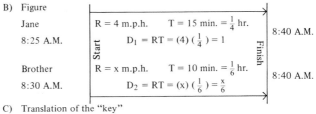

Jane
8:25 A.M.

$R = 4$ m.p.h. $T = 15$ min. $= \frac{1}{4}$ hr.
$D_1 = RT = (4)(\frac{1}{4}) = 1$
8:40 A.M.

Brother
8:30 A.M.

$R = x$ m.p.h. $T = 10$ min. $= \frac{1}{6}$ hr.
$D_2 = RT = (x)(\frac{1}{6}) = \frac{x}{6}$
8:40 A.M.

Start   Finish

C) Translation of the "key"

$D_1 = D_2$

$1 = \frac{x}{6}$

D) Solution

$6 = x$

$\therefore x = 6$ m.p.h. for brother   answer

E) Check

Jane: $D_1 = RT = (4)(\frac{1}{4}) = 1$ mi.

Brother: $D_2 = RT = (^6)(\frac{1}{6}) = 1$ mi.

Figure 5–4 Solution analysis for one problem type (from Francis J. Ranieri, *The Key to Word Problems—Algebra*. Reprinted by permission of Educators Publishing Service, Inc., and Francis J. Ranieri).

ing and realistic problems, hitherto unsuitable for classroom use because of the messy calculations involved.

The first critical incident is designed to provide an application, a real-world problem. Where should you sit in a movie theater? (See Figure 5–6.) The same problem in a different context was presented to a mathematics club in an outstanding high school several years ago:

DRILL PROBLEMS

UNIFORM MOTION – SAME DIRECTION PROBLEMS

1. A car travels south at the rate of 30 miles per hour. Two hours later a second car leaves to overtake the first car, using the same route and going 45 miles per hour. In how many hours will the second car overtake the first car?

2. The sum of the speeds of two trains is 100 miles per hour. The faster train leaves 4 hours after the slower train and overtakes it in 8 hours. Find the speed of both trains.

3. A pick-up truck, while overtaking a large delivery truck, operates at a speed of 15 miles per hour more than the delivery truck and leaves 3 hours after the latter does. If it takes the pick-up truck 7 hours to reach the delivery truck, find the rate of each vehicle.

4. A car traveling at the speed of 30 miles per hour left a certain place at 10:00 A.M. At 11:30 A.M. another car, at 40 miles per hour, departed from the same place and traveled the same route. In how many hours will the second car overtake the first car?

5. A plane flying at "x" miles per hour leaves Logan airport for a certain city. Two hours later, a second plane departs for the same city taking a route which is 50 miles longer and flies at 50 miles per hour faster than the first plane. If both planes arrive at their destination five hours after the departure of the second, find their speeds of traveling.

6. The speed of car "A" and the speed of car "B" are in the ratio of 2 to 3. The slower car, "A," leaves two hours before car "B" for a distant town. Car "B," on a route 20 miles shorter arrives 3 hours later at the town at the same time as car "A." Find the speed of each car.

7. A train in traveling to New York city operates at 60 miles per hour. Five hours later an airplane leaves from the same vicinity as the train to also travel to New York city at the speed of 140 miles per hour. If both arrive at the same time, and if the train travels 20 miles more than the airplane, find the time it takes both to make the trip.

Figure 5–5 Exercises for one problem type (from Francis J. Ranieri, *The Key to Word Problems— Algebra*. Reprinted by permission of Educators Publishing Service, Inc., and Francis J. Ranieri).

(3) Suppose there is an enemy base on the side of a mountain. As intelligence officer you are to obtain the best aerial photograph of the base. A camera is mounted to a plane which will fly over the base in a fixed (straight line) glide path. Where should the picture be taken? (See Figure 5–7.)

Both problems are similar in that, given a line segment (actually some closed area) and a line, we are to select a point on the line to obtain "the best possible picture." When the problem was presented to the mathematics club members, it was approximately two weeks before any student would observe or admit that "best possible picture" needed to be defined. (Such students, typically very successful, find it difficult to admit they do not "understand" the problem.) After some discussion, it was decided to define the "best picture" as the one with the widest angle. (Is the definition reasonable?) Thus the problem reduces to:

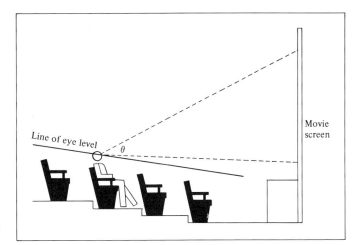

Figure 5-6   Where should you sit in a movie theater?

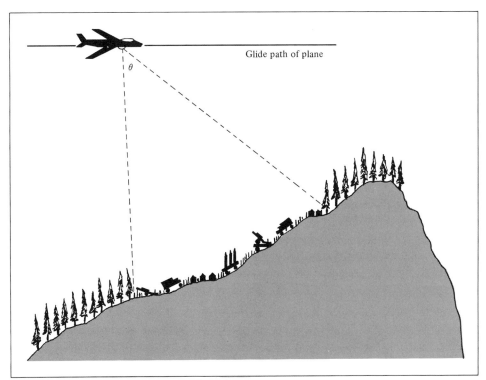

Figure 5-7   Aerial photography of a military base.

(4) Given a line segment and a line, find point B on the line such that the angle formed with a vertex at B and sides passing through the end-points of the line segment is maximal.

The problem will be discussed later (but the reader is invited to find the solution now). There are many sources of mathematical application problems, among them Schiffer (1963) and Bell (1967, 1971, 1972).

Abstract problems may be of as much interest to students as applications. Although we recommend applications of mathematics as a rich source of interesting problems, we must note that students can become just as interested in abstract problems such as Problem (4) above, the final wording of the "movie problem." Students might also enjoy problems we could classify as fantasy (the "bee and sugar" problem in Chapter 2, pp. 37–38).

Open-search problems are becoming more popular in modern mathematics classrooms. Such problems may be stimulating to students in developing their own questions and in exploring extensions of the given problem situation. In solving a good open-search problem a student is likely to uncover some mathematical ideas that are new to him or her or to develop interesting new problems to be solved. Botts (1965a and 1965b), in a two-part article, has described how a simple problem and a lot of imagination can lead to extended mathematical thought. The problem (already cited in Chapter 2) and some of the theorems proposed by Botts follow:

(5) Can you find three (different) positive integers the sum of whose reciprocals is an integer?

EXTENSIONS OF THE PROBLEM

Theorem

The only positive integers a, b, and c such that:

$$a < b < c$$

and

$$\frac{1}{a} + \frac{1}{b} + \frac{1}{c} \text{ is an integer}$$

are $a = 2$, $b = 3$, and $c = 6$

A New Problem

For which positive integers n can we express 1 as the sum of the reciprocals of n (different) positive integers?

Another Problem

For which positive integers m can we express m as a sum of the reciprocals of some number of positive integers, all different?

A Lemma

For every positive integer n, the number 1 can be expressed as a sum of reciprocals of integers, all different from each other and all greater than or equal to n.

Another Theorem

Every positive rational number is expressible as a sum of unit fractions which are distinct (different from each other).

In this example, the original problem has a relatively simple solution—the first theorem in the extension of the problem. However, by posing a new problem and exploring other hypotheses for the original problem, Botts solved a much more general problem. (For a very different approach to this problem see Adkins, 1963.)

In the previous chapter a distinction was made between open search and discovery. In open search the student might, and hopefully will, uncover many generalizations (discoveries). We distinguish between the two to emphasize the teacher's role in discovery lessons. Open search begins simply with a question or situation. A primitive process may not be available, and the student goes through a period of experimentation and searching. Typically, the generalizations uncovered by the student are less predictable in open search and the teacher may stand ready to provide some clues or suggestions at appropriate times to help stimulate the student. The distinction between open search and discovery may not always be clear; it depends on the intent of the teacher and the background of the students.

Projects are another vehicle for problem solving in mathematics. Projects may be applications or abstract; some may involve open search. The distinguishing characteristics of a project are that it should be exploratory and that it is typically investigated over a period ranging from two days to several weeks. Ways to use projects in a classroom—and other techniques for individualizing instruction—are presented in Chapter 8. Here are a few examples of projects:

(6) Determine how much it would cost to seed and fertilize your entire school lot excluding buildings, driveways, and woods.

(7) A point in the coordinate plane may be specified by Cartesian coordinates (x, y) or by polar coordinates (r, θ). Can you find a way to change any function $y = f(x)$ to polar coordinate form without using any of the trigonometric functions?

(8) Write a computer program in BASIC to simulate a tennis match.
(9) Calculate when the next lunar eclipse will occur (Greitzer, 1967, pp. 265–273).

Proof is a special kind of mathematical problem. In Chapter 4 (p. 116), proving a theorem or using it to prove another theorem were identified as ways of demonstrating an understanding of the first theorem (generalization). Proving a theorem may also be a problem-solving experience when students are asked to construct original proofs.

In general, the proof of a statement is an argument based on: (1) statements (axioms) accepted by both the person proving and a listener or reader, (2) definitions, (3) previously proven theorems, and (4) agreed-upon rules of logic. In secondary school, the agreed-upon rules of logic are usually subsumed within courses of study and learned implicitly by the students. A calculus of logic may be (and is recommended to be) studied by teachers, but such a formal structure will probably have little meaning to secondary school students until they have acquired experience in using logic intuitively.

The other elements of a mathematical system (axioms, definitions, and theorems) may be presented and used more or less formally, as in a one year geometry course, or they may be used informally as a way to convince others (students or the teacher) of the validity of a statement. For example, the second critical incident calls for a proof that the sum of twin primes (excluding 3 and 5) is divisible by 12. Let us now consider a proof of this conjecture.

Proof
That the sum of each pair of twin primes, other than 3 and 5, is divisible by 12.
- We may denote a pair of twin primes by p and $p + 2$ and observe that $p > 3$, since 3 and 5 are excluded. Thus, p and $p + 2$ are both prime and odd.
- Thus, $p + 1$, an integer between p and $p + 2$, is even.
- Also, we note one of three consecutive integers must be divisible by 3. But neither p or $p + 2$ is divisible by 3 because they are both prime numbers greater than 3. Therefore, $p + 1$ is divisible by 3.
- Now we have shown that $p + 1$ is divisible by both 2 and 3, so $p + 1$ is divisible by 6 (Fundamental Theorem of Arithmetic).
- But, $p + (p + 2) = 2p + 2$
$$= 2(p + 1)$$
- Since $(p + 1)$ is divisible by 6, $2(p + 1)$ is divisible by 12.

The proof may be presented even more informally, but it still would depend on the definitions of prime number and twin primes, the *Fundamental Theorem of Arithmetic*, and the observation that the middle number is even and divisible by 3.

Let us consider another proof. Recall the first critical incident and the previous discussion of it. The abstracted form of that problem was: "Given a line segment AC, find a point B on line *l* (the glide path of a plane) such that ∠ABC is maximal." (The authors' experience is that most people presented with this problem conjecture that B lies on the perpendicular bisector of $\overline{AC}$, which is the correct solution only in the special case that $\overline{AC} \parallel l$. Such a guess implies an abiding faith that solutions to mathematical problems are always "pretty" in some sense.)

The solution is that B is the point of tangency with the line *l* of a circle containing A and C (Figure 5–8). To find point B analytically is somewhat difficult. However, it is easy to show that, as described, ∠ABC is the maximum angle with a vertex on *l* and sides passing through A and C. If any other point B' on *l* were the vertex, ∠AB'C would intersect the circle in 4 points—say, A, C, A' and C' (Figure 5–9). The measure of ∠AB'C is one half the difference of the intercepted arcs. Since m(AC) is constant, m∠AB'C is less than m∠ABC.

Now that we have examined two instances of proof, we will consider some of the standard classifications of proofs. Every theorem or statement to be proved consists of a *hypothesis* and a conclusion. The hypothesis is a statement of conditions which are asserted in the theorem to be sufficient to guarantee the conditions of the conclusion. In a *synthetic proof* the investigator uses the conditions of the hypothesis and other information (axioms, definitions, and previously proven theorems) to deduce the conditions of the conclusion. An *analytic proof* is constructed by considering the conclusion and exploring what conditions must be demonstrated to justify the conclusion. In like fashion, these conditions are examined to determine what conditions must be demonstrated to justify them. The investigator continues until he or she arrives at the conditions stated in the hypotheses. The following is an example of two proofs of the same theorem.

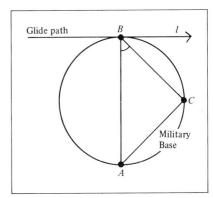

Figure 5–8.

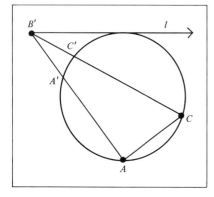

Figure 5–9.

Theorem

In an isosceles triangle, the angles opposite congruent sides are congruent. Suppose $\triangle ABC$ is an isosceles triangle with congruent sides $\overline{AB}$ and $\overline{AC}$ (Figure 5–10).

Synthetic Proof

$\qquad\qquad\quad \overline{AB} \cong \overline{AC}$ Given

$\qquad\qquad\quad \overline{AC} \cong \overline{AB}$ Congruence is reflexive

But $\qquad\quad \overline{BC} \cong \overline{CB}$ A line segment is congruent to itself

Thus $\quad \triangle ABC \cong \triangle ACB$ Side-side-side theorem

and $\qquad \angle ABC \cong \angle ACB$ Corresponding parts of congruent triangles are congruent.

Analytic Proof

I can show that $\angle ABC \cong \angle ACB$ if I can show that the two angles are corresponding parts of congruent triangles. Thus, I need to show that $\triangle ABC \cong \triangle ACB$. Perhaps I can show the congruency by showing that the side-side-side theorem holds. To do this I can show that $\overline{BC} \cong \overline{CB}$, because I know that line segments are congruent to themselves. Thus, I need show only that $\overline{AB} \cong \overline{AC}$, but this follows because $\triangle ABC$ is an isosceles triangle with congruent sides $\overline{AB}$ and $\overline{AC}$.

Proofs may also be classified as *direct* or *indirect*. A direct proof is an argument based on the hypothesis of the theorem. Thus both previous examples of synthetic and analytic proofs are direct. An indirect proof is based on assuming a negation

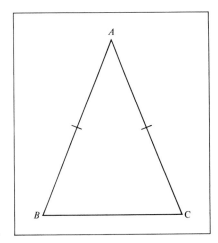

Figure 5–10.

of the conclusion and showing that one can deduce either a negation of the hypothesis or the negation of a (true) statement. A theorem in the form, P implies Q, could be symbolized as

$$P \rightarrow Q$$

where P is the hypothesis and Q is the conclusion. Using $\sim Q$ to mean the negation of a statement Q, an indirect proof is a demonstration of either:

(1) $\sim Q \rightarrow \sim P$, or "not Q implies not P"

or

(2) $(\sim Q$ and $P) \rightarrow (R$ and $\sim R)$, or "not Q and P implies R and not R" where R is some statement.

Case 1 is a proof of the contrapositive of the theorem. Because a theorem and its contrapositive are equivalent statements (that is, they have the same truth value), once we have established the truth of the contrapositive, the theorem is proved. In case 2 the assumption of $(P$ and $\sim Q)$ leads to $(R$ and $\sim R)$ for some statement R. It is easily shown that $(R$ and $\sim R)$ implies A where A is any statement. We note that

$$(R \text{ and } \sim R) \rightarrow R$$

and

$$(R \text{ and } \sim R) \rightarrow \sim R$$

Since we have R, we also have $(R$ or $A)$ where A is any statement. But

$$((R \text{ or } A) \text{ and } (\sim R)) \rightarrow A$$

That is, "R or A and not R implies A." Thus we have shown (R and ~R) → A. Usually we do not state this last argument in a proof; rather, we recognize that P → Q has been proven when we have shown that (~Q and P) implies (R and ~R). We call (R and ~R) a "contradiction."

Some logicians classify case 1, proof of the contrapositive, as a direct proof because direct methods are used to show ~Q → ~P, which is equivalent to P → Q. Although the classification is somewhat arbitrary, we find three reasons for considering case 1 an indirect proof. First, in both case 1 and case 2 we assume (~Q) in our argument. Second, it is acceptable to use the hypothesis P in the argument ~Q → ~P. Thus we are examining a negation of the theorem because ~(P → Q) is equivalent to (P and ~Q). Third, case 1 can be considered a special instance of case 2 because we have shown that (~Q and P) → (P and ~P) and (P and ~P) is a contradiction. (For a more detailed discussion of deductive methods, see Copi, 1954, pp. 40–65.)

To give an example of an indirect proof, suppose we wish to prove the following conjecture: If the product of two natural numbers a and b is odd, then both a and b are odd. The proof: Suppose at least one of the natural numbers, say a, is not odd—that is, a is even. Then by definition of even, there is a natural number n such that $a = 2 \cdot n$.

Thus, $a \cdot b = (2 \cdot n) \cdot b$ substitution
$= 2 \cdot (n \cdot b)$ associative property of multiplication

$n \cdot b$ is a natural number and $2 \cdot (n \cdot b)$ is even. Thus, by assuming that one of the natural numbers, a, was even, we deduce that the product $a \cdot b$ is even. Therefore, a is not even. In the same way, it can be shown that b cannot be even if $a \cdot b$ is odd.

For another example of an indirect proof, consider the following theorem: There is no largest prime number. The proof: suppose there is a largest prime number, p. Since p is known we determine which numbers between 1 and p are prime and call them $p_1, p_2, p_3, \ldots, p_n$. We note that $(p_1 \cdot p_2 \cdot p_3 \cdot \ldots \cdot p_n \cdot p + 1)$ is larger than p and prime. (Why is this number prime?) Thus, by assuming that "p is the largest prime," a statement R, we have shown that "p is not the largest prime," a statement ~R. (R and ~R) is a contradiction. Thus our assumption (R) must be rejected, and we conclude that there is no largest prime number.

PLANNING TO TEACH PROBLEM SOLVING

GOALS

Specifying problem-solving goals is a difficult task. Implicit in developing problem-solving ability is the goal that the student will be able to solve a wide variety of nonroutine problems. Thus goals must be stated in general terms; to list specific problems to be solved by the students is to risk teaching problem types. Alternatively, the teacher could provide some sample problems which are deemed appropriate for the students. However, highly similar problems should be avoided because the consequence may be the teaching of a *skill* of solving problem types.

Every mathematics teacher should collect problems. It is helpful to classify such a collection by the items of knowledge, the skills, and the understandings useful in their solution. A teacher might also want to classify the problems by the categories discussed in the preceding section—applications, abstract problems, open search, projects, and proofs. These classifications are not a partitioning; an individual problem might be assigned to two or three of the categories. Such a file will be useful in selecting a variety of problems for a particular class at a particular time.

PREREQUISITES

To solve almost any mathematics problem a student will need prerequisite knowledge, skill, and understanding. It is important that teachers identify the prerequisite learnings for problems so that only appropriate problems are presented to students. However, some problems may be presented to develop understandings which have not yet been taught. For example, after students have learned to multiply binomial expressions, they might be asked to solve the following problem:

Can you find a general way to expand $(a + b)^n$ as n is assigned different values?

In one sense the problem is open search. In another sense it is a set of exercises based on understanding and skill in obtaining the product of two binomial expressions (and the associative property). The problem may be used in a later class to teach understanding of the binomial expansion, Pascal's triangle, or binomial coefficients. For some students, the problem may become a discovery lesson because a primitive is available—that is, they could carry out the multiplication of $(a + b)^1$, $(a + b)^2$, . . . , $(a + b)^j$ and look for a pattern. However, the major point in emphasizing the need to consider prerequisites for problems is to avoid assigning problems which the students have little likelihood of understanding.

TEACHING PROBLEM SOLVING

To learn to solve problems, students must have an opportunity to solve problems. They should receive rewards and they will need approaches to problem solving. Thus, to develop problem-solving ability in your students, you must schedule time for problem-solving experience, consider a reward system for students, and develop some basic problem-solving approaches for your students.

DELIVERY SYSTEMS

How do you provide students with problems and how do you interest them in solving problems? Teachers have developed several techniques to encourage students in problem solving. First the teacher must have a wide variety of problems available. Some teachers post challenge problems on bulletin boards; others assign a small number of selected problems as homework. One day a week might be set aside as a problem day in which students select problems, work on them, discuss them, and present solutions to the class. Problems may be attacked by individuals or groups of students in the class.

REWARD SYSTEMS

Students who solve problems may be rewarded by bonus grades and/or special recognition, such as being designated a "mathlete" (Hlavaty, 1967, pp. 205–217). Some reward or praise should be bestowed upon the successful solver. It is well known that positive reinforcement is an effective way to stimulate success. However, it should also be noted that many students *enjoy* solving problems—they generate their own rewards by being successful—but teacher praise is also helpful.

A word of caution: the teacher should see that the "reward system" is not punishment. If students who solve problems or who complete assignments quickly are "rewarded" with additional work, such a result may extinguish any desire to succeed.

HEURISTICS

How may a student begin to solve a problem? What strategies could he or she employ? What abilities would be useful for him or her? These approaches, strategies, and abilities are grouped together and called *heuristics*, general student activities to help in solving problems. In a sense heuristics are general approaches, strategies, or abilities which are helpful in solving many problems. As such, they are teachable; and a way to teach problem solving is to teach heuristics. The purpose is to provide students with a repertoire of general approaches so that they can select an activity which holds promise of being productive in obtaining a solution. A teacher's role is to help the students see what heuristics are available to them and how they can be useful in problem solving. In this section we identify several heuristics which are useful in problem solving. We classify the first twelve as *initiating heuristics* because they are ways to begin or progress in the attack on a problem. The last five are called *looking back heuristics* because they are typically employed after the student has found a solution to a problem.*

---

* These heuristics have been adapted from discussions at the Problem Solving Research Workshop, Center for the Study of Learning and Teaching Mathematics, University of Georgia, Athens, Georgia, May 26–29, 1975. Project Director: Larry L. Hatfield. Center Director: Leslie P. Steffe.

1. Select Appropriate Notation   Often the selection of a good notation scheme can help a student achieve a solution. Consider this problem:

(10) Find all rectangles with integral sides whose area and perimeter are equal.

To us it seems apparent that a reasonable notation would be to let l be the length of a rectangle and w be its width. So the perimeter is $(2l + 2w)$ and the area is $l \cdot w$. The problem reduces to finding l and w such that

$$l \cdot w = 2l + 2w$$

However, to a student, who lacks our experience, the selection of such a notation may not be apparent. He might start by letting p be the perimeter and a the area of a rectangle. This choice might delay him.

We might consider selecting an appropriate notation as a skill—practice may help learn it. However, as with most heuristics, it is not entirely a skill; some creativity or artistry may at times be necessary to select a useful notation.

2. Make a Drawing, Figure, or Graph   Initiating heuristics have been chosen to help a student understand the problem. Often a picture is useful in order to abstract important elements in a problem and to identify other helpful heuristics. If a student were to draw a figure in Problem (10), he or she might find it easier to select an appropriate notation.

Some problems may be very difficult to understand unless a figure is drawn. Consider this one:

(11) Construct $\triangle APB$ isosceles with base angles of 15° and the obtuse angle at P. Knowing that $ABCD$ is a square with P in the interior, prove that P, C, and D form the vertices of an equilateral triangle.

Without a figure, this problem is difficult to begin (see Figure 5–11). Here is a hint: locate Q in the interior of square $ABCD$ so that $\triangle BQC \cong \triangle BPA$. Extend $\overleftrightarrow{CQ}$ to intersect $\overleftrightarrow{BP}$.

3. Identify Wanted, Given, and Needed Information   An important ques-

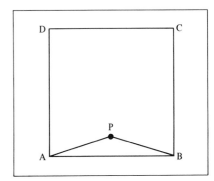

Figure 5–11.

tion is "What am I to find in this problem?" Often the answer will be a hint of how to proceed. If a student who is presented with a problem can identify what information is wanted, what is given, and what is needed, typically that student will be on the way to a solution. Almost any problem could be used as an example of employing this heuristic. In Problem (11) we could determine the following.

Wanted
   Show $\triangle PCD$ is equilateral

Given
   $ABCD$ is a square
   $\triangle APB$ is isosceles
   $m(\angle PAB) = m(\angle PBA) = 15°$
   P is on the interior of square $ABCD$

Needed
   $\overline{PC} \cong \overline{CD} \cong \overline{DP}$ or
   $\angle PCD \cong \angle CDP \cong \angle DPC$
   The hint provides additional information for ''Given'' or ''Needed.''

In teaching this heuristic you may also help students identify relevant and irrelevant information in problems. Relevant information in the problem is part of "Given" but there may be some "Given" information which is irrelevant in the problem.

4. Restate the Problem   Sometimes restating a problem will help a student understand it. Recall how we restated the problem in the first critical incident as Problem (4). In this case the restatement is an abstraction of the problem.

5. Write an Open Sentence   In attacking Problem (10) we used the equation

$$l \cdot w = 2l + 2w$$

Many problems are easy to solve once the student had devised the right open sentence—equation or inequality.

Heuristics 2, 4, and 5 are all translation skills; that is, they involve taking one form of a statement or situation and changing it to a different one. Standard forms used in mathematics problem solving are verbal (written or spoken words), pictorial, and symbolic (in particular, open sentences). Skill in changing from one form to another appears to be closely associated with problem-solving ability (Clarkson, 1977). Using the three forms, there are nine kinds of translations:

v–v	Verbal to verbal	Horizontal
v–p	Verbal to pictorial	Convergent
v–s	Verbal to symbolic	Convergent
p–v	Pictorial to verbal	Divergent
p–p	Pictorial to pictorial	Horizontal
p–s	Pictorial to symbolic	Convergent
s–v	Symbolic to verbal	Divergent
s–p	Symbolic to pictorial	Divergent
s–s	Symbolic to symbolic	Horizontal

Those translations normally used in problem solving are classified as convergent. Those between the same form are classified as horizontal and the others are classified as divergent. Even though the convergent translations are the ones most often used, it appears that skill in all translations is helpful. Teachers can easily develop exercises and questions to aid students in acquiring this skill (see Research Highlight, "What's the Difficulty?").

RESEARCH HIGHLIGHT | WHAT'S THE DIFFICULTY?

Many teachers find that word problems are difficult for students to solve. Computer scientists have noted that the translation from natural language into symbolic formulation is indeed a complicated process. In a study by Kennedy, Eliot, and Krulee (1970), students were asked to solve two sets of algebra problems, some with words and some with numbers only. Twenty-eight juniors from average and advanced classes were asked to read aloud each of six problems and then write down the step-by-step solutions, saying aloud anything which came to mind.

In general, the numerical problems offered little difficulty. The word problems, however, were considerably more difficult for the less able student. Both groups of students recognized equally well the relationships needed for equations. Differences between average and advanced students were a function, however, of their ability to identify or add the logical or physical inferences needed to solve the problem. The researchers suggested that teachers should be less concerned with teaching students to define the relationships between problem elements and more concerned with helping them to identify the logical and physical assumptions made in the problem statement. This ability might be developed as your students learn to restate problems.

6. Draw from a Cognitive Background   Many problems are not solvable without recalling or synthesizing previous learnings. The student might begin by asking "What do I know that is related to this problem?" For example, Problem (11) involves isosceles triangles, and it might be helpful for the student to recall what he or she knows about them. The student might write:

$$\angle PAB \cong \angle PBA$$

$$\overline{AP} \cong \overline{PB}$$

A corollary to this heuristic is to use available resources, such as books, calculators, or resource people, when appropriate. Sometimes it may not be appropriate to use these resources, but many teachers want their students to learn that resources are available and that they can be useful.

7. Construct a Table   In solving some problems a table may provide a way to discover a pattern and gain an insight to the solution. Consider this problem:

(12) Figure 5–12 shows drawings of the five platonic solids. There is an interesting relationship between the numbers of vertices, faces, and edges of these polyhedrons. In fact, this famous relationship is true for any convex polyhedron. Obtain some models of convex polyhedra and see if you can discover what this relationship is.

Figure 5–13 depicts a way of organizing data to solve this problem.

8. Guess and Check   Students must be taught and encouraged to guess. Often they have acquired the idea that guessing should not be part of mathe-

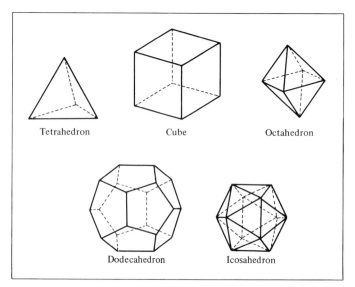

Tetrahedron   Cube   Octahedron

Dodecahedron   Icosahedron

Figure 5–12   The five platonic solids.

Name of Shape	Number of			Rule
	Vertices	Faces	Edges	
Tetrahedron				
Cube				
Square Pyramid				
Triangular Pyramid				
Pentagonal Pyramid				
Hexagonal Pyramid				
Octagonal Pyramid				
Triangular Prism				
Square Prism				
Pentagonal Prism				
Hexagonal Prism				
Octagonal Prism				
Octahedron				
Dodecahedron				
Icosahedron				

Figure 5–13   A table for Problem (12).

matics—but it is! Problem (12) was selected to encourage student guessing. We asked for a generalization, and at some time that generalization would be a guess. A natural follow-up to guessing is checking: does the guess work? Parts of the guess and check heuristic which are sometimes overlooked are to commit the guess and to keep track of guesses. Once a guess is made, try it out. Sometimes students guess a number and then don't check it. Other times they may make a guess but not record it. As they try other guesses they might forget an early guess and try it again—wasted effort. The following is an example of a problem in which the guess and check heuristic would be useful and in which recording guesses would be important. A table of guesses might also help.

(13) Find the digit represented by each letter in the following division problem where each letter stands for a different digit. (*The Mathematics Student Journal*, May, 1969, p. 6).

```
                  X T X T
        T D X ) E D E I T X
                E S X
                S S I
                T D X
                  E X T
                  E S X
                    S N X
                    T D X
                    T S I
```

9. Systematize   In attacking some problems a student may find a way to systematize his or her approach, perhaps even by exhausting all possibilities.

Letter puzzles such as Problem (13) are often solvable by exhaustion—trying all possible digits, say, for the letter "T". Sometimes the system might be one of using successive approximations, at other times it might be to try extreme cases. Consider another problem:

(14) For what n is

$$2^n > n^{10}?$$

This problem might be easier to solve by trying extreme cases. For example, it is easy to see that $n > 10$ and almost as easy to see that $n < 100$. (This problem might be still easier to solve by recalling relevant information from your study of logarithms.)

10. **Make a Simpler Problem** By the very definition we use, problems are perplexing. If a student does not see a way to proceed, he or she might try to change the problem to one that is similar but one that the student can solve. Suppose Problem (12) were stated: "Find the relationship between the numbers of vertices, faces, and edges of convex polyhedrons." A reasonable way to begin would be to examine the relationship between these numbers for a tetrahedron, then for a square pyramid, and then for a cube. That is, it is natural to look at simpler cases rather than start with the general case.

Similarly, consider these problems:

(15) Examine the array of letters in Figure 5–14. Starting at the top, the marked path indicates one way of spelling PROBABILITY. Determine the total number of such paths.

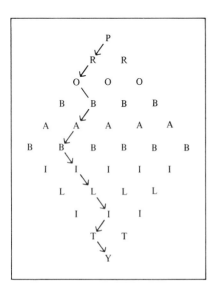

Figure 5–14.

(16) Refer to Figure 5–15. Assuming the truth of the Pythagorean theorem, prove that the area of the figure drawn on the hypotenuse of the triangle equals the sum of the areas of the figures drawn on the legs. The shapes shown are called Gothic arches. Make up other such drawings on your own and prove that the same relationship holds.

One way to start Problem (15) is to change the pattern to one using, say, three letters, like "CAN." Then a student might attempt to solve the simpler problem in a pattern of the word "SOLVE." To solve Problem (16) a student might try to solve the simpler problem of the areas being equal if the Gothic arches are replaced with equilateral triangles.

Here is a final example of a popular problem which can be solved by making it simpler:

(17) How many squares are there in a standard (8 by 8) checkerboard? (The answer is not 64.)

A reasonable approach is to try to solve some simpler problems such as:

How many squares are there in a 1 by 1 square?
How many squares are there in a 2 by 2 square?
How many squares are there in a 3 by 3 square? etc.

Another simpler problem approach is the following:

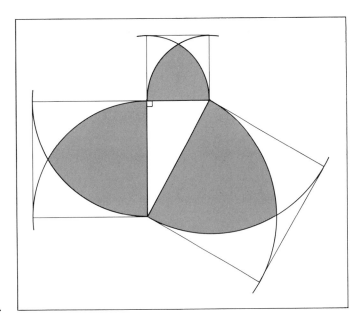

Figure 5–15.

How many 8 by 8 squares are there in an 8 by 8 square?
How many 7 by 7 squares are there in an 8 by 8 square?
How many 6 by 6 squares are there in an 8 by 8 square?, etc.

All these simpler problems may help a student find a strategy to solve the particular problem, or they may help the student see more clearly what he or she can do to solve the given problem.

11. Construct a Physical Model   Problem (17) provides a good example of a problem which could be solved through the construction and use of a physical model. Consider posing the problem to a group of students and providing each of them with a checkerboard and a set of squares from 8 by 8 through 1 by 1. The students can use the squares to answer one of the series of questions given above. Also, it might be helpful if they *constructed a table* to record the results as they *solve simpler problems.* The students then would be using three heuristics.

Physical models might be useful to some students in attacking the next two problems.

(18) If six people are in a room and each person shakes hands with each of the other five people, how many handshakes will occur?
(19) Assume that you have a 3-liter measure and a 5-liter measure. How could you measure out 4 liters? (See Figure 5–16.)

In Problem (18) a group of six students could become the physical model and a table of new handshakes for each person might be made. In Problem (19) a student might use graduated measuring cups or 3- and 5-liter jars to try to obtain 4 liters. Again, a record of the student's procedure might be needed so he or she won't forget the process.

Although some problems are not suited to this heuristic, many are, and sometimes the physical model may need be only imagined to start on a solution.

Figure 5–16.

12. Work Backwards   Sometimes a solution will emerge if the solver can answer two questions repeatedly: "What do I need to solve this problem?" and "Do I know it?"

Often this heuristic would be used with Heuristic 3. For example, see the discussion of an analytic proof on pp. 135–136. Or consider the following problem:

(20) A baseball pitcher has eight baseballs. Seven of them weigh exactly the same, but one is just a bit heavier. Using a balance scale, how can the pitcher find the heaviest ball in just two weighings?

We might think that Heuristic 8, Guess and Check, would be helpful, and perhaps you could try that strategy. However, think of working backwards. "What do I need to solve this problem?" On the second weighing we need to identify the heavy ball. If there are two balls on each side of the balance scale, could we tell which one is heavy? No. Thus we think of one ball on each side. But now, a little insight—what if the scale is in balance? So actually we can tell which of three balls is heavy. Two we put on the balance scale and one we leave off. "What do I need to solve this problem?" We must know that the heavy ball is one of three. "Do I have it?" No. So ask again, "(Now) What do I need to solve this problem?" With a little more insight we might see that for the first weighing we could put three balls on each side of the scale. If one side is heavy, we will know the heavy ball is one of the three on that side. If the two sides balance, we can tell which of the two remaining balls is heavy in one weighing.

These 12 heuristics may be used, one or more at a time, as ways of attacking a problem. A student could try a heuristic before knowing how to solve a problem and thereby gain an insight into the solution by using the heuristic. Thus the heuristics become a repertoire of activities to try. But there are still other heuristics that can be helpful. These are sometimes called "looking back" strategies (Polya, 1973, pp. 14–19). Once a student has a solution he or she may be able to learn more from the problem by trying a looking back heuristic.

13. Generalize   Once a problem is solved, is there a generalization to be discovered? Consider Problem (20) again. If a student has learned to find a heavy ball given three balls with one weighing on a balance scale, the student might also see how to locate the one heavy ball given *nine* balls in two weighings. That student might hypothesize that he or she could find a heavy ball in a group of 3^n balls in n weighings. In this way the student might be able to think of and solve a much more difficult problem.

14. Check the Solution   Students miss a great many problems because they fail to check a prospective solution. When the student has what appears to be a solution, it will usually take little time to find a way to verify the solution. As with the other looking back heuristics, however, it is difficult to teach students to use these heuristics. Often it seems that *finding an answer* is the goal, when in fact we really want students to experience the *process* of solving problems.

15. Find Another Way to Solve It   After the experience of solving a problem and checking the solution, a student may be very close to finding a better way to solve it. The student who thinks about the problem a little longer may identify an alternate approach; he or she can then decide which approach to describe. For example, if you had solved Problem (20) as we described and discovered the generalization that a heavy ball can be identified in a set of 3^n balls in n weighings, could you describe your solution in a different way? (Consider the case of 27 balls and put one-third, or nine balls, aside. Place nine balls on each side of the balance scale. Does the scale balance or not?)

16. Find Another Result Our solution to Problem (20) again exemplifies the heuristic. We found a new, more general result in solving the problem and looking back.

17. Study the Solution Process Finding one or even more solutions to a problem can be satisfying, but it is important for students to see that they may often learn from studying the process of solving a problem. What is important, the *process* or the *product*, when the product is the answer to a problem? The product is a single answer; the process may be the key to solving a great many problems—and obtaining many answers.

We have found it difficult to teach students the looking back heuristics, but it can be done. You may find problems which are exemplary in that new discoveries can be made when students look back. You can also encourage students to use looking back heuristics by providing sufficient time. If students are asked to solve many problems in a short period of time, they may find that they cannot look back and finish the assignment. Finally, you can *reward* students for the process. A teacher can emphasize the importance of the process in problem solving by complimenting those students who find clever and impressive processes. A teacher can help students identify particular processes and show them other applications of these processes (see Research Highlight, "Who Is an Insightful Problem Solver?").

We have approached the teaching of problem solving through teaching heuristics. By emphasizing heuristics you can provide alternative ways for students to attack problems. Surely no one heuristic will work all the time, but the set of 12 initiating heuristics and five looking back heuristics can provide options and general approaches for your students. Each heuristic is teachable in the sense that you can help students apply the heuristic and see that it can be applied in a great variety of problems. You could present the heuristic as questions students might ask themselves: Have I selected an appropriate notation? Should I make a drawing? What is wanted? The important point is that the use of heuristics can help students begin a problem-solving approach. Some will be better solvers than others, but you can help all of your students. We list the heuristics here to summarize them.

INITIATING HEURISTICS
 1. Select appropriate notation.
 2. Make a drawing, figure, or graph.

RESEARCH HIGHLIGHT | WHO IS AN INSIGHTFUL PROBLEM SOLVER?

A study conducted by Dodson (1972) was designed to develop a description of "insightful" mathematics problem solvers. The term "insightful" was used to emphasize that the problems were judged to be challenging and unlike problems that the students had solved before.

Using a data sample from the National Longitudinal Study of Mathematical Abilities (NLSMA), Dodson analyzed student performance on selected problems in relation to scores on mathematics achievement tests and certain cognitive and personality traits. He reported that a typical successful problem solver, aside from being strong in mathematics, (a) scored high on verbal and general reasoning tests; (b) was good at determining spatial relations; (c) was able to resist distractions, to identify critical elements, and to disregard irrelevant elements; (d) was a divergent thinker; (e) had low debilitating test anxiety and high facilitating anxiety; (f) had a positive attitude toward mathematics; and (g) was unconcerned about messiness or neatness.

Dodson suggested that the development of insightful problem solving ability might include such things as (a) student exposure to particular topics in mathematics such as the algebra of inequalities or systems of inequalities; (b) considerable emphasis on solving geometry problems which require students to synthesize a large number of seemingly unrelated geometric ideas; and (c) study in solving routine algebraic equations to provide necessary "tools."

3. Identify wanted, given, and needed information.
4. Restate the problem.
5. Write an open sentence.
6. Draw from a cognitive background.
7. Construct a table.
8. Guess and check.
9. Systematize.
10. Make a simpler problem.
11. Construct a physical model.
12. Work backwards.

LOOKING BACK HEURISTICS
13. Generalize.
14. Check the solution.

15. Find another way to solve it.
16. Find another result.
17. Study the solution process.

THE AFFECTIVE DOMAIN

The affective domain goals are most critical in problem solving. Students who are unsuccessful may be easily led to believe that they cannot do mathematics and that they never will be able to learn. Part of the pattern may be firmly fixed in each student before he reaches middle or secondary school, but many successful teachers have changed students' perceptions of themselves and of mathematics. Careful selection of problems, emphasis on heuristics, helpful hints properly timed, encouragement, and praise can all work wonders in creating positive attitudes and beliefs. Once a student's attitude about himself or herself as a problem solver has become positive, that student will be ready to make progress in attaining problem solving goals.

Critical Incidents Revisited

Each of the critical incidents for this chapter focused on a question or a problem. The fourth critical incident, involving Barb's finding the area of a triangle, was selected to show that not all questions are problems and that a question may be a problem for one student at one time but not a problem for another student or at another time. Collect problems which are appropriate for and of interest to your students. (Remember that the context of a problem may provide differing degrees of interest for boys and girls.)

The first three critical incidents have been used throughout the chapter to demonstrate the characteristics of problems, to show alternative approaches to solving problems, and to present problems to challenge you. You have learned that questions or situations must be perplexing, accepted, timely, and not tricks to be classified as problems in the sense that we use the term. Problems may be classified by types or by other more general characteristics such as an application, an abstract problem, an open search, a project, or a proof. We have considered goals, prerequisites, delivery systems, reward systems, heuristics, and the interaction with the affective domain in the teaching of problem solving. Our emphasis has been on teaching students to use heuristics, and we have explored several examples of the application of heuristics in problem solving.

Appendix B is a collection of problems, including some that are appropriate for solving with a computer. Most of these problems are difficult, but simpler ones are readily available. Many can be adapted for use in secondary school teaching, but we have selected them to represent a wide variety of problems. We hope you will try them.

Activities

1. Select three problems from this chapter or Appendix B, decide at what grade level you could use each, and list the items of knowledge, skills, and understandings you would expect students to employ in solving each.
2. Locate or develop two problems which you could classify as each of the following: applications, abstract problems, open search, projects, and a theorem to be proved.
3. For one of the projects you select in Exercise 2, describe two ways in which different students might develop the project.
4. For one of the open search problems you selected in Exercise 2, describe two ways in which different students might approach the problem.
5. For one of the theorems you selected, construct two proofs.
6. Pick one heuristic and develop a lesson outline to teach students to use it.
7. Design a challenging bulletin board display which contains problems. You might want to provide some solutions which are covered by a "door" that students can open to check a solution. Alternatively, you could provide a space by a problem where students could write or tack up their solutions. Remember, this is a visual display, so plan to use eye appeal.
8. Select a problem you have not solved, solve it, and describe the heuristics you used. Be sure to include at least one looking back heuristic.
9. Choose a heuristic and then present a problem which is solvable with that heuristic. Show how at least three other heuristics could be used in solving that problem.
10. Suppose that there is a student in your general mathematics class who is not a very good problem solver. Suppose also that you have two weeks to work with and no other teaching assignments. What would you try to do?

References

Adkins, Julia. Unit Fractions. In *Enrichment Mathematics for the Grades.* Twenty-seventh Yearbook of the National Council of Teachers of Mathematics. Washington: The Council, 1963.

Albrecht, Robert L.; Lindbert, Eric; and Mara, Walter. *Computer Methods in Mathematics.* Menlo Park, California: Addison-Wesley, 1969.

Bell, Max S., ed. *Some Uses of Mathematics: A Source Book for Teachers and Students of School Mathematics.* Vol. 16, Studies in Mathematics. Stanford, California: School Mathematics Study Group, 1967.

Bell, Max S. Mathematical Models and Applications as an Integral Part of a High School Algebra Class. *Mathematics Teacher* 54: 293–300; April 1971.

Bell, Max S. *Models in Our Everyday World*. Vol. 20, Studies in Mathematics. Stanford, California: School Mathematics Study Group, 1972.

Botts, Truman. Problem Solving in Mathematics, I. *Mathematics Teacher* 58: 496–501; October 1965.

Botts, Truman. Problem Solving in Mathematics, II. *Mathematics Teacher* 58: 596–600; November 1965.

Clarkson, Sandra P. A Study of the Relationships Between Translation and Problem Solving Abilities. (Unpublished doctoral dissertation, University of Georgia, 1977.)

Copi, Irving M. *Symbolic Logic*. New York: Macmillan, 1954.

Dodson, Joseph W. *Characteristics of Successful Insightful Problem Solvers*. NLSMA Report No. 31 (edited by James W. Wilson and Edward G. Begle). Stanford, California: School Mathematics Study Group, 1972.

Greitzer, Samuel L. Computing a Lunar Eclipse: An Exercise in Classical Mathematics. In *Enrichment Mathematics for High School*. Twenty-eighth Yearbook of the National Council of Teachers of Mathematics. Washington: The Council, 1967.

Hlavaty, Julius H. What Every Young Mathlete Should Know. In *Enrichment Mathematics for High School*. Twenty-eighth Yearbook of the National Council of Teachers of Mathematics. Washington: The Council, 1967.

Kennedy, George; Eliot, John; and Krulee, Gilbert. Error Patterns in Problem Solving Formulations. *Psychology in the Schools* 7: 93–99; January 1970.

Polya, George. *How to Solve It*. Princeton: Princeton University Press, 1973.

Polya, George and Kilpatrick, Jeremy. *The Stanford Mathematics Problem Book with Hints and Solutions*. New York: Teachers College Press, 1974.

Ranieri, Francis J. *The Key to Word Problems*. Cambridge, Massachusetts: Educators Publishing Service, 1961.

Runion, Garth E. *The Golden Section and Related Curiosa*. Glenview, Illinois: Scott, Foresman, 1972.

Schiffer, Max M. *Applied Mathematics in the High School*. Vol. 10, Studies in Mathematics. Stanford, California: School Mathematics Study Group, 1963.

Sobel, Max A. and Maletsky, Evan M. *Teaching Mathematics: A Sourcebook of Aids, Activities, and Strategies*. Englewood Cliffs, New Jersey: Prentice-Hall, 1975.

Van Engen, Henry. Strategies of Proof in Secondary Mathematics. *Mathematics Teacher* 63: 637–645; December 1970.

The Man-made World. Engineering Concepts Curriculum Project. New York: McGraw-Hill, 1971.

The *Mathematics Student*. Washington: National Council of Teachers of Mathematics 16: 6; May 1969.

Webster's New Collegiate Dictionary. Springfield, Massachusetts: Merriam, 1975.

Evaluating: How Did
They Do/How Did I Do?

AFTER STUDYING CHAPTER 6, YOU WILL HAVE:

* learned a great many ideas about constructing tests—what items to use, how to use them, and pitfalls to avoid.
* seen how the planning and teaching part of our model leads naturally into the evaluation process.
* learned how to perform simple item analysis on test results.
* considered evaluation of both cognitive and affective goals.

CRITICAL INCIDENTS

1. Richard is a slow learner! "Ironically, schools may be the only treatment centers that blame the patient rather than the treatment when things go wrong" (Schulz, 1972, p. 2).

An earlier version of much of the material in this chapter appears in Suydam (1974) and is reprinted here by permission of the ERIC Information Analysis Center for Science, Mathematics and Environmental Education, Robert W. Howe, Director.

2. If you drop your pencil during one of Mr. Brown's tests, you'll never finish the test.
3. Mrs. Smith's tests are easy—you can guess what she's going to ask.
4. All students' grades were over 85 on Miss Day's quiz. Was it too easy?

Evaluation is a continuing, integral aspect of mathematics teaching concerned with the improvement of instruction. Evaluation ascertains whether the teacher is teaching what the teacher thinks he or she is teaching and the learner is learning what the teacher thinks the learner is learning. Evaluation is qualitative as well as quantitative: it involves appraisal as well as measurement, for it includes the stage of making value judgments.

Evaluation takes a variety of forms, since there is no one technique that is equally appropriate for measuring all aspects of learning. Both cognitive factors and affective factors must be assessed, the knowing and thinking aspects as well as the feeling and doing aspects (see Figure 6-1).

Later in this chapter we will consider nontest evaluation, but because of the significant use of tests and quizzes in evaluating mathematical learnings, the major focus will be placed on test construction.

COMMERCIAL AND TEACHER-MADE TESTS

Since the decade of the 1930s an abundance of commercial standardized tests has been available. Improved printing technology, growing interest in evaluation

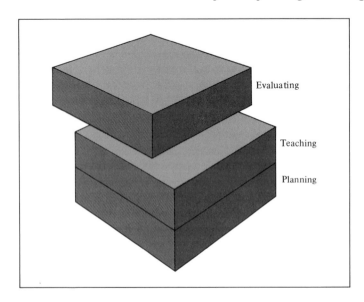

Figure 6–1   The evaluating process in the model of mathematics teaching.

of student performance and in accountability, and advances in statistical theory have all contributed to the publication of a great many tests designed to measure mathematical achievement. Selecting one or more tests from those available is a formidable task.

The late Sheldon Myers observed that mathematics tests may be classified by subject matter, by grade level, by purpose, or by test format (Myers, 1961, pp. 94-95). Here are examples of types of tests in each classification:

CLASSIFICATION	EXAMPLES
Format	Objective, essay, research problem
Purpose	Achievement, survey, diagnostic, prognostic, competitive, or contest
Grade level	Junior high, grade ten, etc.
Subject matter	Arithmetic, general mathematics, algebra I, geometry, algebra II, trigonometry, analytic geometry, or calculus

Usually a teacher, or a group of teachers in a school system, will start with knowledge of the grade level, subject matter, and purpose for which a test is to be selected. They can then determine what tests are available which may meet their need by consulting a reference (such as Buros, 1972) that contains lists and reviews of commercial tests. Copies of those tests which, from their descriptions, appear to meet the purposes should be obtained and reviewed. Particular criteria to be considered in this review are appropriateness of the test, reliability, ease of administration and scoring, and evidence of careful development. Test items should be read to insure appropriateness (validity) for the purpose of the test.

Commercially published tests are almost always standardized. Usually, in the process of constructing the test, the publisher will administer prospective items to a large sample of students. Psychometric properties of the items, such as reliability and validity, are calculated to select those which will appear in the final test. By the use of statistical procedures, the publisher may prepare tables for converting raw scores to grade equivalent scores, to percentile ranks, to stanine scores, or to standard scores. All such tests are called *norm referenced* tests because norms have been established for samples of students in the population for which the test is intended and students' scores are compared to these norms. Most standardized tests are norm referenced.

Criterion referenced tests are designed to determine a student's achievement by ascertaining if he can reach a specified criterion. Such tests are currently growing in popularity because they emphasize what each student has learned, as opposed to how well he achieved compared to other students. Guidelines for constructing criterion referenced tests appear in Gronlund (1973).

It seems safe to state that no students can avoid standardized tests as they progress through school. For that reason it is wise to teach students how to take such tests. Reading aloud the standardized test directions to students as they begin the first test is not enough; the teacher should develop tests that use the same types of items that will be met on standardized tests. (This is particularly necessary for younger students. Many rarely see a multiple-choice item, for instance, until it is met on a standardized test.)

Comparisons with students in other schools by means of standardized testing can help you to attain some perspective on how well your students are doing. The National Assessment of Educational Progress and various state assessment programs attempt to provide such perspective (see Foreman and Mehrens, 1971; Martin and Wilson, 1974; Carpenter et al., 1975). But you are not teaching "other students in other schools." Your goal must be to help each of the students in your classroom to learn and to enjoy mathematics as well as she or he is able (see Research Highlight, "Measuring Progress").

The distinct advantage of a teacher-made test is that the person who has selected the goals for the students is constructing the test to evaluate student achievement in those goals. Often commercially available tests are designed to evaluate some goals of low priority or they fail to evaluate some goals of high priority to a teacher. Teacher-made tests may be either norm referenced or criterion referenced, as the teacher might design a test to achieve a particular distribution of grades or to determine which students have achieved criterion on a set of goals. We believe that criterion referenced, teacher-made tests best serve the needs of teachers and students. Thus we devote the remainder of this chapter to the construction and use of such tests.

DEVELOPING TESTS

In this section some suggestions for developing tests will be considered. These suggestions have been drawn from many sources (e.g., Gronlund, 1968 and 1974). We attempt to be comprehensive, but you must look elsewhere for elaboration and illustrations. Some general procedures will be given first: these apply to the planning and development of all types of instruments.

PLANNING THE TEST

A well-planned test must be designed to accomplish the purpose it is to serve. To construct such a test, you must have the kinds of information that you hope to get from the test clearly in mind.

Review the Goals to Be Assessed Consider what you have taught. What mathematical content and ideas are really important for the students to have learned? Test items should correspond to your instructional goals, but remember that some goals are best measured by other than paper-and-pencil procedures.

RESEARCH HIGHLIGHT | MEASURING PROGRESS

The National Assessment of Educational Progress (NAEP) conducts surveys of the educational attainments, in ten subject areas including mathematics, of learners aged 9, 13, and 17, and adults aged 26–35. Different areas are assessed every year, and all areas are periodically reassessed to measure educational change. The tests are administered to "probability samples" chosen in such a way that the results of their assessment can be generalized to an entire national population.

Many of the items are administered to more than one age group so that response patterns at different ages can be compared. For instance, here is one item and the scores achieved on it in the first assessment:

Weathermen estimate that the amount of water in nine inches of snow is the same as in one inch of rainfall. A certain Arctic island has an annual snowfall of 1,602 inches. Its annual snowfall is the same as an annual rainfall of how many inches?

	Age 13	Age 17	Adult
178 or 178 inches*	31%	53%	58%
Attempted a solution by division— wrong or no answer	14%	19%	15%

* Correct answer.

(NAEP, 1975, p. 26)

This illustrates two points that are also true of some other items: (1) between age 13 and age 17, the percentage of correct answers increases quite dramatically; and (2) between 17-year-olds and adults aged 26–35, differences in scores are small. It is also evident that on some (but not all) items, the percentage of correct answers is disappointingly low.

Decide on the Types of Items to Be Constructed   The type of item depends on the nature of the goal to be measured. Once you have determined that an objective can be measured adequately by a paper-and-pencil item, you need to decide what type of item to use. Some mathematical objectives are measured well by short-answer or completion items, or by multiple-choice items; a few objectives are best measured by true-false or matching items. Such objective-type items (so-called because they can be scored objectively, with independent scorers obtaining the same results) measure knowledge and understanding goals efficiently. They can sample a relatively large field of content, for objective-type items can be answered quickly and one test can contain many questions. This broad coverage helps provide a reliable instrument. For problem-solving goals, however, you will want to consider other types of tests, such as problems and proofs.

Decide on the Number of Items to Be Written for Each Goal   There are no simple rules for determining the "right" number of items to use for measuring each goal. The content of a test should reflect the relative amount of emphasis each objective has received (or will receive) in proportion to the amount of emphasis. And of course you must consider the amount of time available for administration of the test. Tests should measure an adequate sample of the learning outcomes and content included in the instruction. You can never ask all of the questions you might like to; you can only test a sampling of the most important outcomes (see Research Highlight, "What's in a Test?").

WRITING THE TEST ITEMS

The role of each item is to ascertain whether a student has attained the goal or not. There should be nothing about the structure or presentation of an item that leads those who know the correct answer to get the item wrong or those who do not know the answer to get the item right.

1. Select the measurement technique that is most effective for the specific goal.
2. Use clear, simple statements. Use language that students understand. Choose concise vocabulary and sentence construction that is appropriate to the level of your students. Break a complex sentence into two or more separate sentences.
3. Design each item so that it provides evidence that a goal has been achieved. Avoid testing for unimportant details, unrelated bits of information, or irrelevant material.
4. Work with another teacher or group of teachers in reviewing each other's items. Cut out points of doubtful importance and correct unclear wording.
5. Adapt the level of difficulty of a test item to the group and to the purpose for which it is to be used.
6. Initially you may want to write more items than you will need on the final form of the test. You can then discard the weaker items. Many teachers write down items each day for possible inclusion on a test, to help ensure that important points will not be omitted.
7. Number all items consecutively from the first item on the test to the last.
8. Avoid putting part of an item at the bottom of one page and the rest at the top of the next page.

RESEARCH HIGHLIGHT | WHAT'S IN A TEST?

How the tests for the National Longitudinal Study of Mathematical Abilities (NLSMA) were developed is the focus of an article by Romberg and Wilson (1968). The objectives of NLSMA dealt primarily with the relationship of certain variables (such as attitude, textbook, and teacher background) to achievement measures in mathematics. It was assumed that mathematical ability consists of many components rather than being a single unitary trait. The strategy adopted was to develop a short scale or test of 5–15 items to measure an identified component. The authors note that "short scales usually do not provide measures with high reliability, but they are sufficient for differentiating between groups. In this way, many components can be measured within the limited amount of testing time available" (p. 490).

Units of subject matter to be tested were selected and behaviors associated with these units were categorized into four low-cognitive levels (knowing, manipulating, translating, and applying) and three high-cognitive levels (analyzing, synthesizing, and evaluation). A test-writing team then developed items which were critiqued, edited, administered to students, reviewed, screened, edited, pilot-tested again, and finally reviewed for selection. In this way the "best" of 150 items was selected for a 5- to 15-item final form of a scale. Specific items are presented in the article; the entire set of items, with data, is available in the NLSMA Reports (Wilson et al., 1968).

9. If the form of a test or a group of items is unfamiliar, use sample items to clarify the directions. Spend some time teaching students how to take a test.
10. Precede each group of items with a simple, clear statement telling how and where the students are to indicate their answers.
11. When you want students to show their computations, provide adequate space near each item. "Boxing in" this space helps you to locate it quickly.
12. Begin a test with easy items. Difficult items at the beginning of a test are likely to discourage average and below-average achievers. You can then arrange items so that the test gets increasingly more difficult, or you can mix easy and difficult items.
13. Many times you'll need to have more than one type of item on a test (short-answer, multiple-choice, etc.). Place all items of one kind together. Always

have more than one or two items of a particular type (except possibly of the essay type).

14. Avoid a regular sequence in the pattern of responses: students are likely to answer correctly without considering the content of the item at all.
15. Eliminate irrelevant clues and unnecessary or nonfunctional clues, but provide a reasonable basis for responding.
16. Make directions to the student clear, concise, and complete. Instructions should be so clear that each student knows what she or he is expected to do, even though she or he may be unable to do it.
17. Prepare a key containing all the answers that are to be given credit. Make it so that it can be placed beside the answer spaces used by the students.
18. After the test, go over the questions with your students. They can point out ambiguities and other errors, helping you to improve items for future use.
19. Analyze student responses to each item, for diagnostic use.
20. Be sure all drawings are neat and clearly labeled and that the entire test is legible.

Short-Answer Questions and Completion Items   The short-answer item employs a question, an incomplete statement, or a computational example to elicit from the student appropriate words, symbols, or numbers. It is generally limited to questions that call for facts: who, what, when, where, how many. Many classroom mathematics tests are solely of this type. It is frequently used to measure the ability to compute. You can present a number of computational exercises, or you can focus the student's attention on particular aspects of a computation.

In the completion item, certain important words or phrases are replaced by blanks to be filled in by the students. It must be very carefully prepared, or it is likely to measure rote memory or intelligence rather than achievement.

1. State the item so that only a single brief answer is required and possible.
2. Use a direct question; switch to an incomplete statement only when it provides greater conciseness.
3. Words to be supplied should relate to the *main* point of the statement.
4. Blanks should be placed at the *end* of the completion statement.
5. Avoid giving extraneous clues to the answer.
6. If the answer can appear in more than one form, give specific directions about which form to use. Indicate such things as the degree of precision for numerical answers and whether labels must be used.
7. Avoid the use of sentences taken directly from the textbook. Sentences are frequently ambiguous out of context and encourage rote memorization.
8. Do not give clues to the answer by varying the number or length of the blanks.

Multiple-choice Items   The multiple-choice item consists of a *stem*, which is a question or an incomplete sentence presenting a problem situation, followed by several *alternatives*, which are possible solutions to the problem. One of the alternatives is the correct answer; the others are plausible answers, called *distracters* because their function is to distract students who are uncertain of the correct answer. The stem may also be a problem, graph, or diagram followed by the alternatives relating to it.

The ease of scoring undoubtedly plays a big part in the popularity of multiple-choice items. Student answers are easy to read and unambiguous. The use of computer scoring has made the multiple-choice item virtually the only type used when a computer is available or for standardized tests. In general, scores on multiple-choice tests are comparable to those that would be obtained from free-response tests, for the same level of content.

But there are other reasons for deciding to use multiple-choice items: they tend to provide a more adequate measure of many objectives than do other objective-type items; they have high reliability compared with other types of tests; and with careful analysis and development, the multiple-choice item can be adapted to most types of content and to most levels of objectives. It can assess the student's ability to recognize facts or relationships, to discriminate, to interpret, to analyze, to make inferences, to solve problems. Its biggest weakness is that it allows the student to guess, but this affects scores less than on other types of items.

Multiple-choice items should not be used when a simple question is adequate, that is, where there is clearly only one correct answer and there are no plausible distracters. They should not be used when there are only two plausible responses; a true-false item is usually effective in that instance.

Here are other suggestions for constructing multiple-choice items:

1. Make directions explicit so that the student knows exactly what type of response is required. Is more than one answer possible? Is the student to select "the correct answer" or "the best answer"? How should the student record his or her answer? Should the student guess if she or he isn't sure of the correct answer?
2. The stem should present a single worthwhile problem to be solved, expressed clearly and without ambiguity. State the question so that there can be only one interpretation. Check on the clarity of the stem by covering the alternatives and determining whether the question could be answered without the choices.
3. Make each question independent of other questions. Students are often able to select the correct answer to one item because of information gleaned from another item. Where an answer to one item is used in succeeding items, students who miss that item will miss the succeeding items.
4. Make alternative choices as brief as possible. Instead of repeating words in each alternative, include them in the stem.

5. State the stem in positive form whenever possible. When negative wording is used, emphasize it by underlining or by capitalizing.

6. The best alternative choices to the correct answer are those using commonly mistaken ideas, common misconceptions, or errors commonly made by the students. Excellent distracters can be obtained from incorrect responses on short-answer, completion, or essay tests.

7. In general, use the same number of alternatives for each item on a test; but remember that an item is not improved by adding an obviously wrong answer merely to obtain another alternative. Four or five alternatives are used, as a rule, to reduce the chance of guessing the correct answer. It is better to have only four alternatives when five plausible choices are not available.

8. Make all incorrect responses equally plausible or "attractive" to the student who does not know the correct answer. If plausible distractions are difficult to find, use another type of item rather than have ineffective alternatives. The more homogeneous the alternatives, the more difficult the item will be. The correct answer is one which cannot be refuted.

9. Make all alternatives grammatically consistent with the stem and parallel in form. Avoid verbal clues which might enable students to select the correct answer or to eliminate an incorrect alternative: similarity of wording in the stem and the correct answer, for instance, or including two responses that are all-inclusive or two that have the same meaning. Check the structure by reading each alternative with the stem.

10. Do not consistently make the correct response longer or shorter than the distracters. There is a tendency to include unconsciously the greatest amount of detail in the correct answer.

11. Avoid the use of qualifying words such as "always," "never," or "all" as much as possible; they are clues to a test-wise student that an alternative probably is not true.

12. Avoid use of the alternative "all of the above" and use "none of the above" with care. The inclusion of "all of the above" makes it possible to answer the item on the basis of partial information: the student can realize that it is the correct choice by noting that two of the alternatives are correct, or that it is not the correct choice by noting that at least one of the alternatives is incorrect. The chance of guessing the correct answer is thereby increased. The use of "none of the above" may measure only the ability to detect incorrect answers: the student may do this and still not know the correct answer. To reduce the chances of students estimating the answer without doing an entire computation (when that is the objective), use a completion-type item.

13. Avoid using a pattern for the position of the correct response. Students are quick to perceive patterns or apparent patterns and to select their answers accordingly. Use some system of random order for the positions of

the correct answers on each multiple-choice test—and check to make sure that patterns do not inadvertently occur. Many teachers fail to use a, d, and e as often as they use b and c as the correct answer. Students learn that their chances of guessing the correct answer are better if they guess b or c. Be sure the correct response is placed in all positions approximately the same number (but not exactly the same number) of times.

14. Control the difficulty of the item either by varying the problem in the stem or by changing the alternatives.
15. Use an efficient item format.
   a. List alternatives on separate lines, one under the other, making them easy to read and compare.
   b. Use letters in front of alternatives to avoid confusion with numerical answers.

True-False Items   The true-false item can be difficult to construct, for statements must be unquestionably true or false. To construct such items to measure important outcomes is difficult; they adapt best to the measurement of knowledge or understandings and of common misconceptions. They can be used only when there are just two possible alternatives. Because they are highly subject to guessing, true-false items have little value as diagnostic tools.

"Alternative-response" items are variations in which the student must respond "agree" or "disagree"; "right," "partly right," or "wrong"; or with similar words. Other variations include items in which attention is directed to an underlined word or phrase; after deciding that any statement is false, the student must insert the true words in place of the underlined words. Students can also be asked to state why the statement is true or false. Cluster true-false items deal with a single idea; such mathematical content as graphing can be tested with such an item, where students are asked to look at a graph and then respond to a series of true-false items about the data portrayed.

1. Have students circle T and F, or write T and F or + and 0 (rather than t and f or + and −, which cannot be distinguished as readily).
2. State the item clearly and specifically so that it is unequivocally true or false. Avoid the use, however, of specific qualifiers such as "always" or "never"—or use them in both true and false statements. Check for ambiguities.
3. The individual item should deal with a single definite idea. The use of several ideas in each statement tends to be confusing, and the item is then more likely to measure reading ability than achievement. There should be no more than one problem-solving clause.
4. Avoid making true statements longer than false statements.
5. Make the crucial element readily apparent to the student. It is better to have the crucial element come at the end rather than in the early part of a two-part statement.

6. Have an approximately equal (but not exactly equal) number of true and false statements (vary the proportions from test to test).

7. Randomly arrange true and false items; check to be sure there is no inadvertent pattern.

8. Avoid trick statements which appear to be true but are really false because of some inconspicuous or trivial word or phrase.

9. Avoid statements that are partly true and partly false.

10. Avoid the use of statements extracted from textbooks. Out of context, such statements are often unclear or ambiguous.

Matching Items   The matching item measures ability to discriminate among several items of similar material as they are related in a given way with items of another set. The matching exercise is essentially a modification of the multiple-choice form. When all of the responses in a series of multiple-choice items are the same, the matching format is more appropriate. Said another way, unless all of the responses in a matching item serve as plausible alternatives for each premise, the matching format is inappropriate.

Matching items can be used for such content as definitions and words to be defined, measurements and formulas, or geometric shapes and names. They are most appropriate for testing at the knowledge level; it is difficult to adapt them to testing for comprehension and higher-level goals.

1. Place the premise column on the left, the (briefer) responses on the right. Each of the items in the left column should have a test item number; the responses should be preceded by letters. Have the student place the answer to each item in a space to the left of the item number.

2. The items in the two columns must be homogeneous (that is, no responses should be logically excludable as answers by a student who is uninformed). If they are not homogeneous, the student may be provided with clues which will help to match the terms, which results in easier test items.

3. To reduce the effect of guessing, one column should contain more terms than the other. Directions should clearly indicate whether responses may be used once, more than once, or not at all.

4. Do not include too many items in either column; a maximum of twelve items in the premise column should be considered. Longer lists require too much searching time. There should be more responses than items in the premise column when responses are to be used only once, to avoid the selection of the last response on the basis of elimination.

5. Place the items in the response column in some logical order, to enable the student to scan the list quickly to find the term she or he has in mind. Jumbling the terms merely increases searching time.

6. Be sure that there is only one response which is the correct match for each premise when responses are to be used only once.

Essay Items   Essay items are not often used on mathematics tests, but they can and should be. Such items require the student to do more than compute a solution or recall specific facts. The student must think about mathematics and its meaning. Students must organize their own ideas and express themselves effectively in their own words, using both knowledge and reasoning. Purely factual information is not assessed as efficiently as with objective-type items, but higher levels of reasoning can be tapped. Essay questions that assess complex achievement are apt to include such key words as why, explain, compare, relate, interpret, criticize, develop, derive, classify, illustrate, and apply.

There are difficulties in using essay items, as you are aware. The questions on an essay test take so long to answer that relatively few can be answered in a given period of time. A representative sampling of content is not feasible. Essay items are subjective, more difficult to score, and less reliable than objective-type items. Scores are apt to be distorted by writing ability and by bluffing. The student who is fluent can often avoid discussing points of which he or she is unsure. But there are things you can do to minimize these problems, beginning with the writing of clearly defined items that are general enough to offer some leeway but specific enough to set limits.

1. Use essay questions to evaluate achievement only on those goals which are not readily tested by other types of items.
2. Phrase the questions as precisely as possible and be specific in wording, so that the objective of the item is clear and the student is made aware of the specific scope of limits to be included in the answer.
3. Make clear to the student the basis on which the answer will be judged (content, organization, comprehensiveness, relevance, appropriateness, etc.).
4. Require all students to answer all questions so that they will all be taking the same test.
5. Indicate suggested time allotments for each question. Be sure that the student has time to write adequate answers; time must be allowed for thinking as well as for writing. Provide adequate space for answers.
6. Discuss ways of answering essay questions with the students.

Since scoring essay items can be difficult, here are some suggestions which will increase objectivity.

1. List specific goals for each essay question as it is written. Evaluate in terms of the goals. Separate scores may be given for style or writing, or spelling, but they should not contaminate the evaluation of the mathematical objective being assessed.
2. Write out the essentials of a complete answer to each question or prepare a model answer ahead of time. Use it in the same way in scoring each paper. This does not preclude adding other acceptable points made by students.

Determine the number of points to be assigned to each part of the model answer, or determine criteria for levels of expected quality.

3. Keep the identity of students unknown. Have students use coded numerals on their papers or have them write their names on the back or at the end of the test.
4. Read one question through the entire set of papers, scoring each item for all papers before going on to the next item.
5. More uniform standards can be applied by reading the answers twice. At the first reading, sort the papers into several piles. Then reread to check on the uniformity of answers in each pile and make any necessary changes in rating. Assign the same item score to all papers in a pile.
6. Reshuffle the papers after scoring each item so that a paper may not be scored unduly high or low because of its position.

TESTING KNOWLEDGE AND BASIC SKILLS

Both instruction and evaluation are dependent on the choice of selected goals. In Chapters 2 through 5 we saw how the classification of goals is related to the choice of teaching strategies. In the present sections we explore the relationship between the classification of goals and the choice of evaluation items. Almost all our sample items are multiple choice. We selected these items because they clearly exhibit obvious goals and the correct answers may be easily compared to distracters.

Items of knowledge and basic skills are to be learned as automatic responses. As indicated in Chapter 3, these goals are the easiest to evaluate. If students are to learn an item of knowledge, you simply ask them to state it. If students are to learn to perform a skill, you simply provide an exercise which can be solved easily with the skill. The following items are designed to evaluate implicit goals which would be classified as knowledge or basic skills.*

1. Which of the following is not a whole number?
   a. 0   b. 3
   c. $\frac{1}{2}$ d. 4
2. The multiplicative inverse of 5 is
   a. -5 b. $-\frac{1}{5}$
   c. $\frac{1}{5}$ d. 5
   e. don't know

---

* Sample items in this chapter have been selected from: James W. Wilson, Evaluation of Learning in Secondary School Mathematics. In *Handbook on Formative and Summative Evaluation of Student Learning* (Benjamin S. Bloom, J. Thomas Hastings, and George Madaus) © 1971 by McGraw-Hill Book Company. Used by permission of the publisher. Item numbers are those used by Wilson.

3. State the addition algorithm for rational numbers a/b and c/d

$$\frac{a}{b} + \frac{c}{d} = \underline{\hspace{2cm}}$$

4. Which of the following operations is not defined for the real numbers?
   a. $3 + 0$ b. 3×0
   c. $\frac{0}{3}$ d. $\frac{3}{0}$
   e. $0 - 3$
7. Which is the longest?
   a. 34 inches   b. $3\frac{1}{2}$ feet
   c. 1 yard   d. 1 meter
14. Write an example of
   a. a trinomial   b. a binomial
   c. a monomial   d. a polynomial that is none of
                                     the above
16. Draw an example of an isoceles triangle.
20. An axiom is
   a. a proof   b. a corollary
   c. an undefined term   d. a proposition to be proved
   e. an assumption
21. Which of the following is a prime number?
   a. 6   b. 11
   c. 15   d. 39
   e. 51
29. What is the least common demoninator of $\frac{1}{105}$ and of $\frac{1}{70}$?
34. Multiply the following:
   a. $(x + 2)(x + 3)$ b. $(2x + 3)(x + 1)$
   c. $(x - 6)(x + 4)$ d. $(x - 3)(x - 2)$
70. If the intersection of two different planes is not empty, then the intersection is
   a. a point   b. two different points
   c. a line   d. two different lines
   e. a plane
77. If $a \cdot b = 0$, then
   a. a must be zero   b. b must be zero
   c. either a or b must be zero   d. both a and b must be zero
   e. all the choices above are correct

Each of the above examples calls for a particular item of knowledge or a basic skill. The item formats were completion or multiple choice. Other types of format which would be appropriate are true-false questions and matching exercises.

TESTING UNDERSTANDINGS

To demonstrate that she or he has learned an understanding, a student is asked to apply or use the understanding. Although the student may be able to memorize (learn as an item of knowledge) the statement of an understanding, carefully constructed test items will distinguish between knowledge learning and understanding learning. It is more difficult to construct items which will distinguish between skill learning and understanding learning, because in both cases the student is asked to do something, to apply a generalization. One of the underlying distinctions between skills and understandings is that skills must be practiced to remain sharp. We think of understandings as being retained, even in the absence of practice. But it is difficult to tell whether a particular test item is designed to test a skill or an understanding.

The following examples are items designed to test understanding (Wilson, 1971).

88. During the summer a student worked n weeks at k dollars. His expenses for the summer were p dollars. His savings, in dollars, were
   a. $n + k + p$ b. $np - k$
   c. $np \div k$ d. $n \cdot k - p$
   e. none of these
90. A circle whose radius is r is inscribed in a square. Express the area of the square in terms of r.
99. Which of the following numbers, expressed in base 7 numeration, is both prime and odd?
   a. 11   b. 12
   c. 13   d. 14
102. The tens digit of a two-digit number is twice the units digit. If the units digit is represented by x, the number can be represented by
   a. $3x$ b. $12x$
   c. $21x$ d. $30x$
107. A piece of wire 36 inches long is bent into the form of a right triangle. If one of the legs is 12 inches long, find the length of the other leg.
124. If x and y are two distinct real numbers and $xz = yz$, then $z =$
   a. $1/(x - y)$ b. $x - y$
   c. 0   d. 1
   e. x/y
125. The last digit in 4^{10} is
   a. 0   b. 2
   c. 4   d. 6
   e. 8

Example 88 could test skill in translating verbal statements to symbolic expressions, but we take it to be an item to test the understanding that the given statement may be $n \cdot k - p$ (and not the other given expressions). Example 90 is designed to test the understanding that the diameter of a circle inscribed in a square is equal in measure to the side of the square, that the diameter is twice the radius, and that the area of the square is the square of the measure of a side. It seems likely that the last two understandings are immediate and that only the first is tested. Example 99 tests understanding of base 7 numeration, prime number, and odd number. What understandings, items of knowledge, and skills are tested in Examples 102, 107, 124, and 125?

Test items for understandings may be presented in any of several formats: completion, multiple-choice, true-false, matching, exercises, proofs, or discussion.

TESTING PROBLEM SOLVING

Because problem solving is a high-order ability, it is the most difficult to test. Selection of items is strongly dependent on student background. As indicated in Chapter 5, that which is a problem for a student today may be merely an exercise for the student tomorrow. Also, it is difficult to select items to test problem-solving goals because the teacher must attempt to ensure that the student is prepared to solve the problem. Thus the student should have a sufficient background in order to solve the problem, but not so extensive as to reduce the problem to an exercise. The following items could be classified as problems for particular students (Wilson, 1971).

128. What is the largest rational number which is the sum of two rational numbers each having a numerator of 1 and a whole number for its denominator?
133. If $2a + 2b + 5c = 9$, and if $c = 1$, then $a + b + c =$
    a. 2   b. 3
    c. $4\frac{1}{2}$ d. 5
    e. 8
147. Show that $-(-a) = a$.
159. Prove by the methods of analytic geometry that the three altitudes of a triangle meet in a point.

Notice that each problem requires the student to synthesize some previous learning and that in general there are several approaches (heuristics) available for solving each problem. Of course, if a student has memorized the proof required in Example 147 or 159, the item would not test that student's problem-solving ability.

In general, problem-solving goals are evaluated with problems, proofs, and discussions, and students are asked to show their work. Less often, such items may use a completion, multiple-choice, or true-false format.

EVALUATION OF AFFECTIVE DOMAIN GOALS

> The conscientious, skillful, perceptive mathematics teacher is constantly making evaluations of a student's verbal answers, of his blackboard work, of his facial expressions and other non-verbal behavior, but he still needs to give tests in order to have uniform and easily comparable information about the behavior of all his students in the same situation. (Fouch, 1961, p. 172)

Mathematics teachers are evaluating students in every class. Students may not be evaluated formally, but their teachers are watching and listening to them. These observations provide often useful information about students' beliefs and attitudes. Formal evaluation of goals in the affective domain is extremely difficult and often unreliable—however, try! Discussion questions may be used as a way to gain further insight into students' attitudes and beliefs. Such questions should probably *not* be scored, but students could receive extra credit for responses. Here are some examples of questions which might be tried:

1. Describe what you thought when you worked on Problem 133.
2. Can you imagine how you could use the mathematics you learned in this unit during the next week (or year)?
3. Make up a story to describe how you think the first person to prove the theorem of Problem 159 discovered the proof.
4. Pick the question in this quiz you dislike the most and tell why you chose it.
5. Write a dialogue between two famous mathematicians.

Perhaps the most widely used measure of affective goals is the attitude scale. Half a dozen scales have been extensively used; on many of them, items such as those on the scale below appear. The scale attempts to ascertain, less directly than by simply asking "Do you like mathematics?" and therefore hopefully with greater reliability or credibility, how strongly the student likes or dislikes mathematics.

With some practice and hard work, you can construct your own scale to measure specific aspects of mathematics; the procedure is concisely outlined by Corcoran and Gibb (1961).

FORMAT

In a very general way, the formats for items lend themselves to particular kinds of goals. However, this association is very weak and the ingenious reader can probably construct counterexamples of this generalization. By way of summary, Table 6–1 presents the typical, less typical, and seldom used formats for each goal classification.

ATTITUDES TOWARD MATHEMATICS
(Scale Form B)

Marilyn N. Suydam and Cecil R. Trueblood
The Pennsylvania State University

This is to find out how you feel about mathematics. You are to read each statement carefully and decide how *you* feel about it. Then indicate your feelings on the answer sheet by marking:

A — if you strongly agree　　　　D — if you disagree
B — if you agree　　　　　　　　E — if you strongly disagree
C — if your feeling is neutral

1. Mathematics often makes me feel angry.
2. I usually feel happy when doing mathematics problems.
3. I think my mind works well when doing mathematics problems.
4. When I can't figure out a problem, I feel as though I am lost in a mass of words and numbers and can't find my way out.
5. I avoid mathematics because I am not very good with numbers.
6. Mathematics is an interesting subject.
7. My mind goes blank and I am unable to think clearly when working mathematics problems.
8. I feel sure of myself when doing mathematics.
9. I sometimes feel like running away from my mathematics problems.
10. When I hear the word mathematics, I have a feeling of dislike.
11. I am afraid of mathematics.
12. Mathematics is fun.
13. I like anything with numbers in it.
14. Mathematics problems often scare me.
15. I usually feel calm when doing mathematics problems.
16. I feel good toward mathematics.
17. Mathematics tests always seem difficult.
18. I think about mathematics problems outside of class and like to work them out.
19. Trying to work mathematics problems makes me nervous.
20. I have always liked mathematics.
21. I would rather do anything else than do mathematics.
22. Mathematics is easy for me.
23. I dread mathematics.
24. I feel especially capable when doing mathematics problems.
25. Mathematics class makes me look for ways of using mathematics to solve problems.
26. Time drags in a mathematics lesson.

Figure 6–2

Table 6–1. Classification of Test Items.

Goals	Item Formats						
	Short answer	Multiple choice	True-false	Matching	Problems	Proofs	Discussion
Knowledge and basic skills	X	X	X	X	☐	☐	☐
Understandings	X	X	X	X	—	—	—
Problem solving	☐	☐	☐	☐	X	X	X
Affective	X	X	—	—	☐	☐	X

X   Often Used
—   Occasionally Used
☐   Seldom Used

USES OF EVALUATION

There is no reason to give a test unless the results are to be used. Tests are commonly given to determine grades, but there are many other, probably more important uses for the results. In general, tests may be classified as aptitude, achievement, or diagnostic. All cognitive tests are achievement or survey tests; items are presented which students do correctly, incorrectly, or partially correctly. The point is that for each item the student demonstrates (or fails to demonstrate) some degree of achievement.

Aptitude or prognostic tests are those which are used to predict how well a student will succeed at a future learning task. Although such tests are achievement tests, it may well be the case that success on the test has been found to be associated with success in another task. The degree of association (the correlation or multiple correlation) and the test results may be used to select students for special programs or for remediation before attempting the new learning. For this use, aptitude tests are criterion referenced.

Diagnostic tests are designed to identify specific learning needs or strengths of students. Again, the test is an achievement test, but designed and scored in such a way that the teacher can pinpoint priorities for student remediation and/or identify special abilities.

It is conceivable that one test could be used as an aptitude test, say for the next unit of instruction; as an achievement test, say on the previous unit; and as a diagnostic test, again on the previous unit. Because test construction is fairly difficult, few people attempt to design tests for more than one use—and are delighted if they accomplish that use effectively! Often, however, in the school setting teachers can design a test with two purposes in mind. An achievement test designed to measure student success in a unit can also be interpreted to determine specific goals which students have not learned.

LEARNING

Suppose a teacher has listed a set of goals for a unit of instruction, developed a set of lesson plans (samples of these appear in Chapter 7) and constructed a test based on the goals. In constructing the test, the teacher has decided upon behavioral objectives that will be acceptable as criteria for ascertaining student achievement of goals. Thus behavioral objectives are statements of the evidence the teacher will accept to decide that a goal has been attained.

Each goal may be associated with some number of test items. In some cases perhaps one item is sufficient for the teacher to make a judgment, but in other cases the teacher may decide upon five or even ten items. If several items are associated with a goal, the teacher may decide that the goal is attained if a student can correctly respond to some percentage of the items—perhaps the criterion level is 80 percent, or 3 out of 4 items, or 7 out of 10 items. The teacher establishes the criteria to determine students' success realistically, to permit students some degree of flexibility, or to provide an opportunity to include a few items which are somewhat experimental.

"Remember, son, if at first you don't succeed, re-evaluate the situation, draw up various hypotheses for your failure, choose reasonable corrective measures, and try, try again.

After administering a test you can perform an item analysis on the results to obtain much useful information. Performing an item analysis is the process of studying the students' responses to each item. An item analysis can tell you how difficult an item is and how well each question discriminates between high- and low-ranking students. It's especially important if you are going to use the item again; it can indicate whether or not an item needs to be revised. It's also useful even if you don't plan to use the item again, for it can tell you what questions are especially appropriate to test certain goals. Or it can be used simply as part of your diagnostic procedures.

Computer programs are available to conduct an item analysis for tests developed in research studies, for standardized tests, and for other tests that will be used by many groups of students. For most classroom tests, only simple item analysis procedures seem warranted. Here are two suggestions:

1. Look at the test: What items were missed by many students? Were they missed because of a "fault" in the item or a "fault" in the instruction? What do you do next, revise the item or revise the instruction?
2. A simple measure of difficulty is the percentage of students who got the item correct. This gives you an approximation of how difficult the item is. By recording this information for each item in your item pool, you can build a test which will be at an appropriate difficulty level. This is especially helpful when you are developing a test in which you want to rank students; each item should then be of medium difficulty—approximately 40–60 percent. (For mastery tests, your standards will be different.)

You can check the students' papers yourself to obtain the percentages, or you can do an item analysis by show of hands. Call out the item numbers one by one and have each student who has the item correct hold up her or his hand; then count and record the number of hands.

A slightly more extensive item analysis is pictured in Figure 6–3. It consists of a matrix of test items by students. Each item missed by a student is indicated by an X in the student's row under the item column. The teacher can quickly count and record the number of items correct for each student and even calculate a percentage score. By grouping items under goals and noting a criterion (in this case simply the number of correct items to reach criterion within each goal) the teacher can quickly determine which students did not achieve criterion for each goal. Failure to achieve criterion has been indicated by a diagonal line on the student's row under the goals column. Now the teacher can see that all students achieved criterion for goals I and IV, 15 students for goals VI and VII, 14 for goals II and VIII, 13 for goal III, and 7 for goal V.

TEACHING

Further consideration of the item analysis can help teachers make decisions about their teaching and about the kind of follow-up needed by the students. In the

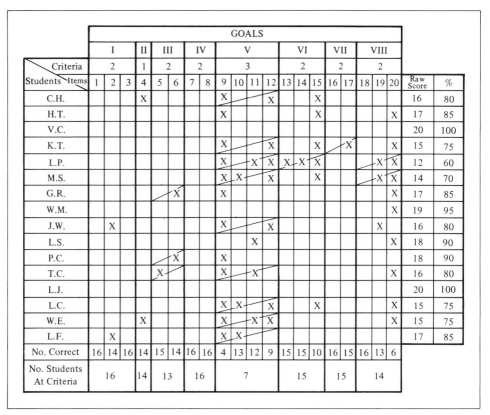

Figure 6-3. Item analysis.

situation analyzed in Figure 6–3, it appears that items 9, 12, and 20 were very difficult. Perhaps the teacher should examine student work for these three items to attempt to identify the difficulty. The teacher may have anticipated that item 20 would be difficult, but the results for items 9 and 12 may be somewhat surprising. Upon reexamination of the items, the teacher might discover that item 9 contained confusing wording. Should his or her criterion be revised?

How can teachers evaluate their teaching? They could determine "grades" for themselves, though such scores might be more or less arbitrary. The important point is that teachers are in the happy position of being able to do more to effect student learning, even after the test has been given. The teacher who reviews the test results with the students might include a reteaching of goal V (not achieved by nine students.) He or she might devise learning packets or hold extra help sessions for the students who did not achieve criteria on goals II, III, V, VI, VII, VIII. Note that only one student would have to work on as many as three goals.

Overall, even a simple item analysis can provide information about student success and teacher success. The data may be used in assigning grades, but more than this, they can be used by a teacher to identify the goals which each student failed to achieve. In this way evaluation may be less of a threat to students and a very useful tool for both teachers and students.

MASTERY LEARNING

Philosophically, we support the notion of criterion-referenced tests and mastery learning. That is, we believe that teachers should determine goals for their students in mathematics, establish criteria for achieving these goals, and set out determined to help students reach criteria. Bloom, Hastings, and Madaus (1971) discuss mastery learning and point out that:

> Given sufficient time and appropriate types of help, 95 percent of the students (the top 5 percent plus the next 90 percent) can learn a subject with a high degree of mastery. To say it another way, we are convinced that the grade of A as an index of mastery of a subject can, under appropriate conditions, be achieved by up to 95 percent of the students in a class. (p. 46)

Further, these authors state:

> As educators we have used the normal curve in grading students for so long that we have come to believe in it. Achievement measures are designed to detect differences among our learners—even if the differences are trivial in terms of the subject matter. We then distribute our grades in a normal fashion. In any group of students we expect to have some small percentage receive A grades. We are surprised when the figure differs greatly from about 10 percent. We are also prepared to fail an equal proportion of students. Quite frequently this failure is determined by the rank order of students in the group rather than by their failure to grasp the essential ideas of the course. (pp. 44-45)

We suggest that the goals of instruction be provided to the students so that they, too, know what they are to learn. Many teachers do this. Some provide alternate forms of each test so that students can identify goals they need to work toward; then the students have another opportunity to demonstrate mastery. Regretfully, very few school systems have adopted this approach, so the teacher using it might have to be prepared for the frustrations typically encountered by innovators.

Critical Incidents Revisited

The critical incidents selected for this chapter indicate some of the common misconceptions about evaluation.

The quotation in the first critical incident emphasizes that too often tests are used to *classify* students—"Richard is a slow learner!" Although you may use test results to assign grades to students, be careful of classifying them. Don't put students in a box with a label on it. Remember, when a student doesn't succeed, it's seldom clear whether the problem was that he didn't learn or that he wasn't taught.

Some mathematics teachers, as indicated in the second critical incident, make tests that are too long to complete in the time allotted. In general, it seems particularly inappropriate for mathematics tests to be speeded (except for knowledge and skill items). Surely students need time for understanding and problem-solving items.

The teachers in the third and fourth critical incidents may be applying some of the ideas presented in this chapter. Mrs. Smith may not necessarily be constructing easy tests; perhaps her students simply have a clear understanding of her goals and thus know what to expect. Miss Day's quiz was not necessarily too easy, either. If it was a criterion-referenced test, the high grades may simply indicate that all students achieved criterion.

This chapter contains a lot of information about constructing and using tests. Emphasis has been placed on teacher-made tests. Suggestions for using several types of items were presented and techniques for performing a simple item analysis were described.

Our aim is to help you see how to construct good tests and how to make effective use of the results.

Activities

1. Select goals you have developed in previous chapters. Establish criteria for student achievement of each goal and construct an appropriate test.
2. Locate a small group of students who would be willing to take a short test. Perform an item analysis on the results and describe your conclusions.
3. Refer to the chart in Table 6–1. Specify a single goal for each row in the table and construct one item for each checked column for that row. See if you can construct items for other classes of items indicated in the table (those marked or those left blank).
4. Locate sources of information on mastery learning and discuss the implications of this idea for teachers and students of mathematics.
5. Carefully examine a standardized test in mathematics, then see if a review of the test is contained in *The Seventh Mental Measurements Yearbook* (Buros, 1972). Evaluate the test.
6. List the characteristics of a good test.
7. Write a set of acceptable instructions for each of the following: a multiple-choice test, an essay test, and a true-false test.
8. Select a topic, list your goals, set your criteria, and write a test for *mastery*.

9. Write a short test in two formats:
   a. multiple-choice
   b. completion
   (That is, use two different formats for the same item). Administer the test to a class, randomly mixing the types of tests among the students. Do an item analysis. Discuss and compare the results.
10. Obtain students' comments on the use of various item formats with mathematics examinations. Evaluate these comments. Try to separate comments about poor items from comments about the qualities inherent in the format.

References

Buros, Oscar K., ed. *The Seventh Mental Measurements Yearbook*. Highland Park, New Jersey: Gryphon Press, 1972.

Bloom, Benjamin S.; Hastings, J. Thomas; and Madaus, George F. *Handbook on Formative and Summative Evaluation of Student Learning*. New York: McGraw-Hill, 1971.

Carpenter, Thomas P.; Coburn, Terrence G.; Reys, Robert E.; and Wilson, James W. Results and Implications of the NAEP Mathematics Assessment: Secondary School. *Mathematics Teacher* 68: 453–470; October 1975.

Corcoran, Mary and Gibb, E. Glenadine. Appraising Attitudes in the Learning of Mathematics. In *Evaluation of Mathematics*. Twenty-sixth Yearbook of the National Council of Teachers of Mathematics. Washington: The Council, 1961, pp. 105–122.

Foreman, Dale I. and Mehrens, William A. National Assessment in Mathematics. *Mathematics Teacher* 64: 193–199; March 1971. (Similar article in *Arithmetic Teacher* 18: 137–143; March 1971.)

Fouch, Robert S. Overview and Practical Interpretations. In *Evaluation in Mathematics*. Twenty-sixth Yearbook of the National Council of Teachers of Mathematics. Washington: The Council, 1961, pp. 167–180.

Gronlund, Norman E. *Constructing Achievement Tests*. Englewood Cliffs, New Jersey: Prentice-Hall, 1968.

Gronlund, Norman E. *Improving Marking and Reporting in Classroom Instruction*. New York: Macmillan, 1974.

Gronlund, Norman E. *Preparing Criterion-Referenced Tests for Classroom Instruction*. New York: Macmillan, 1973.

Martin, Wayne H. and Wilson, James W. The Status of National Assessment in Mathematics. *Arithmetic Teacher* 21: 49–53; January 1974.

Myers, Sheldon. Published Evaluation Materials. In *Evaluation in Mathematics*. Twenty-sixth Yearbook of the National Council of Teachers of Mathematics. Washington: The Council, 1961, pp. 93–104.

Romberg, Thomas A. and Wilson, James W. The Development of Mathematics Achievement Tests for the National Longitudinal Study of Mathematical Abilities. *Mathematics Teacher* 61: 489–495; May 1968.

Schulz, Richard W. Characteristics and Needs of the Slow Learner. In *The Slow Learner in Mathematics*. Thirty-fifth Yearbook of the National Council of Teachers of Mathematics. Washington: The Council, 1972, pp. 1–25.

Suydam, Marilyn N. *Evaluation in the Mathematics Classroom: From What and Why to How and Where.* Columbus, Ohio: ERIC Information Analysis Center for Science, Mathematics and Environmental Education, 1974.

Wilson, James W.; Cahen, Leonard S.; and Begle, Edward G., eds. *NLSMA Reports,* Nos. 1–6. Stanford, California: School Mathematics Study Group, 1968.

Wilson, James W. Evaluation of Learning in Secondary School Mathematics. In Benjamin S. Bloom, J. Thomas Hastings, and George F. Madaus. *Handbook on Formative and Summative Evaluation of Student Learning.* New York: McGraw-Hill, 1971, pp. 643–696.

Math Fundamentals: Selected Results from the First National Assessment of Mathematics. Mathematics Report No. 04-MA-01. Denver: National Assessment of Educational Progress, Education Commission of the States, January 1975.

Good Beginnings

AFTER STUDYING CHAPTER 7, YOU WILL KNOW:

★ specific ways to prevent discipline problems.
★ methods that can be used to handle discipline problems when they do arise.
★ some of the facets involved in planning for teaching a course, a unit, and a lesson.
★ how to plan a daily lesson.
★ some generalizations concerning motivation, homework, record keeping, and the grading of students.

CRITICAL INCIDENTS

1. Each day Mr. Franks asks four or five students to put several of the previous day's homework problems on the chalkboard. Sue, Jan, Marge, Milt, and Bob are usually the ones asked because they always have their homework completed. While they put the problems on the board the other students in class sit and socialize.
2. Although Ed never creates a disturbance, he falls asleep almost every day during geometry class. Ms. Gains is at a loss as to what to do.
3. Mr. Norris is explaining a proof to one of his classes. Ralph, who sits in the

front of the room, crumples up two sheets of paper, gets out of his seat, non-chalantly walks across the front of the room, and throws his paper into the wastebasket.

4. Sheldon is at Mrs. Rice's desk rather vociferously contesting his third-quarter grade in geometry: "But you didn't tell us that our homework grade was going to count the same as a chapter test!"

MAINTAINING DISCIPLINE

Of the many tasks which confront a teacher each day, probably the one which is most talked about and worried about, and which results in more teachers quitting the teaching profession, is maintaining discipline. In our discussion the term *discipline* will mean simply "a system of controlling conduct," or "the ways in which a teacher can maintain order in the classroom." Although teachers vary greatly in the degree of order they maintain in their classes, almost all would agree that some form of discipline is necessary.

Aside from the obvious benefits of insuring the mutual safety of the students (and teacher), providing a classroom atmosphere conducive to learning, and sustaining the sanity of the instructor, discipline serves other important functions. Ausubel (1961) has identified four such functions in the training of youngsters:

> First, it is necessary for socialization—for learning the standards of conduct that are approved and tolerated in any culture. Second, it is necessary for normal personality maturation—for acquiring such adult personality traits as dependability, self-reliance, self-control, persistence, and ability to tolerate frustration. These aspects of maturation do not occur spontaneously, but only in response to sustained social demands and expectations. Third, it is necessary for the internalization of moral standards and obligations or, in other words, for the development of conscience. Standards obviously cannot be internalized unless they also exist in external form; and even after they are effectively internalized, universal cultural experience suggests that external sanctions are still required to insure the stability of the social order. Lastly, discipline is necessary for children's emotional security. Without the guidance provided by unambiguous external controls, the young tend to feel bewildered and apprehensive. Too great a burden is placed on their own limited capacity for self-control. (Ausubel, 1961, p. 132)

Just what are some of the ways available to teachers for disciplining? Generally they fall into two categories, which we will call *preventive* and *corrective*. Although the two sets of behaviors are not totally disjoint, preventive disciplinary procedures refer to those actions which a teacher can take in an effort to prevent disciplinary problems from arising. If a teacher were 100 percent successful in preventing discipline problems, that teacher would, of course, have no need for corrective procedures. Alas, all teachers have discipline problems at one time or another and hence all teachers need a repertoire of corrective actions or at least some guidelines for deciding on appropriate corrective measures.

PREVENTIVE DISCIPLINE

One of the critical times for using preventive disciplinary procedures is on the very first day of class; what transpires during the first few days will set the tone for the entire course. Most teachers want to be liked by their students, but how they go about attaining this goal varies. Let's contrast two approaches.*

Mrs. Likely shuffles papers on her desk as she watches the students come into the room. Then she stands before them with a nervous smile and gives each class a talk which begins, "I am so glad to see all of you! I just know I'm going to love teaching you, and I want so much for you to like me. We're going to have so much fun this year." She continues in this vein for several minutes, stressing how much she wants them to like her. The rest of the time is consumed in checking the attendance list, passing out books, and dealing with other routine matters; almost nothing is said about their purpose in being there: to learn mathematics.

All goes well at first. But after several days Joe decides to find out the answer to what others are questioning, too: how far can we push Mrs. Likely? Instead of going to his seat when she indicates that she's ready to begin, he continues to talk with Pete at the back of the room. Mrs. Likely asks him to go to his seat, but he wanders to the window and stands there. She pretends not to notice that he hasn't done what she told him to do. Everyone in the class is aware of this. The next day, Matt defies her a little more openly, and again she ignores the incident. From such incidents, chaos begins to grow. Mrs. Likely realizes that she has to do something, and she finds that she's doing a lot of yelling every day. Soon no one is happy in her class, and she finds it increasingly hard to get the students' attention for the lesson. They're aware that she doesn't know what to do about the disruptions, and they're aware that the disruptions are keeping them from learning much in her class. She tells them again and again how much she wants them to like her, but they begin to reveal their contempt. Mrs. Likely doesn't know how to regain control; she doesn't even know how she lost it.

Now let's look at how the more experienced and skilled teacher Mrs. Reasoner approaches her class. She, too, wants her students to like her—yes, to have a sincere love for her. But she is also keenly aware of the great responsibility entrusted to her. Here are 30 persons who are hers to nurture, to help develop, and somehow, some way, to assist in learning more about mathematics, about themselves, and about the world. She knows, too, that these 30 individuals will

* These two approaches are adapted from James Dobson, *Dare to Discipline* (1970). Used by permission of Tyndale House Publishers.

vary greatly in their ability to learn, to study, and to be attentive in class. Some will be very cooperative and helpful. Others may not cooperate and may even seem to go out of their way to be uncooperative.

Mrs. Reasoner knows, however, that a large measure of her responsibility will consist of helping students learn to contribute to a productive class situation. She will have to set levels of expectation for each student, and she is aware that many times she will have to work with individual students to help them meet those expectations. At some times she may have to punish students. But at all times she will do her best to be fair in what she expects, firm in continuing to mold those expectations and consistent in her guidelines so that her students know what to expect from day to day.

Mrs. Reasoner may not give her students a long inaugural address; she knows that she herself is soon turned off by people who make long speeches on dull or unpleasant topics. She may state a few simple rules and procedures. But much of what she wants her students to know will come naturally to the surface at the appropriate time. Rather, by her appearance, her tone of voice, and her manner (which often speak much more loudly than words) she communicates a message something like: "I'm glad to be your teacher. I'm looking forward to getting to know each of you. We have a lot to do this year. But we can have some fun, too. So let's get busy!"

Even though Mrs. Likely and Mrs. Reasoner are hypothetical individuals, there are many teachers like them in the classroom today. We are not suggesting that you set up Mrs. Reasoner as a model to be imitated. You will have to decide upon an approach and a style which are most effective for you. However, your chances for success will be greater if you start off the year more like Mrs. Reasoner than Mrs. Likely.

The suggested outline that follows is one that we have found useful in starting the year in high school mathematics classes. You may wish to use some modification of it in your own teaching. Not only will such an outline benefit the students, but it will also help to give you some added security on that frightening first day of class.

1. Welcome students, introduce self. (Write your name clearly on the chalkboard. If you don't want students to call you by your first name, you can reduce the likelihood of that happening by writing simply "Mr. Peterson.")
2. Prepare a seating chart. Remember that your view of the seating arrangement is different from their view. (Learn the students' names as soon as you can.)
3. Distribute needed class textbooks and other materials. Indicate equipment (special pencils, protractors, etc.) which students are expected to bring.
4. Let the students know your expectations. The questions below suggest things you may wish to discuss with your students. You will, of course, have given these items considerable thought prior to discussing them with your classes.

    a. Should students be in their seats and ready for work when the bell rings?

    b. Will students be required to raise their hands before answering questions or making comments in class?

    c. Will gum-chewing or snacking be permitted in class?

    d. Will students be required to remain in their seats until you dismiss class even though the dismissal bell has rung?

5. Discuss grading policies.

    a. How will homework be collected and counted toward final grades?

    b. What about quizzes? (How many? Will they be announced? How much will they count?)

    c. What about exams? (At the end of each chapter? Multiple choice? How long? Can they be retaken?)

    d. Will extra credit problems or projects be assigned? (More will be said about this in the section on motivation.)

    e. Will students have any input into the determination of their grades? (See pp. 222–223 for discussion of a self-evaluation form.)

Have Your Lessons Well Planned   The teacher who has well-defined goals for each class period, who has procedures developed for efficiently expediting routine class chores, who has made some provision for variety in his or her classes, and who has every minute of class time accounted for by meaningful learning experiences, will almost certainly encounter fewer discipline problems than a teacher who does not. We will consider lesson planning in more detail later in this chapter.

Get to Know Your Students, But . . .   One of the most powerful deterrents to discipline problems at a teacher's disposal is a sincere concern and respect for each student as a person. Teachers who are able to convey to their students, by their actions and their words, that they enjoy teaching them, that they are interested in the students' success in the course, and that they want to be helpful, are far more likely to be respected and thus encounter fewer discipline problems than those teachers who view students in an impersonal fashion.

While it is probably true that being friendly is very closely related to one's personality and hence cannot be taught, there are, nevertheless, a few things that teachers can do to help them establish rapport with their students.

1. *Seating Charts*—It means a great deal to most students if the teacher knows the student's name. Seating charts are especially helpful at the beginning of the year for learning names. Seating assignments are also very helpful in judiciously separating prospective troublemakers by isolating them or by assigning them to small groups composed of more conscientious students. (Seating charts also are invaluable to substitute teachers—provided they are kept up-to-date!)

2. *Greeting Students at the Door*—Another technique frequently helpful in getting to know your students is to make it a point to be at the classroom

door as they come into the room. A friendly smile and perhaps some small talk ("How are things going today, Jim?" or "Bill, how did you do in the track meet last night?" or "Gloria, those are cute earrings") can do quite a bit in helping to create good relationships with many of your students.

3. *Using an Information Form*—Information forms (such as the one illustrated in Figure 7–1) or short autobiographies written the first or second day of class can also be effective in helping you to become better acquainted with your students. Knowing a student's interests or hobbies often can be helpful in starting a conversation with him or her. School records can also be a useful source of information on students.

4. *Periodic Conferences*—Individual conferences are quite commonly used by teachers as a corrective disciplinary technique, but less commonly as a preventive measure. As a preventive tool the teacher might set aside the last few minutes of class, during which time he or she talks with one student at a time at his or her desk, or in the hall, or wherever else might be appropriate. The student might be informed of his or her progress in the course, given suggestions on how to improve his or her standing in the class, and given an opportunity to voice complaints or make suggestions to the teacher. Such conferences are intended to let the student know that the teacher is interested in the individual's progress and to inform the teacher of the student's assessment of how the course is going and how his or her learning is progressing. Several such conferences might be scheduled during each grading period.

A word of caution is necessary: There is a fine line between getting to know your students in an effort to establish good rapport and avoid discipline problems and getting to know your students *too* well—and creating discipline problems!

```
Names of brothers/sisters and ages _____
Father's occupation _____
Mother's occupation _____
Home phone number _____
Name of this course _____
Previous high school mathematics courses and grades:

What are your plans after you graduate from high school?

What are your reasons for taking this particular mathematics course?

Mention below any hobbies or special interests that you have:
```

1st	2nd	Sem. Exam	S. Final	3rd	4th	Sem. Exam	S. Final

Name _____   Age ____ Class _____ Date _____

Figure 7–1   Student information form.

It is one thing to show genuine concern; it is quite another to try to become one of them—one of their buddies.

Provide Variety in Your Classes   Most of the secondary mathematics classes with which the authors are familiar follow very closely one particular format. The teacher gives the answers to the previous day's homework, time is allowed for answering questions on the homework, the next section in the textbook is previewed, an assignment is given, and students are given class time in which to begin the homework. Does that sound familiar (see Research Highlight, "Following the Book")? While this routine is not bad per se, if it is practiced day in and day out, as happens with many teachers, it can result in boredom for the students, and that can frequently lead to disciplinary incidents.

RESEARCH HIGHLIGHT | FOLLOWING THE BOOK

The textbook is a valued teaching tool. But how much should a mathematics class revolve around the textbook? This question must be answered on the basis of teaching philosophy; research cannot provide the answer. Yet research indicates that there might be cause for concern about over-reliance on the use of the textbook. Brown (1974) conducted a study of the way in which the textbook is used by teachers and students of geometry and second-year algebra. One class of each of 36 randomly selected teachers and six teachers rated as excellent by their supervisors was observed; the teachers were interviewed and their students given a questionnaire. Results revealed very heavy dependence on the textbook by both teachers and students. The textbook was followed closely for content selection and sequencing; the major objective of lessons tended to be the completion of the exercises at the end of each section of the textbook. Teachers did not make heavy use of such special features as historical and biological material and enrichment exercises. They rarely presented topics not in the textbook. Mathematics was resolved into a sterile sequence of homework-discussion-new homework.

Many ways exist to provide variety in the mathematics classroom. The instructor might vary the way in which homework is discussed, make provision for small-group work, schedule student reports on mathematics-related topics, make use of films or filmstrips, take a field trip, invite a guest speaker, play some mathematics-related game, present some enrichment topic, or occasionally have one or two students be responsible for conducting an entire class period. Other ways of providing some variety in your classes can be gleaned from other sections of this book.

While some variety is desirable, however, the teacher must be sure to maintain enough of a routine to provide a degree of security for the students. This is particularly important for lower-ability classes.

Show Enthusiasm   Teachers who are enthusiastic about their subject, those whose behavior suggests that they enjoy what they are doing, are in a far better position to keep their students interested in what is happening in the classroom than teachers who are unenthusiastic. Interested students are much less inclined to be bored—and consequently less likely to disrupt a class.

The enthusiasm of a teacher can be manifested in various ways. Sometimes it shows itself through a keen sense of humor or through the way in which a teacher modulates his or her voice. In other instances, facial expressions and body movements are the clues. A teacher who has each day's lessons carefully planned so as to provide students with a variety of meaningful learning activities is more likely to be perceived as enthusiastic than one who follows the same routine day after day. Enthusiasm can also be conveyed by the concern which teachers express for their students' academic progress. This concern can be shown through individual conferences, through group discussions, or in written comments on homework and test papers.

Consult Other Teachers   Another source of ideas regarding preventive disciplinary techniques (and, for that matter, corrective as well) is your colleagues in the school where you will be teaching. Their experience can be of value if you are willing to ask for advice and help. In many cases they will have already encountered some of the same problems you are having and can offer suggestions on how to resolve the difficulty. In seeking advice, however, keep two things in mind: not all of your colleagues will be equally successful as teachers; and some advice is not exportable. That is, what works for another person will not necessarily work for you. Also, you should not discard an idea just because it has not worked for another teacher. In other words, weigh carefully the advice you get.

Make Use of Rewards   Rewarding students for behavior which you view as desirable can be a powerful means of both preventing and correcting discipline problems. The variety of rewards available to teachers is extremely wide and includes praising the student, permitting the students to play games, and offering them tokens or points which may be redeemed for prizes or privileges. One of the simpler ways of rewarding students is through the use of sincere praise given for a correct answer or for manifesting a desirable kind of behavior: "That proof was nicely done, Mary" or "Very good, Jim, I never thought of approaching the problem that way." Comments such as these, if offered sincerely, can be a powerful means of maintaining discipline. Avoid, however, the glib use of verbal praise.

© 1977 by Ford Button.

CORRECTIVE DISCIPLINE

Despite all our best efforts, no teacher is ever 100 percent effective in preventing discipline problems. Situations arise in the life of every teacher which require him or her to invoke some sort of corrective action on one or more students. The wide variety of incidents which can arise in a classroom, however, make any attempt at an exhaustive discussion of ways to handle such problems futile. We will, instead, suggest a few of the options available to teachers.

Generally speaking, the discipline problems that most teachers face are minor (or at least begin as minor problems). Such things as blowing bubbles with gum, whispering, and one or two students not being in their seats ready for work when the bell rings are in themselves not major offenses. However, the proverb "White ants pick a carcass cleaner than a lion" certainly has some validity for teachers. In a traditional classroom such small events, can, if left unchecked, multiply and erode the entire system of control until the teacher is faced with a situation where learning is no longer possible (as Mrs. Likely discovered earlier in this chapter). At this point the restoration of control becomes very difficult at best, and impossible at worst. If problems arise from many different students, however, and if those problems recur on a frequent basis, the teacher may be well advised to do some introspection. The problem could be the result of indefinite goals, no attempt at motivation, or other deficiencies in planning. Problems may also be the result of a poor school environment or a weak administration; in this situation, the teacher may discover that attempts at corrective discipline will be to no avail.

Let's consider now some corrective disciplinary suggestions that you may find useful.

Find Out What Goes   Sometime before classes begin for the year, it is wise to consult your department chairman, dean, or principal, to ascertain to what extent you will be expected to handle your own discipline problems. In some schools, for example, teachers are expected to take care of their own problems, while in others they are encouraged to send students to the principal or to some other disciplinarian for corrective action. You should also find out whether you can place students in the hall and under what circumstances you can require them to stay after school. Frequently such information is provided to beginning teachers at orientation sessions at the start of the school year or in a school handbook. If it is not, you should ask for it.

With lawsuits against teachers becoming more commonplace, it is highly advisable to check with your principal or an expert in school law in an effort to ascertain some "dos" and "don'ts" regarding the disciplining of students. The very technical, rapidly changing, and complicated nature of school law makes this one area in which our earlier advice to consult with other teachers may be inappropriate.

Teacher-Student Conferences   Students misbehave for a variety of reasons; for example, to gain attention, as a release from boredom, just because they do not like you, or due to some psychological or physical problem. Many times teacher-student conferences are useful in helping both you and the student (who may not really know why he or she is disturbing you) to identify the problem or its cause. This puts both of you in a better position to resolve the difficulty. If the incident which led to the conference involved a display of temper from either party, however, it is best to allow for a cooling-off period before holding the conference.

Parent-Teacher Conferences   There may be times when a student's behavior interferes enough with his or her own progress or with that of other class members that you feel compelled to notify the student's parents in an attempt to resolve the matter. Usually this should be done only after teacher-student conferences have failed to improve the student's behavior. The following list of guidelines and the evaluation form shown in Figure 7–2 may assist you in conducting effective parent conferences.

1. Responsibility for the success or failure of the conference rests primarily with you, the teacher. Set goals for the conference and have suggestions in mind as to how you and the parents can help the student.
2. Do the best you can to arrange for complete privacy, with no interruptions, during the conference. Have a list of things which you would like to discuss.
3. Your greeting should be friendly and relaxed. Let the parents know that you feel their time is valuable. Spend a few minutes trying to get to know the parents. (Find out their occupations, how long they have lived in the community, etc.) Do not start the conference with an attack.

Pupil' name _____   Subject _____

Parent's name _____

Date of Conference _____

- -

		Yes	No
1.	Did I make the parent feel welcome?	____	____
2.	Did I greet the parent by name?	____	____
3.	Did I introduce myself?	____	____
4.	Were the surroundings: pleasant?	____	____
	private?	____	____
5.	Had I reviewed the pupil's: cumulative record?	____	____
	individual folder?	____	____
6.	Did I let the parent talk all he/she wanted?	____	____
7.	Did I refrain from using technical terms unfamiliar to the parent?	____	____
8.	Did I refrain from giving advice?	____	____
9.	Did we establish good communication?	____	____
10.	Did we help each other understand the pupil better?	____	____
11.	Did our conference help the parent understand better the program of this school and the mathematics department?	____	____
12.	Did we plan any follow-up for the conference?	____	____
13.	Will the home-school relationship developed in this conference benefit the pupil?	____	____
14.	Did the conference help me to know better how to work with the pupil?	____	____

Figure 7–2   Parent-teacher conference evaluation form.

4. Find out what the parent is thinking and how he or she is feeling about the child. Let the parent do the talking.

5. If the parent is concerned about the child's behavior, let the parent explore this. Do not judge, blame, or try to fix responsibility for the difficulties. Remember, parents find it difficult if not impossible to be objective about their children.

6. If the parent gives what he or she thinks is the reason for a child's behavior, accept it. If desirable, present other possible causes for consideration. Try to present several alternatives.

7. Be mindful of what the parent can and cannot do—and if the parent suggests a plan of action, accept it if at all possible.

8. It does not help to argue with a parent. Be accepting of the parent's view.

9. Be early to recognize problems that are too difficult and be aware of possible referral resources.

10. Close the conference with a forward note. Plan for the next conference and any definite action to be undertaken by child, parent, or teacher.

11. Follow up the conference by some statement of what has been subsequently accomplished.

We advise you to have anecdotal records available during such conferences which document the student's misbehavior. (More will be said about anecdotal records in the section on records and grading.) Any other relevant information, such as tests, quizzes, and homework grades, should be at your fingertips during such conferences.

Making Use of Seating Arrangements   The use of seating arrangements has already been mentioned as a means of preventing discipline problems. This can also be very useful as a corrective disciplinary tool. Once the teacher has gotten to know the students, he or she can judiciously reassign seats in an effort to separate troublemakers.

Ignoring Certain Behavior   Sometimes the best way to deal with undesirable behavior is simply to ignore it. In most instances, however, learning to ignore certain undesirable student behavior demands a great deal of restraint and self-discipline on the part of the teacher. And it requires a good deal of judgment on the part of the teacher in regard to which behaviors need ignoring and which require other action. You are urged to examine the studies (cited in References at the end of this chapter) by Clarizio (1971), Dobson (1970), and Meacham and Wiesen (1969) for a more extensive discussion of how to reduce undesirable behavior in students.

A Heart-to-Heart Talk   In some instances where quite a few class members (say, the majority of the class) are misbehaving in some way or other, you might consider discussing the problem with the entire class or with smaller groups within the class. This technique is probably most effectively used after other attempts to solve the problem have failed, and then only when the class can be shown that the teacher is earnestly seeking their help in overcoming the difficulty.

Verbal Reprimands and Teacher Proximity   You will be able to solve many of the disciplinary problems which you will encounter merely with a verbal remark. The majority of students will comply with a sincere request for a change in their behavior. And it is often enough for a teacher to move near students who are misbehaving in order to get them to stop.

Isolating the Disruptive Student   If classroom arrangements permit, it is sometimes effective to separate a misbehaving student from the rest of the class. Isolation is a rather drastic measure and should be done only when verbal reprimands have not had the desired effect, or the offense is serious enough to necessitate removal for the benefit of the other class members. Isolation can be accomplished in a number of ways, depending upon the seriousness of the misbehavior and the physical arrangement of the classroom. In some cases placing the student in a seat away from other students may be all that is required. In other cases, the student may need to be sent to the principal or some other disciplinarian. Whenever you find it necessary to isolate a student, a follow-up conference should be held with the student in order to clarify the reason for the isolation, as well as to explore ways of remedying the cause.

While the above listing in no way exhausts the possible means at a teacher's disposal for dealing with discipline problems, it does give a fair sample of such procedures. We conclude this section by offering a few exhortations which you may find of value in disciplining your students.

1. When you find it necessary to caution students, be certain that you intend and are able to follow through. It can be embarrassing to indicate that a student will receive a certain kind of punishment (for example, staying after school) only to discover that a 48-hour notice must be given.
2. If you find it necessary to accuse a student, be certain you have evidence to justify your accusation.
3. Apply the same standards of behavior to all students. They are very quick to observe the implementation of multiple standards.
4. In some instances it is wiser to hold off punishing a student until you have had an opportunity to consider the most appropriate punishment. Meting out punishment that represents either an underreaction or an overreaction to some infraction can do much to undermine the respect that the students might otherwise have for you.
5. Behave in a mature manner with your students; manifest the kind of behavior that you would want them to exemplify. Arguing with them (about a grade, for example) or resorting to sarcasm will, if done frequently, lower their esteem for you.

THE PLANNING INVOLVED IN TEACHING

We now turn our attention to some of the factors that should be considered when planning for instruction (see Figure 7–3). We cannot separate the control that a

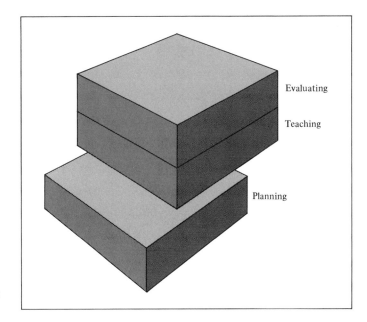

Figure 7-3   The planning process in
the model of mathematics teaching.

teacher has in the classroom from the planning for lessons that he or she does. The effectiveness of corrective disciplinary measures is a function of the rapport that exists between the students and the teacher; and one of the factors influencing rapport relates to how well the teacher has planned the learning activities.

Planning for teaching most commonly is done at three levels: course planning, unit planning, and daily lesson planning. The first two of these will be discussed briefly; the majority of our attention will be devoted to the third and most crucial, daily planning.

COURSE PLANNING

Assuming that a textbook has already been selected for each of the courses you will be teaching, one of the first things to do is to preview the contents of the books. Note the development and sequencing of topics, and consult the teacher's manual to discover the rationale for the particular sequencing used. At this point you may wish to modify the sequencing and some of the development to suit your own preferences. Attempt to develop a rationale for the whole course; that is, you should be thinking of reasons to offer the students as to why the course will benefit them.

Look for answers to questions such as: Are the exercises classified according to difficulty? Are there answers to certain exercises? Does each chapter contain a summary? Are there enrichment topics and a suggested reading list provided at the end of each chapter? What topics are and are not treated in the text? Seeking answers to such questions will assist you in establishing goals for the course.

While it can be argued that ideally goals need to be established before selecting a textbook, a beginning teacher usually is hired after textbook selections have been made. This fact, along with his or her inexperience, usually causes most beginners to rely upon the adopted text for course goals. State or local school district curriculum guides sometimes contain lists of course goals. Regardless of the source of one's goals, it is important that they be established.

Another important decision that you must make relates to how much time will be devoted to the study of various topics. This cannot, of course, be determined precisely; unexpected events will necessitate schedule changes. Planning a tentative course timetable, however, will help you to avoid the embarrassing situation of approaching the end of the school year and discovering that your students have studied only a small portion of what you had intended. Some mathematics departments will have definite syllabi and timetables to which you will be expected to adhere, while in other situations you will be totally responsible for this decision. Carefully constructed pretests which attempt to ascertain how much the students already know about the subject matter can be helpful in the selection of content to be studied as well as in the determination of time allotments.

Other things you will need to consider in making your long-range plans include the number and type of major examinations that you will give. As a general guideline, these examinations are usually given at the completion of each chapter or unit. Depending upon the length of the units, however, you may wish to modify this guideline.

It is also during this period of planning that you will want to give thought to what types of supplementary materials and activities will be appropriate at various points throughout the course. Appendix D should be helpful in making such decisions. Keep in mind that frequently films and filmstrips must be ordered months in advance of their intended use. And do not overlook the possibility of taking field trips or inviting community resource people to speak to your classes.

UNIT PLANNING

In addition to the long-range or course planning, you will also engage in unit planning. A "unit" usually refers to a body of subject matter which deals essentially with a single theme (e.g., congruency, similarity, first-degree equations in one unknown, matrices), requiring anywhere from several days to several weeks to study. In most high school mathematics courses, each chapter in the textbook represents a unit of work. Most of the considerations given to long-range planning are also appropriate at the unit level. The major difference is one of degree rather than kind; at this stage of planning, you are in a position to become more specific.

Goals which you identified at the course-planning level can be stated more specifically at this time. The amount of time to be spent on various topics within the unit can be more precisely delineated. You will be in a much better position to decide upon appropriate learning activities for the unit. Where, for example, might a film be useful, or a field trip or a quiz, or a special project or a report, or a worksheet—and when will quizzes be given?

When introducing the unit to your classes, it is advisable to preview the unit with them. At this time you might discuss some of the goals of the unit with your students and mention any special activities which you might have planned. You should also attempt to show how the unit relates to those which have been or will be studied.

DAILY LESSON PLANNING

A continuum exists in the area of daily lesson planning. At one end of the continuum is the teacher who gives little or no forethought to what he will do in class on a given day. He may believe that to do so would in some way stifle his spontaneity. At the other end of the spectrum is the teacher who has lessons so well planned that they are almost scripted. This teacher may be so bound to a lesson plan that opportunities for meaningful learning experiences are missed because they are not in his or her plan. While neither extreme is desirable, it is more advantageous, in particular for beginning teachers, to be overprepared (while at the same time remaining flexible) rather than underprepared. Let's consider some of the more important components of planning a daily lesson. They are:

1. Goals of the lesson
2. Starting the lesson
3. Key points to be discussed in the lesson
4. The establishment of closure
5. The provision for homework
6. The provision for evaluation
7. Materials needed for instruction

Goals of the Lesson   In Chapters 2–6 we examined the part goals play in the model of teaching mathematics which has been developed in this book. By including your lesson's goals in your daily plan, both you and your students will benefit in these ways:

1. You will subject your educational goals not only to your own critical analysis but to that of your students and some of your colleagues as well.
2. You will be required to delineate clear and definite goals for yourself and your students.
3. You will be in a much better position to select appropriate instructional materials and procedures.
4. You will be better able to evaluate the materials and procedures you use in instruction as well as student achievement.
5. You will be more confident and have a greater sense of purpose by knowing clearly what you wish the students to accomplish.

"Now that I've got your attention . . ."

Starting the Lesson   All of us have been exposed to teachers who begin their lessons in virtually the same way, day in and day out. And yet the first few minutes of a lesson are quite important. They can do much either to "turn off" or "turn on" your students. Every lesson plan should give some indication that you have thought about how you are going to begin the lesson. Some suggestions of ways to begin lessons have been identified by Runion (1973) in what he refers to as "initiating activities." The verbal and nonverbal activities in which a teacher engages when beginning the lesson will be considered initiating activities if:

1. they are relatable to the lessons, and
2. they attempt to serve at least one of the following functions:
    a. to gain the student's attention
    b. to arouse the student's interest in the lesson
    c. to organize the lesson
    d. to show how the content of the lesson is dependent upon previously learned subject matter in mathematics
    e. to establish a common frame of reference between the students and the teacher
    f. to give the student a sense of purpose or direction in studying the lesson

A brief summary of the seven initiating activities which Runion identified is given below. We will briefly discuss and exemplify each activity.

Name of Initiating Activity	In Using this Initiating Activity, the Teacher:
1. Stating goals	Begins the study of the lesson by stating clearly the goal of the lesson.
2. Outlining	Presents either a written or an oral outline of the major points of the lesson.
3. Using an analogue	Uses an analogue within the experiential field of the students that is related to the content of the lesson.
4. Using historical material	Makes use of some piece of historical information (biographical information of a famous mathematician, a problem with an interesting history, etc.).
5. Reviewing subordinate information	Reviews certain topics whose comprehension seems to be a prerequisite for understanding the lesson.
6. Giving reasons	Gives the class reasons why they are studying the lesson.
7. Presenting a problematical situation	Begins the study of the lesson by presenting the students with a problem or a contradictory state of affairs whose resolution is related to the material to be covered in the lesson.

1. *Stating Goals*. Although stating the goals of a lesson is a very simple initiating activity, it seems to be an uncommon occurrence in many mathematics classrooms. But why not let the students in on the lesson's purpose(s)? Would it not give them a sense of direction for you to tell them at the beginning of the class specifically what it was they were expected to know or be able to do or understand at the end of the lesson? If you were teaching a discovery lesson, of course, it would be inappropriate to tell the students what you wished them to discover. (See, for example, the discovery lesson presented in Chapter 4.)

How you would actually present the goals to the class is your prerogative. You might, for example, give the goals to the students verbally, write them out on the chalkboard or overhead projector, or ditto them and pass them out. Numerous examples of goals have already been given.

2. *Outlining*. This, too, is a very simple way to begin a lesson, and one that is not used often. The teacher merely identifies (on a transparency, on the chalkboard, on a handout, etc.) the major points to be covered in the lesson. This can provide direction, organization, and structure to the lesson and, hopefully, will enhance the student's learning. Here is one example of a lesson outline:

General goal

    Know the various relationships between the angles formed by two coplanar lines and a transversal.

    Outline of the Lesson:
  I. The concept of a transversal
 II. The various categories of angles formed by two lines and a transversal
    A. Alternate angles
       1. alternate interior
       2. alternate exterior
    B. Corresponding angles
III. Theorems concerning the various categories of angles and relating them to parallelism
IV. Summary of the major outcomes of the lesson

    3. *Using an Analogue.* In using this initiating activity, the teacher simply tries to relate the unfamiliar content that he intends to present to his students to something with which his students are already familiar. The familiar material, of course, must have characteristics which are similar to the new knowledge or understanding being taught. Here is a description of one instance of using an analogue:

General goal

    Solve, using only tables of trigonometric functions, simple trigonometric equations, that is, equations involving only one function of a single unknown angle.

    The teacher might begin the lesson by displaying to the class the trigonometric equation $3 \sin \theta + 1 = 0$. He or she then could relate the goal of the lesson to the students. At this point the teacher selects as an analogue some algebraic equations involving one unknown, say, $3x + 1 = 0$. He or she then demonstrates the steps necessary to isolate the variable and hence obtain a solution.

    Following this the teacher redirects the student's attention to the trigonometric equation $3 \sin \theta + 1 = 0$. He or she makes clear to the student the similarities in solving this equation for $\sin \theta$ and $3x + 1 = 0$ for x. At this point, the students are informed of a major dissimilarity in the methods of solution, namely, that in solving trigonometric equations one is usually forced to make use of tables of trigonometric functions, a slide rule, or a calculator.

    4. *Using Historical Material.* Although potentially a very powerful means of getting a student's attention and arousing his or her interest in the lesson, this activity is rarely used by high school teachers.

Steinen (1970) provides the following example of an historical anecdote which could be used to focus student attention on the relation between central angles and their intercepted arcs. He writes that the anecdote is "intended to promote interest and respect for mathematics in general" and that the introduction "would probably take the form of a short, informal lecture" (p. 381).*

Most people know that when Christopher Columbus set out to find a new route to India it was commonly believed that the earth was flat. Many feared that Columbus might reach an edge and never return. However, not all people before Columbus's time shared this belief. In fact, more than seventeen hundred years earlier a Greek mathematician named Eratosthenes not only realized the earth was round but also calculated its circumference with an accuracy of within one percent of our present measurement.

A lunar eclipse provided the first clue for Eratosthenes. When the earth's shadow fell on the moon, its shape was clearly curved. In addition, he knew that at noontime of the summer solstice (about June 21) the sun was directly above the city of Syene in Egypt. Evidence of this was that the sun's rays were reflected from the water in a deep well with no shadow from the well's walls. Also, a vertical rod driven into the ground cast no shadow at this time. So, sometime in the latter part of the third century B.C., at the time of the summer solstice, Eratosthenes performed the following experiment. Vertical rods were driven into the ground at Syene and at Alexandria, which was due north of Syene. At noon the rod at Syene cast no shadow at all; the rod at Alexandria cast a short shadow. The shadow's length enabled Eratosthenes to calculate the angle of elevation of the sun, $82\frac{4}{5}°$.

Since the distance from the earth to the sun was very great, Eratosthenes assumed the sun's rays were parallel. Using the alternate interior angle theorem, he concluded that the angle at the center of the earth was $7\frac{1}{5}°$. (In Figure 7–4 the extensions of the rods are represented by dotted lines that meet at the earth's center. Obviously the figure is not drawn to scale.) Since $7\frac{1}{5}° = \frac{1}{50}(360°)$, the distance from Syene to Alexandria must be $\frac{1}{50}$ of the earth's circumference. This distance was known to be 5,000 stadia—roughly 500 miles, in modern terms. Therefore,

$$\begin{aligned} \text{circumference} &\approx 50 \cdot 5{,}000 \text{ stadia} \\ &\approx 250{,}000 \text{ stadia} \\ &\approx 25{,}000 \text{ miles} \end{aligned}$$

(Steinen, 1970, p. 381)

For an excellent collection of topics appropriate for historical initiating activities, you can refer to the Thirty-first Yearbook of the National Council of

* The excerpt that follows is quoted by permission from R. F. Steinen, "An Example from Geometry," in Thirty-third Yearbook of the National Council of Teachers of Mathematics, Washington: The Council, 1970, p. 381.

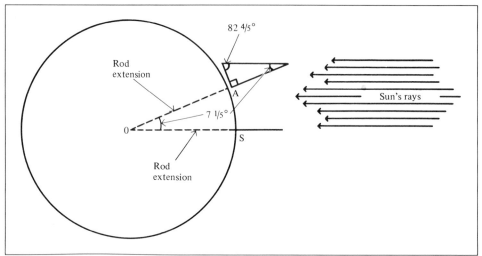

Figure 7–4   Measuring the circumference of the earth.

Teachers of Mathematics, *Historical Topics for the Mathematics Classroom* (1969).

5. *Reviewing Subordinate Information.* This initiating activity is rather common among high school mathematics teachers. Teachers frequently conduct brief reviews of material whose understanding they have judged is essential to a full comprehension of the new lesson. For example, before beginning a lesson on solving first degree equation in one unknown, the teacher reviews the operations that can be performed on an equation that will result in an equivalent equation. The meaning of the term "equivalent equations" is also reviewed.

6. *Giving Reasons.* The activity is primarily used to help give students a sense of purpose in what they are studying and to help motivate them. Here are three examples of how a teacher might proceed:

1. "Since many of you are approaching the age of 17, you will soon be interested in purchasing your own car. This next unit of study related to the cost of maintaining an automobile should be very important to you."
2. (Displaying several newspaper advertisements announcing sales) "Percentages are so much a part of our everyday lives that it is to our financial advantage to know how to work with them. This knowledge could save you many dollars sometime."
3. "Studying what kinds of constructions we can do using only a compass and straightedge is a lot of fun and can help to give us some insight into certain theorems in geometry."

7. *Presenting a Problematical Situation.* This initiating activity can be effective in gaining the student's attention and stimulating interest in the lesson. The

teacher might simply pose a question whose answer will depend in some way upon the content of the lesson. Or the teacher might present the class with a paradoxical state of affairs or some game-type activity which will lead into the lesson. It should be understood that the problematical situation is to be more than just a question posed at the beginning of the lesson and then immediately answered. The intent of this initiating activity is that the problem will serve as sort of a theme around which the teacher presents the new material. Here are two examples:

1. Before studying the Pythagorean Theorem the teacher asks the class "How far is it from one corner of the classroom to the opposite corner?" (See Figure 7–5.) If the student had never seen the Pythagorean Theorem before, this problem could be used very nicely as a lead-in.
2. The Snowflake Curve shown in Chapter 9 (Figure 9–17) could be used in a senior mathematics class to introduce a study of infinite geometric progressions. The teacher might have the various curves on an overhead transparency. They could be exposed one at a time with the same question being asked for each figure: "What is this figure's area and perimeter?" This particular problem situation might very well take more than one class period to resolve.

Key Points to Be Discussed in the Lesson   The daily lesson plan should also contain the key points that the teacher wishes to make during the class period. If the teacher uses an outline as an initiating activity, then this could serve as a reminder of the key points of the lesson. Any example problems or exercises to be used would also be included.

The Establishment of Closure   Simply defined, *closure* refers to "the internal and external organization which students find in the lesson." The internal organization depends on how well the lesson itself is organized, while external organization depends upon how the lesson fits into what has already been studied or will be studied. The degree of closure possessed by any lesson is dependent upon student perceptions: "Achievement of closure means that at the end of the lesson or lesson part the students know where they have been and what they have learned. Furthermore, they are able to integrate the new material into their cognitive structure" (Allen et al., 1969, p. 40).

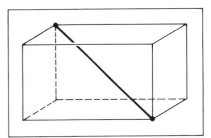

Figure 7–5.

There are a number of ways that are available to you for helping students achieve closure and that have ramifications for planning your lessons.

1. See that your lessons are well organized around a central theme. For example, "Three ways of solving quadratic equations" or "Two methods for simplifying complex fractions."
2. Take advantage of the cueing power of certain statements: "In class today, you will learn how to perform four basic constructions using a compass and straightedge" and "Pay particular attention in today's lesson to the two different methods that will be used to solve a system of equations."
3. As you proceed through the lesson, indicate to the students when you have completed a part of the lesson: "Okay, class; so far today we have seen that one way to solve a system of equations is by the method of substitution."
4. Connect the lesson to previously studied material: "A few days ago we learned two ways to prove triangles congruent, S.A.S. and A.S.A. Today we will study a third way of proving triangles congruent."
5. Let the students demonstrate what they have learned. (This might be done by letting the students practice what they were to have learned—if, for example, the goal of the lesson were to teach some skill—or by letting them summarize the content of the lesson.)

You should notice that there is a close link between the way in which you begin your lesson and the establishment of closure. In particular, the initiating activities of stating goals, outlining, using an analogue, and reviewing subordinate information are useful in helping to achieve closure in the manner suggested in the examples given above.

The Provision for Homework   Homework is very closely linked to the teacher's goals. If the goals for a particular lesson involve the development of skills (for example, being able to factor perfect-square trinomials), then the associated homework would be selected to provide the student with exercises which would give practice in factoring perfect-square trinomials. If the goals were in the understandings or problem-solving category, the assignment would need to be modified. The type of problems assigned would be different, and so would the quantity assigned. As explained in Chapter 2, understanding goals and problem-solving goals are at a higher cognitive level than are knowledge goals (of which skills are a subset). It is therefore usually appropriate to assign fewer problems in the higher level categories.

A major purpose of homework assignments is to provide opportunities to sharpen skills, develop understandings, and improve problem-solving abilities. The previous statement supports our earlier suggestion of letting the students know exactly what the goals of the lesson are. If both teacher and students are aware of the goals, the teacher can relate the homework to the goals and, in so doing, enhance the student's chances of profiting from it.

Another purpose that can be served by homework assignments is to tie in to present work material that has already been studied. This can be done rather easily when the assignment includes a few problems over material already studied that have some relationship to the material now under study. Some textbooks have such problems built into their suggested assignment guides. Besides linking past material to present work, assignments can also effectively be used to set the stage for future topics of study. Thus certain homework problems can quite easily serve as an initiating activity for a new lesson by posing a problematical situation.

Suppose, for example, that in an algebra class students had been working on solving quadratic equations by factoring. As part of one of the latter assignments over this material, you could insert a problem (such as $x^2 + x - 1 = 0$) which is not ordinarily solved by factoring. Assuming that factoring was the first method studied for solving quadratics, such a problem would serve quite nicely to introduce the method of solving quadratic equations by completing the square.

"What shall we do with the homework?" is a frequently asked question. Opinions vary rather widely among teachers as to how much class time should be spent discussing homework problems. Some teachers spend almost all of their class time explaining the homework while others spend almost no time in such discussion. We shall not attempt to resolve this issue; instead, we offer a few alternatives for you to consider.

1. Work every problem in the assignment on the chalkboard or overhead projector.
2. Read the answers to all of the assigned problems and then ask for questions.
3. Have the students turn in their homework papers for you to grade.
4. Let the students exchange papers and correct each other's work.
5. Have certain students come to class with various problems (that you had assigned to them) worked out on an overhead transparency. They would be expected to explain their work.

These are only a few of the options available to teachers for discussing homework. You will have an opportunity to think about these and suggest modifications and alternatives in Exercise 7 in the Activities at the end of this chapter.

Not only must you decide how much class time to devote to homework, but you must also consider whether or not the homework should be a part of the course grade—or whether, indeed, you even want to count the homework at all. Our experience suggests that those students who do homework regularly appreciate receiving some sort of recognition or reward for their efforts.

One method we have found useful and well received by the majority of students involves assigning every class member a digit, 0 through 9. (In most classes, of course, digits will be used repeatedly.) Each day the teacher selects randomly (a table of random digits is very useful) five or six digits, say (for example), 0 1 3 4 7. These are announced to the class, and everyone whose "number is up," so to speak, is expected to turn in homework that day. The teacher then has the option of grading each paper in detail or merely spot-checking and marking it satisfactory (S) or unsatisfactory (U). A record is kept of the number of times the digits are called and the number of times each student turns in homework when his or her digit is "up." Grades can be assigned to students on this basis. This procedure has been found to produce a lot of suspense for the students and quite frequently groans from the "gamblers," a few of whom always seem to inhabit every class!

Some teachers have their students keep notebooks in which they are required to enter class notes and homework assignments. These notebooks are collected periodically and graded.

We know of teachers who collect homework every day from every student. Each problem is carefully graded and the papers are returned the next day. Students who missed or could not work certain problems are expected to redo them and turn them in again. This procedure continues until all problems on each assignment are correctly completed. Obviously such a technique requires a tremendous amount of work on the teacher's part as well as on the part of the students.

In summary, here are several things to consider when planning homework assignments:

1. Have you worked through the exercises or problems you intend to assign, and do they fit in with your intended goals? (It can be embarrassing and time-consuming to get stumped on a problem because you have not worked through it before class.)
2. Have you "set" the students for the assignment by telling them how they should benefit from it or by calling their attention to problems of special interest?
3. Have you been careful to review in class the subordinate information that the students will need in order to work the problems (or at least indicated to them that they will need to conduct their own such review)?
4. Have you attempted to relate the assignment to past work and to future work wherever possible?
5. Have you been sensitive to individual differences? (See Exercise 9 and Chapter 8.)
6. Have you made some provision for supervised study during class time in an effort to provide help for those who need it?
7. Have you consulted other textbooks in an effort to find problems and exercises perhaps better suited to your purposes than those in your textbook (see Research Highlight, "The Effect of Homework")?

The Provision for Evaluation   As we have seen in Chapters 2–6, behavioral objectives are statements of evidence that a teacher will accept in deciding whether the goals of a lesson have been met. There are essentially three major components of every behavioral objective. First, the objective must describe an *observable terminal behavior* which the student will be expected to perform at the end of instruction; second, a statement describing the *condition under which the*

RESEARCH HIGHLIGHT | THE EFFECT OF HOMEWORK

The findings of studies on homework have not been consistent, perhaps because the cumulative effect of homework is not adequately studied in short-term experiments, or because there are so many confounding factors. Several recent studies explored the effect of various ways of distributing the practice provided in homework. Peterson (1970) assigned exploratory homework exercises on a topic three days preceding the teaching of the topic. Classes receiving this type of homework and classes given mathematical puzzles unrelated to the topics each achieved and retained better than did a group not receiving homework. Those who completed at least 50 percent of the exploratory homework had higher retention and transfer scores than students in the other two groups. Laing (1971) reported no significant differences in achievement or retention when practice on a topic was massed in one homework assignment or distributed over several assignments. There was a consistent trend favoring distributed practice, a finding that is supported by research from non-mathematics-education studies. In a third study, Urwiller (1971) assigned homework which contained problems on previously taught topics as well as problems on new material. For another group, assigned homework dealt only with problems on new material. Differences in achievement and retention were found favoring the first group. Only the findings are cited here, but you might find it valuable to look at the actual materials developed for the research. Laing and Peterson (1973) present some illustrations that you can try.

terminal behavior is to be demonstrated; and third, the *minimal acceptable standard of performance* for the terminal behavior. Here are examples of behavioral objectives which a teacher might accept as evidence that the goals listed earlier in this chapter had been achieved:

1. Without using your textbook or notes, define in writing the term *transversal* as stated in your textbook.
2. Without using your textbook or notes, complete the following definition of alternate interior angles as found in your text: If line t intersects line l_1 at M and line l_2 at N _____, then angle AMN and angle BNM are called alternate interior angles.
3. In Figure 7–6, identify, using a numerical designation of angles, all pairs of alternate interior angles and all pairs of corresponding angles. Be certain to label clearly the two categories of angles.

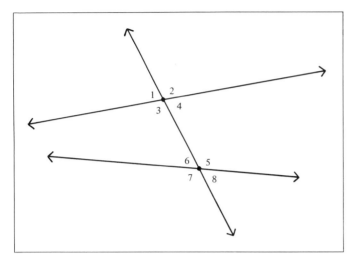

Figure 7–6.

4. For Figure 7–7, prove without using any outside sources, and in a two-column format with the hypothesis and conclusion clearly stated, that *ABCD* is a parallelogram.
5. Using any appropriate proof format and without any outside help, prove the following theorem: If two lines are cut by a transversal so that a pair of alternate interior angles are congruent, then the two lines are parallel.

Observe that each of the foregoing behavioral objectives contains an action verb which is used to describe the sought-after observable terminal behavior. In order of appearance in the objectives, these verbs are "define," "complete," "identify," and "prove." It is characteristic of behavioral objectives that they contain such verbs. Other verbs which are useful in specifying behavioral objectives in mathematics (listed in Gronlund, 1970, p. 55) include: add, bisect, calculate, count, estimate, measure, solve, square, and verify.

Objective 4, although intended to test acquisition of an understanding, could also be used as an evaluation of a problem-solving goal. This is acceptable so long as the behavioral objective is consistent with the teacher's goal.

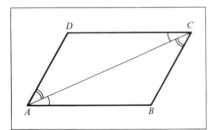

Figure 7–7.

The test items used to evaluate the attainment of a goal and the behavioral objectives associated with a goal do not necessarily represent the same set of items. While it is true that the five behavioral objectives listed above could be used as test items, this is not always the case. Consider this behavioral objective:

▶ Given a quadratic equation of the form $ax^2 + bx + c = 0$, where $a \neq 0$, the student will be able to solve the equation by completing the square, and the student can also check the answer.

Although this is an acceptable behavioral objective, it would not be suitable as a test item. Conversely, the sample test items listed in Chapters 3 and 4 are not suitable behavioral objectives. Behavioral objectives are helpful to a teacher in selecting test items.

Materials Needed for Instruction   It is frustrating as a teacher to begin your class only to discover that you have forgotten to bring some instructional aid with you. Therefore, you should look through your lesson plan carefully to identify and list all equipment (e.g., models, films, colored chalk) and software (e.g., work-sheets, quizzes, tests) that you anticipate using in a given class period.

SAMPLE LESSON PLAN 1

Straightedge and Compass Constructions

Goals:
To learn how to construct, using a straightedge and compass only, the perpendicular bisector of a given line segment.

Starting the lesson:
1. Historical background
    These constructions attributed to Plato, about 400 B.C.
    Restriction to straightedge and compass entirely arbitrary
2. Greeks performed many constructions, but unable to
    • trisect any given angle
    • construct a square having area equal to that of a given circle
    • construct a cube having volume equal to twice that of a given cube

Key points:
1. Familiarization with compass
    • parts of compass (point, pencil, movable part)
    • construct circles on chalkboard (small, then large radii)
    • have students come to chalkboard to construct circles
    • other students use own compasses and draw circles at seats (1 inch radius; 4 cm radius)

- students experiment in producing design (give example on chalkboard)
- students form small groups and share designs, if there is time

2. Constructing arcs
   - arcs: "With a radius of 10 cm and center at A construct two arcs."
   - students construct arcs with given radii
   - intersecting arcs: (points A and B on chalkboard about 30 cm apart) "With center at A construct an arc above AB. With center B construct an arc to intersect the first arc." (Note condition for arcs to intersect.)

3. The perpendicular bisector of given segment
   - points A and B are drawn on chalkboard 40 cm apart. "Construct intersecting arcs above and below $\overline{AB}$ at points P and Q respectively. $\overleftrightarrow{PQ}$ is perpendicular bisector of $\overline{AB}$."
   - hand out dittoed sheet with 5 line segments of various lengths. Students are to bisect each. Move around class to help those with difficulties. Have constructions done on overhead transparency so students can check work.

Closure:

Place two points C and D on chalkboard. Have student construct perpendicular bisector of $\overline{CD}$. "This same approach can be used to bisect an angle and is used in other constructions, too." Also indicate that these basic constructions will be used later in performing more complex ones.

Homework:

Pp. 175–176, Exercises 1, 2, 4, 6, 7, 11, 15 (last 2 for extra credit) The following students do indicated exercises on overhead transparencies to show class tomorrow: Mary #1; Tom #4; Sue #11. Challenge class to prove that the construction does indeed produce the desired perpendicular bisector.

Evaluation:

Provide help for those who have not yet completed handout sheet, and circulate among rest of class. Note those attempting extra credit problems and proof.

Materials needed:

Extra compasses and straightedges; picture of Plato; chalkboard compass and meter stick; colored chalk to use in highlighting parts of constructions; cardboard for students to use under papers to protect desks

This lesson plan suggests how you might introduce the study of geometric constructions. We do not intend it to be followed like a script, but it can help you to recall details and to include the topics you intended. The amount and nature of the detail is a matter of personal preference; generally, during your first lessons you include more detail than you would find necessary later.

Some teachers also find it helpful to indicate roughly how much time they wish to allocate to each part of their lessons. Of course, these limits are not to be strictly followed when unanticipated questions or other developments require more time. But time checks will help you see whether you are ahead or behind schedule and alert you to revise mentally your plan to include extra sidelights or omit certain parts of the lesson.

SAMPLE LESSON PLAN 2

Centroid of a Triangle

Goals:
The student will be able to
1. define centroid of a triangle
2. precisely locate the centroid of any given triangle
3. prove that the medians of a triangle are concurrent at a point which is two-thirds of the distance from a vertex to the midpoint of the opposite side (i.e., prove the Median Concurrency Theorem)

Starting the lesson:
1. Distribute cardboard triangular regions of various sizes and shapes to all students and have them determine by trial and error the "balance point" of their region by using the eraser end of their pencil. Have them trace regions onto note paper and estimate where balance point should be placed. Exchange regions and repeat this procedure.
2. Finding centroid by using plumb line (students work in small groups)
   - insert pin at a vertex so that cardboard figure freely swings
   - using piece of string with a weight on one end, make a plumb line. Hang plumb line from pin (Figure 7–8). Suspend triangular shape by pin. When plumb line comes to rest, draw in the position of the plumb line on shape
   - repeat procedure for other two vertices
   - intersection of drawn-in lines marks center of gravity for region
   - compare the positions of the point found by this method with point found by balancing method

Key points:
1. The balance point of a region is called its *centroid*. Every shape has a centroid. For triangles it is particularly easy to locate.
2. Referring to plumb line activity, elicit from students suggestions as to how centroid might be located. "Since medians bisect sides of triangles, and we want to divide up weights equally, let's see if medians can be used to locate the centroid."

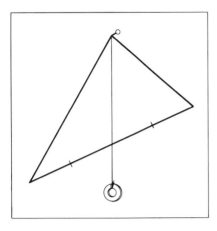

Figure 7–8.

3. Draw triangle on chalkboard. Have student bisect sides of triangle and construct medians. Meanwhile, other students construct medians for their triangular pieces of cardboard.
4. Find point of intersection of medians and compare with balance point found by experiment.
5. Student at chalkboard measures location of centroid (point of intersection of medians) along each median. Is point two-thirds of distance from vertex to opposite side? Other students do same for their figures.
6. State and prove Median Concurrency Theorem (p. 241 of student's text— use different figure).
   a. Locate M, N, and O, the midpoints of $\overline{AG}$, $\overline{BG}$, and $\overline{CG}$, respectively.
   b. Draw $MNA'B'$ and show it is a parallelogram.
   c. Hence diagonals bisect each other and $AM = MG = GA'$ and $B'G = GN = NB$.
   d. Repeat this procedure using medians $\overline{CC'}$ and $\overline{BB'}$ (see Figure 7–9).

Closure:
   1. Review terms *balance point, centroid, concurrent*.
   2. Restate theorem. Ask students to state in own words.
   3. Relate results to concurrency of bisectors of angles of a triangle, and to concurrency of perpendicular bisectors of sides.
   4. Indicate that the results of the Median Concurrency Theorem will be used in several proofs later on.

Homework:
   1. Provide waxed paper. By paper folding, verify Median Concurrency Theorem.
   2. For ten bonus points, prove the theorem using the techniques of analytical geometry.

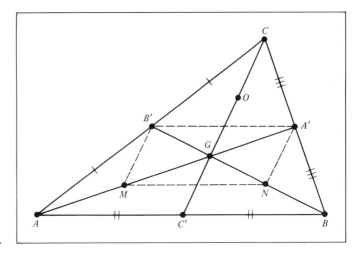

Figure 7–9.

Evaluation:

Quiz will be given tomorrow on finding the centroid of a triangle and proving the theorem.

Materials needed:

Cardboard triangular regions; pins; thread; washers or other weights; extra rulers and compasses; colored chalk; chalkboard compass; meter stick; other shapes for locating centroids if time permits (leaves; other polygons; simple closed curves; horseshoes)

In this lesson, you might notice that we have examples of both knowledge and understanding goals. Goals 1 and 2 are knowledge goals while Goal 3 is intended as an understanding goal.

MOTIVATION

The word *motivation* is a form of *motive*, which denotes an inner drive in a person which causes him or her to do something or to behave in a certain way. Since motives and, hence, motivation comes from within a person, in a strict sense we as teachers cannot actually motivate our students. What we can do, however, is attempt to arrange conditions within our classrooms so that students will be activated to seek those goals that we have selected as worthy of achievement. Holt, in *How Children Fail* (1964), suggests that there are few tasks of greater importance or challenge facing teachers than that of how to capture their students' attention and make their classes interesting.

"Mr. Sproul is always coming up with something new in the way of interest centers."

We ask children to do for most of a day what few adults are able to do even for an hour. How many of us, attending, say a lecture that doesn't interest us, can keep our minds from wandering? Hardly any. Not I, certainly. Yet, children have far less awareness of and control of their attention than we do. No use to shout at them to pay attention. If we want to get tough enough about it, as many schools do, we can terrorize a class of children into sitting still with their hands folded and their eyes glued on us, or somebody; but their minds will be far away. The attention of children must be lured, caught, and held, like a shy wild animal that must be coaxed with bait to come close. If the situations, the materials, the problems before a child do not interest him, his attention will slip off to what does interest him, and no amount of exhortation or threats will bring it back. (p. 158)

The following suggestions should assist you in gaining the attention and interest of your students, thereby increasing the likelihood of their becoming motivated to learn mathematics.

Goals   In the context of motivation, goals are most helpful to students when they can be achieved; that is, when the student is able to view the goals as attainable. Slow learners in particular are less inclined to be motivated by long-range goals. As students enjoy success in achieving daily (short-term) goals, their confidence in themselves and their interest in the material at hand is frequently increased enough to sustain them in a longer activity.

Success   It is usually the case that success breeds success and that people who are successful are self-motivated. This is a characteristic we should like all of our students to have. There are a number of things that you can do to help maximize your students' chances for success. Some of these include encouraging and praising students, presenting clear explanations, and providing appropriate home-work assignments.

Variety   Providing for variety in the classroom has already been discussed in this chapter as a means of reducing boredom and increasing the attention span of students—thus preventing some discipline problems. There is a wide range of things that you can do in your classes to provide variety. One of the more important principles is to provide for several changes of activity during the class period. For example, you might plan a class so that there is time for individual or small-group checking of homework, time for student presentations at the board (maybe by letting them explain a homework problem), time for you to preview new content, time for supervised study, and perhaps some time to play a mathematics-related game or to engage the class in a mathematical puzzle or paradox. On another day, involve students with some mathematics laboratory activity. Do not get in the rut of doing the same thing day after day.

Interests   You should strive to capitalize on the student's own interests in an effort to motivate them. If a student, for example, is interested in nursing, biology, sports, or whatever, attempt to identify mathematically related problems in those areas that would appeal to his or her interest—perhaps even make up a problem in which you use the student's name. Practically, however, it is difficult to find or manufacture problems which would be tailored to the variety of interests which students might have in a typical classroom. It would be to your advantage, though, to start a file of such application-type problems as you encounter them.

An alternative procedure is to attempt to *create* a new interest in all the students. Sobel and Maletsky (1975) suggest four ways in which a teacher might attempt this.

1. *Intuitive Guessing.* This might take the form of a problem to which the students would be encouraged to guess the answer. For example, "Is it possible to stack congruent cubes so that eventually the top cube is completely extended beyond the bottom one?" (See Figure 7–10.) Such a problem could serve as the seed of a very interesting laboratory investigation. The problem could be extended to other shapes.

2. *Mathematical Novelties.* Instructions:

Think of a number	23
Multiply by 5	$23 \times 5 = \quad 115$
Add 6	$115 + 6 = \quad 121$
Multiply by 4	$121 \times 4 = \quad 484$
Add 9	$484 + 9 = \quad 493$
Multiply by 5	$493 \times 5 = 2{,}465$

Ask for the final result and subtract 165 from that number. Drop the zeroes in the units and tens places, and the resulting number is the one started

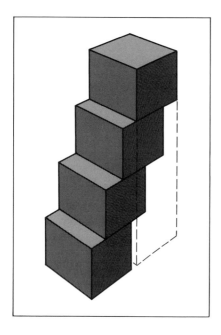

Figure 7–10 Cube-stacking problem.

with. In the example shown we have 2,465 − 165 = 2,300, and the original number thought of was 23 (Sobel and Maletsky, 1975, p. 31).

3. *Computational Curiosities*. Into this category can be placed problems such as the one shown in Figure 7–11. Most students will become involved in finding the message which is spelled out if they correctly work each problem.

4. *Geometric Tidbits*. Into this category Sobel and Maletsky suggest placing all sorts of geometric puzzles and problems "to warm up the class, to gain attention, to involve, to challenge, to maintain interest, or simply to give a change of pace" (p. 39). For example:

Cut a square, measuring 8 cm by 8 cm, from lightweight cardboard. Cut the square into four pieces (as shown in Figure 7–12). Rearrange the four pieces to form a rectangle (illustrated) whose dimensions are 5 cm by 8 cm. The students could then be queried to see if they can discover the apparent discrepancy between the areas of the two figures. Students could also be challenged to explore this puzzle using different dimensions. It could serve as a starting point for discussions of Fibonacci numbers and other additive sequences as well as discussion of other dissection problems.

While the classification scheme used by Sobel and Maletsky is arbitrary, there are scores of books available which contain problems similar to those just described. Some of them are included in the references at the end of this chapter, in Chapter 5, and in Appendix B.

1. $x + m = 2m$ 1. _____

2. $bx = be$ 2. _____

3. $\dfrac{x}{r} = 1$ 3. _____

4. $5x - 2r = r + 2x$ 4. _____

5. $7 + ax = ay + 7$ 5. _____

6. $x - b = -b + c$ 6. _____

7. $lwx = lwh$ 7. _____

8. $gh + mr = gh + xm$ 8. _____

9. $5i + x = 6i$ 9. _____

10. $3x - s = 2s$ 10. _____

11. $\dfrac{vt^2}{g} = \dfrac{vxt}{g}$ 11. _____

12. $m^2x = m^3$ 12. _____

13. $\dfrac{x}{a} = 1$ 13. _____

14. $2x - 2s = 0$ 14. _____

Figure 7–11 Solve each of the equations for x to find the message.

The Grading of Papers   Another simple though potentially time-consuming way of helping to motivate students involves written encouragement on student papers (see Research Highlight, "Your Comments Help!").

Competition   Letting students compete against each other can also serve as a motivational technique. This competition usually assumes the form of some type of game. In using this form of motivational tool, however, do your best to form teams in which students are matched according to ability. Appendix D will be helpful in identifying sources of games for classroom use.

This section on motivation has offered you practical suggestions of things you might do to help motivate your students. One of the more important factors which helps to influence how highly motivated your students will become is the degree of enthusiasm which you demonstrate in the classroom. Your willingness to try different methods and materials as your own creativity leads you can do much to create interest and to make your classes places to which students enjoy coming.

GRADING AND RECORD KEEPING

Grading, like discipline, is one of the more frequently discussed topics among both teachers and students. Hundreds of articles and books have been written on this subject.

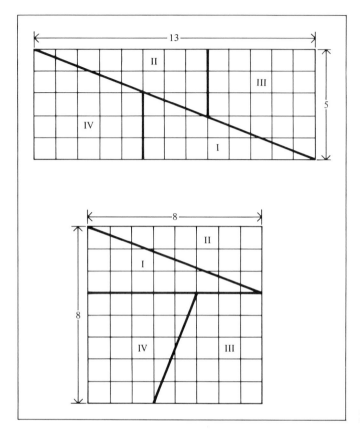

Figure 7–12 A dissection paradox.

SOME UNHEALTHY ASPECTS OF GRADES AND GRADING

Grades Affect the Anxiety Level of Students. Few people have gone through our schools without feeling the tension arising out of some past failure or the fear of a future failure. Anxiety often arises in students when teacher expectations in the area of evaluation are not clear. This consequence of evaluation surely cannot be considered desirable. Anxiety up to a point can be beneficial to the student; beyond this point, however, it can be detrimental (see Research Highlight, "Anxieties in the Mathematics Classroom").

Grades May Become the Object of Learning. The prize of a high grade is often pursued by many students with little regard for what is learned along the way. It is not uncommon to find students who have given up in their studies, thinking that they are no longer in competition for good marks. Hence grades become a substitute for learning; the attainment of a symbol (A, B, etc.) becomes paramount in the student's mind. One's transcript assumes greater importance than what one actually learns.

RESEARCH HIGHLIGHT | YOUR COMMENTS HELP!

Should you write comments on papers that will be returned to students? Will this improve the students' achievement? Are there some kinds of comments that are more effective than others? These are the key questions asked by Page (1958).

He randomly selected 74 secondary teachers, each of whom chose one class for the study: 2139 students were involved. First the teacher administered whatever objective-type test would ordinarily come next for his class. He marked these tests as usual, giving a numerical score and a letter grade. Next he randomly assigned the papers to one of three groups. "No comment" students received only the numerical score and letter grade. "Free comment" students received, in addition, whatever encouraging comment the teacher felt it desirable to make. "Specified comment" students all received comments designated in advance for each letter grade (A, "Excellent! Keep it up." B, "Good work. Keep at it." C, "Perhaps try to do better?" D, "Let's bring this up." F, "Let's raise this grade."). The students were not aware that they were involved in an experiment.

Comment effects were judged by scores achieved on the very next objective-type test given in class. The "specified comment" group achieved higher scores than the "no comment" group. The "free comment" group achieved the highest scores of all. Neither schools nor ability level had an influence on the treatment effect. Comments were as effective for senior school students as for those in junior high school.

Page concluded that, when teachers take the time and trouble to write encouraging comments on student papers, these apparently have a measurable and potent effect on student achievement.

Grading Can Produce Strong Parental Pressures on the Students. Attendant with the "grade fever" which dominates much classroom work and conversation outside the classroom are the parental pressures which haunt many a student. Parent-teacher conferences serve as an excellent place to learn of the multifarious ways in which parents attempt to make students "knuckle down." For instance, "I told Sam if he does not bring his geometry grade up, he will not be able to get into college" or "Linda has been grounded until her grades improve" or "We told John he could not go on the skiing trip until his mathematics grade improved." We know of one teacher who was even asked to write a note to a

RESEARCH | ANXIETIES IN THE
HIGHLIGHT | MATHEMATICS CLASSROOM

The National Longitudinal Study of Mathematical Abilities (NLSMA; see Begle and Wilson, 1970, pp. 386–402) identified and measured two kinds of anxieties which students may experience in mathematics classrooms. "Facilitating" anxiety is that which facilitates or promotes a student's mathematics achievement performance under such stressful conditions as examinations. Here is a sample item from the NLSMA battery for determining the degree of facilitating anxiety possessed by a student:

I keep my arithmetic grades up mainly by doing well on the big tests rather than on homework and quizzes.
a. always
b. usually
c. sometimes
d. hardly ever
e. never
(Travers, 1971, p. 139)

"Debilitating" anxiety interferes with or harms mathematics achievement in stressful conditions; it was measured by NLSMA items such as this one:

When I have been doing poorly in arithmetic, my fear of a bad grade keeps me from doing my best.
a. never
b. hardly ever
c. sometimes
d. usually
e. always
(Travers, 1971, p. 143)

In a study of relationships between these anxieties and actual mathematics achievement performance, Travers (1971) identified students from the NLSMA sample whose performance on the Stanford Achievement Test in mathematics in sixth grade was "much better" (more than 1.0 standard error) or "much worse" (less than 1.0 standard error) than predicted by four intelligence and mathematics achievement scores in fourth grade. Among these students he found those with "high" (greater than 1.0 standard deviation) and "low" (less than 1.0 standard deviation) scores on the anxiety scales. As the resulting classification of the nearly 400 students in each of the two groups shows, strong associations appeared between the achievement classification and anxiety levels (see Tables 7–1 and 7–2). Table 7–1 indicates that of the 200 overachievers found, twice as many students scored high in facilitating anxiety than did those scoring low (133 as compared to 67).

Table 7–1. Two-way classification of students: under- and overachievement versus low and high facilitating anxiety (adapted from Travers, 1971, p. 142).

Facilitating Anxiety

	Low	High	Total
Underachievers	119	65	184
Overachievers	67	133	200
Total	186	198	384

Chi square = 36.05   df = 1 $p < .001$

Table 7–2. Two-way classification of students: under- and overachievement versus low and high debilitating anxiety (adapted from Travers, 1971, p. 146).

Debilitating Anxiety

	Low	High	Total
Underachievers	80	108	188
Overachievers	174	33	207
Total	254	141	395

Chi square = 72.14   df = 1 $p < .001$

mother assuring her that her son's grade had improved sufficiently so he could go out that Saturday night!

It is also common to find many parents who for reasons of prestige are determined that their children attend a college. Under such circumstances, it is common to find students of average or even below-average ability struggling for a spot on some college admissions list. The pressures placed on some students as a result of the need to get a "good-looking" high school transcript are not in the student's best interest.

We do not mean to imply that all forms of parental pressure are undesirable or unnecessary. Indeed, prodding on the part of parents is sometimes needed and beneficial. We also encourage parents to do what they can to provide a quiet study area for their students as well as to help their children establish a regular time of study. It is also wise for parents to check on the completion of homework assignments and to show a genuine interest in their children's work. It is the extreme, almost ruthless, form of pressure to which we object.

Grading Helps to Promote Cheating. Glasser (1969) has noted that William Bowers, a Columbia University researcher, interviewed 6000 students and 600 deans at 99 colleges and universities during a two-year investigation. His study revealed that at least 55 percent of college students in the country cheated to obtain better grades. It probably is not possible to determine the long-range effects of repeated cheating in relation to an individual's personality. We could only conjecture that a loss of self-esteem coupled with guilt feelings may produce an unhappy person later in life.

SUGGESTIONS FOR IMPROVING GRADING PRACTICES

Eliminating the Grading System. It would appear that one of the easiest ways to eradicate the damaging effects of school marks would be to eliminate the grading system altogether. Glasser, in *Schools Without Failure* (1969), gives some details on how to implement such a system.

Give More Weight to Other Factors. Another suggestion for reducing grade pressure on students in high school involves giving more weight to other factors in considering candidates for college entrance and job placement. Recommendations of high school principals or teachers and personal interviews with potential college candidates could be used. While these might reduce some of the pressure for obtaining good grades, other pressures would certainly be felt not only by the student but by teachers, principals, and counselors as well.

Along the same lines, teachers could (and some already do) make use of anecdotal records to replace or at least supplement single-grade marking systems. It is reasonable to suppose that anecdotal records, if kept well, could be of much help to school counselors and psychologists both in placement and in locating and analyzing trouble areas. In addition, such records could be of much help during parent-teacher or student-teacher conferences. Anecdotal records could take a variety of forms; Figure 7–13 illustrates one possibility.

Another example of an interesting computer-assisted anecdotal-type reporting system is illustrated in Figure 7–14. As you can see from the actual student report cards produced by the Computer Assisted Reporting to Parents (CARP) system, parents are given considerably more information on their student's progress and classroom standing than is usually conveyed on conventional report cards.

Let Students Cooperate in the Evaluation Effort. This recommendation involves almost the antithesis of the competition idea so prevalent in most schools and calls for cooperative efforts on the part of students and teachers.

It seems plausible that many undesirable pressures on students and teachers could be relieved in a classroom where cooperation was the key word rather than competition. We have used a limited form of this suggestion in the area of

Name	Class	Date	Situation	Behavior	Comment
Sue	General Mathematics	1/17	group lesson developing meaning of fractions with graph paper	quick to help neighboring students.	
		1/20	computation game	missed most combinations in which she had to multiply by 7 or 9	redevelop and practice multiplication with 7 and 9

Figure 7–13   A sample anecdotal record.

evaluation. During each grading period in some of the secondary classes we have taught, students have been asked to give themselves a grade which they feel they have earned based upon certain criteria. This grade is then averaged in with their other grades (on quizzes, tests, homework). This sort of cooperative evaluation does offer, especially for the slower student who is working hard but having trouble on tests, a kind of salvation. Student comments relative to the use of this procedure have been overwhelmingly positive. A copy of the self-evaluation form is shown in Figure 7–15.

Each grading period, before filling in the form, some class time is spent reminding students that this is a serious endeavor not to be taken lightly. Students are required, on the back of the form, to write a paragraph justifying the grade which they assign themselves. In cases where a student's self-grade is very different from his or her other marks (say he gives himself an A when he has been earnings Ds on tests and quizzes), a personal conference is scheduled with the student. If he or she can justify this discrepancy, the self-grade stands. Otherwise, the student is asked to reconsider.

Use Multi-Area Evaluation. A final suggestion for improving the marking system involves the use of multi-area evaluation. Under such a plan, the student would be given several marks instead of a single grade in a course. Thus the teacher would rate the student, for example, on achievement relative to his or her "peers" in the class, on attitude, on how enterprising he or she is, and so on. The idea underlying such a system is to provide students, parents, and teachers with greater insight into problem areas which the student might have.

JANUARY 15, 1976
TEACHER D. GIANNANGELO
EAST HIGH SCHOOL

PUPIL   SCOTT, JAKE
GRADE 11

** THE FOLLOWING IS A REPORT OF YOUR **
** CHILD'S ALGEBRA PROGRESS **

THE STUDENT'S ALGEBRAIC ACHIEVEMENT IS BELOW THE AVERAGE LEVEL
WHEN COMPARED WITH OTHER STUDENTS' WORK IN THE CLASS.

THE STUDENT EXPERIENCES DIFFICULTY IN SOLVING COMPUTATIONAL
PROBLEMS.

OCCASIONALLY THE STUDENT IS ABLE TO USE BASIC TRANSFORMATIONS TO
SOLVE SIMPLE EQUATIONS. EXAMPLE: $5Y-4=31$

THE STUDENT IS SELDOM ABLE TO SOLVE SIMPLE WORD PROBLEMS REQUIRING
THE USE OF LINEAR EQUATIONS IN ONE VARIABLE. EXAMPLE: THE SUM OF
THE PERIMETERS OF A SQUARE AND AN EQUILATERAL TRIANGLE IS 20 FEET,
AND THE EDGES OF THE TRIANGLE ARE EACH TWO FEET LONGER THAN THOSE
OF THE SQUARE. HOW LONG ARE THE EDGES OF THE SQUARE?

AT TIMES THE STUDENT IS ABLE TO FIND ORDERED PAIRS OF NUMBERS THAT
ARE SOLUTIONS OF A GIVEN LINEAR EQUATION IN TWO VARIABLES. EXAMPLE:
STATE WHETHER THE ORDERED PAIRS ARE EQUAL: (3,5), (5,3)

PRAISE THE STUDENT FOR WHAT HE (SHE) DID WELL.

ENCOURAGE THE STUDENT TO HAVE CONFIDENCE IN WHAT HE (SHE) IS DOING
IN ALGEBRA.

ENCOURAGE THE STUDENT TO COMPLETE THE ASSIGNMENTS.

Figure 7–14 Two sample CARP outputs.

JANUARY 15, 1976
TEACHER D. GIANNANGELO
EAST HIGH SCHOOL

PUPIL   JONES, TOM
GRADE 11

** THE FOLLOWING IS A REPORT OF YOUR **
** CHILD'S ALGEBRA PROGRESS **

THE STUDENT'S ALGEBRAIC ACHIEVEMENT IS BETTER THAN THE AVERAGE
STUDENTS' WORK IN THE CLASS.

THE STUDENT'S WORK REFLECTS ACCURACY AND CAREFULNESS.

THE STUDENT DEVELOPS AN UNDERSTANDING OF NEW CONCEPTS EASILY.

THE STUDENT QUICKLY LEARNED TO SIMPLIFY SUMS AND DIFFERENCES OF
POSITIVE AND NEGATIVE NUMBERS. EXAMPLE: $-15 + (-38) =$

THE STUDENT EASILY LEARNED TO USE BASIC TRANSFORMATIONS TO SOLVE
SIMPLE EQUATIONS. EXAMPLE: $4(2X - 1) + 2 = 5X - 8$

THE STUDENT RAPIDLY LEARNED TO SOLVE SIMPLE WORD PROBLEMS REQUIRING
THE USE OF LINEAR EQUATIONS IN ONE VARIABLE. EXAMPLE: THE SUM OF
THE PERIMETERS OF A SQUARE AND AN EQUILATERAL TRIANGLE IS 20 FEET,
AND THE EDGES OF THE TRIANGLE ARE EACH TWO FEET LONGER THAN THOSE OF
THE SQUARE. HOW LONG ARE THE EDGES OF THE SQUARE?

PRAISE THE STUDENT FOR WHAT HE (SHE) DID WELL.

SOME PRACTICABLE SUGGESTIONS FOR GRADING AND RECORD KEEPING

Here are a few general suggestions which we hope you will find helpful.

1. Keep your grades and anecdotal records clearly labeled and dated. This is especially important if you have many grades in your grade book. It is helpful for purposes of parent-teacher conferences and student-teacher conferences to know exactly what all of your grades mean.

2. Keep your record book confidential. Many students are sensitive about having their grades displayed to their peers. You must take the responsibility of seeing that this confidentiality is not violated. A student, of course, should be permitted to see his or her own grades at any time without being permitted to examine the marks of others. Along this same line, be careful in returning test or quiz papers to students that the marks given are not exposed for all to see. If the students want to reveal their marks, all right; but you should not reveal them.

SELF-EVALUATION FORM FOR THE   STUDENT'S NAME _____
_____ GRADING PERIOD   COURSE_____

This form should be given your careful consideration. Its primary purpose is to cause you to give some serious thought to the manner in which you have been studying your mathematics during this grading period. As you look over the numbered items, bear in mind that your mental answers to these questions ARE NOT NECESSARILY related to your test and quiz grades. In other words, it is hoped that by thinking about the numbered items listed (and others which you might think of) you will be able to determine ways in which you can attempt to improve your understanding of mathematics, and your performance in this class.

Have I:
1. been attentive in class when the teacher or any of my fellow-students is explaining part of the lesson?
2. taken notes in class and reviewed these notes in an attempt to understand the material better?
3. followed good study procedures when I do my homework? For example, each night do I briefly review the previous day's lesson in an attempt to see some continuity in what I am doing?
4. sought extra help on those topics which I do not understand? Do I ask questions in class, do I see the teacher for assistance, do I work with friends on problems which I cannot get?
5. voluntarily taken part in class discussions, and have I volunteered to put problems on the board when I could?

Figure 7–15 Self-evaluation form.

6. given this particular mathematics class a chance or have I concluded that it is too hard for me, I do not like it, and I will never see it, so what is the use of trying very hard?
7. really been spending as much time as I should in studying my mathematics? That is, Have I been STUDYING 30–40 minutes on mathematics almost every night or have I just been putting in time with my book?
8. been regular in my attendance to class and has it been as good as it could have been?

Consider the above items very carefully and then assign yourself a grade which you feel best summarizes your efforts for this nine weeks. Use the grading scheme below, and be certain to write a paragraph on the back of this form justifying your grade selection.

I am:
A. working to the best of my ability. I am trying hard in most of the above areas, and I sincerely feel there is little else I can do to attempt to improve my mathematical understanding.
B. working hard in most of the above areas. I do feel, however, that there are some of the numbered items that I could explore more fully than I have.
C. making an effort in most of the above areas, but definitely have not done as much as I could have to better my understanding.
D. not doing nearly what I could in most of the above areas and have been more or less just trying to get by.
F. very negligent in most of the above areas and have explored few if any of them in any depth.

3. Make grade changes or corrections immediately. There will undoubtedly be times when you will find it necessary to change a student's score on something. For your own protection as well as the student's, these changes should be made as close to the time they are called to your attention as possible.

4. Have your grading policies in writing. You might even wish to have these policies dittoed so you could hand them out on the first day of class.

5. Send both good and bad reports home to parents. Parents generally are interested in keeping abreast of their child's progress in school. Many schools have special forms which teachers may use to alert parents of unsatisfactory work. Some schools have policies which prevent teachers from giving failing grades at the end of a marking period unless the parents have been notified ahead of time via one of these forms. Equally important, although usually much less used, are forms or a brief personal note from the teacher informing parents of their student's improvement or outstanding work in a course. Such letters can be motivating as well as gratifying for the students. Parents are usually pleased to receive such notices.

6. Keep students informed of their progress. This can be done through student-teacher conferences, by jotting a note on a homework or test paper, or perhaps by making a point to see a student during class time to discuss his or her grades.

7. Keep samples of a student's work on file for use in conferences related to the student. It is frequently helpful when talking with parents or students to be able to produce evidence that might be useful in assisting in the correction of a problem. It is a good idea to keep all exams and quizzes on file for the year. And it is wise to have the student's self-evaluation forms on file.

8. Be careful about changing grades. This admonition does not refer to those instances where you have clearly made a mistake in grading; it does apply to times when students approach you seeking additional points for problems or challenging your assignment of partial credit or your grade cut-off points. We are not advocating that you never change your mind on these issues, but that you exercise caution in succumbing to pressure from students. Grading is a subjective matter; you should have given enough thought to how you are grading so that you have some reasonable standards that will permit you to be open to evidence which supports a change while at the same time not being willy-nilly.

Critical Incidents Revisited

The first activity at the end of this chapter will ask you to suggest some alternative ways of handling Mr. Franks's procedure for dealing with homework—which was the focus of the first critical incident at the beginning of this chapter.

The second critical incident found Ed going to sleep almost every day in class. Ms. Gains, of course, has several options available to her. If she knows as little as we do of the situation, she would probably be wise to have a personal conference with Ed in an effort to ascertain the reasons for his classroom slumber. It might be that Ed has a job that does not permit him to get to bed until quite late. If the situation is something like this, Ms. Gains may feel justified in contacting Ed's parents to discuss the problem and attempt its amelioration.

On the other hand, Ms. Gains might discover in the conference that Ed finds mathematics uninteresting and that sleeping in class is a means of escape. In this case she might try to find out more about Ed and his interests and attempt to develop some approaches related to them which might motivate Ed to do some work. Of more far-reaching consequence, Ms. Gains might be led to reexamine what she is doing in her class to motivate the rest of the students. She may discover that Ed is not the only one who is bored. Through this kind of self-evaluation, Ms. Gains could very well alert herself to the need of making additional efforts to motivate her students. (As a matter of fact, Ed was a student in a geometry class taught by one of the authors. What finally turned Ed on was a unit on computers in which class members had an opportunity to type out their own computer programs. Ed snapped out of his lethargy in an astounding fashion and actually led the class in this particular unit of study.)

How might Mr. Norris handle Ralph's behavior in the third critical incident? One option, of course, is simply to ignore Ralph; under some circumstances this might be the appropriate thing to do. In this instance, however, it would probably have been difficult to ignore him (since we are told he walked right across the front of the class). From what we know it seems reasonable to suppose that Mr. Norris ought to have verbally reprimanded Ralph. A comment such as "Ralph, throw that away later, or come in at lunch time" would probably have been enough. To be sure, such an incident as this is going to disturb the class anyway; at least the students would have learned that such behavior is frowned upon. What is important is that Mr. Norris's comment gave Ralph a choice, the option of throwing the paper away later or of coming in at lunch time to serve a detention. Giving students options is frequently beneficial because it allows students a means of saving face. Mr. Norris might have avoided this incident altogether had he included among his beginning-of-the-year remarks something like "Because it is distracting to do so, no one will be permitted to leave his seat while another person is explaining something to the rest of the class."

Mrs. Rice, in the fourth critical incident, could very well have avoided the encounter with Sheldon if she had had a written policy on grading. If each student in her class had been given such a document on the first day of class, no one would have an excuse for not knowing how homework grades were to be counted.

Activities

1. Look at the first critical incident in this chapter. Suggest several ways that Mr. Franks might change his procedure for going over homework so as to have all of his class involved.
2. Possible ways of handling the second, third, and fourth critical incidents were suggested in this chapter. Suggest one other way of handling each of these incidents.
3. Behavioral objectives were described as having three components: an observable terminal behavior, conditions under which the terminal behavior is to be demonstrated, and a minimal acceptable standard of performance. For each of the five objectives listed on page 207, identify the three components.
4. Go back to Chapters 3 (p. 57) and 4 (p. 85). Try writing behavioral objectives for each of the knowledge and understanding goals listed.
5. Select a high school geometry text and an algebra text. Identify three lessons from each book that you might teach. Design appropriate initiating activities for these six lessons.
6. Problematical situations were described as being quite effective in arousing a student's attention. Use Chapter 5 as well as some of the references in Appendix D to identify some problematical situations

which could serve as initiating activities. Indicate for what lesson each problem you describe will be appropriate.

7. Several suggestions were offered dealing with ways in which you could discuss homework assignments. For each of the five suggestions on page 205, discuss what you see as the pros and cons of the procedure. Then offer five more options of your own.

8. What are some of the pros and cons associated with grading homework? Discuss some of the positive and negative features of the three methods of grading homework suggested on pages 205–206. Propose other ways of evaluating student effort on homework.

9. You will soon discover in your teaching that most of your classes contain students who are very rapid learners and students who are very slow. Describe how you will individualize your assignments so that the better students will not be bored in doing what for them are routine problems and the slower students will have additional practice when needed. The plan you suggest should be practicable.

10. Select, from any high school mathematics textbook, five topics that you might teach. For each of these topics, find reasons to give your students for studying it.

11. Use the Thirty-first Yearbook of the National Council of Teachers of Mathematics, *Historical Topics for the Mathematics Classroom* (1969), to see what historical information you can find on three of the five topics you selected in Exercise 10.

12. Using Appendix D to assist you, identify five game-type activities (activities that involve at least two players in some form of competition) that you could use in a high school mathematics class.

References

Allen, Dwight W.; Ryan, Kevin A.; Bush, Robert N.; and Cooper, James M. *Teaching Skills for Secondary School Teachers*. New York: General Learning Corporation, 1969.

Ausubel, David P. A New Look at Classroom Discipline. *Phi Delta Kappan* 43: 25–30; October 1961.

Begle, E. G. and Wilson, J. W. Evaluation of Mathematics Programs. Chapter 10 in *Mathematics Education*. Sixty-ninth Yearbook of the National Society for the Study of Education, Part I. Chicago: University of Chicago Press, 1970.

Besvinick, Sidney L. An Effective Daily Lesson Plan. *Clearing House* 34: 431–433; March 1960.

Boehm, Ann E. and White, Mary Alice. Pupils' Perception of School Marks. *Elementary School Journal* 67: 237–240; February 1967.

Briggs, Frances M. Grades: Tool or Tyrant? A Commentary on High School Grades. *High School Journal* 47: 280–284; April 1964.

Brown, John Kenneth, Jr. Textbook Use by Teachers and Students of Geometry and

Second-Year Algebra. (University of Illinois at Urbana-Champaign, 1973.) *Dissertation Abstracts International* 34A: 5795–5796; March 1974.

Chansky, Norman M. Resolving the Grading Problem. *Education Forum* 37: 189–194; January 1973.

Clarizio, Harvey F. *Toward Positive Classroom Discipline*. New York: Wiley, 1971.

Crosswhite, F. Joe. Implications for Teacher Planning. In *The Teaching of Secondary School Mathematics*. Thirty-third Yearbook of the National Council of Teachers of Mathematics. Washington: The Council, 1970.

DeRoche, Edward F. What Do Marks Mean to Your Pupils? *Instructor* 73: 79; May 1964.

Dobson, James. *Dare to Discipline*. Wheaton, Illinois: Tyndale House Publishers, 1970.

Elder, Florence L. Using "Take-Home" Tests. *Mathematics Teacher* 50: 526–528; November 1957.

Epstein, Marion G. Testing in Mathematics: Why? What? How? *Arithmetic Teacher* 15: 311–319; April 1968.

Farwell, Gaylord H.; Nelson, Robert H.; and Thompson, Michael L. Pressures Behind the Grade. *Clearing House* 38: 462–466; April 1964.

Fast, Julius. *Body Language*. New York: Pocket Books, 1970.

Freitag, H. T. and Freitag, A. H. Using the History of Mathematics in Teaching on the Secondary School Level. *Mathematics Teacher* 50: 220–224; 1957.

Gagné, R. M. Educational Objectives and Human Performance. In *Learning and the Educational Process* (edited by J. D. Krumboltz). Chicago: Rand McNally, 1965, pp. 1–24.

Glasser, William. *Schools Without Failure*. New York: Harper and Row, 1969.

Gronlund, N. E. *Stating Behavioral Objectives for Classroom Instruction*. New York: Macmillan, 1970.

Hall, Edward T. *The Silent Language*. New York: Fawcett, 1959.

Harnueling, H., Jr. Using Historical Stories to Stimulate Interest in Mathematics. *Mathematics Teacher* 57: 258–259; 1964.

Holt, John. *How Children Fail*. New York: Dell, 1964.

Johnson, David R. The Element of Surprise: An Effective Classroom Technique. *Mathematics Teacher* 66: 13–16; January 1973.

Kirschenbaum, Howard; Simon, Sidney B.; and Napier, Rodney W. *Wad-Ja-Get? The Grading Game in American Education*. New York: Hart Publishing Company, 1971.

Laing, Robert Andrew. Relative Effects of Massed and Distributed Scheduling of Topics on Homework Assignments of Eighth Grade Mathematics Students. (The Ohio State University, 1970.) *Dissertation Abstracts International* 31A: 4625; March 1971.

Laing, Robert A. and Peterson, John C. Assignments: Yesterday, Today, and Tomorrow—Today. *Mathematics Teacher* 66: 508–518; October 1973.

Lorber, Michael. *Precise Instructional Objectives*. Urbana, Illinois: University of Illinois. Mimeo.

Mager, Robert F. *Developing Attitude Toward Learning*. Belmont, California: Fearon, 1968.

Meacham, Merle and Wiesen, Allen F. *Changing Classroom Behavior: A Manual for Precision Teaching*. Scranton, Pennsylvania: International Textbook Company, 1969.

Page, Ellis B. Teacher Comments and Student Performance: A Seventy-four Classroom Experiment in School Motivation. *Journal of Educational Psychology* 49: 173–181; August 1958.

Peterson, John Charles. Effect of Exploratory Homework Exercises Upon Achievement in Eighth Grade Mathematics. (The Ohio State University, 1969.) *Dissertation Abstracts International* 30A: 4339; April 1970.

Rosenberg, Herman. The Art of Generating Interest. In *The Teaching of Secondary School Mathematics*. Thirty-third Yearbook of the National Council of Teachers of Mathematics. Washington: The Council, 1970.

Runion, Garth E. The Development of Selected Initiating Activities in the Teaching of Mathematics. (University of Illinois at Urbana-Champaign, 1972.) *Dissertation Abstracts International* 34A: 653; August 1973.

Shain, Robert L. *Discipline: How to Establish and Maintain It*. Englewood Cliffs, New Jersey: Prentice-Hall, 1961.

Shain, Robert L. and Polner, Murray. *Using Effective Discipline for Better Class Control*. Englewood Cliffs, New Jersey: Prentice-Hall, 1964.

Sobel, Max A. and Maletsky, Evan M. *Teaching Mathematics: A Sourcebook of Aids, Activities, and Strategies*. Englewood Cliffs, New Jersey: Prentice-Hall, 1975.

Steinen, R. F. An Example from Geometry. In *The Teaching of Secondary School Mathematics*. Thirty-third Yearbook of the National Council of Teachers of Mathematics. Washington: The Council, 1970.

Travers, Kenneth J. *Non-intellective Correlates of Under- and Over-Achievement in Grades 4 and 6*. NLSMA Reports, No. 19 (edited by James W. Wilson and Edward G. Begle). Stanford, California: School Mathematics Study Group, 1971.

Urwiller, Stanley LaVerne. A Comparative Study of Achievement, Retention, and Attitude Toward Mathematics Between Students Using Spiral Homework Assignments and Students Using Traditional Homework Assignments in Second Year Algebra. (University of Nebraska, 1971.) *Dissertation Abstracts International* 32A: 845; August 1971.

Walbesser, Henry H. *Constructing Behavioral Objectives*. College Park: Bureau of Educational Research and Field Services, College of Education, University of Maryland, 1970.

Evaluation in Mathematics. Twenty-sixth Yearbook of the National Council of Teachers of Mathematics. Washington: The Council, 1961.

Historical Topics for the Mathematics Classroom. Thirty-first Yearbook of the National Council of Teachers of Mathematics. Washington: The Council, 1969.

Individualizing

AFTER STUDYING CHAPTER 8, YOU WILL BE ABLE TO:

★ specify various degrees of individualization which can be provided in the mathematics classroom.

★ give examples of activities designed for students having a wide range of interests and abilities.

★ discuss relative advantages and disadvantages of various individualizing techniques.

★ list several ways in which individualizing can be implemented in the classroom.

CRITICAL INCIDENTS

1. Mr. Greenwood, who has been teaching mathematics for three years, has become very discouraged. His classes are always "spread out" in achievement. He has a terrible time challenging the faster students, and the slower students seem to demand all his attention.

2. Miss Ferguson is dissatisfied with the way her classes are going. It seems as if she spends most of her time at the front of the classroom writing and talking. Her students have little input into her lessons; they just sit there. Isn't there more to education than this?

233

3. Bob Brown went to a mathematics teachers' conference and heard an exciting talk on individualizing instruction. The speaker had many ideas about how to teach in more interesting ways. But the following week, when he finally got enough courage to try a totally new approach to teaching, Bob's classroom became chaotic. Everybody milled around and nobody seemed to pay any attention to what he had to say. This week Bob has gone back to his old way of teaching.

INDIVIDUAL DIFFERENCES

Stop for a moment and look at any group of children, or adults, at a playground or in a park. Some are slim and some are stocky. Some are lethargic, some hyperactive. Some are attractive, others plain. Differences in physical characteristics are readily observed and are familiar to all of us. Differences in mental characteristics—intelligence, aptitude, achievement, attitudes—are less readily observable, but they are nonetheless a fact of life to be dealt with in teaching.

Project TALENT, a nation-wide survey of mathematics achievement, provides data to illustrate the existence of wide differences between high school students. On the introductory mathematics test, for example, scores for the boys increased 0.7 of a standard deviation from ninth to twelfth grades. That is, according to this test, about one-third of the ninth graders knew as much mathematics as one-half of the twelfth-grade boys. For the girls, the differences between ninth and twelfth grades were even smaller, with the researchers stating: "We may tentatively conclude that a large proportion of girls forget more mathematics than they learn from the end of the ninth grade to the end of the twelfth grade" (Project TALENT *Bulletin*, 1965, p. 8). Thomas and Thomas (1965), in their study of individual differences in the classroom, point out that these differences increase from elementary to high school. In what they called a "typical fifth grade class," they found achievement test scores, based on national norms, ranging from third to ninth grade (p. 5). In the junior high school which these students would be attending, mathematics scores for ninth graders ranged from third grade to the junior college level, with "clusters of students at all levels of the ladder." Those authors conclude, "It would hardly seem possible that . . . one mathematics class designed for the 'average ninth grader' could suit the needs of this typical collection of students" (pp. 30–31).

Students vary a great deal in their interests as well. Some of them will be highly interested in your favorite subject, mathematics. Others will not. A survey of 15,000 high school juniors showed that, out of 12 school subjects, mathematics ranked sixth in interest among boys and tenth in interest among girls. Industrial arts and physical science were ranked as having lowest interest for girls and highest interest for boys (*ETS Developments*, 1970, pp. 1–2).

How will you make provisions for such wide differences in your own classes? Like Mr. Greenwood in our first critical incident, you are likely to find this task one of your greatest challenges. We regard individualizing more as a style of approaching your day-to-day teaching than as a distinct program of instruction which is "done" on certain days with certain classes. This chapter presents a variety of ways in which you can begin individualizing activities so that your teaching will reflect the needs, abilities, and interests of individual students, even though they have been assigned to you as a class (see Figure 8–1).

MASTERY LEARNING

One important component of the current activity in individualization is that of mastery learning, already referred to in Chapter 6, p. 178. Bloom and his collaborators (Bloom, Hastings, and Madaus, 1971) point out that, in the typical classroom, all students receive the same kind of instruction and the same amount of class time to complete the assignments. The result is what we have learned to expect. Those students who have done well in the past will continue to do well and the poorer students will continue to do poorly. Statistically speaking, we say there will be a high correlation between student aptitude scores and achievement before and after participation in a course. Furthermore, the achievement scores will be distributed normally (see Figure 8–2). That is, the frequencies of scores, when graphed, will approximate the familiar bell-shaped curve, with most scores close to the mean and few very high or very low scores.

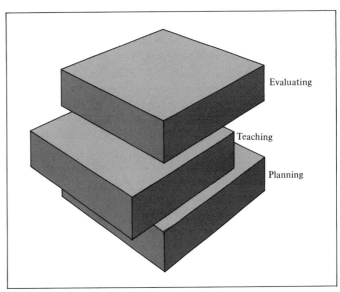

Figure 8–1   The teaching process in the model of mathematics teaching.

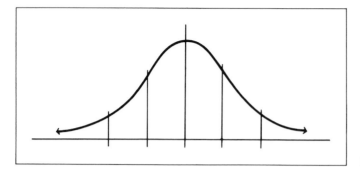

Figure 8–2   A normal distribution.

However, Bloom's argument continues, this sort of outcome is not desirable—or inevitable. Instead of receiving uniform instruction, students might receive individualized instruction in which teaching strategies are chosen on the basis of the needs and interests of each student. Instead of a fixed period of time for completing assigned tasks, children could have as much time as required for mastery. Feedback and corrective procedures can be used to determine what a student has learned and to help that student learn those goals not mastered. As a result all students will achieve at a high level, even the "low achievers." Hence achievement scores will not be normally distributed, but will be skewed to the left (see Figure 8–3). Such a graph indicates that most of the children did quite well on the particular measure being reported. Some might even argue that such a distribution of scores suggests more effective teaching than does a "normal" (bell-shaped) curve.

Mastery learning has been the subject of considerable research in the past few years (see, for example, the reviews by Block, 1971 and 1974). One study (Collins, 1971) involved teaching eighth-grade mathematics classes using various combinations of mastery-learning strategies. Collins found that when students were pro-

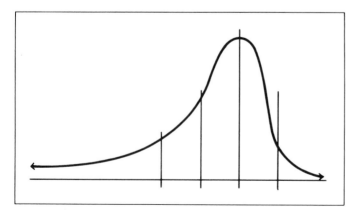

Figure 8–3   A negatively skewed distribution.

vided lists of instructional objectives, test problems to review the objectives, and extra assistance in learning those topics not mastered, 80 percent of the students attained the mastery criterion of A or B grades while D and F grades were "practically eliminated" (Block, 1971, pp. 111–112).

PROVIDING FOR INDIVIDUAL DIFFERENCES

How to provide for individual differences among students has been a concern of educators for many years. In 1894, Search published a report on his individualized high school in Pueblo, Colorado, which permitted the student to proceed at his own pace. Some 70 years later, two administrators of a high school in New York State wrote:

> We generally are willing to acknowledge that in every group of youngsters there can be found without exception wide differences in physical and intellectual potential, achievement, desire, ability, and motivation to learn. Many of the practices found within secondary schools, however, announce much more vividly that we are not willing, or that we are unable to put into practice what we so fervently preach. Many of the more common practices within secondary schools deny emphatically that youngsters are in fact unique, individualistic entities; that they are motivated educationally in an inordinate number of ways. Such educational practices as singular curriculum plans, standard grading systems, strict, age-graded promotional criteria, and standardization of learning assignments completely dispel any belief that students have different capacities, abilities, or motivations for learning. (Wiley and Bishop, 1968, p. 136)

Those authors go on to describe a program of independent study at their high school which has as its main objective, "to bring to the student the realization that he can learn something *almost anytime* and nearly *anywhere* with or without the school" (p. 137).

The widespread interest and activity in individualizing instruction has led one researcher (Gibbons, 1971) to conclude that we have so many programs, based on such a wide variety of theories and philosophies, that the very term *individualized program* is not a useful category of instructional methods (p. 2). A further difficulty is that it is very difficult actually to bring about significant changes in teaching, despite the best intentions of teachers (see Research Highlight, "Who Are the Good Students?").

In spite of the many problems surrounding the individualizing of instruction, we present in this chapter some basic information which we believe will be of help as you work at providing an optimal educational experience for each of your students. Figure 8–4 presents four major classifications of management schemes for working with students: large group, single class, small group, and individual

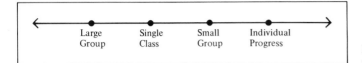

Figure 8–4 Four ways of organizing students for instruction.

progress. The figure is intended to suggest roughly that, as one proceeds from left to right, the amount of individualizing which takes place increases. We recognize, however, that there are exceptions. One could envision large group instruction which, under the direction of a skilled teacher, permits considerable individualization, while a small group poorly handled allows virtually no attention to the needs and interests of individual students.

LARGE GROUPS

Large groups of students assembled for instructional purposes have become rather common in high schools with the advent of a school scheduling practice called *modular scheduling*, initially developed by Allen, Bush, and their associates (see Bush and Allen, 1964). Operationally, we define *large group* as one consisting of numbers of students which are multiples of the typical single class size of 25 or so students. Groups of up to 150 or 200 students are not uncommon in a school operating on a modular schedule.

Two fundamental assumptions of modular scheduling are:

Assumption 1. Each subject, when properly taught, will include four basic types of instruction: large group instruction, small group instruction, independent instruction, and laboratory instruction.
Assumption 2. Class size, length of class meeting, and number and spacing of classes ought to vary according to the nature and aim of the subject, the type of instruction, the level of ability and interest of the pupils, and the aim and purpose of teaching.

Modular scheduling is designed to make possible the use of different types of instruction and variations in the length and frequency of class meetings by providing class schedules for both teachers and students which are tailored to their goals and interests. A teacher, for example, may wish to meet an algebra class for one and a half hours once a week for presentation of new material and a half hour twice a week for discussion and review.

Modular scheduling makes use of considerable "unassigned" time for students, who must learn to use their time constructively. The availability of abundant resources such as books and videotapes, and of training in their use, appears to be an essential component of successful modular-scheduling programs.

Although large-group lectures typically do not feature much student interaction, it is possible to achieve some involvement. Consider the following example

---

RESEARCH | WHO ARE THE GOOD
HIGHLIGHT | STUDENTS?

Thelen (1963) found that teachers seldom agree on who are good and who are bad students (for example, those who the teachers think get a great deal or very little out of class). In one study, three teachers worked with the same 50 students for three hours a day. One of the teachers named seven students and the other two teachers named nine students as getting a great deal out of class. But the three teachers agreed on only three students (p. 84).

In the same report, Thelen cites the work of Ekstrom, who studied all published research on ability grouping (grouping together students of comparable academic ability) over the preceding 50 years. Ekstrom found little conclusiveness in the research, stating that student achievement was higher in the ability-grouped classes only about one-third of the time. The interpretation given for the inconclusiveness of the research was that only a few teachers knew how to adapt their teaching to particular groups of students (Thelen, 1963, p. 84). Most teachers, apparently, tend to teach in the same way regardless of whether their groups are high- or low-ability students. Furthermore, Thelen found that although teachers generally perceive their "good" classes as being brighter than their difficult ones, a comparison of student I.Q. scores may reveal no differences (p. 84).

---

concerning the teaching of a trigonometry class of 140 students. The topic under study was the solution of right triangles. To break the monotony of the lecture and to arouse the students' interests (it was an eight o'clock class!), the teacher told the students something about the large, manlike, apelike creature known as Big-Foot which allegedly roams the forests of the northwestern United States and southwestern Canada. The teacher related background information concerning the size of the creature and projected a photograph of it which had appeared in a national magazine a few years earlier.

After capturing the students' attention, the teacher asked each student to imagine himself or herself in the forest sitting on a log. Suddenly the student looked up and there, a short distance away, was a strange creature. At this point, the overhead transparency shown in Figure 8–5 was flashed on the screen. The student (being conditioned to think mathematically!) immediately made an estimate of the creature's distance from the student and then quickly estimated the angle of elevation from ground level to the top of the animal's head.

In order to make this example more concrete, several students were asked to

Figure 8–5.

take a ball of twine and form a large right triangle, extending the length of the lecture hall, as illustrated in Figure 8–6. Students then measured the length of twine from student A to student B, and used a chalkboard protractor to find the angle of elevation α. The teacher then lets the students determine which of the trigonometric functions would be the most expedient to use to determine the height of the strange creature.

Should you have the responsibility of teaching a large group, you will find Berger's chapter in the Thirty-fourth Yearbook of the National Council of Teachers of Mathematics (1973) to be helpful. Berger notes that for maximum effectiveness, large-group instruction should be followed by small groups which encourage discussion, clarification of issues, and independent study of new material (p. 18).

SINGLE CLASSES

The single class of about 25 students and one teacher characterizes what we usually think of as typical schooling. This approach of putting teacher and stu-

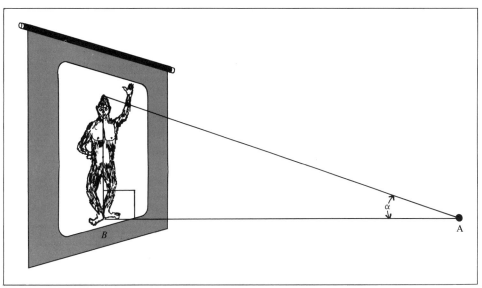

Figure 8–6.

dents together has changed little over the years and is likely to be the most prevalent pattern in the foreseeable future. As you begin your teaching career, the
single-class approach is likely to be the most viable method of classroom management. It is simply much easier to work with one class of 25 or so students than
with several groups, even though they may be smaller. Indeed, for some teachers
in some teaching situations, the single-class approach should be used most of the
time.

We encourage you, however, to consider opportunities within the context of
the single class for dealing with students as individuals. The following approaches
provide examples of what you might do.

Individualized homework assignments are one way in which to respond to
students as individuals. As Bradley (1968) has observed, there is some question
about the educational value of giving "blanket-type assignments" to all members
of a given class. Too often each student is assigned the same page of exercises
regardless of whether those exercises are so difficult for that student as to be
totally frustrating or so easy as to be hopelessly boring. Bradley experimented
with "individualized homework assignments," which he defined as

> assignments related to the work of the classroom but based primarily on the needs
> of the individual child with full recognition that the difficulty, amount and
> nature of the work may at times be similar to and at other times quite different
> from that of other members of the class. (p. 353)

The experimenter found that when he took pains to give homework assignments individually, these assignments took on remedial and enrichment characteristics as well as reinforcing the work of the day. Furthermore, he found that the students receiving the individualized assignments showed higher mean gains on mastery tests than did students receiving the conventional, "blanket-type" assignments.

Independent study and self-selection can be used in the mathematics class with selected students. For high-achieving students, this involves working at their own pace, participating in class only as they choose, and taking chapter tests as they feel ready for them. So long as the student maintains good progress and outstanding grades for the course, he or she is permitted to continue in the program. Brannon (1962) gives details on such a program for advanced high school students. Self-selection may be used for low achievers (see Research Highlight, "If They Choose. . . . " This study indicates the need for the teacher to motivate and encourage students to make wise choices).

Programs of independent study in high schools around the country have been surveyed by Alexander and Hines (1967). They describe a geometry program at one high school in which students contract with their teacher to study certain topics, deciding how they will learn the materials and the degree to which they will learn on their own. A variety of textbooks and other supplemental materials is available. End of unit tests are taken on a honor basis and turned in to the teacher for grading (p. 29). Alexander and Hines found that independent programs could be successful with both slow and fast students (p. 76).

Tutorial programs are a variation of independent study programs which, according to Polos (1966), have been used with success in high schools in California. Under the tutorial plan, students do programs of study in accordance with their academic strengths and interests under the direction of the teacher. Meetings are scheduled to discuss the readings and to exchange ideas. Emphasis is on oral rather than written work. The final examination is oral and evaluation is based on the student's "ability to correlate areas of knowledge and to show an understanding of the relationships of interdisciplinary knowledge" (p. 405).

SMALL GROUPS

There are times when it is beneficial to organize a class into five or six groups consisting of a half dozen students each. This practice is standard in elementary school, but its potential at the secondary level does not appear to be widely recognized.

There are many benefits of the small-group approach to teaching. Perhaps the most noteworthy is that teachers report being able to spend much more time with more students than in the more formal "teacher in front of the class" mode. Students tend to respond enthusiastically to the less formal structure due to greater accessibility to the teacher and to increased social contacts with each other.

RESEARCH HIGHLIGHT | IF THEY CHOOSE . . .

Six pre-algebra classes for ninth-graders who were low achievers in mathematics participated in a study by Baker (1971). Students in the first treatment group in each class were permitted to make weekly choices of activities and indicated these activity choices on cards. Activity choices included teacher-made assignments, curriculum activity sheets, a computational skills development kit, programmed textbooks, and independent study. The second treatment group in each class chose problems from a page assigned by the teacher and indicated these choices on cards. The third treatment group served as a control group and were allowed no choice of activities or assignments. Pretest and posttest data were obtained on a testing instrument devised by the investigator to yield measures of confidence, interest, and achievement.

No significant differences in achievement, confidence, or interest were found between the three groups. Thus low achievers did as well when they were allowed to make choices as when they were directed by the teacher. Significant interaction effects were noted, however, between treatments and classes, suggesting that the efficacy of providing activity choices to low achievers depends on the teachers and/or classes involved. The teacher-made assignments were the activities most often chosen by the experimental group students, indicating a strong dependence of the low achiever on the guidance and direction of the teacher.

What choice would *your* classes make?

Small Group Instruction for Low Achievers A general mathematics course has used small group instruction with success for several years at Arlington High School, Arlington Heights, Illinois. The teacher, Phyllis Ferrel, has developed a program for a class of about 15 students who have a long history of difficulty with mathematics. The students are organized into four or five groups, with each group at a work table (see Figure 8-7). Assignments are given to each group on a rotating basis, as indicated in the organizational chart in Figure 8-8. Within the week, each group will have had several opportunities to work on each of the activities involving a catalogue unit, calculators, and the abacus. The teacher concentrates her efforts on the groups working on the abacus, since it is a new activity. Each student is expected to reach a specified level of mastery on both the calculator and the abacus. Since the class period of 45 minutes has proved to be too long for the students to work productively, the period is broken into two parts and assignments are rotated at half time. The management plan for this

Figure 8–7 Small group instruction in mathematics at Arlington High School, Arlington Heights, Illinois.

class is rather complex, so you are encouraged to work through it in detail, observing how each of the small groups is to spend its time.

The activities for one week involve the following:

> *Catalogue Unit:* Students are provided with mail order catalogues and a hypothetical budget, say $40.00. They are to select items they wish, complete the order form, calculate costs (including tax) and keep within

Goals: To review multiplication skills and to use them in a practical situation (mail orders).

Week of November 27

Groups remain the same all week.

Mrs. Ferrel will work with each group on the abacus (place value).

Activity	Catalogue	Catalogue	Calculators	Abacus
Monday	Group I	Group II	Group III	Group IV
1st half	Rich Steve Marge Betty	Ron Sally Sue Sharon	John Pete Kris Karyn	Mary Sue Ellen Rob
2nd half	Groups I and II use calculators, Groups III and IV work on catalogue unit.			
Tuesday 1st half	Group II	Group III	Group IV	Group I
2nd half	Groups II and III use calculators, Groups IV and I work on catalogue unit.			
Wednesday 1st half	Group III	Group IV	Group I	Group II
2nd half	Groups III and IV use calculators, Groups I and II work on catalogue unit.			
Tuesday 1st half	Group IV	Group I	Group II	Group III
2nd half	Groups I and IV use calculators, Groups II and III work on catalogue unit.			
Friday				
1st half	All groups complete catalogue unit.			
2nd half	When finished catalogue unit, all groups do cross number puzzles.			

NOTE: When working on catalogue unit, use the mail order forms. Fill them out, then find totals when you go to the calculators. When you complete your order forms, using the calculators, do the exercises in the calculator workbook, pages 15 to 20.

Figure 8–8 Organizational chart for working with small groups.

their budget. Apart from the arithmetical skills which this activity requires, it provides the teacher a great deal of information about the student in the selection of items ordered!

Calculators: They are used in calculating the costs on the mail order forms and in other activities requiring computation. They are also used to check manual computation.

Abacus: Place value and numeration are reviewed.

Teaching Trigonometry Using Small Groups   In south suburban Chicago, Homewood Flossmoor High School has used small-group instruction in trigonometry classes. The rationale for small-group instruction was that, by working in groups, students can get help when and where they need it. According to Carlton Bodine of the mathematics department, "They learn how to ask for help and where help can be found. This is a real-life situation. The students not only learn how to work with each other, they socialize."

The role of the teacher in a small-group setting is quite different from the traditional one. Here the teacher consults with individuals, identifies learning difficulties, makes and modifies assignments, and in general facilitates the learning climate. Such an approach demands a great deal of preparation. Topics for the day must be identified, examples selected, materials prepared, and patterns of grouping decided upon. From some points of view, this approach to teaching is very structured.

Here is a typical pattern for managing the class. As students enter the classroom, they obtain the answer key for the previous day's assignment and check their work in the same groups they were in the day before. The answer key differs from the typical answer book provided by publishers in that exercises are worked out in considerable detail, suggesting alternate directions in which the solution may have gone. The teacher is available to answer questions concerning the assignment.

After the previous day's work has been dealt with, the teacher typically makes a presentation to the entire class. The teacher outlines the goals to be learned, discusses each, then assigns exercises to be done in small groups. The group to which a student goes usually is determined by the teacher; hence it is called one of "teacher's groups." There may be particular reasons why certain students should be grouped together—to discuss individual projects, to assist one another with particular difficulties, to help an individual who has missed work because of absence, and so on. In the meantime, the teacher is kept busy moving from group to group, either responding to questions from individual students or dealing with a matter of concern to an entire small group.

At times the students will form their own groups. These groups consist of students who prefer to work together, and usually they are productive as well. Toward the close of the lesson, the teacher typically brings the class together again to consider common difficulties encountered, to review main points of the lesson, and to give homework assignments.

In another example of small-group instruction, Davidson (1974) describes a technique he has used with success in dividing his elementary algebra classes into four or five groups with four to six students in a group. The students in each group were mixed in achievement and chosen on the teacher's assessment of their ability to work together. Davidson's primary rationale for using groups was that students would receive more help because the teacher is better able to assist five groups than 30 individuals, and often students within a group can help each other.

INDIVIDUAL PROGRESS

As Figure 8–4 indicates, at the other end of the scale from teaching a single class of students is the teaching of "classes" consisting of one student each. Under such a plan, each child is allowed to progress at his or her own rate. One writer describes the technique:

> Thus the teacher may be seen working with individuals. He counsels a child regarding the scheduling of time. He uses a firm hand or a gentle touch depending upon each child's needs. The talents of the teacher may be better utilized with greater efficiency in a one-to-one relationship than in a one-to-ten or one-to-30 relationship. One minute spent with a child at his moment of need may be more worthwhile than 15 minutes spent with the entire class. (Duker, 1972, p. 25)

A variety of programs for individual progress has appeared in recent years. Gibbons (1971) provides a comprehensive summary of these programs.

Small Group–Individual Study Program in High School Geometry This individualized geometry program was devised by Gretchen Potter, a mathematics teacher at Central High School, Champaign, Illinois. The program was designed to allow students to study topics in geometry of most interest to them. A statement about the course, prepared by the teacher for her students, outlined these basic assumptions for the program:

> 1. Learning is an individual process. Each student works and learns at a different rate. In this type of program, each student can spend as much time on a topic as he feels is necessary.
> 2. Each student must take the responsibility for his own learning. It is important that students learn to determine their own interests and limitations. This is a chance for you to have a say in what you study and how long you study it. I will make every effort to provide you with all the help you need. Sometimes I may suggest homework assignments and outside work to help you.
> 3. It is the teacher's role to create a classroom atmosphere where students are free to learn and have the desire to learn. This means I am to help every student as he needs help. No one will be deserted and left to stumble through assignments by himself.

4. Insight and understanding depends upon the freedom to discover concepts and relationships. Just because the book solves a problem one way doesn't mean it is the only way the problem can be solved. You should feel free to make up your own ways to solve problems.

Using these basic assumptions, it should be possible for you to choose who you want to study with and what topics you want to study.

The class was organized in small groups comprised of students who wanted to study the same material. Some other students chose to work alone. Each group decided what to study and discussed its plans with the teacher. In some cases the teacher made assignments to a group and gave introductory instruction to help students begin their study.

The students were informed about the record-keeping and testing procedures as follows:

Each student will have an individual assignment sheet. On this sheet you will keep track of all work completed and the date on which it was handed in to be checked and discussed. This way I can keep track of your progress, too.

Since this is your own study program, you may take a test over a certain chapter or section whenever you feel you are ready. If you look at the test and decide that you need more time to study before taking it, I will postpone the test and let you try again when you are prepared. In certain cases, I may suggest that you take a test when it seems that you know the material.*

Topics for the course were outlined under headings such as "basic terms," "constructions," and "loci," and page references in the textbook were provided. Possible sequences of study were proposed in a chart such as that in Figure 8–9 (starred topics are those required to be completed during the course). For each topic, a detailed assignment sheet and study guide was prepared.

Individual Progress in Junior High School Mathematics   This program, devised by mathematics teacher Jana Rotz, has been in operation at Brookens Junior High School, Urbana, Illinois, for several years. The content of the curriculum is somewhat conventional: basic skills in the fundamental operations on whole numbers, common fractions, and decimal fractions; place value and systems of numeration; coordinate graphing; ratio and percentage; and so forth.

No single textbook is used for the class. Instead, five or ten copies of each of several textbooks are available for student use. Assignments are provided by worksheets, games, or selections from the textbooks, which may be checked out overnight. If the available assignments are not suitable for a particular student, or if additional study is deemed necessary by either student or teacher, then exercises are given which are suited to that student. Frequent use is made of

* The authors wish to thank Gretchen Potter for providing these examples.

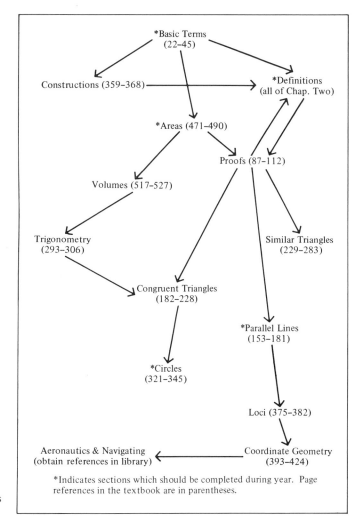

Figure 8–9 Proposed study sequences for a geometry course.

supplementary materials such as filmstrips, puzzles, games, pamphlets, and books.

Goals are determined for the student by the teacher in consultation with the student, his or her parents, and previous teachers. Also considered are present teacher assessments based upon classroom observation and discussion. A checklist indicating level of attainment of these goals and a comment sheet for more general assessments of learning are kept by the teacher and used in determining future directions for the student.

The checklist and comment sheets are useful in evaluating student progress and in reporting to parents. Each student is responsible for keeping all of his or her work in a folder so it can be periodically reviewed by the student and teacher. Each student is provided with a weekly calendar on which the student keeps a record of daily activities and thus informs the teacher of his or her progress. The rates at which the students work vary. Most students set their own pace; some, however, work under teacher-imposed deadlines or according to study contracts between student and teacher. Throughout the program, students are encouraged to become aware of their own strengths and weaknesses and to work with the teacher in reaching the desired instructional goals.

Study contracts have been used with success in some schools. The work for each semester is divided into topics, and agreements between individual students and the teacher are arrived at concerning the amount of work to be covered. Agreements are also made concerning home assignments to be completed and tests to be written. Grades for the course are assigned on the basis of the amount of work in the contract actually completed. Often a mastery-learning approach is taken, so that students are required to reach a predetermined level of achievement before moving on to the next topic. Records for each student consist of such information as contracts completed, degree of mastery of each topic, home assignments completed, and performance on standardized achievement tests.

THE MATHEMATICS LABORATORY

The mathematics laboratory in a high school is more likely to represent the approach to instruction at that school than to be simply a particular room in the building, a set of equipment, or a group of students. Every classroom can become a mathematics laboratory, if by this we mean a place where students are actively involved in mathematics—practicing skills, applying understandings, or solving

"Jeff, just because the school is on an individualized program doesn't mean your answers should be different from everybody else's.

© 1977 by Ford Button.

problems. For some activities there may be a need for special equipment such as calculators, and there may be justification for holding the laboratory in a special room. But much of the equipment needed for mathematical investigations is rather simple and often can be constructed by students or teachers. By and large, we would encourage you to establish a mathematical laboratory within your own classroom rather than in a separate location in the school. Many of the more effective activities will arise naturally out of ongoing instruction. And what a pity if you had to postpone the pursuit of an intriguing hypothesis because "We don't have mathematics lab until Thursday!" The laboratory in your own classroom is always at your and your students' disposal.

WHY USE LABORATORY ACTIVITIES?

Your goals for a mathematics laboratory will depend upon your view of what a laboratory is and the facilities you have available. The following are some purposes of high school laboratory activities.

Maximize student participation. In a good laboratory activity student participation is required; each student should produce his or her own set of results.

Provide appropriate level of difficulty. Activities should be such that the student is challenged, yet can be successful. The laboratory situation makes it feasible to provide for varied completion times required by students.

Offer novel approaches. Through laboratory activities, familiar topics can be approached in new ways. Practice in working with fractions can result from a field trip to a local supermarket to determine best buys; factoring can be reviewed in an introductory computer-programming activity; geometric figures can be studied through a kite-building project.

Enrich the curriculum. As students work independently on topics which interest them, they are likely to be led in new directions. A study of mathematics in relation to postage stamps leads to an examination of mathematics in other cultures; a film-making project involves research in the history of mathematics; and a computer mathematics unit leads to cryptography—the study of secret codes and codebreaking.

Improve attitudes toward mathematics. Many positive feelings about mathematics, and about learning in general, can result from interesting, rewarding laboratory activities.

KINDS OF LABORATORY ACTIVITIES

The following are sample laboratory activities for a variety of grade levels of the secondary school. The descriptions given here are not intended as lesson plans; they are merely suggestions which you will adapt to suit your students, your style of teaching, and your instructional goals. We expect that these activities could be conducted in a variety of ways: Some will be appropriate for large-group or single-class formats, others will lend themselves to small-group activities, and still others will be best handled on an individual-study basis.

We commend to your attention the use of the *learning station* as one means of incorporating laboratory activities into your classes. A learning station is some portion of the classroom where students can go and do laboratory activities. There they will find activity cards (see Figure 8–10), materials (paper, pencils, rulers, etc.), needed equipment for the activities, and—hopefully—work space (tables or desks).

The daily newspaper can provide many opportunities for mathematical ex-

Pretend that you have $750 with which to take a trip anywhere you want.
Using the travel section of your newspaper, plan your vacation spot, transportation, accommodations, etc.
Draw up a budget, and remember to include <u>all</u> trip expenses (But don't spend more than $750).

(a)

Turn to the "Houses for Sale" section of the classified ads.
Figure out the average cost per home for four different cities (If there are more than ten homes for sale in a city, only use the first ten listings).
Draw a bar graph which charts the average cost per home in the four cities.

(b)

CLASS ACTIVITY
Have everyone in the class select a stock to follow for a week.
Each student follow the progress of his stock and draw a line graph depicting its fluctuation.

(c)

Turn to the sports page and find the standings for any professional sport.
Add up the <u>total</u> number of victories and defeats for each division of the league.
Figure out the won-lost percentage of each division.
Which division appears to be the strongest? Which appears to be the weakest?

(d)

Figure 8–10 Activity cards from the *Chicago Sun-Times Newspaper in the Classroom Program* (reprinted by permission of *Chicago Sun-Times*).

periences, as in these examples from the *Newspaper in the Classroom Program* of the *Chicago Sun-Times:*

▶ Decimals can be computed from batting averages and percentage of pass completions for a favorite quarterback (p. 7).

▶ Common fractions are found in recipes. Find a recipe serving six persons and have students convert the quantities so the recipe will serve 15 persons (p. 5).

▶ The metric system can be introduced or reviewed by reading and interpreting articles, especially from foreign news services. "For example, the strength of a nuclear blast is measured in megatons and kilotons, small arms ammunition is given in millimeters, and distances are given in kilometers" (p. 10).

▶ Consumer economics can be learned from advertisements. "Automobile batteries are advertised at different prices with guarantees for different numbers of months. Students can figure the cost of each battery on a per-month basis to determine the most economical purchase" (p. 17). Students might become more interested in current legislative debates, such as those concerning truth-in-packaging bills, by attempting to determine actual value of the "economy" size of one product compared with the "giant" size produced by another company (p. 15).

▶ Graphs are common in most newspapers, and students can draw their own from data they find in articles. Bar graphs may be constructed from sports statistics such as runs scored by each of the major league baseball teams. Election results can be depicted on a circle graph in which sectors indicate the percentage of vote received by each candidate. Line graphs can record daily temperatures or hours of daylight (plot from November 22 to February 22 to demonstrate the last month of decreasing hours of daylight before the winter solstice followed by the first two months of increasing hours of daylight after the winter solstice) (pp. 19–21).

Field Work   Field work—finding heights of buildings, areas of playing fields, distances between far-off landmarks—provides opportunities for applications of classroom knowledge. Preparation for field work might involve research on such instruments as the hypsometer, angle mirror, sextant, transit, and planimeter (see Shuster and Bedford, 1935, or the Nineteenth Yearbook of the National Council of Teachers of Mathematics, *Surveying Instruments: Their History and Classroom Use*, 1947). Simple versions of the hypsometer and planimeter can be made by students and used as suggested below.

Hypsometer

The word *hypsometer* comes from Greek words meaning height measure. A hypsometer is a simple instrument used to determine heights of trees, buildings, and other objects. The version shown in Figure 8–11 includes a protractor and thus can serve as a clinometer for finding angles of elevation and declination as well.

Materials needed:
> Graph paper fastened to firm cardboard or plywood; plumb bob; soda straw
> a. Soda straw serves as a sight. Affix to upper edge of graph paper so that top of object whose height is to be measured can be sighted as Figure 8–12 suggests.
> b. Suspend plumb bob from point A at upper right corner of graph paper.

To use hypsometer:
> a. Sight top of object to be measured, holding hypsometer with plumb bob suspended (see Figure 8–12). Clamping hypsometer to a pole at the desired angle will improve accuracy.

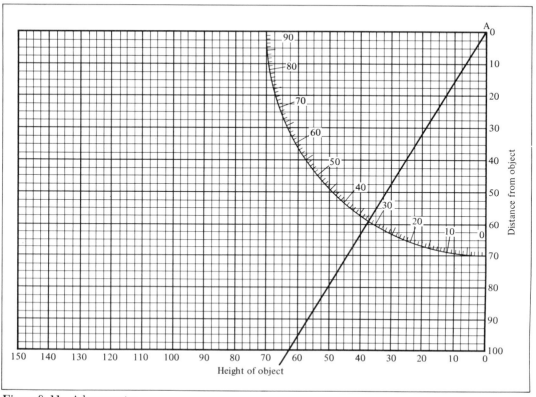

Figure 8–11   A hypsometer.

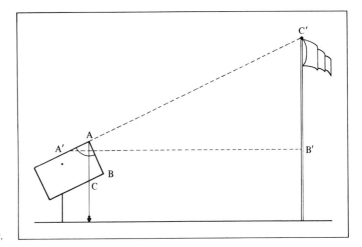

Figure 8–12   Using a hypsometer.

b. The sides of the graph and the plumb line form the triangle ABC which is similar to the triangle $A'B'C'$.
Note: Hypsometer can be graduated (as shown in Figure 8–11) so that height can be read directly from the graph.

Exercises:
a. Find the height of a tree by sighting treetop from one position.
b. Find the height of the school's flagpole by recording the angle of elevation to the top of the pole from one location, then proceeding a given distance directly towards the pole and recording the new angle of elevation.
c. Learn how to find the latitude of your location by using the hypsometer (with protractor). Caution: Never look directly at the sun.
d. Learn how to "level," a technique in surveying to find the difference in elevation between two points.

Hatchet Planimeter*

A planimeter is a device which can be used to measure the area of an irregular region such as a leaf or a lake shown on a map. The hatchet planimeter is a simplified version of the device. It can be made by students (especially if they have access to a few tools in the school's shop) and is fairly easy to learn how to use. At lower grade levels the planimeter would be useful when dealing with measurement concepts. In the senior grades it could be regarded as a mechanical integrator.

* The authors wish to acknowledge the contributions of Roger K. Brown of the School Science Curriculum Project, University of Illinois, to this section. The written material is adapted from George R. Frost, *The Hatchet Planimeter* (1965) with the permission of the author.

Description:

Two versions of the hatchet planimeter are shown in the figures. The one in Figure 8–13 is made from the list of materials given below. Another version, made from a wire coat hanger, is shown in Figure 8–14.

The hatchet planimeter consists of three main parts: the hatchet shaped blade H, the tracing point P, and the arm of length R. The arm separates the center of the hatchet blade H and either of the tracing points P by a known, fixed distance. The hatchet blade H controls the direction in which the free end of the arm will move as the tracing point P is guided around the outline of the unknown area of an irregular plane figure.

The hatchet blade H allows the free end of the arm R, which bears the blade, either to pivot about the blade's center or to move parallel with the blade's edge. All sideways motion of the arm is resisted by the sharp blade's contact with the surface on which it rests. As a result, the hatchet blade moves over a restricted path for each irregular area traced.

The sharp edge of the hatchet blade also allows the recording of the beginning and the end of the path over which it moves. Pressure put on the blade by the finger will indent a mark in the surface on which the blade rests.

Between the starting and stopping of the tracing operation, the hatchet blade will be laterally displaced a distance s in moving over its restricted path (see Figure 8–15). It is this distance s that, when measured and multiplied by the known length R of the arm, produces a very close approximation of the area of the traced figure.

Materials needed (for plastic version of planimeter):

1 piece transparent plastic, $11\frac{1}{2}$ inches by 2 inches, approximately 0.02 inches thick (thickness of light cardboard)

1 brass machine screw, $1\frac{1}{2}$ inch by $\frac{1}{4}$ inch

2 brass hexagonal nuts, $\frac{1}{4}$ inch

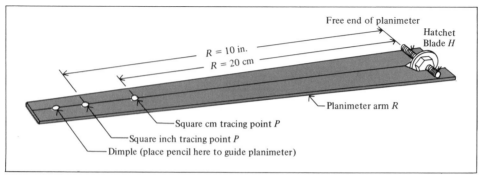

Figure 8–13 A plastic hatchet planimeter. (The drawings here and in the following figures are adapted from George R. Frost, *The Hatchet Planimeter*, with the permission of the author.)

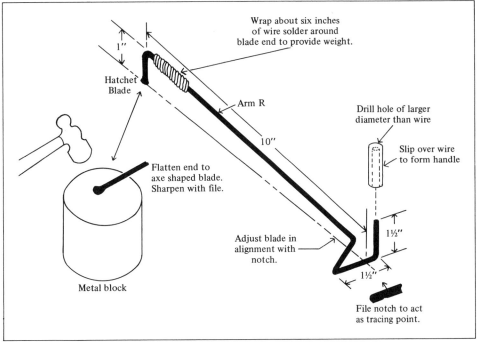

Figure 8–14 A hatchet planimeter made from a coathanger.

1 iron washer, $\frac{3}{16}$ inch
1 tube plastic cement (obtainable at hobby shops or hardware stores)
button thread

To use hatchet planimeter:

On a sheet of smooth paper, trace the outline of the area under investigation (as shown in Figure 8–16). Locate and mark the approximate center of the area. Draw a straight line from the center to the outline of the area. Fasten the paper on which the outline is traced on a table top or to a drawing board. Fasten a second sheet of paper directly above the traced pattern as shown. A piece of paper towel placed under the second sheet will allow the start- and stop-marks to be clearly impressed into the second sheet.

Place the hatchet planimeter on the two sheets of paper in a position such that either the square-inch tracing point or the square-centimeter tracing point is directly over the marked center of the area. Press firmly downward on the top of the hatchet blade to indent a starting mark on the upper sheet of paper. Be sure that the hatchet blade is not accidently moved out of the indentation during this operation.

Allowing the hatchet blade freedom to move, carefully guide the tracing

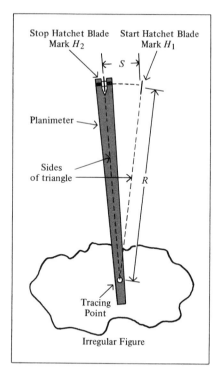

Figure 8–15   How the planimeter
works.

point upward along the straight drawn line to the outline of the pattern.
Continue to guide the tracing point along the outline until one complete
circuit is made. Return the tracing point along the straight drawn line to the
center of the pattern. Stop at this point. Press firmly on the hatchet blade
to indent a stop mark in the upper sheet of paper.

Measure the shortest distance s between the indented start- and stop-
marks made by the hatchet blade. Multiply this distance s by the length R
of the planimeter arm. If the square-inch tracing point is used, distance
s is measured in inches. The 10-inch length R of the planimeter arm is used
as the multiplier, and the product is given in square inches. For example,
if the measured distance

$$s = .25 \text{ inch},$$

the area of the traced figure is

$$.25 \text{ inch} \times 10 \text{ inches} = 2.5 \text{ square inches.}$$

However, if the square-centimeter tracing point is used, distance s is measured in centimeters. The multiplier is now the 20-centimeter length R of
the arm, and the answer is in square centimeters. If the measured distance

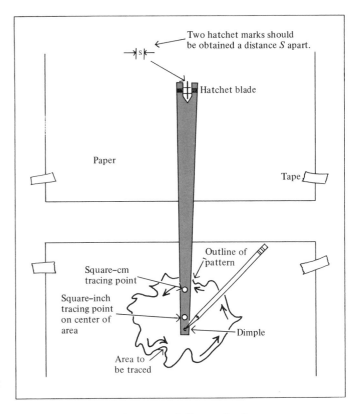

Figure 8-16 Using the hatchet planimeter.

s is 6 centimeters, for example, the area of the traced figure is 6 cm × 20 cm = 120 square centimeters.

Repeat the process of tracing and multiplying several times, tracing first in a clockwise and then in a counterclockwise direction. Average the results obtained from the trials to minimize error introduced by inaccurate tracing. Have several students trace and compute the area. Average these computations to obtain a solution.

Exercises

Geography: Which of the Great Lakes is the largest in area? Which is the smallest? How many times greater is the largest than the smallest? Which state in the United States is the largest? And so on. Tracings of the entities to be compared can be kept to the same scale by making the outlines from the same map.

Social Studies: The population of any state in the United States can usually be found in a dictionary or an encyclopedia. What is the population density (persons per square mile) of a particular state? The number of persons divided by the number of square miles will give the density.

What is the population density in persons per hectare of the region in which you live?

Biology: What change in area occurs in an individual leaf of a fast-growing plant as the growing season advances? (Tracings can be made of the same leaf over several weeks.) What is the rate of change (as found from a simple graph of area versus time)? How does the rate of change in area compare with the rate of change in the length of the leaf?

What is the average area of the tracings of the boys' hands and of the girls' hands in the classroom? (Each student may make a tracing of his or her own hand and find its area.) Which have the greatest average area, the boys' or the girls' hands?

Muskrats thrive best when the population has a density of approximately one muskrat per 86 acres. Identify an area which forms an ecological niche for muskrats. What would be the ideal number of muskrats distributed over this area?

If the number of stomates on one square millimeter of a leaf is counted with the aid of a calibrated microscope and found to be n in number, how many stomates does the leaf contain?

A certain woods, well defined on a scaled map, is known to receive an annual rainfall of 80 centimeters. Sixty percent of the rainfall is absorbed by the earth. How many liters of water are annually absorbed by the earth in those woods?

Physics: The lift of a certain biplane wing shown on a scaled drawing is known to be 8 pounds per square foot. What lift will be developed by the airplane?

The distributed electric current through a metal strip of uniform but irregular cross section is one ampere. What is the current density per square centimeter across the cross section?

A petrie dish containing a small depth of water is floated on an accurately mapped pond. The petrie dish is found to have lost a certain weight of water through evaporation over a 24-hour period. What is the approximate loss of water from the pond in kilograms through evaporation over this same period?

Polyhedra   Polyhedra can form the basis for quite a number of interesting investigations, some of which are outlined below. The references will provide further sources of ideas.

1. Many students will enjoy constructing polyhedra using posterboard and patterns (nets) like the one shown in Figure 8–17 for the octahedron. In order to get well-formed models, this type of project requires much attention to detail. Such models, once completed, are attractive and can be painted and displayed in the classroom. Wenninger (1971) gives complete details for the construction of such models.

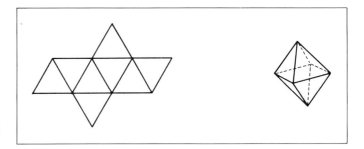

Figure 8–17   The net and completed model of an octahedron.

2. Explorations related to the symmetry of the various polyhedra can be stimulated by asking students to identify all of the axes of rotational symmetry or all of the planes of symmetry that a particular polyhedron has. It is interesting for many students to discover, for example, that a cube has 13 axes of symmetry and nine planes of symmetry. (Can you draw them all? Try before looking at Figure 8–18.) Such investigations can lead very nicely into discussions of isometries and a consideration of the properties of a group. (The film "Dihedral Kaleidoscopes" beautifully illustrates in animated form these axes and planes of symmetry. See Appendix D.)

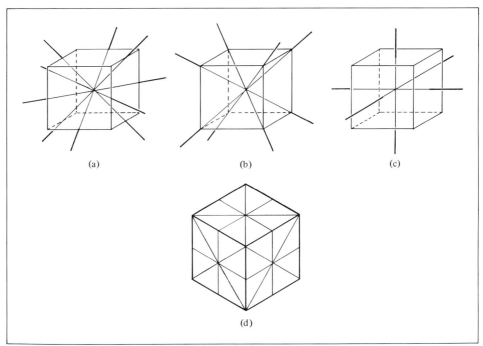

(a)   (b)   (c)

(d)

Figure 8–18   The axes and planes of symmetry of a cube (a. six 2-fold axes; b. four 3-fold axes; c. three 4-fold axes; d. the nine planes of symmetry).

3. Problem 12 of Chapter 5 (p. 144) suggests another appropriate inquiry related to polyhedra. Students usually find it relatively easy to discover Euler's formula $V + F - E = 2$. A proof of this formula can be found in Lines (1965, p. 135).

Tessellations We use the term *tessellation* to denote "the covering of a plane by planar regions such that (1) none of the planar regions overlap, and (2) there is no portion of the plane which is left uncovered." A study of tessellations lends itself nicely to laboratory-type investigations at a variety of difficulty levels.

Investigations of this topic which require very little in the way of mathematical sophistication include:

1. Letting the students identify and record (either by drawing or photographing) occurrences of tessellations in their environment (e.g., in walls, ceilings, floors).
2. Having students draw in the various lines of symmetry that exist in most tessellations.
3. Letting the students color tessellations which you might distribute to them on dittos.
4. Having your students create (by drawing or by making use of colored pieces of construction paper) some tessellations which you might display in your room.

At a higher level of complexity, you might let your students try to answer the question "What kinds of polygons will tessellate?" Investigating this question is best done if you restrict your students to using just one kind of polygon at a time. Students should be provided with various cardboard cut-outs or templates of polygons to assist them in conducting their investigation. Such shapes might include those shown in Figure 8–19. While students will not usually be able to answer the above question completely, many will discover that any parallelogram can be used to tessellate the plane. This discovery will then lead to the generalization that any triangle will tessellate. Figure 8–20 suggests the rationale for this generalization.

It usually requires considerably more experimentation for students to conclude that *any* quadrilateral can be used to generate a tessellation. As Figure 8–21 shows, however, we can always form a tessellation from any given quadrilateral merely

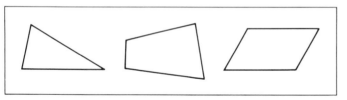

Figure 8–19 Templates of polygons.

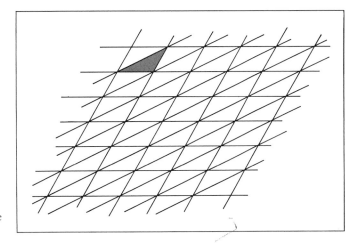

Figure 8–20 Any triangle will tessellate the plane.

by performing half-turns about the midpoints of the sides of the original quadrilateral. Continuing to perform half-turns about the midpoints of the sides of the newly generated quadrilaterals will yield the desired tessellation.

After students have had an opportunity to begin to answer the question "What kinds of polygons tessellate?" you might narrow the investigation somewhat by asking "Which of the regular polygons can be used to tessellate a plane?" Again cardboard models or templates can be used to assist the student. Some students will undoubtedly demonstrate that equilateral triangles, squares, and regular hexagons will tessellate. Some may even be able to prove, using information about

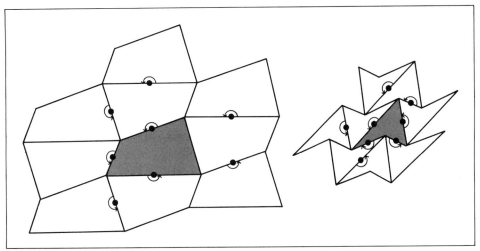

Figure 8–21 Any quadrilateral will tessellate the plane.

the angle sizes of regular *n*-gons, that the tessellations in Figure 8–22 are the *only* three regular tessellations. (A regular tessellation is one which is formed by using only one kind of regular polygon.)

INTERIOR ANGLES OF REGULAR *n*-GONS

3 sides	60 degrees
4 sides	90 degrees
5 sides	108 degrees
6 sides	120 degrees
7 sides	$128\frac{4}{7}$ degrees
8 sides	135 degrees
9 sides	140 degrees
10 sides	144 degrees
11 sides	$147\frac{3}{11}$ degrees
12 sides	150 degrees

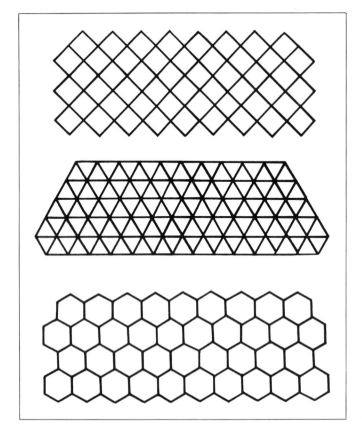

Figure 8–22   The three regular tessellations.

In such a proof, students would note that there can be no fewer than three regular polygons arranged around a point and no more than six so that no overlapping or gapping occurs (Why?). This means that there can be but 3, 4, 5, or 6. Dividing each of these values into 360° yields 120°, 90°, 72°, and 60°, respectively. There are no regular polygons, however, that have interior angles of 72°. Hence, there are only three types of regular polygons that can be used to surround a point. They are hexagons, squares, and equilateral triangles (see Figure 8–23). It is rather easy to see that each of these patterns can be extended to form the tessellations shown in Figure 8–22.

Here are suggestions for further investigations with tessellations:

1. Explore the realm of semi-regular tessellations. These are tessellations which are formed by using two or more different regular polygons in such a way that the arrangement of polygons about each vertex is identical. Students are usually surprised to learn that there are only eight such patterns (see Figure 8–24).
2. Prove that there exist only eight semi-regular tessellations.
3. Using three mirrors, construct a trihedral kaleidoscope as illustrated in Figure 8–25.
4. Use the trihedral kaleidoscope and construction paper to generate the three regular and as many of the eight semi-regular tessellations as possible. (see Figure 8–26).
5. Explore the kinds of three dimensional shapes that can be used to fill space with no gapping. (Cubes, for example, will work, but what other shapes will? This investigation might best be handled after your students have studied various kinds of polyhedra.)

Probability and Statistics   Probability and statistics provides a rich source of interesting, real-life problems. Initially the only laboratory equipment needed is ordinary coins and dice. As problems become more varied, use can be made of polyhedral dice, that is, dice formed from each of the five regular polyhedra (available from Creative Publications—see Appendix D). Another source of chance outcomes is a table of random digits. Even young students can readily

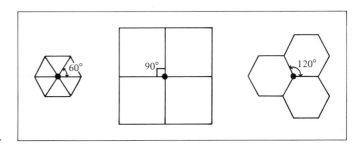

Figure 8–23.

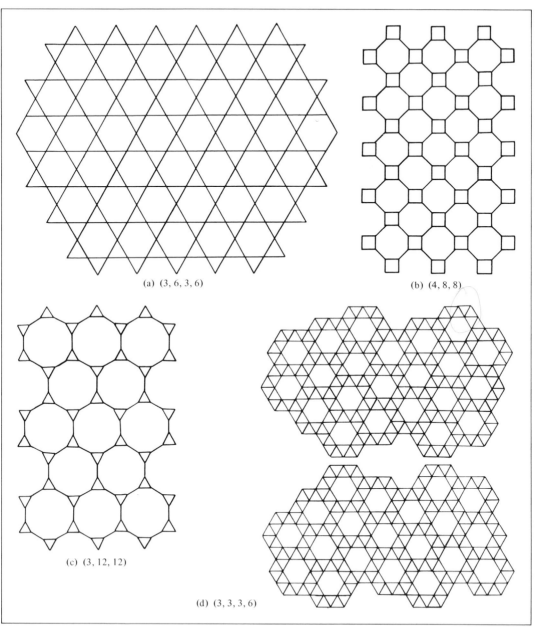

(a) (3, 6, 3, 6)

(b) (4, 8, 8)

(c) (3, 12, 12)

(d) (3, 3, 3, 6)

Figure 8–24   The eight semi-regular tessellations. Semi-regular tessellations contain at least two different regular polygons and are such that the arrangement of these polygons around each vertex is always the same. This regularity in vertex arrangement provides a convenient means of identifying the various patterns.

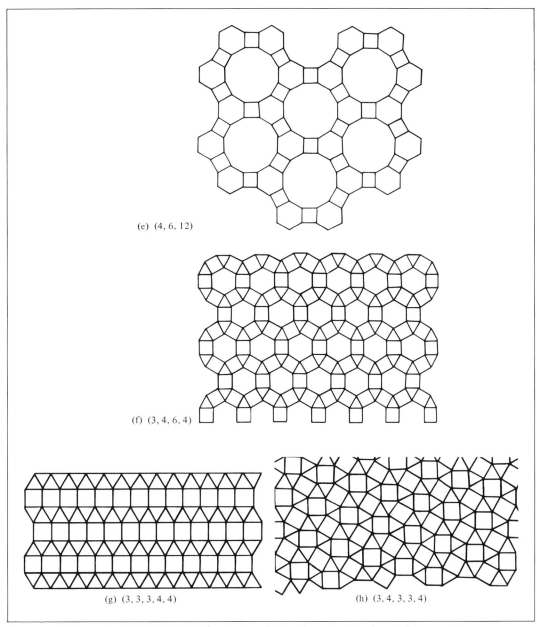

(e) (4, 6, 12)

(f) (3, 4, 6, 4)

(g) (3, 3, 3, 4, 4)

(h) (3, 4, 3, 3, 4)

For example, the symbol "(3,6,3,6)" identifies the tessellation shown in (a), where "3" represents an equilateral triangle and "6" a regular hexagon. While the two (3,3,3,3,6) tessellations shown in (d) appear identical, a closer inspection reveals that one is a mirror image of the other. They are said to be *enantiomorphic*.

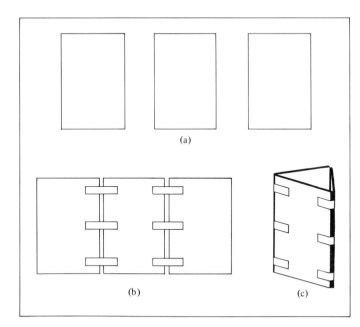

(a)

(b)   (c)

Figure 8–25 Making a three-mirror kaleidoscope.
(*a*) Place three congruent mirrors with their reflecting sides down on a table.
(*b*) Tape the mirrors together as shown. Be sure to separate the mirrors slightly so they can be "folded" as shown in (*c*).
(*c*) The mirrors are set upright to form the kaleidoscope.

learn the use of such a table, provided in most statistics books (such as Mosteller et al., 1970). A BASIC computer program to generate random digits, and the resulting output of 500 such digits, appears in Appendix C.

A sample problem to solve using random outcomes is the *simplified epidemic problem* (adapted from Mosteller et al., 1970, p. 15):

> An infectious disease has a one-day infectious period, and after that day the patient is immune. Six hermits (numbered 1, 2, 3, 4, 5, 6) live on an island, and if one has the disease he randomly visits another hermit for help during his infectious period. If the visited hermit has not had the disease, he catches it and is infectious the next day. How many hermits on the average can be expected to get the disease?

We could solve the problem by rolling an ordinary die, but we will use instead the table of random digits in Appendix C (pp. 546–547). We will use digits 1–6 to correspond to the outcomes of rolling a die. Digits 0, 7, 8, 9 will be ignored. Arbitrarily, assume that hermit 1 has the disease today, and the rest have not had it. We determine which hermit he visits by going to row one of the table of random digits. The first digit is "2" (corresponding to rolling a two on a die). Hermit 2 is infectious tomorrow and we indicate this in Figure 8–27. We go to the next digit in the table to see whom hermit 2 visits and obtain a "6." The next digit is a "2," and since hermit 2 has already had the disease the epidemic dies out. In Figure 8–27 we note that three hermits had the disease during the course of the

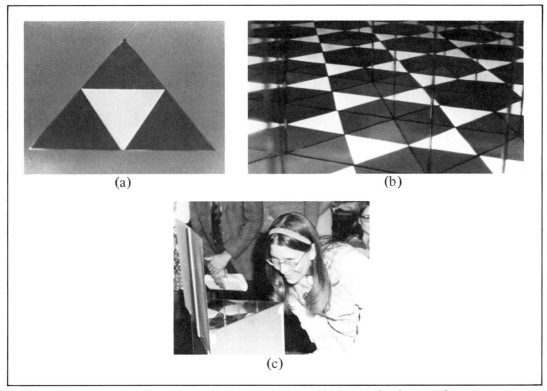

(a)   (b)

(c)

Figure 8–26 An unlimited variety of tessellations can be formed by fitting colored pieces of construction paper together to form an equilateral triangle that will just fit in the bottom of a trihedral kaleidoscope. The triangle shown in (a), for example, will generate the (3,6,3,6) semiregular tessellations as seen in (b). Figure (c) illustrates how the kaleidoscope is viewed to obtain the tessellation effect. O'Daffer and Clemens (1976) suggest further activities using the trihedral kaleidoscope (see Appendix D). (Photos used by permission of Stanley R. Clemens.)

epidemic. We perform the experiment 12 times to find the average (or mean) number of persons getting the disease. We continue in the random digit table where we left off. If the same digit appears a second time in succession, it is ignored (a hermit must visit someone else). The problem may be extended to twelve hermits by using a die which is a dodecahedron (has twelve faces). The average number of persons getting the disease can again be computed.

Now consider a variation on the same problem: What is the probability that, in a group of five people, at least two share the same birthmonth? An approach similar to the one used in the hermit problem, involving the use of the dodecahedra die, can provide an estimate of the required probability. Statistical theory tells us that the estimate will improve in accuracy as the number of tosses of the die increases, and provides guidelines as to the number of tosses needed for the desired degree of confidence in our results.

Epidemic No.	Hermit visited						Total Hermits Getting Disease
	1	2	3	4	5	6	
1	X	X				X	3
2							
3							
4							
5							
6							
7							
8							
9							
10							
11							
12							
Average No. Getting Disease							

Figure 8–27.

REPORTS AND PROJECTS

Reports and projects can take on the characteristics of good laboratory activities. They permit students to pursue individual areas of interest. They open up new dimensions of study for the student, and often for the teacher as well. They can provide excellent opportunities for you to get to know your students better as you learn their interests and work with them on an individual basis.

The results of student work generally should be shared with the rest of the class. Oral or written reports may be given. Bulletin board displays may be prepared, which may include charts or graphs. A model might be displayed or an experiment performed, with the entire class involved in the presentation and the ensuing discussion.

The difference between a report and a project is not always clear or even important. We will make the distinction that a *report* involves doing background research on some topic, organizing the information in some way, and preparing some form of presentation. A *project* goes beyond a report in that it actually involves the student in solving some problem. We present the following topics as suggestions for reports and projects. In some cases the way in which the assignment is handled by the student will determine whether the result is a report or a project.

Evaluation of reports and projects is difficult, but in most cases it should be given; students are entitled to your judgment of the quality of their work. In some cases either "fail" or "pass," with some commentary from you, will be sufficient.

Here are several subject areas that offer many possibilities for reports and projects.

The number pi has fascinated man for centuries. The Old Testament records a mathematical experiment on the value of pi (I Kings 7:23 and II Chronicles 4:2). The computer program PI (Appendix C, pp. 531–537) demonstrates six methods of computing pi. Details on the computation of pi can be found in the Thirty-first Yearbook of the National Council of Teachers of Mathematics (1969, pp. 148–154) and the School Mathematics Study Group Reprints, *Nature and History of Pi* and *Computation of Pi*. See also the account in Chapter 3 of this book of the high school student who memorized pi to 5000 places (Figure 3–2). An interesting computer exercise would be to find common fraction approximations to pi which are correct to a given number of decimal places. A method for solving this problem is given in Feng (1969, pp. 49–51).

The Pythagorean Theorem has over 300 proofs (see, for example, Loomis, 1968). The intriguing secret order of the Pythagoreans will interest some students (see Shulte, 1964).

The mathematics of the honeycomb was studied by the fourth-century Alexandrian mathematician Pappus. No other shape of cell is as economical of space and material as is the hexagonal structure of the honeycomb; recall the illustration of honeycomb structure in Chapter 1 of this book (Figure 1–5). See also Siemens (1965 and 1967).

Mathematics and postage stamps will interest some students as a project. Schaaf, in a beautifully illustrated article (1974), studied postage stamps from various countries and found examples of mathematics in use for measurement, computation, cartography (mapmaking), and design, and he also noted examples of mathematics in nature (pp. 16–24).

"Gas Station Map Mathematics" is the title of an article by Allison (1973) in which he points to ways in which road maps can be used in mathematics classes. Some activities he suggests are:

1. Trace a highway across a state or a region of the country and count the towns it passes through.

2. Measure the distance between two cities shown on the map, then compute how far apart these cities are (in miles or kilometers).
3. Use the mileage chart on the map to find distances between cities. Then use yarn or string to find distances, using the map's conversion scale. Why are these distances not the same? In a trip from Philadelphia to New York, how much less land does the airplane have to cover than a car?
4. Find areas of states and compare results with official figures in an atlas or encyclopedia (pp. 328–329). (A booklet suggesting road map activities is available from the Illinois Office of Education.)

Kite building can lead to the study of geometric shapes (such as Graham Bell's famous tetrahedral kites), to methods of measuring heights (such as using the hypsometer), and to studies of the physics of flight. And kite building can result in a product which actually works—great motivation for many students. Greger (1975) provides details on the history of kites, on how to make several different kinds, and on how to present the topic in the classroom.

Graphing pictures is instructive and entertaining. Students find solution sets for equations and plot them; the results are pictures of a variety of objects. The activity shown in Figure 8–28 was devised by Kathy Vick, a high school student in a class of one of the authors.

Many trigonometries could be the subject of extensive study. In addition to the familiar trigonometry of the right triangle, there is a trigonometry of the square, for example (see Biddle, 1967). Podbelsek (1973) surveyed the literature and identified many models for trigonometries, including:

Vector trigonometry model (see Copeland, 1962; Amir-Moez, 1958)
Vector-complex number model (see Hillman and Alexanderson, 1971)
Wrapping function model (see Yandl, 1964)
Riemann integral model (see Zaring, 1967)

ADDITIONAL TOPICS FOR LABORATORY ACTIVITIES

Full listings for all references cited in this section will be found in Appendix D. (Further illustrations of student mathematics projects appear in Figures 8–29 and 8–30.)

ACTIVITY	REFERENCES
Anecdotes and stories	Eves 1971, 1972
Calculation (manual)	Asimov 1964
Calculators	Feldzamen and Henle 1973; Hunter 1974; Judd 1974; Mullish 1973

1. $3x - y = 3$
    $0 \leqslant x \leqslant 1$

2. $2x + y = 2$
    $0 \leqslant x \leqslant 1$

3. $y = 15$
    $3 \leqslant x \leqslant 10$

4. $y = -8$
    $8 \leqslant x \leqslant 10$

5. $x = -10$
    $9 \leqslant y \leqslant 10$

6. $y = 10$
    $-10 \leqslant x \leqslant -9$

7. $x = 2$
    $10 \leqslant y \leqslant 11.5$

8. $x - 9y = -99$
    $-9 \leqslant x \leqslant 0$

9. $3x + y = 37$
    $13 \leqslant x \leqslant 14$

10. $5x + y = 76$
    $13 \leqslant x \leqslant 14$

11. $5x - y = 67$
    $-2 \leqslant y \leqslant 3$

12. $\frac{x}{3} - y = 11\frac{1}{3}$
    $-8 \leqslant y \leqslant -7$

13. $x - y = -8$
    $10 \leqslant y \leqslant 11$

14. $3x + y = -19$
    $-9 \leqslant x \leqslant -8$

15. $x = 2$
    $4 \leqslant y \leqslant 6$

16. $4x - y = 12$
    $-4 \leqslant y \leqslant 0$

17. $\frac{x}{3} + y = -5\frac{1}{3}$
    $-8 \leqslant y \leqslant -7$

18. $2x - y = 33$
    $-7 \leqslant y \leqslant -5$

19. $\frac{x}{2} + y = -\frac{9}{2}$
    $-7 \leqslant y \leqslant -6$

20. $x + y = -1$
    $8 \leqslant y \leqslant 9$

21. $2x + y = 37$
    $12 \leqslant x \leqslant 13$

22. $x + y = 25$
    $13 \leqslant y \leqslant 15$

23. $x - y = -12$
    $14 \leqslant y \leqslant 15$

24. $x - y = -11$
    $0 \leqslant x \leqslant 1$

25. $x - y = -8$
    $4 \leqslant x \leqslant 5$

26. $\frac{x}{2} + y = \frac{25}{2}$
    $1 \leqslant x \leqslant 3$

27. $x = -9$
    $8 \leqslant y \leqslant 10$

28. $x = 14$
    $3 \leqslant y \leqslant 6$

29. $4x - y = 2$
    $2 \leqslant x \leqslant 3$

30. $2x + y = 0$
    $-6 \leqslant y \leqslant -4$

31. $\frac{x}{6} + y = 2$
    $-6 \leqslant x \leqslant 0$

32. $4x + y = 12$
    $2 \leqslant x \leqslant 3$

33. $2x - y = -4$
    $10 \leqslant y \leqslant 12$

34. $x + y = -3$
    $-8 \leqslant x \leqslant -6$

35. $2x - y = -10$
    $1 \leqslant x \leqslant 2$

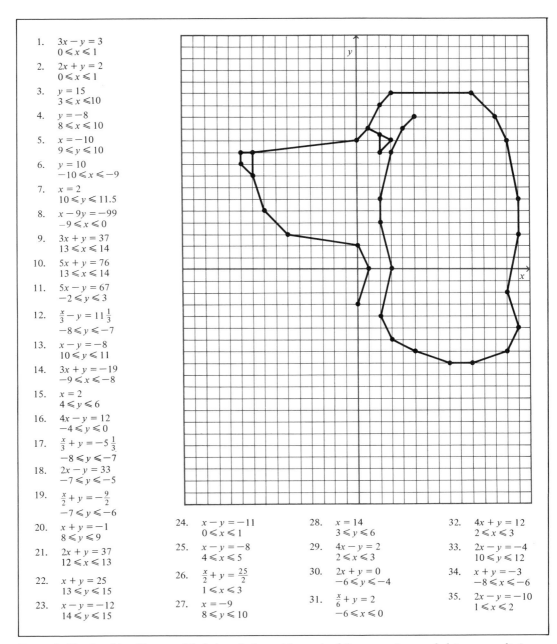

Figure 8–28 Graphing Pictures. (Find the solution set for each of these equations and plot your results.)

Figure 8–29   Constructing the five Platonic solids (shown in the lower left of the figure) and the thirteen Archimedean solids can be an interesting, profitable, and painstaking activity for students.

Computers in classroom	Ahl 1974; Crowler 1967; Koetke 1971; Post 1970; Spencer 1973
Computer dictionaries	Sippl and Sippl 1974; Spencer 1973
Computer games	Ahl 1973
Computer mathematics textbooks	Feng 1969
Computer problems	Allison 1971; Nievergelt et al. 1974
Computer programming	Crawford 1969; Corlett 1972; Dorn 1965; Gear 1969; Newey 1973
Contests and clubs	Dalton and Snyder 1973; Gruver 1968
Cryptography	Brooke 1963; Peck 1961; Phillips 1961b
Dictionaries	Bendick and Livin 1965; Marks undated
Fantasy	Burger 1965
Games	Gellis 1971; Henderson and Glunn 1972; Holt and Dienes 1973; Rice 1973
Geoboards	Niman and Postman 1974
Golden mean	Beard 1973; Huntley 1970; Runion 1972
Graphing pictures	Boyle 1971, 1972
Graphs	Brant and Keedy undated; Gelfand undated;

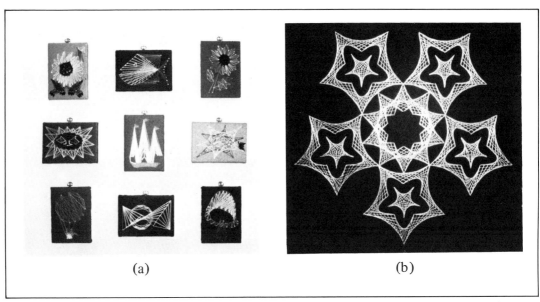

(a)   (b)

Figure 8–30   String art projects by students of Richard Deptuch of Oak Park-River Forest High School, Oak Park, Illinois.

	Hogben 1955; Johnson and Glenn 1960–1963
History	Aaboe 1964; Barnard 1968; Bell 1937; Eves 1964; Kline 1972
Laboratory manuals	Hooten and Mahaffey 1973; Krulik 1971, 1972; Zelenik undated
Libraries	Hardgrove and Miller 1973; Schaaf 1973
Logic and proof	Del Grande et al. 1967; Johnson and Glenn 1960–1963; Summers 1972
Magic, tricks	Fults 1974; Gardner 1955; Gellis 1973; Meyer 1972
Models	Cundy and Rollett 1961; Holt and Marjoram 1973; Wenninger 1966, 1971
Metric system	Glaser 1974; Henderson and Glunn 1974; Hopkins 1974; Ross 1974; Smart 1974
Number theory	Barnett 1961; Lockwood 1967
Outdoor mathematics	Johnson 1974; 19th NCTM Yearbook 1947
Paperfolding	Cundy 1961; Olson 1975
Polyhedra	Coxeter 1973; Laycock 1970; Wenninger 1966, 1971
Probability and statistics	Adler 1963; Freund 1967; Johnson and Glenn 1960–1963; Huff and Geis 1959

Problems	Charosh 1965; Hill 1974; Kordensky 1972; Lindgren 1972; Posamentier and Wernick 1973; Salkind 1973
Puzzles	Bakst 1954; Brooke 1963; Emmet 1972; Fujii 1966; Gardner 1959a, 1959b, 1969a, 1969b; Greenblatt 1965; Hunter 1974; Kordemsky 1972; Phillips 1961a
Recreation	Ball 1939; Barnard 1965; Friend 1972; Fadiman 1958; Gellis 1973; Golomb 1965; Lindgren 1972; Schaaf 1970a, 1970b, 1973
Slide rule	Merrill 1961
Space exploration	Ahrendt 1965
String sculpture	Jayne 1962; Sharpton 1974; Winter 1972
Tangrams	Read 1965
Two dimensional geometry	Burger 1965; Hilbert and Cohn-Vossen 1952

MATHEMATICS CLUBS

The various activities suggested in the preceding sections on mathematics laboratories can lead to sustained student interest and work on a project over the period of a semester or even longer. A mathematics club can promote such interest and provide opportunities for developing new interests as well. While it is important that the mathematically talented students are involved in such clubs, we believe that this is an opportunity to make mathematics interesting to all students, not just those who are interested to begin with. In many ways, teaching is a selling job—and through the mathematics club you can sell mathematics.

Dalton and Snyder's *Topics for Mathematics Clubs* (1973) contains many suggestions for club activities (see Appendix D). One format which has been used successfully is to study topics as a group, identifying subtopics for individual members. At club meetings the topic is presented by a "panel of experts," those who have been studying and preparing material on their subtopics for considerable time. Suppose, for example, the topic of using Monte Carlo methods to solve probability problems were to be studied. Possible subtopics for investigation would include: history of the Monte Carlo method; applications of Monte Carlo methods in industry; calculating π by Monte Carlo methods; solving well-known problems (such as the Birthday Problem) by Monte Carlo methods; and using Monte Carlo methods to solve problems too difficult (or even impossible) to solve by analytical methods—for example, the heat flow problem (computer program TEMP, Appendix C, p. 537) and the percolation problem (Hammersley and Handscomb, 1964).

Suggestions for organizing a mathematics club are contained in a booklet edited by Carnahan (1958), which considers such topics as: purposes of mathematics clubs, how to begin, a model constitution, and sample club programs. Possible activities for clubs include writing articles, preparing bulletin boards, speaking to school groups, presenting dramatic productions, taking field trips, and having social activities (pp. 14–16). Another useful publication is Ransom's *Thirty Projects for Mathematical Clubs and Exhibitions* (1961), which contains suggested activities relating mathematics to astronomy, music, maps, and many other areas.

The National Council of Teachers of Mathematics publishes the *Mathematics Student Journal*, which provides articles of interest to high school students and contains articles by students. Communication between mathematics students is also promoted by Mu Alpha Theta, a national honor society cosponsored by the National Council of Teachers of Mathematics and the Mathematical Association of America. In 1975, Mu Alpha Theta had chapters in approximately 1300 high schools and junior colleges.

Critical Incidents Revisited

Our first critical incident dealt with Mr. Greenwood's problem of differences in student achievement. We discussed this problem, pointing out that such differences, at times very large ones, are a fact of life to be recognized at the outset of teaching. We then suggested ways for dealing with these differences.

Miss Ferguson was worried because she had little student participation in her classes. She wants to involve the students more, but she has not been able to do so. We have tried to point out in this chapter that the "single-class" approach can have a negative effect on students because it may not recognize their individuality. In the extreme, students become nameless, faceless entities "somewhere out there" (recall the case of Cliff Evans in the article, "Cipher in the Snow," pp. 18–20); everyone gets the same teaching presentation, the same examples, the same assignment.

We have found, on the other hand, that by (1) recognizing students as individuals, (2) giving them tasks specifically geared to their needs and interests, and (3) providing opportunities for their interaction with the teacher and with each other, classes will come alive and will become productive mathematically.

We have also found, though, that it takes a lot of judgment to know how to give recognition for work well done. For example, a reward for solving a problem quickly and elegantly does not mean assigning the student more of the same. A more appropriate reinforcement might be the posting of the solution on the departmental bulletin board, sending it to a newsletter for possible publication, or allowing the student class time to pursue further information on the subject and prepare a report. We have learned that what are interesting, rewarding activities for us may not be regarded as such by our students. Let's let our students have a say as to how they should be recognized! Little offerings, like an ice-cream cone as a prize (one that can realistically be won by all members of the class), can mean much to many students. You might well ask yourself from time to time how many of your students have *ever* received any sort of recognition for accomplishments in mathematics class.

We close with a note of caution about the third critical incident as well: Don't expect overnight changes in pupil expectations or behavior. If you have a class of tenth graders who have been trained for a decade to be part of "one class," it is not reasonable to expect that, if you suddenly change the rules, everyone in your class (or even your department chairman or school principal) will necessarily see the wisdom of your enlightened educational approach! Change is a slow process and fruitful, worthwhile change does not come about easily. In terms of our teaching model, determine your instructional objectives carefully; then plan painstakingly and in consultation with others. Take into account the advice of others, but if your approach is educationally and pedagogically sound, don't be discouraged by others who feel uncomfortable about change. Bob gave up too easily. Experienced teachers know that significant modifications take months, even years, to carry out successfully. The major problem is often not with "those other people" (students or teachers) but with us. We have to experiment with approaches, ways of saying and doing things, which will work for us in our own unique classroom situation.

Activities

1. You have been assigned to teach a topic of your choice to a large group of high school seniors. You have one hour for the presentation. What provisions could you make for pupil involvement in the lesson?
2. Consider the factors to take into account in assigning individualized homework for a class of 25 students. How would you go about selecting the assignments and how would you handle their grading?
3. Figure 8-8 is an organizational chart for managing four small groups of students in the classroom. Work through the chart to see how each group has been assigned. Determine the amount of time each group can spend on each of the activities and the amount of this time which is under direct teacher supervision.
4. The reporting of student achievement in an individual progress program raises certain unique difficulties. What are some of these difficulties and how might they be handled by the teacher?
5. Devise a set of activity cards for one of the topics listed on pages 272–276. Assume the cards are to be used at learning stations in an eighth-grade class.
6. Frost (1966) has made the following comment about how the hatchet planimeter works:

> In the process of tracing an outline of a two dimensional figure, the hatchet planimeter mechanically displaces its blade over a distance s. . . .

This distance s, when multiplied by the fixed distance R between the center of the hatchet blade and the tracing point P produces a close approximation of the area A of the traced pattern. (p. 33)

Verify the author's assertion (refer to Figure 8–15).

7. Choose a topic for a mathematics laboratory activity for a ninth-grade general mathematics class. Prepare a plan of how you would present the activity. What introductory material would you need? How would you prepare the students to participate in the activity? What sort of finished product would you expect from the class and how would you evaluate it?
8. The topic of individualization is controversial. The May 1972 issue of the *Mathematics Teacher* featured a presentation of various points of view on individualizing instruction. Read the various articles on individualizing, summarize the points of view, and then determine how you would respond to the major issues which appear to be subject to debate.
9. Prepare a report on Mu Alpha Theta. Find out its history, the nature of its organization, and its present activities.
10. Torrance (1965), an expert in the study of exceptional children, has said:

> To me, by far the most exciting insight which has come . . . (is) that different kinds of children learn best when given opportunities to learn in ways best suited to their motivations and abilities. (p. 253)

Choose a classroom with which you are familiar. In two or three paragraphs, tell how this classroom provides opportunities for children to learn mathematics in ways best suited to their motivations and abilities.

References

Alexander, William M. and Hines, Vynce A. *Independent Study in Secondary Schools*. New York: Holt, Rinehart and Winston, 1967.

Allison, William M. Gas Station Map Mathematics. *Arithmetic Teacher* 20: 328–329; May 1973.

Amir-Moez, Ali R. Teaching Trigonometry Through Vectors. *Mathematics Magazine* 32: 19; September-October 1958.

Baker, Betty Louise. A Study of the Effects of Student Choice of Learning Activities on Achievement in Ninth Grade Pre-Algebra Mathematics. (Northwestern University, 1971.) *Dissertation Abstracts International* 32A: 2895; December 1971.

Bell, E. T. *Men of Mathematics*. New York: Simon and Schuster, 1937.

Berger, Emil J., ed. *Instructional Aids in Mathematics*. Thirty-fourth Yearbook of the National Council of Teachers of Mathematics. Reston, Virginia: The Council, 1973.

Biddle, John C. The Square Function: An Abstract System for Trigonometry. *Mathematics Teacher* 60: 121–123; February 1967.

Block, James H., ed. *Mastery Learning: Theory and Practice*. New York: Holt, Rinehart and Winston, 1971.

Block, James H., ed. *Schools, Society and Mastery Learning*. New York: Holt, Rinehart and Winston, 1974.

Bloom, Benjamin S., Hasting, J. Thomas; and Madaus, George F. *Handbook on Formative and Summative Evaluation of Student Learning*. New York: McGraw-Hill, 1971.

Bradley, Richard Moore. An Experimental Study of Individualized Versus Blanket-Type Homework Assignments in Elementary School Mathematics. (Temple University, 1967.) *Dissertation Abstracts* 28A: 3874; April 1968.

Brannon, M. J. Individual Mathematics Study Plan. *Mathematics Teacher* 55: 52–56; January 1962.

Bush, Robert N. and Allen, Dwight W. *A New Design for High School Education*. New York: McGraw-Hill, 1964.

Carnahan, Walter H., ed. *Mathematics Clubs in High Schools*. Washington: National Council of Teachers of Mathematics, 1958.

Collins, Kenneth M. A Strategy for Mastery Learning in Modern Mathematics. In James H. Block, ed. *Mastery Learning: Theory and Practice*. New York: Holt, Rinehart and Winston, 1971, pp. 111–112.

Copeland, Arthur H. *Geometry, Algebra, Trigonometry by Vector Methods*. New York: Macmillan, 1962.

Cundy, H. Martyn and Rollett, A. P. *Mathematical Models*. 2nd edition. New York: Oxford University Press, 1961.

Davidson, Dennis. Learning Mathematics in a Group Situation. *Mathematics Teacher* 67: 101–106; February 1974.

Duker, Sam. *Individualized Instruction in Mathematics*. Metuchen, New Jersey: Scarecrow Press, 1972.

Feng, Chuan C. *Second Course in Algebra and Trigonometry with Computer Programming*. Boulder, Colorado: Colorado Schools Computing Science Curriculum Development Project, University of Colorado, 1969.

Frost, George R. *The Hatchet Planimeter*. Urbana, Illinois: School Science Curriculum Project, University of Illinois, 1966. (Mimeo)

Gibbons, Maurice. *Individualized Instruction: A Descriptive Analysis*. New York: Teachers College Press, Columbia University, 1971.

Greger, Margaret. The Complete Kite Curriculum. *Learning* 3: 84–88; March 1975.

Hammersley, J. M. and Handscomb, D. C. *Monte Carlo Methods*. London: Methuen, 1964.

Hess, Adrien L. *Mathematics Projects Handbook*. Boston: Heath, 1962.

Hillman, Abraham P. and Alexanderson, Gerald L. *Functional Trigonometry*. 3rd ed. Boston: Allyn and Bacon, 1971.

Kidd, Kenneth P.; Myers, Shirley S.; and Cilley, David M. *The Laboratory Approach to Mathematics*. Chicago: Science Research Associates, 1970.

Lines, L. *Solid Geometry with Chapters on Space-Lattices, Sphere-Packs and Crystals*. New York: Dover, 1965.

Loomis, Elisha Scott. *The Pythagorean Proposition*. Washington: National Council of Teachers of Mathematics, 1968.

Mold, Josephine. *Tessellations*. New York: Cambridge University Press, 1969.

Mosteller, Frederick; Rouke, Robert E. K.; and Thomas, George B., Jr. *Probability with Statistical Applications*. 2nd ed. Reading, Massachusetts: Addison-Wesley, 1970.

O'Daffer, Phares G. and Clemens, Stanley R. *Geometry: An Investigative Approach*. Menlo Park, California: Addison-Wesley, 1976.

Podbelsek, Allen R. A Study of the Various Deductive Models for Developing and Teaching Plane Geometry Including an Investigation of the General Nature of Trigonometry. (University of Illinois, Urbana, 1972.) *Dissertation Abstracts International* 33B: 4916; April 1973.

Polos, N. C. Tutorial Adapted for High School. *Clearing House* 40: 404–405; March 1966.

Ransom, William R. *Thirty Projects for Mathematical Clubs and Exhibitions*. Portland, Maine: Walch, 1961.

Schaaf, William L. Mathematics in Use, As Seen on Postage Stamps. *Mathematics Teacher* 67: 16–24; January 1974.

Search, Preston W. Individual Teaching: The Pueblo Plan. *Educational Review* 8: 154–170; February 1894.

Shulte, Albert P. Pythagorean Mathematics in the Modern Classroom. *Mathematics Teacher* 57: 228–232; April 1964.

Shuster, Carl N. and Bedford, Fred L. *Field Work in Mathematics*. New York: American Book Company, 1935.

Siemens, David F., Jr. The Mathematics of the Honeycomb. *Mathematics Teacher* 58: 334–337; April 1965.

Siemens, David F., Jr. Of Bees and Mathematicians. *Mathematics Teacher* 60: 758–761; November 1967.

Sweet, Raymond. Organizing a Mathematics Laboratory. *Mathematics Teacher* 60: 117–120; February 1967.

Thelen, Herbert A. Grouping for Teachability. *Theory Into Practice* 2: 81–89; April 1963.

Thomas, R. Murray and Thomas, Shirley M. *Individual Differences in the Classroom*. New York: David McKay, 1965.

Torrance, E. Paul. Different Ways of Learning for Different Kinds of Children. In E. Paul Torrence and Robert D. Strom, eds. *Mental Health and Achievement: Increasing Potential and Reducing School Dropout*. New York: Wiley, 1965.

Wenninger, Magnus J. *Polyhedron Models*. New York: Cambridge University Press, 1971.

Wiley, W. Deane and Bishop, Lloyd K. *The Flexibly Scheduled High School*. West Nyack, New York: Parker, 1968.

Yandl, Andre L. *The Non-Algebraic Elementary Functions, A Rigorous Approach*. Englewood Cliffs, New Jersey: Prentice-Hall, 1964.

Zaring, Wilson M. *An Introduction to Analysis*. New York: Macmillan, 1967.

Chicago Sun-Times. *Newspaper in the Classroom Program: Activities for Mathematics*. Undated.

ETS Developments. 18: 1–2; October 1970.

Historical Topics for the Mathematics Classroom. Thirty-first Yearbook of the National Council of Teachers of Mathematics. Washington: The Council, 1969.

Illinois Office of Education. *Road Map Math*. Springfield, Illinois: Illinois Office of Education, undated.

National Council of Teachers of Mathematics. *Mathematics Teachers* 65: May 1972.

Project TALENT. *Bulletin* 4: February 1965.

School Mathematics Study Group. *Nature and History of Pi*. Stanford, California: School Mathematics Study Group, undated.

School Mathematics Study Group. *Computation of Pi*. Stanford, California: School Mathematics Study Group, undated.

Surveying Instruments: Their History and Classroom Use. Nineteenth Yearbook of the National Council of Teachers of Mathematics. Washington: The Council, 1947.

Using Computers

AFTER STUDYING CHAPTER 9, YOU WILL BE ABLE TO:

★ write simple computer programs in the BASIC programming language for use in teaching mathematics.

★ list four instructional applications of computers in mathematics classrooms and give an example of each.

★ select a computer activity appropriate for reaching a given instructional goal.

★ suggest ways of supplementing noncomputer-oriented mathematics textbooks so that computers may be used to help attain the desired goals.

★ list steps for beginning a computer-oriented mathematics program in your school.

CRITICAL INCIDENTS

1. Central High School has a new computer and the mathematics department chairman has announced that every teacher should be doing some "computer math." Bill is a first-year teacher who thinks he has enough to do already, just teaching his classes "regular math." What should Bill do?

2. Sue, a beginning teacher, is concerned about the computer mathematics course at her school. The syllabus seems to be filled with topics like FOR/

NEXT loops and subroutines. But she wants to emphasize mathematics, not computer programming. What should she teach?

3. It was great to get a computer terminal at West High. But Gary Williams tries to use the computer with his class and concludes it is more trouble than it is worth. Students are frustrated because the computer is complicated to run, and they don't like having to wait in line to use it.

4. This year Miss Kenyon is teaching in a school which has a computer center. She quickly finds that she can use mathematics problems in class which she previously had to omit because too much calculation was involved. Now her students can find hundreds of terms in a series and look for convergence. They find several ways of calculating an estimate for π and determine which methods give the most accurate estimates. They do independent study projects on estimating roots of equations of higher powers. Even her general mathematics students (who had shown little interest in mathematics) have become enthused about preparing simple computer programs and obtaining "their very own" results. Some students actually beg to be allowed to stay after school and do mathematics on the computer!

COMPUTERS AND SCHOOLS

Just as computers are playing important roles in many aspects of society—for airline reservations, billing, payrolls, traffic control—computers are becoming important in education. Figure 9-1 shows that computer uses in education may be classified in two major divisions: administrative and instructional. Presently, the

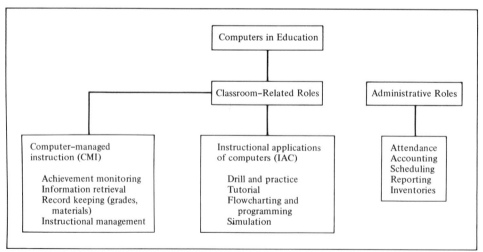

Figure 9-1   Uses of computers in education.

primary uses of computers in education are for administrative functions such as keeping pupil attendance records, preparing pupil schedules and grade reports, and maintaining personnel and financial records. However, as computers have become easier to use, more compact, more dependable, and less expensive, their use for instructional purposes has increased. Surveys conducted by the American Institutes for Research for the National Science Foundation (Darby et al., 1972, and Korotkin and Bukoski, 1975) provide data on the growth of computer usage in United States secondary schools between 1970 and 1975. It was found, for example, that the percentage of secondary schools using computers grew from 34.4 percent in 1970 to 58.2 percent in 1975, an average increase of 4.8 percent per year (Korotkin and Bukoski, p. 17). Computers were used for instructional purposes in 12.9 percent of the schools in 1970 and 26.7 percent in 1975. Furthermore, it was projected in the study that by the mid-1980s all secondary schools in the country will be making some use of computers in their educational program and that over one-half of the schools will have some type of computer-based instructional application (Korotkin and Bukoski, pp. 16–17).

This chapter is devoted to considering a variety of ways in which computers can be used for instructional purposes in the mathematics classroom. In Figure 9–1, these uses are divided into two subcategories: computer-managed instruction (CMI) and instructional applications of computers (IAC).

COMPUTER-MANAGED INSTRUCTION

Computer-managed instruction denotes those uses of computers in classroom situations which relate to the managerial aspects of teaching. For example, the computer can be helpful in record-keeping duties. Scores on daily or weekly quizzes can be recorded on the computer, transformed to standard scores, and weighted; then students can be ranked at the end of a grading period according to their weighted scores. The computer program SCORES (Appendix C) is such a program. Several projects in computer-managed instruction have been developed in the past few years. The following examples suggest some of the possibilities.

COMPREHENSIVE ACHIEVEMENT MONITORING (CAM)

This management program was designed to test and report student achievement every few weeks during the school year. The computer scores the tests and prepares printouts of results for the student and for the teacher, indicating total scores as well as performance on individual course objectives. Developmental work on CAM was done at the Evaluation Center for the schools of Hopkins, Minnesota. The Evaluation Center, an ESEA Title III Project, devised training materials to explain CAM, its capabilities, and the procedures for using it in the classroom.

BASIC MANAGED STUDY (BASMS)

BASMS is a computer-managed study guide program, written in the BASIC language and easily implemented on most time-sharing systems (see Daykin et al., 1975). The teacher wishing to use the program for a class is required to prepare a series of instructions and questions on the material to be studied. The teacher then supplies this information to the program, step-by-step, as asked by the system. Very little familiarity with computers is required for operation of the program. Once the material is prepared and entered, the student goes to the teletype and receives instructions on what to study—and perhaps is given a pretest. The student then leaves the terminal, does the assigned work, and returns for further instructions and testing. Students who reach the desired level of mastery are given the next study assignment. Those doing less well may be given a remedial assignment on the computer or directed to the instructor for tutorial work.

Here is a portion of the printout from the BASMS program:*

```
IF IN THE FOLLOWING MATRIX EQUATION WE HAVE AB=0, THEN
    1) A MUST EQUAL THE ZERO MATRIX
    2) B MUST EQUAL THE ZERO MATRIX
    3) EITHER A OR B MUST EQUAL THE ZERO MATRIX
    4) BOTH A AND B MUST EQUAL THE ZERO MATRIX
    5) ALL OF THE ABOVE ARE CORRECT
    6) NONE OF THE ABOVE IS CORRECT
THE CORRECT ANSWER IS? 3
YOUR ANSWER IS WRONG.
PLEASE CONFER WITH YOUR INSTRUCTOR.
APPOINTMENT TIMES ON MONDAY ARE AVAILABLE AS FOLLOWS:
1) 8:00 AM   2) 10:00 AM   3) 1:30 PM   4) 3:00 PM
CHOOSE A TIME BY TYPING ITS PRECEDING NUMBER? 4
YOUR APPOINTMENT TIME WITH PROFESSOR HICKS ON MONDAY, APRIL 15
AT 3:00 PM IS RECORDED.
DO YOU WISH TO SIGN OFF NOW? YES
```

INSTRUCTIONAL APPLICATIONS OF COMPUTERS

The various ways in which computers may be used for mathematics instruction are discussed here under four headings: (1) drill and practice, (2) tutorial, (3) flowcharting and programming, and (4) simulation.

---

* Numerals or other symbols entered by the person using the computer program are indicated throughout this chapter by italics and they always follow a question mark. For example, in the BASMS printout, the program user types "3" in line 8 and "4" in line 13.

DRILL AND PRACTICE

The importance of practice in skill learning has been discussed in Chapter 3. A computer is in many ways ideally suited for giving practice in skills because it can provide individual attention to the learner, virtually instantaneous feedback, and tireless repetition. A simple example of a drill and practice program is the following, which provides practice on the basic facts of addition. Recall that in this text we have adopted these conventions:

1. The computer program is enclosed like this:
2. The output printed by the computer is enclosed like this:

```
10    PRINT
20    LET X=INT(10*RND(1)+1)
30    LET Y=INT(10*RND(1)+1)
50    PRINT X;"+";Y;"=";
60    INPUT A
70    IF A=X+Y THEN 100
80    PRINT "NØ,";X;"+";Y;"=";X+Y
90    GØ TØ 10
100   PRINT "RIGHT! NØW TRY THIS ØNE"
110   GØ TØ 10
200   END
```

```
5+9=? 14
RIGHT! NØW TRY THIS ØNE

6+4=? 9
NØ, 6+4=10

5+7=? 12
RIGHT! NØW TRY THIS ØNE

8+3=?
```

In statements 20 and 30 of the program above, two integers between 1 and 10, inclusive, are randomly generated. (The RND function is explained in Appendix C.) With a few changes, the program is readily extended to provide practice with three addends, as shown below. Arithmetic operations other than addition can also be used (see Activity 1).

```
10    PRINT
20    LET X=INT(10*RND(1)+1)
30    LET Y=INT(10*RND(1)+1)
40    LET Z=INT(10*RND(1)+1)
50    PRINT X;"+";Y;"=";
60    INPUT A
70    IF A=X+Y THEN 100
80    PRINT "NØ,";X;"+";Y;"=";X+Y
90    GØ TØ 10
100   PRINT "RIGHT! NØW TRY THIS ØNE";
110   PRINT X;"+";Y;"+";Z;"=";
120   INPUT B
130   IF B=X+Y+Z THEN 160
140   PRINT "NØ,";X;"+";Y;"+";Z;"=";X+Y+Z
150   GØ TØ 10
160   PRINT "WAY TØ GØ...!"
170   GØ TØ 10
200   END
```

```
10+4=? 14
RIGHT! NØW TRY THIS ØNE 10+4+8=? 18
NØ, 10+4+8=22

9+2=? 12
NØ, 9+2=11

1+8=? 9
RIGHT! NØW TRY THIS ØNE 1+8+7=? 16
WAY TØ GØ...!

10+9=? 19
RIGHT! NØW TRY THIS ØNE 10+9+10=? 29
WAY TØ GØ...!
```

To be sure, this drill and practice program is simple and has limited applications in actual classroom situations. However, elaborate programs have been written which have a great variety of features. For example, Hewlett-Packard (1972) has a system of programs which provides individualized drill and practice in arithmetic for grades one through six. The programs have been used with success in remedial instruction at the secondary school level as well: Each child is assigned a difficulty level by his teacher, and the system then provides exercises of an appropriate level based on the child's performance on pretests and lessons. The Hewlett-Packard system also provides the teacher with reports of student progress, time spent on the system, and achievement gains for both individual students and for classes (see Research Highlight, "Computers in Remedial Mathematics").

RESEARCH HIGHLIGHT | COMPUTERS IN REMEDIAL MATHEMATICS

Woodrow Wilson High School in San Francisco has been using computers with entering students having severe deficiencies in reading and mathematics. Approximately one-third of the 1500 students in the school receive instruction in some form from a computer. Some students receive as much as three hours of instructional time per week by computer. One study on a four-month period of computer-assisted mathematics showed average gains for the students of .91 and 1.03 years for arithmetic applications and computation, respectively, as measured by the Stanford Achievement Test. Corresponding gains for students not using the computer were .31 years for applications and .5 years for computation. (Hewlett-Packard, 1972)

TUTORIAL APPLICATIONS

In the tutorial mode, the computer teaches a block of subject matter, be it a single concept, a unit of material, or an entire course. In the sample printout below, for example, the student is being taught how to reduce a common fraction to its simplest terms. (The complete program, REDUCE, is given in Appendix C.)

```
REDUCE THIS FRACTION: 24/12

TO REDUCE THIS FRACTION YOU LOOK FOR ALL THE COMMON
    FACTORS OF THE NUMERATOR AND THE DENOMINATOR.
    WHAT IS ONE FACTOR THAT THE NUMERATOR AND
    DENOMINATOR HAVE IN COMMON?
    ? 7
TRY ANOTHER NUMBER, THAT NUMBER IS NOT A FACTOR
    OF BOTH THE NUMERATOR AND THE DENOMINATOR.
    ? 4
```

```
YØU ARE CØRRECT! THAT NUMBER IS A FACTØR ØF BØTH THE
    NUMERATØR AND THE DENØMINATØR.
    NØW YØU NEED TØ DIVIDE BØTH THE NUMERATØR AND
    THE DENØMINATØR BY THAT NUMBER. WHAT IS THE
    RESULT? (EXPRESS ANSWER N/D AS N,D)
    ? 8,3
YØU HAVE MADE A MISTAKE DIVIDING NUMERATØR BY
    THAT FACTØR. TRY AGAIN.
    ? 6,2
YØU HAVE MADE A MISTAKE DIVIDING DENØMINATØR
    BY THAT FACTØR. TRY AGAIN.
    ? 6,3
YØUR RESULT IS CØRRECT SØ FAR, BUT
    CAN YØU FIND A GREATER CØMMØN FACTØR? (TYPE 1=YES, 2=NØ)
    ? 1
WHAT IS THE CØMMØN FACTØR?
    ? 2
2 IS NØT GREATER THAN 4
```

The programs in this book are written in BASIC and can be used on an ordinary teletypewriter. However, the possibilities for devising interesting lessons can be increased by the availability of special computer languages for teaching purposes and by special equipment for presenting information to the learner. One such language is Coursewriter, available since the early 1960s on IBM (International Business Machines) computers and more recently implemented on other computers as well. Another instructional language is IDF (Instructional Dialogue Facility) developed by the Hewlett-Packard Corporation. A third such language is TUTOR, developed at the Computer-Based Educational Research Laboratory at the University of Illinois, Urbana, for use on their teaching computer system, PLATO (Programmed Logic for Automatic Teaching Operations). The PLATO system has been under development and testing since 1960. The programming language TUTOR enables the teacher to present messages to the student using a single command. When the student answers by typing in a response, PLATO checks for spelling errors and marks unacceptable words. One of the outstanding features of PLATO is its display panel which enables the teacher to present figures, diagrams, and text on a 20-centimeter square screen at the rate of 180 characters per second (see Figure 9–2). A slide projector permits the superimposing of color photographs on the screen. Other auxiliary equipment includes audio systems for giving spoken commands to the student and a touch panel which can detect that portion of the display which has been touched by the user.

Figure 9–2   Student using a PLATO terminal. (Photo courtesy Computer-based Education Research Laboratory, University of Illinois at Urbana-Champaign.)

In the PLATO system the teacher, the computer, and the students are all members of an interactive team. The teacher designs the instructional materials and prepares computer programs for teaching the materials. The computer presents the materials, monitors the student's performance, and collects data which can be used to evaluate student learning and the effectiveness of the lessons. The programs may then be modified on the basis of feedback. To date, thousands of students from elementary grades through graduate and professional school have studied on PLATO such diverse subjects as mathematics, foreign languages, physics, medicine, and social welfare. PLATO terminals are located not only on the University of Illinois campus at Urbana, but in some public schools in the state, on campuses of the City College System of Chicago, at locations as far away as the east and west coasts of the United States, and in various European countries (see Figure 9–3).

FLOWCHARTING AND PROGRAMMING

Flowcharting provides a useful tool for analyzing and solving problems because it provides a solution in pictorial form. Although professional programmers may use a bewildering array of flowchart symbols, most programming problems encountered in secondary school mathematics can be flowcharted using the four symbols shown in Figure 9–4.

Elementary skills of flowcharting can readily be learned by most junior high school students, and such skills can then provide an effective introduction to computer programming. Once a correct flowchart has been prepared, the writing of the corresponding computer program should be an easy task.

As an introduction to flowcharting, you may wish to have students analyze

Figure 9–3   Location of PLATO terminals in North America.

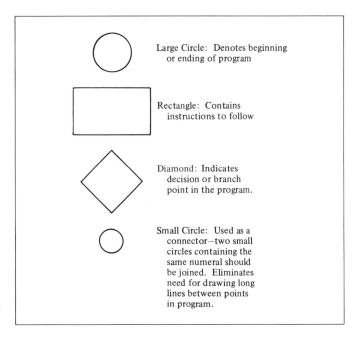

Figure 9–4 Four elementary flowchart symbols.

a real-world situation, such as determining whether a pay telephone is in working order. Figure 9–5 gives one analysis of a component of this situation using the four symbols of Figure 9–4. Extensions of this flowchart can be made to take into account other aspects of the task of placing a call from a pay telephone. (For example, what is the procedure to follow should the phone be found out of order, yet the coin not be returned?)

Programming Languages In order to communicate with a computer (to give it instructions, for example) we need to know what languages it "understands." Every computer knows at least one language, the language which gives instructions broken down into steps which it can follow, such as "store a number," "subtract a second number from the first," or "test to see if a certain stored number is negative." These instructions are often written in the form of binary digits. Since these "machine languages" are difficult for most people to learn, over the years a number of "user languages" have been written. Many of these languages are simple, from the point of view of the user. For example, the instruction SQR(N) in a user language can be translated into a complex sequence of small steps which the computer is able to follow (that is, machine language instructions) in order to calculate the square root of N. This translation from the user language to the machine language is done by another set of instructions in the computer called the "compiler." Fortunately for those of us who wish to use computers with a minimum of effort, it is necessary only to learn a user language, several examples

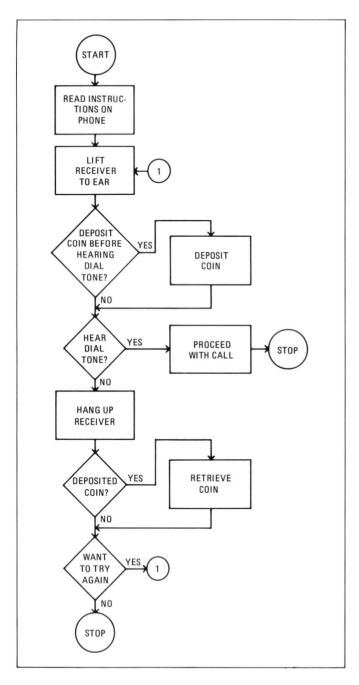

Figure 9–5 Flowchart for determining whether a pay telephone is in service.

of which follow. However, for those students who will enter various fields of computer science, it is important that they learn a machine language as well. One source of information about machine language suitable for high school students is Hagelbarger and Fingerman (1968).

FORTRAN (FORmula TRANslator) is one of the oldest and most commonly used languages in the scientific community. FORTRAN was developed for use in complex mathematical computation. It is relatively easy to learn, but implementations of the language often involve cumbersome input and output formatting requirements. Therefore, although some secondary schools use FORTRAN, it has not become widespread.

FOCAL (FOrmula CALculator) is a language somewhat easier to learn than FORTRAN. This language consists of only a few simple instructions and requires very little output or input formatting. However, the language is somewhat hard to follow for those inexperienced with computers.

APL (A Programming Language) is an extremely powerful language developed by Iverson at the IBM Corporation. One unique feature of the language is that it is designed to handle vectors as easily as it handles scalar quantities. For example, the expression $X = A + Y$ in the APL language might refer either to addition of vectors or to addition of real numbers in the usual sense. Iverson (1971) has written an introductory algebra course based on the use of the APL language. The course provides an example of how novel approaches to elementary algebra are made possible by computers.

COBOL (COmmon Business Oriented Language) is a special-purpose language for business uses such as accounting and keeping inventories.

PL/I (Programming Language/One) is designed to process numerical as well as string data, although some of the other languages mentioned here can handle string data as well. String data are made up of characters such as letters or numerals. (All of the lines of print on this page, for example, constitute a set of string data.)

BASIC (Beginner's All-purpose Symbolic Instructional Code) is probably the computer language most often used today in schools—and is gaining popularity in colleges as well. You can learn BASIC by studying Chapters 3–4 and Appendix C, where the essentials of the language are summarized, and then working through the various programs given there and in this chapter. For a more extensive introduction to BASIC, see Kemeny and Kurtz (1971), Coan (1970), or People's Computer Company (1972).

Divisors: A Sample Flowcharting and Programming Activity   Let's say that you are dealing with the topic of divisors. You want your students to find the divisors of several numbers, and you wish to enlist the help of a computer to check their work. A flowchart of a procedure for obtaining the divisors of a number appears in Figure 9–6. The flowchart suggests that a key programming task is to

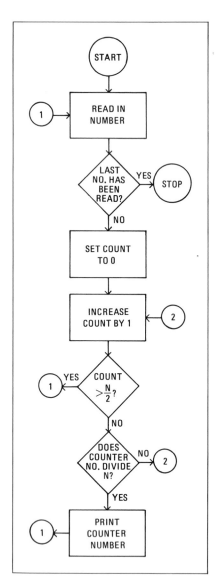

Figure 9–6 Flowchart for finding the
divisors of a number.

check whether a number I is a divisor of the given number N. The result is a
programming exercise which demonstrates the use of the BASIC function
INT(A). The INT function simply yields the greatest integer value in A. For
example, INT (3.36) = 3 and INT (.237) = 0. Now, what is meant by "divisor?"

I is a divisor of N if N/I has a remainder of zero

But by the division algorithm, our definition can be restated:

I is a divisor of N if N = I*INT(N/I)

Let's try this definition on a few examples to see how the rule works:

N	I	INT(N/I)	I*INT(N/I)	I divisor of N?
6	2	3	2*3=6	yes
9	2	4	2*4=8	no
21	3	7	3*7=21	yes
21	6	3	6*3=18	no

All that remains now is to write a program from the flowchart using our rule for checking divisibility. Notice that in statement 60, where the rule is applied, it is more convenient to use the "not equal" relation written in BASIC either as "<>" or "#".

```
10    PRINT "WHAT NUMBER",
20    INPUT N
30    IF N=-999 THEN 110
40    PRINT "THE DIVISØRS ØF ";N;"ARE";
50    FØR I=1 TØ N/2
60    IF N#I*INT(N/I) THEN 80
70    PRINT I;
80    NEXT I
90    PRINT
100   GØ TØ 10
110   END
```

```
WHAT NUMBER   ? 6
THE DIVISØRS ØF    6    ARE 1 2 3
WHAT NUMBER   ? 7
THE DIVISØRS ØF    7    ARE 1
WHAT NUMBER   ? 28
THE DIVISØRS ØF   28    ARE 1 2 4 7 14
WHAT NUMBER   ? 211
THE DIVISØRS ØF   211   ARE 1
WHAT NUMBER   ? 1728
THE DIVISØRS ØF   1728 ARE 1 2 3 4 6 8 9
   12 16 18 24 27 32 36 48 54 64 72 96
   108 144 192 216 288 432 576 864
WHAT NUMBER   ? -999
```

In this program, we have assumed "divisor" to mean "proper divisor"; that is, all divisors of N excluding N itself. However, the program could easily be modified to include N as a divisor if desired.

Now let's think about how this program might be used in class. You could prepare a worksheet on which there are some numbers for which the divisors are easy to find and some for which it is more difficult. This list might be appropriate as a start:

6   7   21   28   84   120   129   200   211   1640   1728   1991

You could have the class begin by finding as many divisors as they can using pencil and paper. Then students could go to the computer either individually or in groups of two or three to check their answers and to see the answers obtained by the computer for those which the students could not work. The class might also enjoy comparing times—theirs and the computer's—for doing the same exercises. (This exercise has several possibilities for future investigation; some whole numbers have many divisors while others have only a few, and this leads to classifying numbers as prime or composite.)

Another possibility is to classify numbers according to the sum of their proper divisors. The program above could be modified to compute the sum of the divisors. Students could then use this information to tell whether a number is perfect, abundant, or deficient, as shown in Table 9–1. Prielipp in the *Mathematics Teacher* (1970) discusses this topic in some detail. (See also the problem on amicable numbers, Appendix B, p. 510.

SIMULATION

Simulation can be very useful in teaching. In the training of aircraft pilots, for example, considerable time may be required in a simulated cockpit before actual flight is attempted. Such experience is so highly rated that the Federal Aviation Administration will accept time spent in a simulator up to a specified limit as credit toward required flight time for licensing of pilots. And new maneuvers in the handling of a spacecraft may be simulated on the ground before they are attempted in actual spaceflight. Simulation is the representation or imitation of actual events or organizations for the purpose of studying how they operate under

Table 9–1. Numbers Classified According to Sums of Their Proper Divisors.

Kind of Number	Definition	Example
Perfect	sum of its divisors equals the number itself	$6 = 3 + 2 + 1$ 6 is a perfect number
Abundant	sum of its divisors exceeds the number itself	$9 + 6 + 3 + 2 + 1 > 18$ 18 is an abundant number
Deficient	sum of its divisors is less than the number itself	$4 + 2 + 1 < 8$ 8 is a deficient number

certain conditions over a period of time. With the availability of computers, some very complex simulations, such as the behavior of the voting public, can be developed. We will consider a simple example first and then turn to more advanced simulations which have been prepared for school use.

Let's assume we are studying quadratic equations and wish to investigate the way in which a function $f(x)$ varies as x increases. We might begin with the function

$$s(t) = \tfrac{1}{2}gt^2$$

and consider the model of a falling ball.

Example: The Falling Ball, or Galileo Revisited   One of the best known anecdotes in science is that of Galileo and the Leaning Tower of Pisa (see Gamow, 1961). The story goes that by watching swinging pendulums, Galileo formulated the laws of falling bodies. He then verified the laws by dropping from the tower objects of equal sizes but different masses and observing that they hit the ground simultaneously. A simple version of the law he investigated is

$$s(t) = \tfrac{1}{2}gt^2$$

where s is the distance which the object falls in t seconds and g is the acceleration due to gravity, which we take here as 10 meters per second per second (see Figure 9–7).

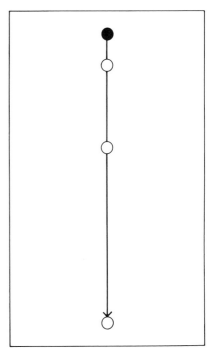

Figure 9–7   Acceleration of the falling ball.

With the help of this computer program, we simulate a "visit" to the Leaning Tower and "drop" a ball. The program tells how far a free-falling body will have dropped at the end of t seconds. Or we could use the program to obtain an estimate of how many seconds it would take the ball to reach the ground if the tower were 200 meters high.

```
10   DEF FNS(T)= .5*G*T↑2
20   PRINT "NØ. SECØNDS";
30   INPUT K
40   PRINT "VALUE ØF G";
50   INPUT G
60   PRINT
70   PRINT " T", "DISTANCE"
80   FØR T=1 TØ K
90   PRINT T, FNS(T)
100  NEXT T
110  END
```

```
NØ. SECØNDS ? 12
VALUE ØF G ? 10

    T              DISTANCE
    1              5
    2              20
    3              45
    4              80
    5              125
    6              180
    7              245
    8              320
    9              405
    10             500
    11             605
    12             720
```

An interesting feature of this program is the use of the DEF function in BASIC to define a function in one variable. Once a function, say FNS(T), is defined, it can be evaluated later in the program for the current value of t. Let's see how this feature can be used to generate another kind of activity. Suppose we drop a ball not from the tower in Pisa but from a tower on some other planet.

The problem now is to determine the value of g, the acceleration due to gravity, on this unknown planet.

```
10   DEF FNS(T)=.5*G*T↑2
20   PRINT "NØ. SECØNDS";
30   INPUT K
40   LET G=INT(20*RND(1))+2
50   PRINT
60   PRINT " T", "DISTANCE"
70   FØR T=1 TØ K
80   PRINT T, FNS(T)
90   NEXT T
100  END
```

```
NØ. SECØNDS ? 10

    T               DISTANCE
    1               2.85
    2               11.4
    3               25.65
    4               45.6
    5               71.25
    6               102.6
    7               139.65
    8               182.4
    9               230.85
    10              285
```

From the output of this program, can you calculate the value of g? You are at an advantage since you can see from statement 40 of the program that a random integer between 2 and 21 is generated as a value for g. But you might not want to let students see the program; just the output.

If the relationship between time and distance also takes the initial velocity of the falling object into account, it is a more challenging problem to calculate the value of g—that is, to define

$$s(t) = \tfrac{1}{2}gt^2 + v_0 t$$

where v_0 is initial velocity, which may also be randomly generated by the program.

Finally, we must be aware that our function $s(t)$ assumes that the object is falling in a vacuum; that is, air resistance is ignored. A program which takes air resistance into account is given on pp. 332–333.

The Huntington II Simulation Project This project has developed instructional packages for using computer simulations in secondary school classes. The materials were devised at the Polytechnic Institute of Brooklyn and field tested in schools throughout the United States. Teacher manuals for the materials were developed by the Northwest Regional Educational Laboratory in Portland, Oregon. The computer programs are written in BASIC and should easily adapt to most computer systems which accept BASIC. Sample materials include:

BALPAY   A program for use in economics classes. It permits development of understanding of balance of payments in foreign trade.

SLITS   A physics program which simulates Young's double-slit experiment in demonstrating interference patterns in light.

WHEELS   A simulation of the operation and maintenance problems associated with automobiles.

POLUT   A simulation of water pollution which yields effects of pollutants on a river, pond, or lake. Graphical output of the various factors involved is provided.

Each package of materials consists of a punched paper tape of the BASIC program, a teacher's manual describing briefly how to use the program, a list of preparatory materials to provide for the students, some questions to use in discussing results of the simulation, and some sample runs to give the teacher an understanding of how the simulation works. The Huntington II simulation materials are available through such educational computer companies as Hewlett-Packard and Digital Equipment Corporation (see Appendix D for addresses).

COMPUTER HARDWARE AND SOFTWARE

The terms *hardware* and *software* are commonly used by computer specialists. *Hardware* refers to the computer itself, to its size, its capabilities, and its components; *software* is the programming needed to run a computer. Hardware typically is specified by the manufacturer—that is, the computer is equipped with a certain memory size and speed and with such equipment as teletypes, printers, and so forth. Software, on the other hand, can be changed, depending upon such factors as the skill of the computer programmer and the limitations of time and of the computer's hardware.

Three major kinds of computer usage are available for schools: interactive time-sharing, quick batch, and slow batch. *Interactive time-sharing* permits several different users (the more sophisticated the computer facility, the more users may be permitted) to interact with the computer virtually simultaneously, usually through a teletypewriter or a cathode-ray tube terminal. Since computer operations are so fast, even though each user is being responded to in turn, the impression is usually given that each user has the entire machine at his or her disposal. *Batch* is a form of computer use which involves inputting a *job* (often a computer program and a set of data to be processed), executing it, and providing the results, typically printed out on a sheet of paper. For *quick batch*, the time between reading in the program (perhaps on cards or punched tape) and receiving the output varies from a few seconds to several minutes. *Slow batch* may require a day for processing, since the jobs may be held until late at night when the computer is less busy (hence this form of usage may be less expensive) or it may be necessary physically to transport the jobs to a central location where the computer can be accessed (see Research Highlight, "To Batch or Not to Batch").

RESEARCH HIGHLIGHT | TO BATCH OR NOT TO BATCH

A question frequently asked is "How should the computer be used?" Pack (1971) compared three ways of using the computer as a problem-solving tool: interactive time-sharing; quick batch, requiring eight minutes; and slow batch, in which solutions were returned to students the next day. Following two weeks of instruction in BASIC, 36 high-ability students rotated through the three modes, spending ten hours in each. No significant differences were found in the number of problems solved in each mode. However, student questionnaires revealed a marked preference for interactive time-sharing, with quick batch a second choice and slow batch least preferred.

The answer to the question of which mode to use is one that must be answered on the basis of philosophical considerations plus cost factors. It appears that batch processing of student programs, without computer access, is at least as effective as direct access. The important factor may be experience in writing programs rather than the time it takes to receive computer solutions.

Many experienced computer users agree on the value of some form of "hands on" experience for students working with computers. That is, there should be some opportunity for students to run their own programs, enter them on a teletypewriter, or in some way become at least minimally familiar with the operation of the computer itself. But if this is to be effective, efforts should be made to permit easy access to the machine, including the minimizing of queues of students waiting to use the computer. One way to maximize student involvement is to use portable computer terminals which can be moved easily from classroom to classroom. In some cases these have successfully been utilized in an arrangement permitting the terminals to be taken home for evenings and weekends.

The length of student queues can be minimized by use of a mark-sense card reader. When a computer has such a reader, each student can prepare a program at his or her own desk. The student simply uses a special card (see Figure 9–8), one card per program instruction, and marks appropriate boxes on the card with a soft lead pencil. For example, to give the instruction PRINT, the student darkens the box labeled "PRINT" in column six of the card shown in the figure. When the entire program has been prepared, the resulting stack of cards is dropped into the reader, and the program is fed into the computer for execution. In this way, 50 or more short programs could easily be run in a class period.

Off-line teletypes (terminals not connected to the computer) can help in maximizing computer usage and student access. It is, after all, wasteful of computer time to have the computer wait for a person to write a program on a teletype, especially if that person is composing the program as he or she goes along. At the same time, other students are being prevented from getting to the machine. Off-line teletypes, which are relatively inexpensive, allow the preparation of programs on punched paper tapes. Once the tape has been prepared, the student goes to the on-line terminal and uses a tape reader on the teletype to feed the program into the computer at a relatively high speed. In such a computer setup, student usage of on-line terminals might be reserved for minor debugging of programs, at least during peak usage hours.

Minicomputers A recent development in computer technology which has important implications for schools is that of the minicomputer. These machines, small enough to fit on an ordinary desk, have many features of the large computers. They can read cards, print at high speed, and process most programs

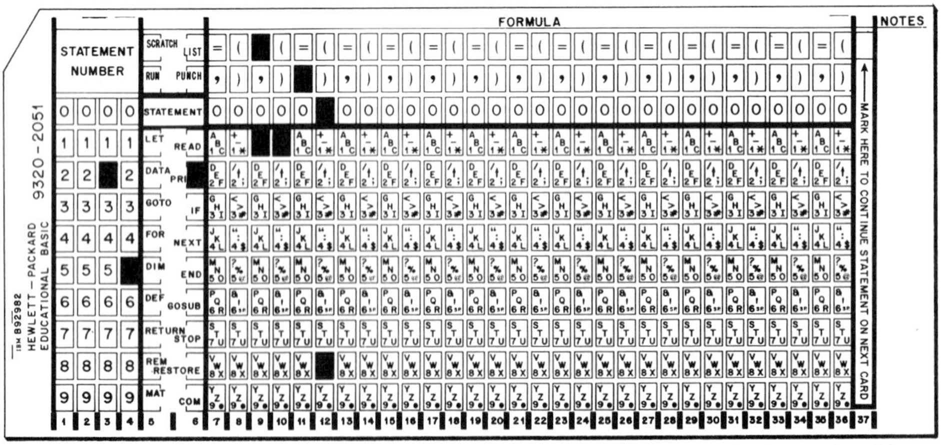

Figure 9–8 Mark-sense card showing BASIC statement 25 PRINT A1,X. (Card reproduced by permission of Hewlett-Packard.)

likely to be encountered in schools. They have a powerful version of a programming language such as BASIC and can store data electronically on regular cassette tapes. In terms of dollar value, many minicomputers offer an enormous computing capability at relatively low cost.

The array of hardware and software available is vast, and it is rapidly increasing in size and diversity. You may be well advised to seek help from specialists in educational applications of computers should you be in the position of making decisions for your school. One such source of information is the newsletter of the People's Computer Company. (The March 1974 issue, for example, features reviews of hardware and software available from the major manufacturers of instructional computers.)

WHY USE COMPUTERS?

We have already noted that the amount of computer programming done by students will depend upon a number of factors, including that of the teacher's instructional objectives. If your objectives are to teach programming concepts, then obviously you will emphasize the details of programming. However, if your objectives relate more closely to mathematics, then you should be careful not to lose sight of those objectives by excessive emphasis on programming. Many students will find computer programming, particularly BASIC, very easy. But you may have to make special provisions for helping those students who do not have an aptitude for programming with the programming tasks. For example, programs can often be written ahead of time and stored in the computer. This allows students to use the program for problem solving without worrying about the programming details involved. Or you may wish to provide students with partially prepared program decks. The student is then required only to supply one or two missing statements and his or her own data. Consider the following exercise, using the program developed on pp. 295–298.

```
10    (missing)
20    (missing)
30    IF N=−999 THEN 110
40    PRINT "THE DIVISØRS ØF ";N;"ARE";
50    FØR I=1 TØ N/2
60    IF N#I∗INT(N/I) THEN 80
70    PRINT I;
80    NEXT I
100   END
```

You may want to tell the student that the missing statements are READ and DATA, and have him decide what variable to use in conjunction with READ and what values to provide for DATA.

Remember that the computer is another teaching aid. How you use it is a matter of your own ingenuity and creativity. Considerable experience and experimentation will be required before you find ways of using the computer which are most effective in helping your students learn (see Research Highlight, "Does Computer-Use Help?").

RESEARCH HIGHLIGHT | DOES COMPUTER-USE HELP?

Behaviors while solving nonroutine problems were studied by Foster (1973). Four treatment groups were distinguished by the supplementary aids made available to the students: G1 used neither computer nor flowcharts, G2 used flowcharts only, G3 used the computer only, and G4 used both computer and flowcharts. Mean performance of the treatment groups on the Problem Solving Abilities Test developed for the study were: G1 < G2 < G4 < G3. The significant difference between the computer-only group (G3) and the no-computer group (G1) was due primarily to performance on four behaviors: specifying conditions a datum satisfies, selecting a relevant solution, proposing a hypothesis, and constructing an algorithm. It was concluded that computer programming, and to a lesser extent flowcharting, tends to support the development of selected problem-solving behaviors.

In a study of the feasibility of a system for using the computer with low achievers, King (1972) used three treatments: mastery learning (ML), mastery learning and flowcharting (ML + FC), and mastery learning, flowcharting, and computer access (ML + FC + CA). On an achievement post-test, the ML + FC group scored significantly higher than the other groups.

From studies such as these it appears evident that problem-solving skills are enhanced by the use of computer programming. Moreover, flowcharting may be a valuable experience even when computers are not available.

STUDENT CONTROL VS. COMPUTER CONTROL

As you consider which instructional applications of the computer to use for a particular lesson, Figure 9–9 may be of assistance. It suggests that the four applications which we have described vary in the amount of control which the student has over her or his instruction. In our drill and practice example, every student goes through essentially the same sequence of exercises except for random generation of the actual number facts presented. If the student provides the

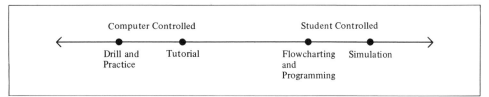

Figure 9–9 Control element in instructional applications of computers.

number making the sentence true, the student is presented with another randomly generated number sentence. If the student does not provide the correct response, the student is told what that number is.

Moving along the continuum of student control we find that many variations of this program would be possible, and probably essential, for maximizing the instructional potential of drill and practice. For example, the range of integers used in the number sentences might be determined by the student's past performance record. In this case students would sign onto the drill program under their own names and the computer would look up their past performance records, select number facts at a level of difficulty which is perhaps a little easier than the individual student is capable of, and then successively increase the difficulty level as the student works through the exercises.

More elaborate instruction would provide a diagnosis of student errors and branch the student to a remedial program if a particular weakness persists. (If you are interested in learning about a relatively elaborate instructional computer program which has been operational in schools for several years, see Suppes et al., 1968.)

Flowcharting and programming offer the student the opportunity to approach a problem from a novel point of view. Some students will pride themselves (and often surprise you) with elegant and efficient programming methods. One of our favorite anecdotes is about a group of high school students who, unbeknownst to their teacher, entered a national competition for professional programmers and walked away with one of the top prizes. The students had earned grades of C+ or B at school.

At the far right of the continuum, simulation offers students the opportunity to create their own world, such as (in our example of the falling ball) one in which acceleration due to gravity is very small or very large. The student then can experiment by "dropping" an object and comparing the distance it falls during the same time period, but with differing values of gravity. More elaborate simulations can be more open ended; there the student can establish a variety of conditions and test hypothesized relationships between variables. In POLUT, the Huntington II simulation, for example, the student specifies a certain kind of body of water (lake, river), its size, and types of pollutants being put into that body of water. The student then is provided detailed output as to the condition over a period of time of the body of water being polluted.

USING COMPUTERS TO ATTAIN GOALS

We will now consider examples of computer activities which may be useful in attaining certain of the mathematical goals already discussed in Chapters 2–6 of this book. (Other examples of computer programs for selected instructional purposes will be found in Appendix C.) In terms of our model, we are here con-concerned with the teaching slice in the goals and content dimensions (see Figure 9–10).

We will provide examples of the four instructional applications of computers by filling in each cell on the main diagonal of the matrix in Figure 9–11. The classification of the examples is not absolute; justification could be given for placing them in different cells. However, we believe this scheme provides a useful starting place for mapping out an instructional program involving computer applications.

(1) Instructional application: Drill and practice
     Goal: Knowledge-skills
     Topic: Difference of two squares

This program randomly generates a B value between 1 and 11 in the expression X^2-B^2. This value can be modified by making changes in statement 90. (See Appendix C for further information on the RND function.)

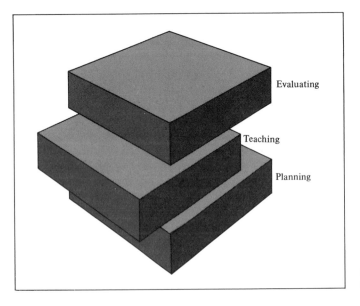

Evaluating

Teaching

Planning

Figure 9–10 The teaching process in the model of mathematics teaching.

		Goals of Instruction			
		Know Skills	Know Facts	Understanding	Solve Problems
Instructional Application of Computers	Drill and Practice	▓			
	Tutorial		▓		
	Flowchart-ing and Programming			▓	
	Simulation				▓

Figure 9–11   Matrix of instructional goals and applications of computers.

Program:

```
10   PRINT "I WILL GIVE YØU PRACTICE IN FACTØRING THE DIFFERENCE"
20   PRINT "ØF TWØ SQUARES"
30   PRINT
40   PRINT "REMEMBER THAT AN EXPRESSIØN ØF THE FØRM"
50   PRINT "X↑2−B↑2"
60   PRINT "HAS TWØ LINEAR FACTØRS ØF THE FØRM (X+B)(X−B)"
70   PRINT "EXAMPLE: X↑2−25=(X+5)(X−5)"
80   PRINT
90   LET A=INT(11*RND(1)+1)
100  LET C=A↑2
110  PRINT "X↑2−";C;"=(X+B)(X−B)"
120  PRINT "WHAT IS B";
130  INPUT B
140  IF A=B THEN 170
150  PRINT "NØ, FØR YØUR VALUE ØF B, B↑2=";B↑2;" TRY AGAIN."
160  GØ TØ 110
170  PRINT "RIGHT ØN....X↑2−";C;"=(X+";B;")(X−";B;")"
180  GØ TØ 80
190  END
```

Sample Output:

```
I WILL GIVE YØU PRACTICE IN FACTØRING THE DIFFERENCE
    ØF TWØ SQUARES


REMEMBER THAT AN EXPRESSIØN ØF THE FØRM
            X↑2−B↑2
HAS TWØ LINEAR FACTØRS ØF THE FØRM (X+B)(X−B)
EXAMPLE: X↑2−25=(X+5)(X−5)


X↑2−64=(X+B)(X−B)
WHAT IS B? 8
RIGHT ØN....X↑2−64=(X+8)(X−8)


X↑2−16=(X+B)(X−B)
WHAT IS B? 6
NØ, FØR YØUR VALUE ØF B, B↑2=36      TRY AGAIN
X↑2−16=(X+B)(X−B)
WHAT IS B? 4
RIGHT ØN....X↑2−16=(X+4)(X−4)
```

(2) Instructional application: Drill and practice
     Goal: Knowledge-skills
     Topic: Complementary and supplementary angles

Two novel features are introduced in this program. First, the variable N\$ in statement 40 stores the student's name so that she or he can be addressed personally. Second, positive reinforcers (VERY GOOD, WAY TO GO, etc.) are chosen randomly in statements 500–535 so that the student is not always given the same response for a correct answer. Negative reinforcers (statements 600–650) are handled in the same way.

Program:

```
10    PRINT "THIS PRØGRAM WILL GIVE YØU PRACTICE IN FINDING"
20    PRINT "CØMPLEMENTARY AND SUPPLEMENTARY ANGLES."
30    PRINT "BEFØRE WE START, WHAT IS YØUR FIRST NAME";
40    INPUT N$
50    PRINT "THANKS, ";N$;", HERE WE GØ ! !"
60    PRINT
70    PRINT "REMEMBER THESE TWØ DEFINITIØNS:"
80    PRINT "   CØMPLEMENTARY ANGLES ARE TWØ ANGLES WHØSE SUM IS"
90    PRINT "      90 DEGREES. EXAMPLE: IF ØNE ANGLE IS 40 DEGREES"
100   PRINT "      THEN ITS CØMPLEMENT IS AN ANGLE ØF 50 DEGREES"
```

```
110   PRINT
120   PRINT "    SUPPLEMENTARY ANGLES ARE TWØ ANGLES WHØSE SUM IS"
130   PRINT "       180 DEGREES. EXAMPLE: IF ØNE ANGLE IS 20 DEGREES"
140   PRINT "       THEN ITS SUPPLEMENT IS AN ANGLE ØF 160 DEGREES."
190   PRINT
200   GØ TØ INT(2*RND(1)+1) ØF 210,300
210   LET C=INT(91*RND(1))
220   PRINT "IF ØNE ANGLE IS ";C;" DEGREES THEN ITS CØMPLEMENT IS"
230   INPUT X
240   IF X=90−C THEN 500
250   GØSUB 600
260   GØ TØ 220
300   LET S=5*INT(37*RND(1))
310   PRINT "IF ØNE ANGLE IS ";S;" DEGREES THEN ITS SUPPLEMENT IS"
320   INPUT Y
330   IF Y=180−S THEN 500
340   GØSUB 600
350   GØ TØ 310
500   GØ TØ INT(3*RND(1)+1) ØF 510,520,530
510   PRINT "VERY GØØD, ";N$
515   GØ TØ 200
520   PRINT "FAR ØUT, ";N$;", YØU GØT IT !"
525   GØ TØ 200
530   PRINT "WAY TØ GØ, ";N$
535   GØ TØ 200
600   GØ TØ INT(3*RND(1)+1) ØF 610, 620, 630
610   PRINT "SØRRY, ";N$ ;", TRY AGAIN"
615   GØ TØ 650
620   PRINT "NØ, CHECK YØUR ARITHMETIC"
625   GØ TØ 650
630   PRINT "UH UH. REMEMBER THE DEFINITIØN?"
650   RETURN
1000  END
```

Sample Output:

THIS PRØGRAM WILL GIVE YØU PRACTICE IN FINDING
CØMPLEMENTARY AND SUPPLEMENTARY ANGLES.
BEFØRE WE START, WHAT IS YØUR FIRST NAME?   JØ
THANKS, JØ, HERE WE GØ! !

REMEMBER THESE TWØ DEFINITIØNS:

COMPLEMENTARY ANGLES ARE TWO ANGLES WHOSE SUM IS
90 DEGREES. EXAMPLE: IF ONE ANGLE IS 40 DEGREES
THEN ITS COMPLEMENT IS AN ANGLE OF 50 DEGREES.

SUPPLEMENTARY ANGLES ARE TWO ANGLES WHOSE SUM IS
180 DEGREES. EXAMPLE: IF ONE ANGLE IS 20 DEGREES
THEN ITS SUPPLEMENT IS AN ANGLE OF 160 DEGREES.

IF ONE ANGLE IS 165   DEGREES THEN ITS SUPPLEMENT IS
? 15
VERY GOOD, JO
IF ONE ANGLE IS 11   DEGREES THEN ITS COMPLEMENT IS
? 79
FAR OUT, JO, YOU GOT IT !
IF ONE ANGLE IS 52   DEGREES THEN ITS COMPLEMENT IS
? 28
NO, CHECK YOUR ARITHMETIC
IF ONE ANGLE IS 52   DEGREES THEN ITS COMPLEMENT IS
? 38
WAY TO GO, JO
IF ONE ANGLE IS 9   DEGREES THEN ITS COMPLEMENT IS
? 81
FAR OUT, JO, YOU GOT IT!
IF ONE ANGLE IS 24   DEGREES THEN ITS COMPLEMENT IS
? 56
SORRY, JO, TRY AGAIN
IF ONE ANGLE IS 24   DEGREES THEN ITS COMPLEMENT IS
? 66
VERY GOOD, JO
IF ONE ANGLE IS 48   DEGREES THEN ITS COMPLEMENT IS
? 132
UH UH. REMEMBER THE DEFINITION?

(3) Instructional application: Tutorial
   Goal: Knowledge-facts
   Topic: Equation of a line
The student provides one of the following sets of information about a straight line:
a. its slope and one point on the line
b. two points on the line
c. its slope and Y-intercept
d. the Y-intercept and one point on the line

Program:

```
10    PRINT "THE EQUATIØN ØF A STRAIGHT LINE HAS THE FØRM Y=M∗X+B"
20    PRINT "WHERE M IS THE SLØPE ØF THE LINE"
30    PRINT "AND B IS THE Y-INTERCEPT. FØR EXAMPLE, THE LINE Y=3∗X+5"
40    PRINT "HAS A SLØPE ØF 3 AND CRØSSES THE Y AXIS AT 5."
50    PRINT "IF YØU WILL GIVE ME CERTAIN INFØRMATIØN ABØUT A LINE,"
60    PRINT "I WILL GIVE YØU ITS EQUATIØN."
70    GØ TØ INT(RND(1)∗4+1) ØF 100,200,300,400
100   PRINT "SUPPØSE YØU KNØW THE SLØPE AND ØNE PØINT (X1,Y1)"
110   PRINT "ØN THE LINE."
120   PRINT "WHAT IS THE SLØPE";
130   INPUT M
140   PRINT "WHAT ARE X1,Y1";
150   INPUT X1,Y1
160   GØ TØ 700
200   PRINT "SUPPØSE YØU KNØW TWØ PØINTS ØN THE LINE."
210   PRINT "GIVE ME X1,Y1 FØR THE FIRST PØINT";
220   INPUT X1,Y1
230   PRINT "GIVE ME X2,Y2 FØR THE SECØND PØINT";
240   INPUT X2,Y2
250   IF X1−X2=0 THEN 280
260   LET M=(Y2−Y1)/(X2−X1)
270   GØ TØ 700
280   PRINT "THE LINE IS PARALLEL TØ THE Y AXIS."
290   PRINT "ITS EQUATIØN IS X=";X1
295   GØ TØ 810
300   PRINT "SUPPØSE YØU KNØW THE SLØPE ØF THE LINE"
310   PRINT "AND ITS Y-INTERCEPT."
320   PRINT "WHAT IS THE SLØPE";
330   INPUT M
340   PRINT "WHAT IS THE Y-INTERCEPT";
350   INPUT B
360   PRINT "THIS ØNE IS EASY...."
370   GØ TØ 800
400   PRINT "SUPPØSE YØU KNØW ØNE PØINT (X1,Y1) ØN THE LINE"
410   PRINT "AND THE Y-INTERCEPT."
420   PRINT "WHAT ARE X1,Y1";
430   INPUT X1,Y1
440   PRINT "WHAT IS THE Y-INTERCEPT";
450   INPUT B
460   LET M=(Y1−B)/X1
```

```
470   GØ TØ 800
700   LET B=Y1−M*×1
800   PRINT "THE EQUATIØN ØF THE LINE IS Y=";M;"*X+";B
810   PRINT "WANT TØ TRY ANØTHER? (0=NØ, 1=YES)";
820   INPUT K
825   PRINT
830   IF K=1 THEN 70
840   END
```

Sample Output:

```
THE EQUATIØN ØF A STRAIGHT LINE HAS THE FØRM Y=M*X+B
WHERE M IS THE SLØPE ØF THE LINE
AND B IS THE Y-INTERCEPT. FØR EXAMPLE, THE LINE Y=3*X+5
HAS A SLØPE ØF 3 AND CRØSSES THE Y AXIS AT 5.
IF YØU WILL GIVE ME CERTAIN INFØRMATIØN ABØUT A LINE,
I WILL GIVE YØU ITS EQUATIØN.
SUPPØSE YØU KNØW THE SLØPE AND ØNE PØINT (X1,Y1)
ØN THE LINE.
WHAT IS THE SLØPE? 1.5
WHAT ARE X1,Y1? 2, −1
THE EQUATIØN ØF THE LINE IS Y=1.5      *X+−4
WANT TØ TRY ANØTHER? (0=NØ, 1=YES)? 1

SUPPØSE YØU KNØW ØNE PØINT (X1,Y1) ØN THE LINE
AND THE Y-INTERCEPT.
WHAT ARE X1,Y1? 2,3
WHAT IS THE Y-INTERCEPT? −1
THE EQUATIØN ØF THE LINE IS Y=2      *X+−1
WANT TØ TRY ANØTHER? (0=NØ, 1=YES)? 1

SUPPØSE YØU KNØW THE SLØPE AND ØNE PØINT (X1,Y1)
ØN THE LINE.
WHAT IS THE SLØPE? −.5
WHAT ARE X1,Y1? 2,0
THE EQUATIØN ØF THE LINE IS Y=−.5      *X+1
WANT TØ TRY ANØTHER? (0=NØ, 1=YES)? 1

SUPPØSE YØU KNØW THE SLØPE ØF THE LINE
AND ITS Y-INTERCEPT.
WHAT IS THE SLØPE? −1
```

```
WHAT IS THE Y-INTERCEPT? 3
THIS ØNE IS EASY....
THE EQUATIØN ØF THE LINE IS Y= −1      *X+3
WANT TØ TRY ANØTHER? (0=NØ, 1=YES)? 0
```

(4) Instructional application: Tutorial
    Goal: Knowledge-facts
    Topic: Cartesian product of a set with itself
This program might be useful in reviewing the concept of Cartesian product. An interesting modification of the program consists of regarding the ordered pairs of integers as fractions. The question is then "How many nonequivalent fractions are thus produced in the Cartesian product?" (Note, for example, that (3,3),(4,4), (5,5), and (6,6) are all equivalent fractions.)
    Program:

```
10    PRINT "I WILL GIVE YØU ALL THE ØRDERED PAIRS WHICH"
20    PRINT "CAN BE MADE FRØM A GIVEN SET ØF CØNSECUTIVE INTEGERS."
30    PRINT "WHAT IS THE FIRST MEMBER ";
40    INPUT M
50    PRINT "WHAT IS THE LAST MEMBER ";
60    INPUT N
70    PRINT "THE ØRDERED PAIRS ARE:"
80    FØR I=M TØ N
90    FØR J=M TØ N
100   PRINT "(";I;",";J;")",
110   NEXT J
120   PRINT
130   NEXT I
140   END
```

Sample Output:

```
I WILL GIVE YØU ALL THE ØRDERED PAIRS WHICH
CAN BE MADE FRØM A GIVEN SET ØF CØNSECUTIVE INTEGERS.
WHAT IS THE FIRST MEMBER? 3
WHAT IS THE LAST MEMBER? 6
THE ØRDERED PAIRS ARE:
( 3   , 3   ) ( 3   , 4   ) ( 3   , 5   ) ( 3   , 6   )
( 4   , 3   ) ( 4   , 4   ) ( 4   , 5   ) ( 4   , 6   )
( 5   , 3   ) ( 5   , 4   ) ( 5   , 5   ) ( 5   , 6   )
( 6   , 3   ) ( 6   , 4   ) ( 6   , 5   ) ( 6   , 6   )
```

(5) Instructional application: Flowcharting and programming
Goal: Understanding
Topic: Rational numbers
Every rational number can be written either as a quotient of integers

$$\frac{a}{b}, \ (b \neq 0)$$

or as a repeating decimal fraction. With the help of a computer we can easily find the decimal name for any rational number. The program consists simply of following the steps for dividing the numerator a by the denominator b (see Figure 9–12). Check the steps by using the rational number

$$\frac{22}{7}$$

Students may then examine relationships between the number of digits in a repeating block (called the period) and determine whether the denominator is a prime or composite number.

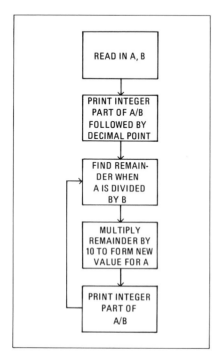

Figure 9–12   Flowchart for dividing the numerator a by the denominator b.

Sample Output:

22/7 =											
3	1	4	2	8	5	7	1	4	2	8	5
7	1	4	2	8	5	7	1	4	2	8	5
7	1	4	2	8	5	7	1	4	2	8	5
7	1	4	2	8	5	7	1	4	2	8	

Output of the program can be made more compact by printing out numerals from a character string. This is accomplished by creating the array N$ as in statement 5 below, then printing out the Zth character of the array by using PRINT N$(n,n) in statement 50.* An added feature to this program would be to calculate the length of the period (that is, the number of digits in the repeating block). Note that 1 is added to Z in statement 45 since the numeral 3 is in the fourth position in N$, and so on.

Modified Program:

```
4    DIM N$(10)
5    LET N$="0123456789"
10   READ A,B
20   PRINT INT(A/B);".";
30   LET R=A−B*INT(A/B)
40   LET A=R*10
45   LET Z=INT(A/B)+1
50   PRINT N$(Z,Z);
60   GO TO 30
70   DATA 22,7
80   END
```

Sample Output:

```
22/7 =
3      .14285714285714285714285714285714285714285714285714285714285
714285714285714285714285714285714285714285714285714285714285
```

---

* This use of an array is not available in all versions of BASIC. Some require that statement 5 be effected by defining individually each of the ten characters (LET N$(1)="0", LET N$(2)="1", and so forth); ten different statements would be needed to do the work of statement 5.

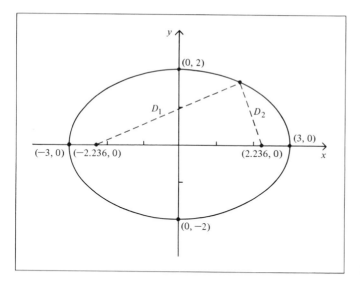

Figure 9–13 An ellipse.

(6) Instructional application: Flowcharting and programming
   Goal: Understanding
   Topic: The ellipse
   An ellipse is the locus of points in a plane such that the sum of the distances of each point from two fixed points (called the foci) is a constant. This programming activity illustrates that points satisfying the equation for the ellipse also satisfy the conditions of the locus. The program requests values for $A, B, C,$ and F where the equation of the ellipse is

$$AX^2 + BY^2 = C \quad \text{for} \quad C = A \cdot B$$

and the foci are $(-F, O), (F, O)$. The values of $D1$ and $D2$ (see Figure 9–13) are then computed:

$$D1 = \sqrt{(X + F)^2 + Y^2}$$

and

$$D2 = \sqrt{(X - F)^2 + Y^2}$$

For these calculations, X is initiated at $-\sqrt{B}$ and incremented until $X = \sqrt{B}$ (see Figure 9–14). One variation of this exercise would be to increment on Y rather than X. In our example, the equation $4X^2 + 9Y^2 = 36$ is used.
   Program:*

---

  * This program uses the PRINT USING feature (statements 130, 140) which is not available in all versions of BASIC. A combination of variables and desired symbols, such as parentheses, in quotation marks can achieve the same effect. See, for example, statement 100 of the program for Example (4).

```
10    DEF FNE(X)=SQR((C−A∗X↑2)/B)
20    PRINT "IN A∗X↑2+B∗X↑2=C WHAT ARE A,B,C";
30    INPUT A,B,C
40    PRINT "WHAT IS FØCUS X IN (X,0)";
50    INPUT F
60    PRINT
70    PRINT "     (X,Y)     D1     D2     D1+D2     "
80    PRINT
90    LET X=−SQR(B)
100   LET Y=FNE(X)
110   LET D1=SQR((X+F)↑2+Y↑2)
120   LET D2=SQR((X−F)↑2+Y↑2)
130   PRINT USING 140;X,Y,D1,D2,D1+D2
140   IMAGE "(",DD.DD,",",DD.DD,")",3(DD.3D6X)
150   LET X=X+.5
160   IF X<=SQR(B) THEN 100
170   END
```

Sample Output:

```
IN A∗X↑2+B∗X↑2=C WHAT ARE A,B,C, ? 4,9,36
WHAT IS FØCUS X IN (X,0)? 2.236


    (X,Y)        D1        D2       D1+D2

(−3.00, 0.00)  0.764     5.236     6.000
(−2.50, 1.11)  1.137     4.863     6.000
(−2.00, 1.49)  1.509     4.491     6.000
(−1.50, 1.73)  1.882     4.118     6.000
(−1.00, 1.89)  2.255     3.745     6.000
(−0.50, 1.97)  2.627     3.373     6.000
(  0.00, 2.00)  3.000     3.000     6.000
(  0.50, 1.97)  3.373     2.627     6.000
(  1.00, 1.89)  3.745     2.255     6.000
(  1.50, 1.73)  4.118     1.882     6.000
(  2.00, 1.49)  4.491     1.509     6.000
(  2.50, 1.11)  4.863     1.137     6.000
(  3.00, 0.00)  5.236     0.764     6.000
```

(7) Instructional application: Flowcharting and programming
      Goal: Understanding
      Topic: Probability

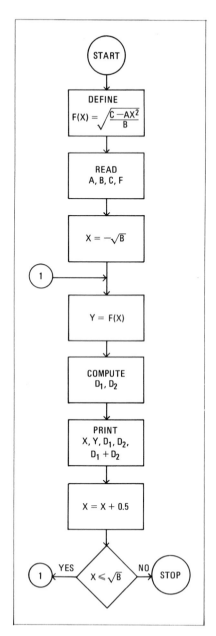

Figure 9–14

The computer makes possible some novel approaches to teaching probability. Consider the problem: "What is the probability that, in a family of three children, all are girls?"

One approach to this problem would be to teach the product rule for the probability of the occurrence of independent events. That is

$$P(3 \text{ girls}) = \tfrac{1}{2}(\tfrac{1}{2})(\tfrac{1}{2}) = \tfrac{1}{8}$$

However, a more intuitive approach, which will apply to a great variety of complex problems (see, for example, Simon and Holmes, 1969) is to "sample" a number of three-child families and see how many consist of three girls. An analogous problem is to toss three coins many times and find the proportion of tosses producing three heads. Here we are using the rule that upon the toss of a coin:

tails corresponds to outcome "child is a boy"
heads corresponds to outcome "child is a girl"

The BASIC program generates zeros and ones each with probability of 0.5 by using the RND function (see statements 30–50). One execution of statements 30–50 corresponds to one toss of three coins or the surveying of one three-child family:

zero corresponds to outcome "child is a boy"
one corresponds to outcome "child is a girl"

Probability theory (see, for example, Blum and Rosenblatt, 1972, pp. 92ff) gives these results:

$$P(0 \text{ girls}) = .125; P(1 \text{ girl}) = .375; P(2 \text{ girls}) = .375; P(3 \text{ girls}) = .125$$

Compare these values for the sample output of the program below for 1000 tosses of three coins.

Program:

```
  5   DIM A(4)
 10   MAT A=ZER
 20   FØR I=1 TØ 1000
 30   X1=INT(RND(1)+.5)
 40   X2=INT(RND(1)+.5)
 50   X3=INT(RND(1)+.5)
 60   A(X1+X2+X3+1)=A(X1+X2+X3+1)+1
 70   NEXT I
 80   PRINT "P(0 GIRLS)=";A(1)/1000
 90   PRINT "P(1 GIRL)=";A(2)/1000
100   PRINT "P(2 GIRLS)=";A(3)/1000
110   PRINT "P(3 GIRLS)=";A(4)/1000
120   END
```

Sample Output:

```
P(0 GIRLS)=.124
P(1 GIRL)=.383
P(2 GIRLS)=.363
P(3 GIRLS)=.13
```

(8) Instructional application: Flowcharting and programming
   Goal: Understanding
   Topic: Inscribed polygons, perimeters, and limits
This program computes the perimeter of a polygon inscribed in a unit circle as the number of sides increases. It may not be self-evident that the program does the required task. One useful activity would be to justify each statement in the program (Figure 9–15 provides a hint). Another use of the program might be to give it to a student with the direction, "Here is a program, tell me what it does. Do not look at the statements, only at the output. What seems to be going on?" Note that $\pi \cong 3.14159265$ and $2\pi \cong 6.28318530$. (See also Figure 9–16.)
   Program:*

```
10    PRINT "N SIDES";"1 SIDE";"PERIM"
15    LET Q=2
20    FOR J=2 TO 16
25    LET S=SQR(Q)
30    LET N=2↑J
35    LET P=N*S
45    PRINT N,S,P
50    LET Q=2-SQR(4-Q)
55    NEXT J
100   END
```

Sample Output:

N SIDES	1 SIDE	PERIM
4	1.414213562	5.656854249
8	.765266865	6.122134920
16	.390180644	6.242890304
32	.196034281	6.273096992
64	.098135349	6.280662336
128	.049082457	6.282554496

* This program should be executed in double-precision BASIC.

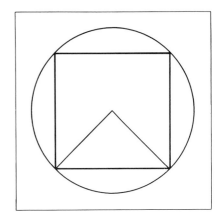

Figure 9–15

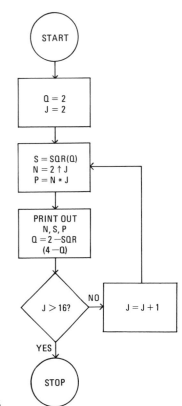

Figure 9–16

256	.024543077	6.283145728
512	.012271769	6.283145881
1024	.006135914	6.283175936
2048	.003067960	6.283182080
4096	.001533981	6.283186176
8192	.000755990	6.283185154
16384	.000383495	6.283182080
32768	.000191748	6.283198464
65536	.000095874	6.283198464

(9) Instruction application: Simulation
   Goal: Problem solving
   Topic: Population models

The problem is to find a function yielding the population N_k of a species (such as humans) at the end of the kth time period. The example here is developed fully by Dorn (1971). If the population growth is P percent during each time period then

$$N_{k+1} = N_k + PN_k$$

This program gives populations for 20 periods, given an initial population of 1000.
   Program:

```
10    REM DATA VALUES: P IS PERCENTAGE GRØWTH PER PERIØD
20    REM N IS INITIAL PØPULATIØN: M IS NØ ØF PERIØDS
30    READ P,N,M
40    DATA .3,1000,20
50    PRINT "PERIØD","PØPULATIØN"
60    FØR I=1 TØ M
70    PRINT I,N
80    LET N=N+P*N
90    NEXT I
100   END
```

Sample Output:

PERIØD	PØPULATIØN
1	1000
2	1300
3	1690
4	2197

5	2856
6	3713
7	4827
8	6275
9	8157
10	10605
11	13786
12	17922
13	23298
14	30288
15	39374
16	51186
17	66542
18	86504
19	112455
20	146192

For many reasons (for example, eventually the species will fill all available space) the function used in statement 80 does not provide a satisfactory model for population growth. A major improvement would be to have the population growth depend upon the population itself. For example,

$$N_{k+1} = N_k + (P - Q*N_k)*N_k$$

Using values for P and Q suggested by Dorn (pp. 6–7) and an initial population of 62.948 million for the United States in the year 1890, we obtain these predicted values.

Modified Program:

```
10    REM DATA VALUES: P IS PERCENTAGE GRØWTH PER PERIØD
20    REM N IS INITIAL PØPULATIØN; M IS LAST YEAR
25    REM Q IS DEPENDENCY FACTØR (SEE LINE 80)
30    READ P,Q,N,M
40    DATA .168,1.77E−04,62.948,2060
50    PRINT "PERIØD","PØPULATIØN (MILLIØNS)"
60    FØR I=1890 TØ M STEP 10
70    PRINT I,N
80    LET N=N+(P−Q*N)*N
90    NEXT I
100   END
```

Sample Output:

PERIØD	PØPULATIØN (MILLIØNS)
1890	62.9480
1900	72.8219
1910	84.1173
1920	96.9966
1930	111.627
1940	128.175
1950	146.800
1960	167.648
1970	190.838
1980	216.453
1990	244.524
2000	275.021
2010	307.836
2020	342.780
2030	379.570
2040	417.836
2050	457.131
2060	496.941

This program gives reasonably accurate results (for the 48 contiguous states) up to 1970. Furthermore, the population values level off at about 950 million by the year 2500.

(10) Instructional application: Simulation
  Goal: Problem solving
  Topic: Convergence

One snowflake will not make a snowman. But a figure constructed with the help of a computer does produce a fascinating picture. Look at the output to see what appears to be happening to the perimeter and area of the resulting polygon.

The snowflake curve (see Figure 9–17) begins with an equilateral triangle having a side of length K. If you trisect each side and on the center section construct a new equilateral triangle, you will have an equilateral polygon of 12 sides. If you repeat this construction procedure a figure resembling a snowflake emerges. Fehr (1951, pp. 140ff) attributes this problem to Koch (as presented by Boltzmann in 1898), stating that the curve is used in the theoretical study of gases. While the perimeter of the curve grows without bounds, the area has the limit

$$\frac{2\sqrt{3}}{5} \cdot K^2$$

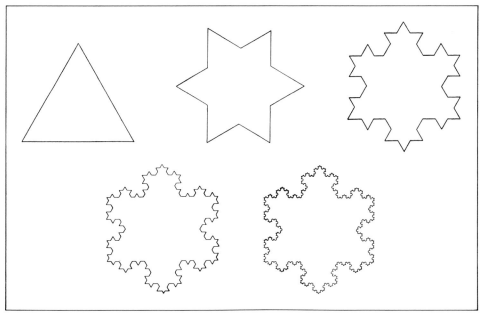

Figure 9–17   Steps in constructing the snowflake curve.

Program:

```
10    PRINT "HØW MANY CØNSTRUCTIØNS";
20    INPUT J
30    PRINT "LENGTH ØF SIDE ØF ØRIGINAL EQUILATERAL TRIANGLE";
40    INPUT K
50    PRINT
60    A=K*K*SQR(3)/4
70    P=3*K
80    S=3
90    PRINT "NCØNST     NSIDES     PERIM     AREA"
100   PRINT USING 110;1,S,P,A
110   IMAGE 3D,4X10D,10X4D.4D,5X4D.4D
120   FØR N=2 TØ J
130   LET P=P*4/3
140   A=A+(A/3)*((4/9)↑(N−2))
150   S=S*4
160   PRINT USING 110;N,S,P,A
170   NEXT N
180   END
```

Sample Output:

HØW MANY CØNSTRUCTIØNS? *15*
LENGTH ØF SIDE ØF ØRIGINAL EQUILATERAL TRIANGLE? *2*

NCØNST	NSIDES	PERIM	AREA
1	3	6.0000	1.7321
2	12	8.0000	2.3094
3	48	10.6667	2.6515
4	192	14.2222	2.8261
5	768	18.9630	2.9088
6	3072	25.2840	2.9467
7	12288	33.7119	2.9637
8	49152	44.9492	2.9713
9	196608	59.9323	2.9747
10	786432	79.9098	2.9762
11	3145728	106.5464	2.9769
12	12582912	142.0618	2.9772
13	50331647	189.4158	2.9773
14	201326588	252.5544	2.9774
15	805306350	336.7391	2.9774

(11) Instructional application: Simulation
    Goal: Problem solving
    Topic: Areas of irregular regions

Suppose we have a curve which encloses a region as suggested in Figure 9–18. If we use the rectangle *ABDE* as a dart board and throw darts in such a way that each point on the board has equal likelihood of being hit, then the probability of hitting the shaded region, which we call *P*(hit), is

$$P(\text{hit}) = \frac{\text{Area of shaded region}}{\text{Area of } ABDE}$$

Using a computer, we simulate the tossing of many darts at the rectangular region by generating pairs of random numbers. We then keep score on how many of these "darts" hit the region whose area we are attempting to determine. The resulting score, in terms of number of hits per total darts tossed, together with our knowledge of the area of the rectangle *ABDE*, enables us to solve the equation for the area of the shaded region. This Monte Carlo method, called the "hit and miss" approach (for obvious reasons), is one of the least efficient but one of the easiest to understand. A discussion of Monte Carlo methods is available in Hammersley and Hanscomb (1964) or in Sobol (1974).

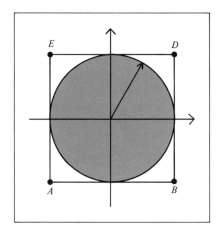

Figure 9–18

The first program finds the area enclosed by the circle $X^2 + Y^2 = 1$ (see Figure 9–18). But since we know the area of this circle to be π we have:

$$P(\text{hit}) = \frac{\pi r^2}{\text{Area of } ABDE} = \frac{\pi}{4}$$

Hence, by finding $P(\text{hit})$ for many tosses we can estimate π. (Appendix C gives a program for estimating π by several different methods.)

Program:

```
10    LET H=0
89    REM AREA ØF SQUARE IS 2*2
90    LET A=4
100   PRINT "N TØSSES","EST. ØF PI"
110   FØR I=1 TØ 1500
120   REM GENERATE RANDØM X AND Y IN INTERVAL (−1,1)
130   LET X=2*RND(1)−1
150   LET Y=2*RND(1)−1
160   REM CHECK WHETHER PØINT(X,Y) IS HIT
170   IF X↑2+Y↑2>1 THEN 190
180   LET H=H+1
190   REM PRINT ØUT VALUE EVERY 100 TØSSES
200   IF I#100*INT(I/100) THEN 220
210   PRINT I, (H/I)*A
220   NEXT I
230   END
```

Sample Output:

N TOSSES	EST. OF PI
100	3.2
200	3.06
300	3.02667
400	3.09
500	3.136
600	3.11333
700	3.11429
800	3.08
900	3.09778
1000	3.1
1100	3.09455
1200	3.11667
1300	3.08923
1400	3.08286
1500	3.088

The second program finds the area under the graph of Y = SIN(X) between X = 0 and X = π (see Figure 9–19). Hence, we have an estimate for:

$$\int_{x=0}^{x=\pi} \sin (x) \, dx$$

Note that BASIC trigonometric functions assume the argument to be in radians.
  Modified Program:

```
10    LET H=0
20    DEF FNA(X)=SIN(X)
30    PRINT "WHAT ARE ENDPOINTS OF INTERVAL FOR X";
40    INPUT X1,X2
50    PRINT "MAXIMUM VALUE OF FUNCTION IN THIS INTERVAL";
60    INPUT M
70    PRINT
80    REM AREA OF RECTANGLE IS M*(S2-X1)
90    LET A=M*(X2-X1)
100   PRINT "N TOSSES","AREA"
110   FOR I=1 TO 1000
120   REM GENERATE RANDOM X IN INTERVAL X1,X2
130   LET X=(X2-X1)*RND(1)+X1
140   REM GENERATE RANDOM Y IN INTERVAL 0,M
```

```
150   LET Y=M*RND(1)
160   REM CHECK WHETHER PØINT (X,Y) IS HIT
170   IF Y>FNA(X) THEN 190
180   LET H=H+1
190   REM PRINT ØUT AREA EVERY 100 TØSSES
200   IF I#100*INT(I/100) THEN 220
210   PRINT I,(H/I)*A
220   NEXT I
230   END
```

Sample Output:

```
WHAT ARE ENDPØINTS ØF INTERVAL FØR X? 0,3.1415926
MAXIMUM VALUE ØF FUNCTIØN IN THIS INTERVAL? 1

N TØSSES                          AREA

    100                          1.82212
    200                          1.85354
    300                          1.92684
    400                          1.97135
    500                          1.95407
    600                          1.98444
    700                          1.97023
    800                          1.98706
    900                          2.0176
   1000                          2.0169
```

(12) Instructional application: Simulation

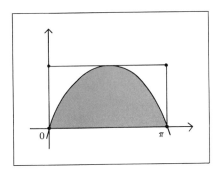

Figure 9–19 The sine curve.

Goal: Problem solving

Topic: Falling bodies in air*

We have already discussed the problem of simulating falling bodies under ideal conditions (i.e., no air resistance). The more realistic situation is more complex and is often neglected because of the calculations involved. With the help of a computer, however, much more satisfactory approximations of reality are often easily obtained. Two models are presented here: Model I, where air resistance is taken to be proportional to the velocity of the falling body, and Model II, where air resistance is taken to be proportional to the square of the velocity.

Model I (air resistance proportional to velocity)

$$a = g - \frac{kv}{m}$$

$$v = \left(\frac{mg}{k}\right)(1 - e^{-kt/m})$$

$$y = \left(\frac{mg}{k}\right)\left(t + \frac{me^{-kt/m}}{k} - \frac{m}{k}\right)$$

where:

a = acceleration of body

v = velocity of body

y = vertical distance of body from rest

g = acceleration due to gravity (10 m/sec/sec)

m = mass of body in kilograms

k = constant of proportionality (assume here to be 10.0)

t = time elapsed in seconds

e = 2.71828

Model II (air resistance proportional to square of velocity)

$$a = g - \frac{kv^2}{m}$$

$$v = \sqrt{\frac{mg}{k}} \tanh\left(t\sqrt{\frac{gk}{m}}\right)$$

$$y = \frac{m}{k} \ln\left(\cosh\left(t\sqrt{\frac{gk}{m}}\right)\right)$$

where variables have same meaning as for Model I, but here take the constant $k = .125$, and recall that

$$\tanh(x) = \frac{e^x - e^{-x}}{e^x + e^{-x}} \quad \text{and} \quad \cosh(x) = \frac{e^x + e^{-x}}{2}$$

---

* The authors are indebted to David Thiessen, physics instructor at Wheeling High School, High School Township District 214, Arlington Heights, Illinois, for providing this example.

The program below provides values based on Model II. Students might also prepare a program using Model I and determine which model provides the best simulation of actual falling bodies in air. Graphs of each of Y, V, and A plotted against time T will be helpful in studying the characteristics of each model.

Program:

```
10    REM IN DATA STATEMENTS
20    REM M: MASS ØF FALLING ØBJECT IN KILØGRAMS
30    REM G: ACCELERATIØN DUE TØ GRAVITY M/SEC/SEC
40    REM K: CØNSTANT ØF PRØPØRTIØNALITY
50    REM X: NØ. ØF SECØNDS ØF FALL: X1: TIME INCREMENT
60    READ M,G,K,X,X1
70    PRINT "T SEC","  Y METERS","  V M/SEC","  ACCEL M/SEC/SEC"
80    FØR T=0 TØ X STEP X1
90    U=SQR(G*K/M)*T
100   Y=(M/K)*LØG((EXP(U)+EXP(−U))/2)
110   V=SQR(M*G/K)*((EXP(U)−EXP(−U))/(EXP(U)+EXP(−U)))
120   A=G−K/M*V↑2
130   PRINT T,Y,V,A
140   NEXT T
145   DATA 80,10,.125,120,10
150   END
```

Sample Output:

T SEC	Y METERS	V M/SEC	ACCEL M/SEC/SEC
0	−0.00011	0.00000	10.00000
10	406.87561	67.86270	2.80415
20	1160.68383	78.92915	0.26592
30	1956.74022	79.91154	0.02210
40	2756.41550	79.99274	0.00182
50	3556.38865	79.99940	0.00015
60	4356.38570	79.99995	0.00001
70	5156.38666	80.00000	0.00000
80	5956.38666	80.00000	0.00000
90	6756.38665	80.00000	0.00000
100	7556.38568	80.00000	0.00000
110	8356.38665	80.00000	0.00000
120	9156.38663	80.00000	0.00000

COMPUTERS AND THE CURRICULUM

In this chapter we have attempted to depict the computer as an instructional tool with a role similar perhaps to the one a microscope plays for the biologist in the laboratory. The biologist would not be using the microscope intelligently if its use did not promote his research goals or other purposes. Bill, in our first critical incident, wants to avoid doing the computer mathematics just because it is fashionable. But we use computers because they motivate, they challenge, and they permit new perspectives and approaches to mathematics. Therefore, we select computer activities on the basis of what goals they can help us reach and what mode is most appropriate for a given situation in our classroom.

The extent to which we spend time and energy teaching programming and other computer science topics was important to Sue, in our second critical incident. To respond to this issue, let's consider further the analogy of the microscope and the biologist. While the biologist in his training may have learned something about lenses, magnification, and the physics of light, he probably learned only enough to enable him to make effective use of the instrument. Similarly, while using the computer as an instructional device, we would do well to learn something about computer word size and memory core. But the primary objective is to learn and teach mathematics. Therefore, as teachers, we may have to pare away excess material that does not bear directly on enhancing mathematics instruction. That is a professional decision, one of many we have to make.

The question of what computer topics are relevant to mathematics instruction also arises with flowcharting. We believe flowcharting is a useful tool to facilitate the analysis of problems and the writing of computer programs. We do not recommend teaching flowcharting in mathematics class as an end in itself. The most efficient way to teach flowcharting is probably to use flowcharts as a natural part of programming exercises. If students see flowcharting as a genuine aid to programming, they will be motivated to learn the techniques. If students do not see flowcharting as useful, there would seem to be little value in teaching it.

You will find yourself having to make decisions about how much time you spend on teaching computer programming per se. Factors to take into account in making such decisions include the ease of the programming language itself, the availability of the computer for student use, and your own instructional goals. As a general guideline, we suggest that you teach only enough programming to facilitate attainment of your mathematical goals. But in individual situations there will be great variation in the emphasis which ought to be placed on computer programming.

There are managerial problems to worry about, too, as Gary Williams in the third critical incident discovered. The following hints may be helpful as you go about using the computer.

1. Avoid line-ups by having groups work on a program and assigning one person to enter the program on the computer, by having a mark-sense card reader which permits each person to prepare a program on mark-sense cards at her or his desk, or by providing an off-line terminal for preparing program tapes.
2. Maximize success by using simplified instructions and programs.
3. Encourage investigation. Young people pick up computer skills and concepts very quickly, often faster than do their teachers. Don't hesitate to let students attack problems and projects on their own. You'll be impressed with what some of them can do. A computer can be a great asset to mathematics teaching, as Miss Kenyon discovered (in the fourth critical incident). But in order for this to happen, planning, imagination, and effort are required.

Activities

1. Modify the drill and practice program on pages 287–288 so that it will give practice on basic multiplication facts. (Hint: Change statement 50 to read 50 PRINT X; "*"; Y; "="; and make a change in statement 70 as well.) How would you change the range of integers in which numbers are generated so that you could adapt the program to various difficulty levels? Keep in mind that negative as well as positive integers can be randomly generated by appropriate use of the RND function.
2. Read the article by Prielipp (1970) on abundant, deficient, and perfect numbers. Write programs which could be used to test his conjectures, such as (a) every prime number is a deficient number, and (b) all abundant numbers are even numbers.
3. How could the drill and practice program for basic number facts be modified to teach basic skills in algebra such as factoring? See Example (1), on factoring the difference of two squares, for some suggestions. Devise a program to provide drill for a skill of your choice.

4. Large units of subject matter are now being taught by teaching computer systems such as the PLATO system. What are some advantages you can foresee of being taught an entire course by a computer? What are the disadvantages?

5. Some persons have described problem solving in terms of programming: "If you can write a program to solve a problem, you have solved the problem." Go back to Chapter 5. Select a problem there that lends itself to solution by computer. Write a program (including a flowchart) to solve this problem. What steps did you go through in solving the problem using computer programming? Were they analogous to how you solved the problem without programming?

6. Flowchart some actual situation, like instructing a person how to leave the classroom. As a preliminary activity to flowcharting, you might try blindfolding a person and giving him instructions on how to leave his desk and exit the classroom.

7. Modify Example (6), pp. 317 ff., on the ellipse, so that it will calculate the foci.

8. Modify the program in Example (7), pp. 320 ff., on probability, so that it will compute the probability that in a three-child family, at least one child is a girl. How could the program be changed to solve the general question "What is the probability that in an N child family, at least X are girls $(0 \leq X \leq N)$?"

9. Suppose that in the falling ball example (pp. 299 ff.) the ball is allowed to hit the ground and bounce. Suppose further that each time it bounces it rebounds to 40 percent of its previous height. Write a program to calculate the total distance traveled by the ball before "coming to rest." You may want to refer to the article by Young (1970). Note that the solution involves twice the sum of a geometric series.

10. Add a touch of realism to the problem in Activity 9 by tossing the ball straight outward from the top of the tower with a speed of 2 m per second so that it will clear the base of the tower. Write a program to calculate the total time until the ball reaches the ground after the fifth bounce and calculate the total horizontal distance the ball travels in that time (adapted from Gruenberger and Jaffray, 1965, p. 45).

11. Identify a simulation program which has already been written, and have it put on your computer. (If this is not practical, go through the sample output of the program to determine what results might be expected.) What problem-solving skills could be learned by this simulation program?

12. Develop a "computer guide" to a mathematics textbook with which you are familiar. Select a chapter and identify places in the chapter in which you could use various instructional applications of computers.

13. Write to several computer companies to determine what educational

services they provide. One such service is the user's group—a group which shares computer ideas, programs, and class activities. Some of these groups also conduct workshops at meetings of the National Council of Teachers of Mathematics and the School Science and Mathematics Association. (Addresses appear in Appendix D.)

Hewlett Packard Corporation
Digital Equipment Corporation
Honeywell
Wang Calculators
IBM

References

Ahl, David. *101 BASIC Computer Games*. Maynard, Massachusetts: Digital Equipment Corporation, 1974.

Albrecht, Robert L. et al. The Role of Electronic Computers and Calculators. In *Instructional Aids in Mathematics* (edited by Emil J. Berger). Thirty-fourth Yearbook of the National Council of Teachers of Mathematics. Reston, Virginia: The Council, 1973.

Allen, D. W. and Gorth, William. *Comprehensive Random Achievement Monitoring*. Technical Memorandum AR-2, Stanford University, 1968.

Barrodale, Ian; Roberts, Frank D. K.; and Ehle, Byron L. *Elementary Computer Applications in Science, Engineering and Business*. New York: Wiley, 1971.

Bradford, John W.; Jones, Kenneth; and Larsen, Dean C. *Computer Extended Mathematics: Complex Numbers*. Denver: Computing and Mathematics Curriculum Project, University of Denver, 1970.

Blum, Julius and Rosenblatt, J. *Probability and Statistics*. Philadelphia: Saunders, 1972.

Coan, James S. *Basic BASIC*. New York: Haydn, 1970.

Darby, C. A., Jr.; Korotkin, A. L.; and Romanshke, T. *Survey of Computing Activities in Secondary Schools*. New York: Praeger, 1972.

Daykin, P. N.; Gilfillan, J. W.; and Hicks, B. L. Computer Managed Study for Small Computer Systems. *Educational Technology* 15: 46–50; March 1975.

Dorn, William S. Notes for Mathematical Modeling and Computing. Denver: University of Denver, 1971. (Mimeo)

Doyle, F. J. and Goodwill, D. Z. An Exploration of the Future in Educational Technology. Bell Canada, January 1971.

Fehr, Howard F. *Secondary Mathematics: A Functional Approach for Teachers*. Boston: Heath, 1951.

Feng, Chuan C., project director. *Second Course in Algebra and Trigonometry with Computer Programming*. Boulder, Colorado: Colorado Schools Computing Science Curriculum Development Project, 1969.

Foster, Thomas Edward. The Effect of Computer Programming Experiences on Student Problem Solving Behaviors in Eighth Grade Mathematics. (The University of Wisconsin, 1972.) *Dissertation Abstracts International* 33A: 4239–4240; February 1973.

Gamow, George. *Biography of Physics*. New York: Harper & Row, 1961.

Gruenberger, Fred and Jaffray, George. *Problems for Computer Solution*. New York: Wiley, 1965.

Hagelbarger, David and Fingerman, Saul. *CARDIAC: A Cardboard Illustrative Aid to Computation*. Bell Telephone Laboratories, 1968.

Hammersley, J. M. and Hanscomb, D. C. *Monte Carlo Methods*. London: Methuen, 1964.

Hicks, Bruce L. and Hunka, S. *The Teacher and the Computer*. Philadelphia: Saunders, 1972.

Iverson, Kenneth E. *Elementary Algebra*. Technical Report 320-3001. Philadelphia: IBM, 1971.

Johnson, David C. et al. *Computer-Assisted Mathematics Program (CAMP)*. Glenview, Illinois: Scott, Foresman, 1969.

Katz, Saul M. A Comparison of the Effects of Two Computer Augmented Methods of Instruction with Traditional Methods upon Achievement of Algebra Two Students in a Comprehensive High School. (Temple University, 1971.) *Dissertation Abstracts International* 32A: 1188–1189; September 1971.

Kemeny, John G. and Kurtz, Thomas E. *BASIC Programming*. New York: Wiley, 1971.

King, Donald Thomas. An Instructional System for the Low-Achiever in Mathematics: A Formative Study. (The University of Wisconsin, 1972.) *Dissertation Abstracts International* 32A: 6743; June 1972.

Korotkin, Arthur L. and Bukoski, William J. *Computing Activities in Secondary Education*. Washington: American Institutes for Research, 1975.

Lee, J. A. N. *Computers*. Rev. ed. North Brunswick, New Jersey: Boy Scouts of America, 1973.

Pack, Elbert Chandler. The Effect of Mode of Computer Operation on Learning a Computer Language and on Problem Solving Efficiency of College Bound High School Students. (University of California, Los Angeles, 1970.) *Dissertation Abstracts International* 31A: 6477; June 1971.

Post, Dudley, ed. *The Use of Computers in Secondary School Mathematics*. Newburyport, Massachusetts: Entelek, 1970.

Prielipp, Robert. Perfect Numbers, Abundant Numbers and Deficient Numbers. *Mathematics Teacher* 63: 692–696; December 1970.

Rosenbaum, Sema Joy Marks. A Course in Computer Simulation for High School Students. (Harvard University, 1970.) *Dissertation Abstracts International* 31A: 5676; May 1971.

Sage, Edwin. *Problem-solving with the Computer*. Newburyport, Massachusetts: Entelek, 1969.

Simon, Julian and Holmes, Allan. A Really New Way to Teach Probability Statistics. *Mathematics Teacher* 62: 283–288; April 1969.

Sobol, I. M. *The Monte Carlo Method*. Chicago: University of Chicago Press, 1974.

Suppes, Patrick; Jerman, Max; and Brian, Dow. *Computer-Assisted Instruction: Stanford's 1965–1966 Arithmetic Program*. New York: Academic Press, 1968.

Young, Worth J. The Bouncing Ball DOES Come to Rest. *Mathematics Teacher* 63: 391–392; May 1970.

Computer-based Educational Research Laboratory. *PLATO*. Urbana: Computer-based Education Research Laboratory, University of Illinois, May 1975.

Hewlett-Packard Corporation. Computer System Helps Motivate High School Students. Application Note 145-7, Hewlett-Packard, March 1972.

People's Computer Company. *My Computer Likes Me (When I Speak in BASIC).* Menlo Park, California: People's Computer Company, 1972.

Simulation Packages, Huntington II Project. Maynard, Massachusetts: Digital Equipment Corporation.

Using Research

AFTER STUDYING CHAPTER 10, YOU WILL KNOW:

* what you'll find in a research report.
* how to read and analyze research reports.
* how to locate research reports.
* some research terms and types.
* some research findings that can be applied.
* how research may be done simply.
* one point of view about research and researching.

CRITICAL INCIDENTS

1. Sue's students are having a hard time doing computation. She wonders if calculators or computer programming might help them.
2. In the teachers' lounge, John argues, "Individualize! That's all I hear. But will it really help my students achieve better? And will they like it?"
3. Bob wants to try something different. He asks, "How do I find out whether my students might learn more in mathematics laboratories?"
4. Ann has found a new way to teach problem solving. It seems to work well and the students like it. But she doesn't know how to determine whether her new methods are really any better than the old ones.

RESEARCH: WHO NEEDS IT?

The answer to that question is simple: every teacher needs research! Teachers profit from research every day. The textbooks they use, the way they teach, the appearance of their classrooms—all of these are, to some extent, based on research.

Through the reading of research you will obtain ideas that you can apply in your teaching. The focus of research studies varies widely. Some studies are oriented toward developing a theory of instruction or learning; these may merely suggest some tentative guidelines for teachers to test. From other research studies come ideas that are very practical, and some of these can be applied directly. For example, Austin and Austin (1974) suggest that a teacher try grading only half of each set of homework papers: you'll save time without affecting student achievement adversely. Hanna (1971) describes how multiple-choice test items can be made better. Other research studies indicate what you might expect from certain types of students. Baker (1971) found that low-achieving ninth graders most frequently chose teacher-made assignments when they had a choice of activities. Some studies have implications for your program. For example, Dodson (1972) lists characteristics of successful problem solvers and states some things a teacher can do to help all problem solvers. Lankford (1972) presents a list of the actual errors students made as they attacked computation examples; you could use the list to design teaching techniques which might help your students to avoid making some of these errors.

Among the many other topics which research has explored are:
- what can be taught (e.g., Hoffman, 1973)
- how to teach (e.g., King, 1972)
- the use of time (e.g., Zahn, 1966)
- the use of materials (e.g., Kuhfittig, 1974)
- what teachers do in a classroom (e.g., Strickmeier, 1971)

As you can see, the range of research interests is wide. Some of the findings will confirm practices you already believe to be appropriate; others may make you want to try new approaches to teaching.

Many references to applicable research findings and *what* you can learn from research have been scattered throughout this book. This chapter focuses attention on the research process itself and on *how* you can learn from research.

WHAT IS RESEARCH?

Research is both a product and a process. You can be involved with both aspects of research: as a consumer—a reader and a user of research findings—and as an active participant—doing research yourself or taking part in a research study being conducted by someone else.

Research is controlled inquiry, an application of scientific methods. It is not independent of teaching and learning but is derived from and applied to the teacher-learning process. Every teacher is involved in one kind of research every time new ideas are tried out; this is a type of action research. You are constantly trying to find the methods, the materials, and the procedures which will work best for you and your students. You are assessing what students have learned and using your findings as you plan what to do next. You are concerned with what will help *you* to teach better and help *your* students to learn better. This is a vital component of your concern with the evaluation process.

Research involves carefully established conditions with precise controls. Frequently we attempt through research to secure information which can be generalized to many other teachers and to many different situations. We maintain greater control of many variables than is usually true in the classroom; we want a more precise measurement of the status or level of learning.

Research can provide a foundation on which to make decisions about *what* to teach and *how* to teach. It provides us with guidelines that aid us in making decisions. Teachers are involved in a continuous decision-making process; on what should these decisions be based? Certainly, many are based on the teacher's own experience in the classroom and on the experiences of other teachers with other children. Research provides a broader base of information, combining "what to do" with "why do it," so the teacher has a rationale on which to make decisions. However, not all problems are amenable to research. Some decisions must be made on the basis of your philosophy and values. For instance, research can provide an answer to "*Can* we teach _____?", but this is not an answer to the question "*Should* we teach _____?"

Research has a valid role to play in assessing and improving the quality of instruction. In fact, merely being involved in research helps in achieving this latter goal; participating in research is motivational for both teachers and students. Teachers often see evidence akin to this "Hawthorne effect" when we say "Sure, it worked in the research study, but will it work in a 'real' classroom?" (see Research Highlight, "What Is a 'Hawthorne Effect'?").

Research can also help us in understanding how children learn. The more we know about how learners think or process information, the better we can structure experiences which will help them to learn. The interaction of the cognitive domain with the affective domain is still relatively unexplored, but as a research area it may provide information to help us as teachers to improve students' beliefs and attitudes about themselves and mathematics.

That research is a professional activity in which teachers *should* engage cannot be overlooked. Each teacher can make a contribution which can help all mathematics teachers and students. And there is also an educational benefit for each individual. A researcher may start with constructs or ideas of "mental ability" or "mathematical achievement" or "discovery teaching," but before very long, the researcher must carefully define these constructs for that particular study. For

> ## RESEARCH HIGHLIGHT | WHAT IS A "HAWTHORNE EFFECT"?
>
> Some years ago an experiment was conducted in a plant in Hawthorne, New Jersey. The purpose was to determine whether a coffee-break would have any effect on worker productivity. A midmorning coffee-break resulted in increased productivity. So the experimenters tried more than one coffee-break and then extended the time of each coffee-break. Productivity increased with each change: the workers produced more as they worked fewer minutes but had more coffee-breaks. Then the experimenters cut out all coffee-breaks. Productivity continued to increase! The explanation seems to be that the workers were highly motivated, not by the coffee-breaks, but merely by being involved in an experiment. Thus the term "Hawthorne effect" is used to describe this affective variable—the presence of the research itself—which can influence the results of a study. Researchers must attempt to control it or take it into account; teachers can exploit it to promote achievement.

example, *creativity* may be defined as "the ability to give unusual solutions to a specific problem." You can learn operational definitions of mental constructs and gain sensitivity to an understanding of the constructs through participation in research.

HOW TO READ RESEARCH

How exciting it would be if the news of educational research discoveries were to be splashed in headlines across the pages of newspapers! (See Figure 10–1.) Alas, only rarely is research given a headline. Too often we find the news of a research finding buried in a jumble of erudite words. Researchers have been conditioned to present their research reports in a rather static format from which all the joy and woes of the actual research process are removed. Rarely does a researcher tell what it's really like to do research; instead he or she reports only the matter-of-fact why, what, who, when, where, how. There are reasons why the emphasis is on the reporting of facts, of course; we want to know exactly what was done in order to be able to replicate the study or apply it in other situations. And in the limited amount of space available to a journal article, there seems to be little room to provide the "human-interest" descriptions.

The structure of a research report is often made clear by subheadings. Even

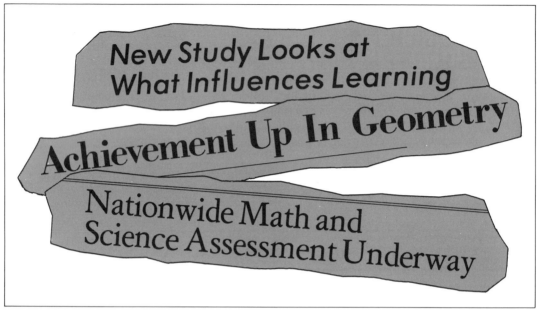

Figure 10-1

when subheadings are missing, you will find that most research reports include ten basic components. We will illustrate each of these components with excerpts from an actual research report by Bright and Carry (1974) which studied the effect that two professional groups, mathematics educators and mathematicians, have on the decisions made by pre-service mathematics teachers. Notice how the report tells you not only *why* and *how* they conducted the study, but also what the findings may indicate.*

1. A *rationale* for the importance of the research:

> Almost universally, teacher education programs for mathematics teachers include a course titled Methods of Teaching Secondary School Mathematics. Some professionals believe that methods courses have negligible impact on the behavior of the teacher when he finally arrives in his own classroom. Others argue a different point of view, strongly supporting the importance of such a course. There seems to be relatively little empirical evidence to support either position. (p. 87)

* The quotations that follow are reprinted from George W. Bright and L. Ray Carry, "The Influence of Professional Reference Groups on Decisions of Preservice Secondary School Mathematics Teachers," *Journal for Research in Mathematics Education*, March 1974 (vol. 5, pp. 87–97), © 1974 by the National Council of Teachers of Mathematics. Used by permission.

2. A *statement of the problem* or question:

> Much of the "methods" instruction consists of exposing the future teacher to the
> opinions of professors and authors of books and journal articles, so an interesting
> question is, Will the opinions of these professionals influence the future classroom
> decisions of the teacher? Furthermore, since the opinionating professionals
> can be generally categorized as either mathematicians or educators, is one of
> these groups more influential than the other? (p. 87)

3. A presentation of *hypotheses*, which state what is expected to be ascertained from the research:

> This study was an experiment to test the following hypotheses:
> 1. The professional reference group of mathematicians and the corresponding
> group of educators influence the decisions of preservice secondary school teachers
> in projected classroom situations.
> 2. The influence of mathematicians as a reference group is stronger than the
> influence of educators on the decisions of preservice secondary school mathematics
> teachers in projected classroom situations. (p. 87)

4. A description of precisely what *procedures* or treatment were used:

> The instruments were distributed to S s [subjects, in this case preservice teachers]
> during regular class sessions in October 1970. . . . When the S s finished, the
> booklets were collected. The choices of each S were recorded as an ordered triple
> . . . . (p. 91)

5. Information on the *students* who were involved in the study:

> Preservice secondary school mathematics teachers at The University of Texas at
> Austin were randomly assigned to an experimental group (N = 28) [N or n means
> number] and a control group (N = 33) from two methods classes and one student-
> teacher seminar . . . . (p. 90)

6. An indication of the type of *tests* or other measurement instruments or methods used:

> A set of 36 hypothetical classroom situations were [sic] developed through a
> series of three pilot studies and printed in booklets . . . Four categories of nine
> situations each were developed . . . For each S, the computer randomly ordered
> the three resolutions for each situation . . . . (p. 90)

7. Presentation of *data*, often in tables:

Table 10–1. Mean Response Distributions.

	Mean Number of Agreements with Label		
Group	Mathematics	Education	Blank
Experimental (N = 28)	10.1	10.5	3.4
Control (N = 33)	8.5	7.9	7.6

The mean number of S's agreements with the particular labels are presented in Table 10–1. (p. 92)

[Such data are then analyzed, using statistical procedures or tests.]

8. Discussion of the *findings*:

The significant values yielded by the Kolmogorov-Smirnov test [a statistical test applied to the data] supported Hypothesis 1. That is, the attaching of labels did influence the responses of the Ss. The nonsignificance of the Wilcoxon statistic indicated no support for Hypothesis 2. That is, for the available sample, no differential effects of the labels were observed. (p. 94)

9. Delineation of the *limitations* of the study:

Several limitations of the study need to be mentioned. The sample may have been atypical of the entire population of all prospective secondary school mathematics teachers. The observed behaviors of Ss were not classroom behaviors —what a teacher will do in a classroom may differ from what he thinks he will do before he begins teaching. . . . (pp. 95–96)

10. Discussion of the *conclusions* the researcher has drawn on the basis of the data and of the implications which she or he believes may be derived from the study:

The fact that preservice mathematics teachers are influenced by opinions associated with professional groups has implications for several kinds of activities. . . . It also seems evident that . . . the teaching of content and methods should be closely coordinated. . . . (p. 96)

It should be noted that all but the last component of a research report are fact-oriented, scientifically objective factors. The researcher is presenting an account which presumably could be verified and replicated: others can trace the route the researcher has taken and repeat the experiment to see if the same results can be attained. But in the last component the researcher is granted the right to become subjective, to talk about what he or she thinks the research means: this is the researcher's opportunity to speculate. Some of the speculations may turn out to be true; the researcher who has been immersed in a study, who has thought of it night and day for a period of time, is in a position to have some insight. But the researcher may also be carried away by hopes and expectations.

The component of a research study which deals with conclusions, interpretations, and implications is probably the section most frequently read, especially by the novice at reading research reports. Perhaps this is because the writing sometimes seems clearer and more interesting in this section. Perhaps it is because there is a different sense of meaning—of "putting it all together"— in this part of the report. Perhaps it is because there is no threat from tables of data and statistical formulas. Perhaps it is simply an attempt to save time. But one should say BEWARE to the reader: BEWARE because this is speculation.

There is nothing wrong with speculation; it has a valid purpose in a research report and, not infrequently, it presents a thought that sparks further research. The reader must, however, beware of believing that this section contains the basic information about the study. The reader must beware of accepting a conclusion or interpretation as if it were a finding. In the report cited above, the authors conjecture that "the teaching of content and methods should be closely coordinated." This may be plausible, but the reader must remember that the research was not directed to this point and that it is not a finding but a speculation.

A finding is data-related and hypothesis-related. The researcher collects the data by administering a test or, as in the study cited, a specially developed instrument. The researcher subjects the data to statistical procedures with the intent of ascertaining the significance of the data in terms of the probability of that data occurring purely by chance. The findings are stated in terms of the hypotheses: because the statistical procedure indicated that the data were significant, the hypothesis that was the basis for the data being collected is rejected or not rejected. There are limitations even to the findings, but the findings are expected to be more objective than the conclusions, interpretations, and implications.

To ascertain the validity of a research study—that is, to determine whether or not a finding is to be accepted—it is necessary to read the entire study. Many inexperienced research readers find that things don't go too badly until they hit the portion which presents data and statistics. Unless the reader has a background in statistics, this *is* likely to be a problem. Perhaps you must accept this temporarily: you may not understand it all, but you should at least skim it. You will want to acquire more background for understanding research design and statistics from courses and from books. You can get the reactions of those more knowledgeable about research in such journals as *Investigations in Mathematics Education* (a sample article appears at the end of this chapter).

You should concentrate your attention on the remainder of the report. The entire research report must meet the test of plausibility. The most important facet to be considered is: Was the research worth doing? Studying the influence of mathematicians and educators on prospective mathematics teachers is plausible; studying the influence of the President of France on them would not be plausible. Perceptions of what is worthwhile will differ, but you can answer the question in terms of your needs and the scope of the study. Look at the question being posed in the study: Does it need to be answered? Look at the statement of the problem: Is this a realistic way in which to go about answering the question? Look at what the researcher actually did, the procedures used and the treatments administered: Are they really plausible as a way of attacking the problem? Is the sample chosen to be representative of the population of students to whom the research question is applicable? Are the tests or other measuring instruments appropriate for the treatments and thus for providing an answer to the research question? Does the research tell you enough so that you know what is going on? Are there details of the investigation which are not included in the report, but which you need to know? Are the conclusions warranted by the findings?

These questions may be summarized in another way, to provide a guide as you read a research report:

1. Is the problem practically and/or theoretically *significant?*
2. Is the problem clearly *defined?*
3. Is the *design* of the study appropriate to answer the research question?
4. Does the design facilitate the *control* of variables?
5. Is the *sample* properly selected for a specified population and for the design and purpose of the research?
6. Are the measuring instruments *valid* and *reliable?*
7. Are the techniques of *analysis* of the data appropriate?
8. Are the *interpretations* and *generalizations* valid conclusions from the results?
9. Is the research adequately *reported*, or do you need more information about the study?

Then you come to the most important question of all: What do the findings mean to you as a teacher? Do they suggest that you might try something different in your teaching? A research study can only indicate that a procedure has been effective with some teachers and some students; it cannot tell you whether it will be effective for you and your students. A research study can tell you what others have done; it cannot tell you what you should do. The decision is up to you: to try or not to try, to change or not to change. This is a part of the challenge of research (see Research Highlight, "Perhaps You Might Have Guessed It!").

HOW TO DO RESEARCH

Suppose you decide to change some of your teaching procedures. Why not do some research to ascertain whether or not a procedure is effective for you and your students? Maybe you have a question for which you cannot find the answer in previous research. Why not do some research to find an answer? Here's an

RESEARCH HIGHLIGHT | PERHAPS YOU MIGHT HAVE GUESSED IT!

A recent study assessed the relationships among secondary school teachers' knowledge of and attitudes toward educational research. A total of 204 teachers from all subject areas volunteered to complete a test on research knowledge and an attitude measure.

Mathematics teachers had significantly higher scores on the research knowledge test than either social studies teachers, English teachers, or a combination of teachers from other subject areas; mathematics teachers also scored higher than science teachers, though this difference was not significant. The correlations of attitude and knowledge scores also were not significant; that is, no strong relationship between attitude and knowledge scores was found. Scores indicated that attitudes were positive, however. A previous study was cited in which knowledge of educational research correlated highly with a favorable attitude toward educational research (Short and Szabo, 1974).

account from one teacher who was asked to describe what was involved in doing research in a classroom.

> I'd read a number of research reports during a course I had taken, but I never thought that it might be possible for *me* to do some research. And then I became interested in individualizing instruction for my general math students. I checked a review of research on the topic and found that research hadn't come up with much of an answer as to how effective individualizing instruction is. The studies didn't indicate that it was noneffective; "no significant differences" between an individualized approach and other approaches were found in many instances. I'd read enough articles to know that many people have favorable things to say about individualizing—and I was more than ready to give it a try.
>
> It occurred to me that the important thing was not how successful *other* teachers and students had been with individualizing instruction, but how effective it would be for *me* and *my* students. So I decided to become a researcher!
>
> I spent quite a bit of time deciding what questions I really wanted to answer and *could* answer. I knew I had to delimit the problem so that I could do the research—and feel afterward that the research had really given me an answer. The research questions I posed were for an experimental research study using a pretest and posttest design:

1. How do my general math students achieve with certain individualized instruction procedures when compared with my "usual" procedures?
2. What is the attitude of the students toward the two types of procedures?
3. Does achievement or attitude toward the two types of procedures differ at different ability levels?

Then I had to decide exactly what to do. Even though I usually plan carefully for teaching, I found that my research concerns made me aware of some aspects of teaching that I'd never given much thought to before this. I had to:

1. Decide what content to include.
2. Define exactly what "certain individualized instruction procedures" I would use to define how these differed from "usual" procedures. (I knew from reading research reports that these were called "independent variables" or "treatments.")
3. Decide when to begin and how long the study would be. I had to plan for a long enough period so that individualized instruction would be given a fair chance of showing how effective or ineffective it might be.
4. Plan what materials to use and what to do every day.
5. Decide how to assign the students to the two treatments. I decided to assign them randomly to the two treatments after first dividing them into three groups on the basis of scores on previous standardized tests. I put the names of those in each group in a box; then I assigned them to first one group and then the other as the names were drawn. (With such random assignment you can expect that extraneous variables will be evenly distributed among the groups.)
6. Decide what measurement instruments to use. I chose an appropriate attitude scale and made up a test to measure the "dependent" variables of achievement and attitude. I knew that I couldn't use a standardized achievement test because such a test includes only a few items on topics we'd explore during the study, so I had to develop a test to do this. I knew that a study may result in "no significant differences" when in fact differences were present but unmeasured. (Actually, I would have developed a test for my class even if I hadn't wanted to use it for the research—but I was a lot more careful in developing the test to make sure that it covered only what the students would actually study. And I selected and wrote the items carefully.)
7. Decide what to say to the students about being involved in research.

I also had to identify other variables which might affect the research; such relevant variables must be controlled. For instance, I knew that the time at which I taught each treatment was a variable. So instead of teaching the two always in the same order—the individualized instruction group first, followed by the usual instruction group—I changed the order every three days. I made sure the time I spent with each group was exactly the same. I made sure that each group was exposed to the same content—though in different ways. During the experiment, the groups involved should have common experiences except for the treatment. Since I was trying to find out about the effect of a procedure, I wanted the content or mathematical topics to be the same for both groups; only then could differences at the end of the study be attributed to the treatments.

I haven't mentioned the step at which I had to decide how to analyze the data. At first I was going to do this very simply: I would find the means for the two groups and see whether one was greater than the other, and then I would find the means for the students in different ability groups and see if they were different. But I decided that it might be fun to do this more scientifically—as a researcher really would. So, since I know so little about analyzing data, I called our school's research department. A woman from the department went over my plans with me. (She suggested the procedure for the random assignment of the students to the two treatments.) Since I was only trying to find an answer for my own use, and therefore only using my one general math class, I could use a simple statistic to analyze the differences. She showed me what to do and offered to help me further when I had collected the data.

Sometime I might want to do this study on a larger scale, with some other teachers in my school or with teachers from other schools. Many of the things I'd done for my one-class study would be appropriate in a larger study, too—the definitions of the treatments, the materials, the tests. One of the few changes might be in the way the data would be analyzed, since this depends on the scope and limitations of the study. Yet the basic purpose of the simple statistical test that I used is just that of some of the more sophisticated tests, to analyze the variance of differences between or among scores and determine whether or not there are "significant" differences.

(Differences, I learned, are determined to be "significant" or "statistically significant." This means that there is a specified likelihood that such differences would not have occurred by chance. Frequently the level of significance is set at the probability of .05, or .01, or .001—thus the results would occur by chance only 5 times in 100, or only 1 time in 100, or only 1 time in 1000. "No significant differences" means that a specified level of significance was not reached—thus the results could occur frequently by chance. In some instances, this may mean that the effect of two procedures on a given measure was so similar that you could select either one. Researchers set a level which seems appropriate to them in terms of the content and design of the study.)

If I had been doing a study with more than just me and my one class involved, I probably would have done a pilot study. This gives you the chance to find out what problems may arise and allows you to resolve them before they affect the results of the main study. You do everything in a pilot study that you would do in the main study—but with fewer students and with the knowledge that this was only a trial, that you could make whatever changes seemed necessary. (The "rules" for a major study would be more restrictive—that is, you shouldn't expect to make any changes as the study progressed.)

I kept a log of what I did each day during the study and anecdotal records of particular incidents and reactions. Whenever I did something different from what I'd expected to do, or the students didn't react as I'd expected, I noted it. These records could be useful, and they can help me to improve my teaching.

I won't write up my study—except for this account—because the findings are limited in generalizability since I was only one teacher teaching one class. But I found out what I wanted to know, and now I'm trying to get some of the other teachers in my building to try some research, too.

Incidentally, I enjoyed doing the research tremendously—and so did my class!

Your reaction to this report may be "What a lot of work!"—and indeed it was. Yet consider how many activities were performed that the teacher would have engaged in even without a research goal: outlining the content, determining exactly how to individualize, planning the materials, deciding what to do each day, developing instruments to assess progress, scoring the student's papers. You might notice that some things not always done by teachers are helpful in improving teaching skills, such as keeping anecdotal records. And increased understanding of research methodology might lead to better teaching. Most important, this is an example of a scientific approach to evaluating the effect of a teaching approach, individualizing instruction, on achievement and attitude. The teacher now has some evidence, derived from "personalized" research, to aid in making a decision.

WHERE TO FIND RESEARCH

Sources of information about research in mathematics education have become increasingly easy to locate. One of the better known sources is the *Journal for Research in Mathematics Education*, published by the National Council of Teachers of Mathematics. This journal presents reports on a range of mathematical topics from preschool through college levels. The style is generally formal, but researchers are encouraged to include discussion of data-related implications for teachers. Occasionally articles about developing or designing research in mathematics are published.

Two other journals of the NCTM include research reports written in a less formal style. The *Arithmetic Teacher* has articles on using research in the classroom, in which research is reported almost entirely in terms of its potential for classroom application. Although the major focus of the journal is at the elementary school level, some of these reports involve junior high school students. The *Mathematics Teacher* also presents research reports using basically the same applicability criterion. Most of the articles involve secondary school students.

Once a year the *Journal for Research in Mathematics Education* publishes an annotated list of research published during the previous year. From this listing you can ascertain other journals in which reports of research on mathematics education can be found. The annotations should provide some indication of whether or not a given report is one that you might want to obtain and read completely. Not only articles appear on the list; dissertations in mathematics education which have been listed in *Dissertation Abstracts International* are also included. *Dissertation Abstracts International* is a monthly publication of abstracts of doctoral dissertations in all disciplines which have been completed at universities in the United States or in several other countries. *School Science and Mathematics*, the official journal of the School Science and Mathematics Association, also publishes a list of dissertation abstracts as well as research and other types of articles.

One journal concentrates on publishing critiques of selected reports of research in mathematics education. The entries in *Investigations in Mathematics Education* are prepared by knowledgeable mathematics education researchers. A sample entry appears at the end of this chapter; you will note that, like other entries, an abstract of the report is followed by the "abstractor's notes," in which questions are raised about facets of the research.

The *Encyclopedia of Educational Research* and the *Handbook for Research on Teaching* (see Dessart and Frandsen, 1973, and Henderson, 1963) also include reviews of research on mathematics education.

HAVE YOU MET *ERIC?*

Another source of research information—plus a variety of nonresearch information—is ERIC. The letters stand for Educational Resources Information Center. ERIC is a national information system that is actually a network of centers or "clearinghouses," each contributing to a collection of educational materials in a particular educational area.

The documents which form the ERIC collection come from a variety of sources, including federally funded research and development projects, school systems and other educational agencies, educational organizations, individuals, various meetings and conferences, and educational journals. In the clearinghouses the information is indexed and abstracted in preparation for listing in one of two ERIC reference publications, *Resources in Education* (RIE) and *Current Index to Journals in Education* (CIJE), both of which are available in most college libraries. (See Figure 10-2 for sample listings.) *RIE* affords easy access to abstracts of innovative programs, curriculum materials, conference proceedings, bibliographies, outstanding professional papers, and reports of significant efforts in educational research and development. Copies of most of

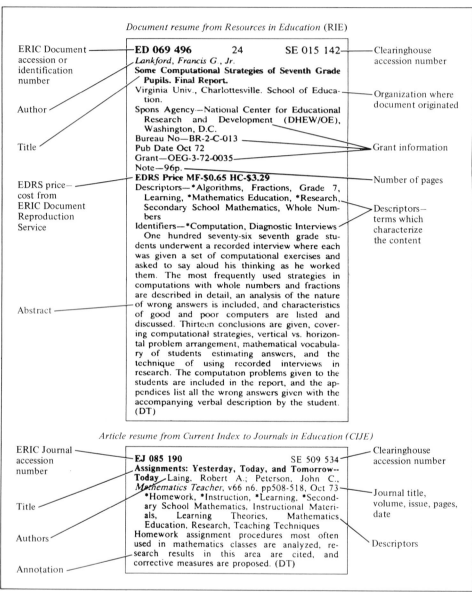

Figure 10–2

these documents may be purchased from the ERIC Document Reproduction Service. *CIJE* contains indexed and annotated articles found in approximately 700 journals. Computer searches of the ERIC document base may also be conducted through public and commercial agencies.

The clearinghouse for mathematics education materials is the ERIC Information Analysis Center for Science, Mathematics, and Environmental Education (ERIC/SMEAC), currently located at The Ohio State University and directed by Robert W. Howe. In addition to the documents available from the ERIC system, several types of publications are developed at and distributed by ERIC/SMEAC:*

1. Bibliographies on both general and specific topics. Listings of research reports are published, as well as listings of research on topics such as unpublished mathematics tests.
2. Research reviews. Bulletins are published which give overviews of research on secondary school mathematics. There have also been reviews of research on such topics as the use of computers in mathematics education and the role of attitudes.
3. Interpretive summaries. These are generally longer publications which attempt to provide a broader analysis of a body of research with extensive discussion of the meaning and applicability of the research.
4. Background information. Publications on such topics as evaluation and metrication provide information on specific matters of concern to teachers.

THE VARIETIES OF RESEARCH

There are all kinds of researchers. Some are interested in basic research, on topics such as why and how mathematical ideas are attained. Some are interested in applied research, on topics such as "the effect of computer programming experiences on student problem-solving behaviors" or "student attitudes toward geometry." Some work in a clinical or laboratory setting, where variables can be controlled comparatively readily. Some work in a classroom setting, where reality makes it impossible to control all variables but where all the forces that affect teaching and learning are at work. Most researchers are intensely interested in contributing to better understanding of the learning process and the teaching process. Some will do only one controlled research study. Some will continue to be involved in research either directly through setting up studies or indirectly through applying and testing research findings all their lives. Hopefully you will become one of the latter—if not one of the former.

It takes all kinds of research, too, to help us in better understanding the teaching-learning process. How to categorize research has been of some concern ever since the beginning of the century and the origins of scientific educational research. Many categorization schemes have been devised, none of which meets the needs of all users. Arbitrarily, we might consider this set of categories:**

---

* To request a list of current publications, or for other information, write to ERIC/SMEAC, The Ohio State University, 1200 Chambers Road, Columbus, Ohio 43212.
** Descriptions of the studies cited here are given in Appendix A.

1. *Descriptive*—research in which the researcher reports on records which may have been kept by someone else; includes reviews, historical studies, and textbook analyses or comparisons. An example of a study of this type is Aiken (1970).
2. *Survey*—research which attempts to find characteristics of a population by asking a sample through the use of a questionnaire or interview; includes also the status study, in which a group is investigated as it is to ascertain pertinent characteristics. An example: Dutton (1968).
3. *Case study*—research in which the researcher describes in depth what is happening to one designated unit, usually one child. An example: Burke (1974).
4. *Action*—research which uses nominal controls; generally teacher or school originated; procedures of actual practice may be described. An example: Bierden (1969).
5. *Correlational*—research which studies relationships between or among two or more variables; uses correlational statistics primarily. An example: Smith (1974).
6. *Ex post facto*—research in which the independent variable or variables were manipulated in the past; the researcher starts with the observation of a dependent variable or variables and then studies the independent variables in retrospect for their possible effects on the dependent variables. An example: Norland (1971).
7. *Experimental*—research in which the independent variable or variables are manipulated by the researcher to quantitatively measure their effect on some dependent variable or variables, to test a logically determined hypothesis. An example: Sherrill (1973).

Experimental research is, perhaps, the type most often involved in most writing about research. The terms in which we write of research probably apply most fully to experimental designs. In experimental research we must be highly concerned about such things as the control of variables or the explicitness of the procedures. Yet to some extent most of these terms can be applied to other types of research as well. In descriptive research it is important to keep a record of what variables enter into the situation, and in action and ex post facto research it is important to keep the lack of control of variables in mind. In survey research the selection of the sample and the type of instrument used is of vital concern.

We need and can get important information from all types of research. We should be aware of what each type can contribute as well as the limitations of each type. Descriptive research may provide a thoughtful analysis of trends and patterns—but it is by nature subjective and speculative. Case studies provide detailed information on how a few specific children learned or reacted—but this information may not be generalizable to other children. Correlational research indicates the degree of relationship between or among certain factors—but it may not involve all important factors or it may involve extraneous factors. Ex post facto research provides information on the effectiveness of some procedures that occur over a period of time which may not be studied otherwise—but the lack of control over, or research awareness during, the treatment period means that one cannot be entirely sure what variables might have influenced the outcome. Action research allows one to obtain a picture of research in the real environment—but the lack of control of variables leaves generalizability in question. Survey research indicates what practices have been followed or how well students have done—but it is firmly welded to one point in time and what was true at that point in time may not be true at a later point in time. Experimental research provides an indication of the effect of using certain procedures—but the indication can be influenced by an array of variables.

For any given topic, we need to take into consideration the sum of all types of research and of many studies. We need to consider the "value" of the research as well as the findings. We need to consider the implications for the particular situation to which we want to apply those findings. In short, research provides a key to better teaching: as teachers we need to decide when it is appropriate to use that key.

We need research on many different types of questions. For the individual teacher it may be important to know what has been ascertained about effective procedures for teaching transformational geometry. For the school it may be important to know what types of geometry most favorably affect achievement in geometric concepts. Local school systems may at times need to engage in their own research, for generalized findings may not be applicable when the unique characteristics of any one system are considered. For the researcher it may be important to know which geometric concepts have been attained so that new procedures may be tested.

The search for one "best" answer is one of the most misleading expectations of research. Rather than searching for the one best procedure, we need to know what procedures are effective for what children. Determining what and how to teach is a professional task for which research can provide guidelines.

USING RESEARCH TO IMPROVE INSTRUCTION

Research can be confusing if you expect to find an unequivocal answer to whatever question you might have. In the first place, not all of the questions which it may be appropriate for research to answer have been tackled. In the second place, almost without exception, when there have been several studies on a topic, the findings are divergent. You must do more than just look at the outcome. You must look at the "goodness" of a study—at how well it was designed and executed—and you must consider the reasons for the divergence. How were two studies alike and how were they different? What were the treatments? What were the tests? Who were the students?

The significant influence of research does not come piecemeal, study by study. Instead it comes cumulatively, as we put the pieces together. When the research evidence is equivocal, we must almost "read between the lines" or look for factors which might be affecting the research. An example of this is given in an answer to the question "How do class organization patterns affect achievement?"

> The studies are frequently of the action-research type, which implies less-firm control of variables; when firmer controls are attempted, the problem of extraneous variables is still present. This causes an analysis problem—for it remains true that any organizational pattern *will* be affected by a multitude of variables. And it seems evident that one generalization can be made: *any* organizational pattern can be effective, can result in better achievement, depending on the variables involved. Chief among these is the teacher factor: if the teachers are committed to a particular pattern, they can make it work. Conversely, some teachers can make any pattern work. . . . Thus it seems clear that there is no *one* organizational pattern which is *best*. (Suydam, 1972, p. 5)

You might consider this statement in relation to the findings of some of the studies cited in Appendix A: Bachman (1969), Baley and Benesch (1969), Beul (1974), Buchman (1972), Morrison (1967), Paige (1967), Steere (1967), and Willcutt (1969).

A second example of equivocal findings is provided by evidence on the question "What is the role of inductive and deductive strategies in the teaching-learning process?" Among the studies in Appendix A which might be checked are: Ballew (1966), Eldredge (1966), Henderson and Rollins (1967), Hirsch (1973), Lackner (1968), Maynard and Strickland (1969), Meconi (1967), Michael (1949), Price (1967), Ray (1961), Sobel (1956), and Volchansky (1969). It will be interesting to see what conclusions you reach!

At times the evidence overwhelmingly supports the value of one approach. This was the case with research on the role of teaching with meaning: Almost all of this research was conducted in the elementary school, and the evidence was so clear that there apparently seemed little need to verify it in the secondary school. We simply began to apply it, moving from a "here's-what-to-do" pattern to one that included "here's-why-you-do-it."

Another example is provided in the research attempting to answer the question "Do students like mathematics?" Consider the findings of surveys by Dutton (1956), Dutton (1968), Dutton and Blum (1968), and Callahan (1971). What do you conclude?

The critical incidents presented at the start of this chapter have not been answered. That is a task you can tackle, using in part the studies cited in Appendix A. Studies which will help you in reaching a decision about "whether calculators and computers will really help them" (the first critical incident) include: Boyd

(1973), Cech (1970), Durall (1972), Durrance (1965), Gaslin (1972), Katz (1971), Keough and Burke (1969), Mastbaum (1969), Ostheller (1971), and Pack (1971).

The teacher's account of a research project that appears in this chapter suggests one way in which an answer for the second critical incident might be obtained. Other studies which might be investigated are: Beul (1974), Bull (1971), Drake (1935), Fitzgerald (1965), Gadske (1933), Hirsch (1973), Nix (1970), Olson (1971), Pearl (1967), and White (1972).

Mathematics laboratories are found in many schools. And some teachers have wondered about their general effect, as does Bob in the third critical incident. For some findings, you might see: Baker (1971), Cohen (1971), Schippert (1965), Silbaugh (1972), Vance and Kieren (1971, 1972), and White (1972).

Perhaps reading *this* chapter might help Ann, the teacher in the fourth critical incident. Many researchers have explored various problem-solving techniques, and some of their findings will be of interest to Ann and to others. Some of these are: Ashton (1962), Bechtold (1965), Denmark (1965), Dodson (1972), Kennedy, Eliot, and Krulee (1970), Kilpatrick (1968), Martin (1964), Sekyra (1969), Sherrill (1973), Shoecraft (1972), and Wilson (1968).

Research can be an important guide for better teaching; it's up to the individual teacher to make use of it.

Activities

1. Consider the following three questions. For each, what data might you collect that would help answer that question?
   a. Are students' attitudes toward mathematics related to the students' achievement in mathematics?
   b. What procedures are effective with low achievers?
   c. Is grouping for mathematics instruction a useful technique?
2. Select a subject area (such as general mathematics or geometry) and see what the studies in Appendix A indicate about the teaching and learning of this area. You might also want to investigate some research reviews: Henderson (1963) or Dessart and Frandsen (1973).
3. Select a title from Appendix A. Think how *you* would design a study with that title. *Then* read the research report and see how the researcher investigated the topic.
4. An actual research report (Coppedge and Hanna, 1971), published in the *Journal for Research in Mathematics Education*, appears on pages 361–365. Read the report, identifying the components listed on pages 344–346. Develop answers for the nine questions on page 348. Then read the review of the article by Nelson (1973) on pages 365–368 (published in *Investigations in Mathematics Education*). What agreement is there between your analysis and Nelson's?

5. Read other research reports in the *Journal for Research in Mathematics Education*. Begin a card file of abstracts of these articles. (You might also want to read reports from other sources.)
6. (If you have studied elementary statistics, you probably have an understanding of the statistical tests used in this exercise. If you have not yet studied statistics, you may nevertheless want to try to apply the statistical formulas.) The data which follow are from the study described on pages 349–352. To simplify the exercise, only the scores of the "high-ability group" for each treatment are given (X, individualized; and Y, usual), from a 25-item test on rational numbers.

X	Y
17	18
18	15
21	16
19	12
20	23
23	20
22	17
16	19

Using these data, compute the following:
   Mean ($\overline{X}$):

$$\overline{X} = \frac{\Sigma X}{n_x}$$

(the sum (Σ) of the scores divided by n_x, the number of scores)
   Standard deviation (s_x):

$$s_x = \sqrt{\frac{\Sigma (X - \overline{X})^2}{n_x - 1}}$$

   t-test:

$$t = \frac{\overline{X} - \overline{Y}}{\sqrt{\frac{\Sigma (X - \overline{X})^2 + \Sigma (Y - \overline{Y})^2}{n_x + n_y - 2} \left(\frac{1}{n_x} + \frac{1}{n_y} \right)}}$$

(To ascertain whether the resulting *t* is significant, you might consult a table in a statistics book.)

Comparison of Teacher-Written and Empirically Derived Distractors to Multiple-Choice Test Questions

FLOYD L. COPPEDGE and GERALD S. HANNA

The preference of many mathematics teachers for completion test items over multiple-choice tests for numeric and algebraic problems probably reflects their judgment that test validity is greatest when students *produce* right answers rather than *identify* them. Yet considerations such as speed and reliability of scoring stimulate use of multiple-choice questions.

Teachers who prefer completion items but find it more practical to use multiple-choice items may wish to define operationally the best multiple-choice item distractors, or wrong answers, as those that discriminate best between good and poor students in *completion-format* test administrations. However, teachers usually write multiple-choice item distractors in an attempt to maximize their discrimination between good and poor students in multiple-choice test administration. Scores on multiple-choice tests built to discriminate in the same way that completion tests discriminate probably correlate more highly with actual completion test scores than do scores on multiple-choice tests built with multiple-choice format discrimination in mind. Although the discriminatory power of items with the two kinds of origins may not necessarily differ, the mental processes they measure, or their content and concurrent validity, may be different. Therefore, teachers, and authors of standardized tests, concerned with students' ability to produce rather than recognize answers might well consider building their multiple-choice tests by first administering the items in completion form then selecting for multiple-choice item distractors those student responses that discriminate best between good and poor students. Similar procedures have been suggested by Furst (1958), Adams (1964), and Lore (1948). However, these authors utilized *frequency* of pupil responses to completion items rather than *discrimination* power.

The purpose of this study was to determine how congruent the multiple-choice distractors that experienced and student teachers supplied to geometry questions, in completion format, having numeric and algebraic answers were with the discriminating errors that students actually made in completion format.

Reprinted from *Journal for Research in Mathematics Education*, Nov. 1971 (vol. 2, pp. 299–303), © 1971 by National Council of Teachers of Mathematics. Used by permission. This research report is used in Activity 4 of the present chapter.

PROCEDURE

A 33-item completion geometry test of numeric and algebraic content was taken by 357 students in 15 classes in three midwestern high schools late in April of 1968. These widely used geometry textbooks were used by the five cooperating teachers.

An item analysis was performed in which the responses of the top 27% of the students were compared with the responses of the bottom 27%.[1] The percent of each group that gave each response to each item was computed. The fraction of the low group that gave each response was then subtracted from the fraction of the high group that gave the same response. This difference was the discrimination index for each response (the index is usually positive for correct answers and negative for incorrect answers).

In May of 1968, 13 secondary mathematics teachers, 11 of whom had taught five or more years and had supervised student teachers, and 18 university seniors who had completed student training in secondary mathematics, were asked to supply, without access to the item-analysis data, the three best distractors for each item, i.e., those they thought would most negatively discriminate (as defined above) if these distractors were to be used in a multiple-choice format.

FINDINGS AND DISCUSSION

Table 1 presents selected findings. The first section reports data for the geometry students. The first column gives the percent that passed each item. The next column reports the discrimination index of each item's right response. The third column reports for each item the sum of the discrimination indexes of the three best-functioning wrong answers; this figure indicates the extent to which the item's efficiency was attributable to the three best-discriminating student errors. Item 29, for example, has no greater total discrimination than does its three best distractors while item 19 has three times as much discrimination power as its three best distractors.

The next section of Table 1 reports the extent to which the experienced teachers created for multiple-choice format the responses that best discriminated in completion format. The first column reports for each item the percent of teachers who included among their three distractors the best single option. The second column gives the sum of the discrimination indexes, determined by analysis of student responses,

---

[1] The item-analysis procedure used is not as sophisticated as those using point-biserial correlations or X50-beta statistics. While more refined techniques are feasible for the development of tests, the method used in this paper possesses sufficient computational ease for use by classroom teachers.

of the three most frequently supplied distractors. The last section of Table 1 reports the corresponding data for the student teachers.

Both the experienced teachers and the student teachers varied greatly from item to item in their collective efficiency in providing the most discriminating distractors. In items like numbers 2, 16, 19, and 21, little

Table 1. Item-Analysis Findings

Item	Student ($N = 357$)		
	% Passing	Disc. index	Total disc. of 3 best distractors
2	52	+.83	−.49
14	30	+.43	−.11
16	49	+.72	−.47
19	40	+.63	−.21
21	40	+.51	−.39
28	28	+.47	−.15
29	10	+.15	−.15
Mean ($N = 33$)	38.3	+41.4	−.204

Item	Teachers ($N = 13$)	
	% Giving best distractor	Total disc. of 3 most-provided distractors
2	62	−.41
14	08	+ 0.8
16	31	−.38
19	54	−.21
21	46	−.30
28	00	+.11
29	08	+.01
Mean ($N = 33$)	23.6	−.057

Item	Student teachers ($N = 18$)	
	% Giving best distractor	Total disc. of 3 most-provided distractors
2	61	−.39
14	06	+.08
16	67	−.40
19	50	−.21
21	61	−.39
28	00	+.11
29	00	−.04
Mean ($N = 33$)	27.0	−.066

discriminatory loss would result from using the composite judgment of either group. On the other hand, the collective judgment of both groups actually discriminated in the wrong direction in items like 14 and 28. This situation arose when items had wrong answers that were more attractive to capable students than to poor students. Both experienced teachers and student teachers appeared unable to differentiate popular distractors from best-discriminating distractors.

The bottom row of each section in Table 1 reports the mean data for the 33 items. The loss of discriminatory potential of completion-format student errors that resulted from the respective use of experienced teachers' and student teachers' judgment is 72% and 68%. Similarly, the single best distractors to items were identified by each group only about one-fourth of the time. Contrary to expectations, the experienced teachers anticipated the best discriminating student responses slightly less accurately than did the student teachers. This may have resulted from (1) chance factors, (2) the role expectation of greater conscientiousness of student teachers in research activities, (3) the student teachers' greater identification with the mental processes of examinees, and/or (4) the student teachers' more accurate perception of the distinction between discrimination and popularity of distractors.

CONCLUSIONS AND RECOMMENDATIONS

In this study neither experienced teachers nor student teachers produced multiple-choice numeric and algebraic geometry item distractors that were very similar to the discriminating errors that students made when responding to the same questions in completion format. Consequently, if one defines the best multiple-choice item distractors as those which discriminate best in completion format administration, there appears to be merit in utilizing the following procedure:

1. Administering items in completion format,
2. Item analyzing the results,
3. Selecting the most discriminating student errors as the multiple-choice item distractors.

There is, however, a need for empirical comparison of the correlation of completion items with parallel multiple-choice items of each kind of origin.

While practical considerations may prevent teachers from following the more time-consuming developmental procedures for preparing multiple-choice items indicated in this paper it may be, however, that they would increase their skill in writing items by occasionally using such a technique in conjunction with their standard procedures (a hypothesis, of course, open to research). This would give individual teachers a way to study the kinds of items, if any, for which poorer than necessary multiple-choice distractors are produced. In addition, such a procedure may provide a setting for improving skill in anticipating student errors (again, a researchable question). The competencies attained in learning to write item distractors that would efficiently discriminate in completion-format administrations may also increase teachers' insights into the mental processes of high and low ability students alike.

REFERENCES

Adams, G., & Torgerson, T. *Measurement and evaluation in education, psychology, and guidance.* New York: Holt, Rinehart, & Winston, 1964.

Furst, E. J. *Constructing evaluation instruments.* New York: Longmans, 1958.

Loree, M. R. A study of a technique for improving tests. Unpublished doctoral dissertation, University of Chicago, 1948.

Review of Coppedge and Hanna's "Comparison of Teacher-Written and Empirically Derived Distractors to Multiple-Choice Test Questions"

L. D. NELSON

EJ 046 714   520   SE 504 303
COMPARISON OF TEACHER-WRITTEN AND EMPIRICALLY DERIVED DISTRACTORS TO MULTIPLE-CHOICE TEST QUESTIONS. Coppedge, Floyd L.; Hanna, Gerald S., *Journal for Research in Mathematics Education,* v2 n4, pp299–303, Nov 71
    Descriptors—*Geometry, *Mathematics Education, *Multiple Choice Tests, *Test Construction, Objective Tests, Research, Teacher Developed Materials, Testing

Expanded Abstract and Analysis Prepared Especially for I.M.E. by L. D. Nelson, University of Alberta

1. PURPOSE

To determine the ability of experienced teachers and student teachers to select the most discriminating distractors for multiple-choice items in evaluating certain aspects of geometry.

2. RATIONALE

The authors point out the dilemma of teachers who prefer to use completion-type items in constructing tests but who often use a

Reprinted from *Investigations in Mathematics Education,* Winter 1973 (vol. 6, pp. 14–19). Used by permission of the ERIC Information Analysis Center for Science, Mathematics, and Environmental Education, The Ohio State University.

multiple-choice format instead because of ease and reliability of scoring. One way that teachers can attempt to improve the quality of any particular multiple-choice items is to administer the item to pupils first in completion form and then select from among the actual errors those distractors which have the most discriminating power. Teachers often try to do this but instead of choosing for distractors those erroneous responses that are the best discriminators they choose the responses which occur most frequently. The authors imply that the most frequent erroneous response is not necessarily the most discriminating.

3. RESEARCH DESIGN AND PROCEDURE

To check the ability of experienced and student teachers to construct the most discriminating distractors for multiple-choice test items, a 33-item geometry test in completion format was first administered to 357 students.

An item analysis (or more correctly a response analysis) was then done in which the responses of the top 27% of the students were compared with the responses of the bottom 27%. There were among these, of course, some correct answers and some incorrect answers. In any case the percent of each group that gave each response to each item was calculated. A *discrimination index* for each response was obtained by subtracting the percent of the low group giving the response from the percent of the high group giving that response. This index would normally be positive for correct responses but negative for incorrect responses. The best distractors among the incorrect responses would presumably be those with the highest negative values. Sometime after this analysis was done 13 experienced secondary mathematics teachers and 18 university seniors (student teachers) were supplied with the 33-item test and the responses of the 357 geometry students. They were not supplied the item analysis data, however.

From this information they were each asked to change the completion form to multiple-choice form and to supply what they considered to be the three best distractors for each item.

The results of the item analysis for a few selected items (7 out of 33) are given in a table. There is a separate section in the table for the analysis of the students' responses, another for those of the experienced teachers, and a third for those of the student teachers.

4. FINDINGS

In the student section of the table there is a column which shows the percent of students passing each of the 7 items, the next gives the discrimination index of the correct response and the last gives the total of the discrimination index of the best three distractors.

For both the experienced teachers and student teachers there is one column which shows the percent choosing the best distractor. Another column gives the total of the discrimination index of the three distractors they choose. Whether the authors selected items which showed as great a variety as possible or not, the tabulated information indicates there was a great deal of variation in the efficiency with which experienced and student teachers chose distractors. On some items there was little difference between the collective judgment of the teachers and the actual results of the response analysis. On others they chose responses which discriminated in the wrong way. The authors state that both groups of teachers found it difficult to separate the popular distractors from best discriminating factors.

They also give further information that does not appear in the table. They state that there is a "loss in discriminatory potential of completion format student errors" of 72% for experienced teachers and 68% for student teachers. They also report that each group of teachers selected the single best distractor for items only about once out of four.

The experienced teachers are reported to have selected the best distractor slightly less accurately than did the student teachers. This finding was attributed to either 1. chance factors, 2. the student teachers' greater conscientiousness in research activities, 3. the student teachers' greater identification with the mental processes of examinees, 4. the student teachers' ability to distinguish between most popular distractors and the most discriminating ones.

5. INTERPRETATIONS

The authors recommend that if one defines the best multiple-choice item distractors as those which discriminate best in completion test format one might follow this procedure:

1. Administer items in completion format
2. Analyze the items
3. Select the most discriminating errors as distractors for multiple-choice items

They point out the need, however, of determining the relationship between completion items and parallel multiple-choice items.

Finally they suggest that even though teachers might not follow the recommended procedure all the time in constructing multiple-choice items, it might be done from time to time. The advantages would be that teachers might learn to spot items with very poor distractors, they might increase skill in anticipating student errors, and they might improve their insight into the mental processes of high and low ability students alike.

ABSTRACTOR'S NOTES

This is a neat little study which points out the possibility of selecting multiple-choice distractors on the basis of discriminatory power rather than frequency. The underlying assumption in the whole study is that the procedure outlined would produce scores on multiple-choice tests that would probably correlate higher than usual with parallel tests in completion form. It is a pity that the investigators did not use the information they already had to go on and test this assumption. One still has the uneasy feeling that the most discriminating errors which show up in completion format may not discriminate in the same way if they are used as distractors in multiple-choice format.

This is a short report (only 4 pages in the journal) and probably was designed that way. However, more specific information about what data were provided to the teachers would have made the report easier to read. One wonders, too, why the complete analysis is shown for only 7 of the 33 items. An example or two of a completion test item, the kinds of response students made to them, and the kind of distractors chosen by teachers (were they the most frequent errors?) in making multiple-choice items would have improved the report immeasurably.

References

Austin, Joe Dan and Austin, Kathleen A. Homework Grading Procedures in Junior High Mathematics Classes. *School Science and Mathematics* 74: 269–272; April 1974.

Baker, Betty Louise. A Study of the Effects of Student Choice of Learning Activities on Achievement in Ninth Grade Pre-Algebra Mathematics. (Northwestern University, 1971.) *Dissertation Abstracts International* 32A: 2895; December 1971.

Bright, George W. and Carry, L. Ray. The Influence of Professional Reference Groups on Decisions of Preservice Secondary School Mathematics Teachers. *Journal for Research in Mathematics Education* 5: 87–97; March 1974.

Coppedge, Floyd L. and Hanna, Gerald S. Comparison of Teacher-Written and Empirically Derived Distractors to Multiple-Choice Test Questions. *Journal for Research in Mathematics Education* 2: 299–303; November 1971.

Dessart, Donald J. and Frandsen, Henry. Research on Teaching Secondary-School Mathematics. In *Second Handbook of Research on Teaching* (edited by Robert M. W. Travers). Chicago: Rand McNally, 1973, pp. 1177–1195.

Dodson, Joseph Wesley. *Characteristics of Successful Insightful Problem Solvers*. NLSMA Report No. 31. Stanford, California: School Mathematics Study Group, 1972.

Hanna, Gerald S. Testing Students' Ability To Do Geometric Proofs: A Comparison of Three Objective Item Types. *Journal for Research in Mathematics Education* 2: 213–217; May 1971.

Henderson, Kenneth B. Research on Teaching Secondary School Mathematics. In *Handbook for Research on Teaching* (edited by N. L. Gage). Chicago: Rand McNally, 1963, pp. 1007–1030.

Hoffman, Nathan. Geometry in Mathematics: A Survey of Some Recent Proposals for the Content of Secondary School Geometry. (University of Montana, 1973). *Dissertation Abstracts International* 34A: 3026; December 1973.

King, Donald Thomas. An Instructional System for the Low-Achiever in Mathematics: A Formative Study. (University of Wisconsin, 1972.) *Dissertation Abstracts International* 32A: 6743; June 1972.

Kuhfittig, Peter K. The Relative Effectiveness of Concrete Aids in Discovery Learning. *School Science and Mathematics* 74: 104–108; February 1974.

Lankford, Francis G., Jr. *Some Computational Strategies of Seventh Grade Pupils.* Final Report. Charlottesville: The Center for Advanced Study, The University of Virginia, October 1972. ERIC Document No. ED 069 496.

Nelson, L. D. Review of Coppedge and Hanna (1971). *Investigations in Mathematics Education* 6: 14–19; Winter 1973.

Short, Byrl G. and Szabo, Michael. Secondary School Teachers' Knowledge of and Attitudes Toward Educational Research. *Journal of Experimental Education* 43: 75–78; Fall 1974.

Strickmeier, Henry Bernard, Jr. An Analysis of Verbal Teaching in Seventh Grade Mathematics Classes Grouped by Ability. (University of Texas at Austin, 1970.) *Dissertation Abstracts International* 31A: 3428; January 1971.

Suydam, Marilyn N. *A Review of Research on Secondary School Mathematics.* Columbus, Ohio: ERIC Information Analysis Center for Science, Mathematics, and Environmental Education, March 1972.

Zahn, Karl George. The Optimum Ratio of Class Time To Be Allotted to Developmental Activities and to Individual Practice in Teaching Arithmetic. (University of Colorado, 1965.) *Dissertation Abstracts* 26: 6459; May 1966.

Students, Teachers, and Student Teachers

AFTER STUDYING CHAPTER 11, YOU WILL HAVE:

- ★ reviewed comments by student teachers about their student teaching experiences.
- ★ obtained hints on how to get the most out of visits from supervisors.
- ★ examined a sample student teacher evaluation form and learned what supervisors look for when they visit your classes.
- ★ received suggestions for a successful student teaching experience.

CRITICAL INCIDENTS

1. Walter has wanted to be a teacher since he was in elementary school. But now that he is about to do his student teaching, he has mixed feelings. His big problem is that he can't see himself "on the other side of the desk."
2. Elaine is nervous. It's her first day of student teaching and she stands in front of her class. She is sure that she will make a mistake in the proof she is teaching, even though she knows it thoroughly and has prepared her lesson fully. But if she messes it up, the students will think she's stupid!

3. Andrew had a fantastic student teaching experience. His cooperating teacher, Ms. Baker, was super. She had a great way with her students, and they all seemed to enjoy being in her classes. He learned a lot about presenting different topics, using various teaching aids, and preparing and grading assignments. He has stacks of sample worksheets, teaching plans, and tests to which he can refer when he starts teaching. Andrew really feels prepared to teach.

STUDENT TEACHING: TIME OF TRANSITION

You have been preparing for teaching for quite a while now, maybe for as long as you can remember. You have heard a lot of people talk about what teaching is like—for them. Now you will be able to find out for yourself.

Most teachers agree that student teaching was the highlight of their teacher education program. They probably have different reasons for saying so, but most of the reasons likely center around the fact that in student teaching they learned a great deal about teaching itself. Teaching requires a great variety of skills, and skills are learned by doing. Student teaching provides the first significant opportunity for doing what you have been preparing to do throughout college.

In this chapter we present topics which we have found to be important to student teachers. Our first choice for discussing student teaching would be to arrange for you to talk with former student teachers (the more recent their experience, the better) about what they did—the good times they had, the friendships they established, the many things they learned. Since we can't arrange such interviews for you, we have obtained written commentaries by students about their student teaching.

FROM STUDENT TO TEACHER: CHANGING ROLES

The first commentary from our students has to do with setting goals for student teaching. How do these goals compare with yours?

> *February* 8. Anyone who goes into a branch of education naturally expects to be placed in a *real* school and in front of a *real* bunch of students at some time before his formal education ends. This time has come. I can truthfully say that I feel a bit apprehensive when I remember some of the student teachers I had as a high school student. I can also remember a lot of the "good times" (at least for the students) that we had with them. It is difficult to list precise objectives for teaching when one has been peering only one way over the teacher's desk for so long. However, I have a few "ideas" which may possibly turn into "objectives" as the teaching experience becomes more real.

1. I would like to develop such an enthusiastic attitude toward the material presented that it is contagious to the students themselves. This may be difficult since many students are only taking math courses to "work off" requirements. However, I know that this inspiring atmosphere does exist. My family is not rich, but we certainly have never needed the added income from my mother's teaching salary. Yet she goes to school every day and is happiest when she is in front of a class of children. She is even going back to school in the evenings to keep herself from becoming molded into the same patterns.

2. At the end of this semester I would like to be able to say that I had not taught for my own self-flattery. Teachers who call the students "my kids" or say "See how easily Johnny learned what I taught him" are often only boosting their own self-esteem. The attitude shouldn't be "See how well the students learn when I teach," but "See how well the students have developed their learning abilities." Of course, there is no more wonderful feeling for a teacher than to know that his teaching has been effective. I have given flute lessons for years, and I know the excitement I feel when a young musician finally masters a certain technique.

3. After this semester, I want to feel definitely that teaching is right for me. This implies there may be some uncertainty in my thoughts about becoming a teacher, but this is not the case! I am convinced now that this is what I want to do; I only ask that this semester confirm my feelings.

The next account relates to the problem faced by Walter in our first critical incident—that of defining a new role for oneself.

March 7. One thing we as student teachers are able to observe is the difference—almost a class difference—between students and teachers. When I was in high school, I had such a firm belief that the teachers were superbeings that I was never able to talk personally with them. Teachers had their own lounges, restrooms, and large desks with chairs which swiveled. I would never have gone into a lounge to find a teacher. If only I had known then what I know now—that teachers are human beings. They don't mind being interrupted with questions, and they can understand when something comes up so that a student cannot complete his homework.

As I think back now, perhaps I was trying to please the teacher when I should have been trying to learn for myself. I hope the students in my classes will want to learn for their own satisfaction and pleasure.

We will not try to add to what the student teacher has said about changing roles from that of student to that of teacher, except to say that we believe the best way to learn a new role is to enter into it and learn by experience. Student teaching provides that opportunity.

STUDENT TEACHER MEETS STUDENTS: THAT FIRST DAY

If you're nervous about your first day of student teaching, welcome to the club! It's a very common feeling for several reasons. You don't know exactly what to expect (all of us to some degree are perturbed about the unknown). You want to do a good job, and you need that assurance of actually teaching to prove to yourself, and to others, that you can do it. You have waited and prepared for this moment—and you want to "get on with the show."

The story is told of an opera singer who, if she did not feel nervous prior to a performance, would refuse to go on stage. She knew that without the resource of that extra nervous energy she would not be at her best. This account may be of some assurance to those who suffer from that common affliction, butterflies in the stomach, and who feel they have far more nervous energy than they really need! Here is how one of our student teachers put it:

> *February 15.* Today I taught—my first actual day of teaching. I was scared. I'm glad now that I didn't have time right before class to review. I think I would have been very nervous then. I think the class went well. I don't know if I talked as slowly as I wished. There's other room for improvement. I must remember to do a few things so that the kids might follow more easily. I think they followed most of my proofs reasonably well. I think I included everyone. Naturally, the kids are slow at responding at first. Towards the end of the discussion problems, the kids were getting into them quite well. Once I know I fished for an answer—"does everyone agree with him?" That might have made him feel bad. I was alert after that. So I think I handled the rest of it pretty well. Overall, I'm pretty pleased for my first lesson. It wasn't as bad as I expected. Calmness, easier flow of the lesson, better feedback, are some of the things I'd like to work on. Friday is going to be my first day before the general mathematics class. That's going to require a different approach. Well, now's not the time to worry about that.

This student teacher seemed to approach her class rather well; she was able to focus on the class instead of on herself. A good principle to follow when in front of the class is to concentrate on your students. Rather than ask yourself "How am I doing?" ask yourself: "How are my students doing?" "Is Mary following this example?" "Was this explanation too difficult for Jon?" As you become more occupied with your students and their progress, you will find that your concerns about yourself will decrease.

You also may be helped during your first days of teaching by getting together with other student teachers and talking over your experiences—the successes and the disappointments. The student just quoted wrote the following after taking part in a discussion group with other student teachers.

February 22. This afternoon was the first meeting of our group. I didn't know how it would turn out. From what we knew, the kids (the other student teachers) didn't feel like they would want to go in and talk about their fears, hopes, etc. I felt that in this first session, Earl (our leader) hoped that we would get to know each other better, not just state facts about hometown, age, etc., but what we like, are aiming for, etc. I was looking forward to it and was a little apprehensive. It takes me a pretty long time to open up, usually. I'm better, but I can be awfully difficult. Well, I think the session went great. I loved it! Earl is great, the kids are great! I wasn't afraid to talk! I think seeing them sitting there just as scared as I, helped me to start. Wow! I can't believe it.

THE IMPORTANCE OF PREPARATION

Next to developing a healthy concern for your students as individuals, the most important factor for success in student teaching is preparation. It is indeed reassuring to know that when you go in front of a class you have ample material, a well-worked-out teaching plan, and worthwhile activities—come what may. Part of this preparation includes, of course, having done your best to learn the student's names (recall the hints given for learning names, pp. 63–64, and for planning, pp. 194–197). You will quickly learn, as did the student teacher of the following commentary, that while preparation is extremely time-comsuming, it pays dividends in terms of student learnings and attitudes.

April 27. I am amazed by the amount of time I am putting into class preparation and general study for teaching. Student teaching—although I love it and am getting a lot of very good experience—is a 24-hour job! For the first year or two —until I am more familiar with my material—I will have to prepare materials as well as techniques. After a few years, the material will be more familiar, I will have more self-confidence, and then I can work more with techniques and style. I don't feel I will ever be "perfect" enough not to prepare or experiment with new and improved ideas for teaching. But right now I am overwhelmed with preparation!

STUDENT TEACHING: BUILDING PROFESSIONAL RELATIONSHIPS

So far we have dealt with your encounters with your students. Teachers spend most of their time working with students and learn much from their students in return. But your relationships with the teachers in your school are also important. For most of you, this will be your first experience in relating to another person on a professional basis—one professional to another (soon-to-be) professional. You should find this relationship to be an essential source of learning, too.

Your learning situation while student teaching is different from that in the college classroom, however. To a large extent, what you learn will depend upon what you do to see that opportunities to learn actually arise. Your teachers are not likely to give you formal assignments; they may suggest assignments for your students, but not for you. However, we do know of one teacher who, when first meeting his student teachers, routinely assigns them the reading of Fawcett's *The Nature of Proof* (1938).

The number and variety of professionals you will work with will vary, depending upon your own situation. We assume there will be one classroom teacher with whom you will work most closely, and we call this person your "supervising teacher." You may work with other classroom teachers as well; this is usually very helpful. You will thus have opportunities to compare approaches, to explore points of view, and perhaps to note how similar topics can be handled well in different ways. In most high schools there also is a department chairperson (sometimes this is your supervising teacher). You may not be scheduled to interact with the department chairperson, but you should make attempts to get to know this person and to make it clear that you are willing to help and to learn. Usually student teachers are welcome to attend department meetings, to work in the mathematics resource room as a tutor, to assist students in the mathematics laboratory, to give presentations for the mathematics club, to help in the operation of the computer center—the possibilities are many, and you will profit from the efforts you make. Finally, the faculty member from your college or university is another key member of the team of professionals concerned with making your student teaching a success.

WORKING WITH YOUR SUPERVISOR

Supervisors are primarily concerned with doing what they can to see that you learn as much as possible while you are student teaching. Typically, supervisors are themselves excellent teachers, have had considerable experience in secondary school teaching, and have done graduate study in mathematics education. But for them to most effective, they need your help. Perhaps nothing impresses supervisors more than a student teacher who is eager to learn, who asks opinions of others, and who attempts to work out methods which are best for that student teacher. The following suggestions may be helpful in developing an effective relationship with your supervisor.

1. Get to know your supervisor. Confer about his or her ideas on teaching, on mathematics, on schools. Your supervisor will be flattered that you seem interested in what he or she has to say!
2. Find out your supervisor's expectations for you. You have a right to know what criteria will be used for your evaluation. Write down these points. Use them as you assess your own performance. Compare your assessments with your supervisor's and try to determine reasons for any differences.

"Elwood, I do not consider 'having them stop throwing things at me' as a significant behavioral objective."

© 1977 by Ford Button.

3. Take the supervisor's comments seriously. It is frustrating for a supervisor to spend a great deal of time with a candidate, only to gain the impression that his or her words of wisdom fell on deaf ears! In some cases you may disagree with your supervisor's recommendations—and you probably should say so—but in any event there should be some indication on your part that you have thought about what has been said.

4. Capitalize upon this opportunity to get feedback on your teaching. Strange as it may seem at first, very few opportunities to analyze your teaching—to regard it from other points of view, to discuss alternatives for improvement —will be available once you receive your certificate. Current teacher-teacher and teacher-administrator roles do not tend to promote analysis and feedback for improving classroom instruction.

5. Help your supervisor set up a schedule for visiting you. Provide a bell schedule and floor plan for the school, and exchange telephone numbers so that you can keep each other informed about last-minute changes in plans.

6. Don't hesitate to take the initiative at times during your supervisory conference (though you will want to avoid completely dominating the discussion!). You might think of questions to raise with the supervisor, such as "What did you think of the examples I used today?" or "Am I asking better questions this week than I did last week?"

7. Make a record of your supervisory conferences. Such a record will be

valuable to you as you look back over your student teaching experience: you will see a map of your own professional growth. You also will be reminded of points on which you were to concentrate. Share these notes with your supervisor; she or he will be encouraged to see your interest in self-analysis and improvement.

WHEN SHOULD I ASK FOR HELP?

At times student teachers are hesitant to ask for help, believing that they may be regarded as weak, dependent, or ill prepared. Nothing in our experience as supervisors bears out such a belief. Never have we known a student teacher to be downgraded because she or he asked questions or sought assistance. But we have, on the other hand, known student teachers who have not received highest ratings because they (1) seemed unwilling to learn, (2) did not attempt to learn from the experience of others, or (3) failed to build up their own reserve of techniques, hints, and materials.

Almost all student teachers worry about discipline. "Will I be able to handle my classes?" "Will my supervisors think my students are too noisy?" The possibilities for questions and doubts are limitless. But this common concern provides us with a good object lesson. Why not talk about this problem with your supervisor? Rather than thinking less of you because of it, your supervisor will probably think more highly of you—for being open, for showing a willingness to learn, for seeking opinions, and for valuing them. The student teacher with the following commentary had a similar concern (Mr. Phillips is the supervising teacher).

> I wonder how Mr. Phillips thinks I'll do as a teacher. I told him that I'm from a small school and had some pretty strict teachers. He asked me if I thought I'd be the same. His questions seemed to imply I wouldn't be authoritative enough, which might lead me into trouble because I look so young. I know that will be a problem, and I think that's a major concern of mine—can I get the material across?— and can I command their respect and compliance? I'm going to go in and approach it the way I think best. I'm scared to death. Scared to fail, but I'll try.

WHAT IF I DISAGREE WITH MY SUPERVISOR?

As you may have learned already, teaching involves many points of view and ways of doing things. You may see your supervisor use a technique in class that you think you would *never* use. On the other hand, some approaches which you believe to be sound and helpful may not meet the approval of your supervisor.

These are real-life situations. There simply is little unanimity on many aspects of teaching. The important point to keep in mind here is that you are developing criteria which will be useful in helping you make your own decisions about teaching. What are the characteristics of good planning? What is an effective teaching technique for you? What sorts of organizational patterns lend themselves to profitable use in the classroom?

One situation which sometimes arises, and which could be difficult to handle, is when a staff member at your school, trying to be helpful, wants to fill you in on "all the troublemakers you are going to have in class." This person may well give you detailed biographies of the worst offenders (as he or she sees them), pointing out that he or she has taught all of the children in a given family—and that they are all losers. You have to use your own imagination and tact in dealing with a would-be assistant of this kind. We simply want to say (in the words of a well-worn cliche) that a person should have a reputation to live up to, not to live down. We should do our best to approach our classes expecting the best of them —and usually they will live up to our expectations (see Research Highlight, "Teachers' Expectations and Pupil Performance").

WHEN CAN I HAVE MY OWN CLASS?

One of the more common sources of frustration for a student teacher arises from the temporary nature of the position. The student teacher may feel that he or she is only a visitor in the school and classrooms, here today and gone tomorrow. And the students may act like that, too. "If only I had my own class, things would be different. . . . "

We have found that most supervising teachers will readily assign to the student teacher as much responsibility as they feel can be assumed. In other words, as a general rule, student teachers will get as much responsibility as they demonstrate they can handle. But the burden of proof is usually on the student teacher. You are obliged to show that you are capable of assuming responsibility before you can realistically expect to be put in a responsible position.

TEACHER OBSERVATION

Most student teachers are expected to spend considerable time observing other teachers at work. Unfortunately, in many cases, this time is not well spent. There is a knack to observing; it takes skill to observe profitably. Here are some points to remember:

1. Talk with the teacher of the observation class. Depending on your purposes for observing, this may be before or after your visit to the class. You may wish to discuss such matters as the teacher's goals for the lesson, his or her views on teaching, and his or her philosophy of education.
2. Determine your own goals for your observation. Will you concentrate on teacher strategies or learner behavior? Will you attempt to identify kinds of questions asked? Will you examine the nature of the questions which evoke the most interesting responses from the students?
3. A particularly useful technique is to concentrate on one student alone.

RESEARCH HIGHLIGHT | TEACHERS' EXPECTATIONS AND PUPIL PERFORMANCE

In one of the better-known research studies in education of the 1960s, Rosenthal and Jacobson (1968) investigated the effects of teachers' expectations on the performance of their pupils. The researchers hypothesized that when a teacher expects superior academic performance from certain students, those students will show greater intellectual growth than those for whom such performance is not expected. The study involved classes designated as fast, medium, and slow in reading at each grade level from first through sixth in a South San Francisco school.

During May 1964 these children were administered a test called the "Harvard Test of Inflected Acquisition" by researchers Rosenthal and Jacobson. This test was described to the teachers as a new instrument purported to identify "bloomers" who would probably experience unusual gains in school achievement during the following year. Actually, the test was a little-known test of intelligence, the Flanagan Test of General Ability. In the following fall, a randomly chosen 20 percent of the students were designated "spurters." Each teacher was provided a list of those "spurters" who would be in her or his class.

The I.Q. test was readministered in January 1965, May 1965, and May 1966. Many of the students designated as "spurters" showed remarkable gains in I.Q. scores, allowing the researchers to conclude "that teachers' favorable expectations can be responsible for gains in their pupils' I.Q.s and, for the lower grades, that these gains can be quite dramatic" (p. 98).

It should be pointed out that this study has been the subject of a great deal of controversy. One attack on the study has been made by Thorndike (1968), who wrote, "In spite of anything I can say, I am sure it will become a classic—widely referred to and rarely examined critically. Alas, it is so defective technically that one can only regret that it ever got beyond the eyes of the original investigators!" (p. 708). See also Elashoff and Snow (1971) and Mendels and Flanders (1973).

Mentally put yourself in the student's shoes. How would you have responded in that situation were you the student? Why did the student respond, or fail to respond, to questions or problems posed by the teacher? In this connection, Holt (1964) has commented with insight:

A teacher who is really thinking about what a particular child is doing or asking, or about what he, himself, is trying to explain, will not be able to know what all the rest of the class is doing. And if he does notice that other children are doing what they should not, and tells them to stop, they know they have only to wait until he gets back, as he must, to his real job. Classroom observers don't seem to see much of this. Why not? Some of them do not stay with a class long enough for the children to begin to act naturally in their presence. But even those who are with a class for a long time make the mistake of watching the teacher too much and the children too little. Student teachers in training spend long periods of time in one classroom, but they think they are in there to learn *How to Teach*, to pick up the tricks of child management from watching a *Master at Work*. Their concern is with manipulating and controlling children rather than understanding them. So they watch the teacher, see only what the teacher sees, and thus lose much of what could be a valuable experience. (pp. 21–22)

In order to point up the difference between "just looking" at a class, which many of us may do when we first visit teaching situations, and doing some significant observing, we provide two accounts. The first is fictitious; the second is an actual report provided by one of our student teachers.

We visited a freshman mathematics class today. It was very dull. The teacher didn't do anything interesting and the kids hadn't completed their homework. I hope I don't ever teach like that.

It was interesting to watch the students take their final computer test. Michele worked confidently on her test and was finished very quickly. She checked each of her answers and ran them through the computer as a double check. Maureen worked quickly, but without the confidence of Michele. Maureen took a couple of quick glances into her notebook to see if she could find hints on how to work the problems. I saw her do this and she saw me watching her, so she quickly closed her notebook and went on with the test. Mike and Ellyn were by far the slowest. The expressions on Mike's face were giveaways for what was going through his mind. He was the last one to finish, but he checked his work and even wrote out two computer cards, instead of the one which was required. The results of the test were (out of 40):

Michele	38
Maureen	35
Mike	33
Ellyn	33

Several points can be made about these two sets of comments. The first account reveals very little about what went on in the classroom. It is of no help or value to simply record that bad teaching took place. Why was the teaching dull? What did the teacher do? How might the teaching have been improved? What was the role of the students? The second account reveals a great deal, not only about the class observed, but about the observer! Even though the class was "only taking a test," many important cues were picked up—about the students, why they acted the way they did, the nature of the learning climate, how the students responded to observers, and about seeing class as the students saw it.

STUDYING YOUR OWN TEACHING

When you become accomplished at observing, you can, through the help of audio and video recordings, apply these skills to your teaching. (This is the subject of Chapter 12.) In the meantime, however, you can profit simply from writing your own analysis of what you are doing. Here is one such self-analysis by a student teacher:

> *February 17.* Today was a good day. I worried whenever I wasn't thinking about something else for my algebra class tomorrow. I've settled down a little now though. I think I've learned some things to look out for. Yesterday, H. (a fellow student teacher) used the book an awful lot. Today he didn't and I think that's much better. I'll have to be careful about not talking into the board either. I hope people tell me if I'm going about something a hard way or doing something wrong. I hadn't noticed that I talked to the board quite a bit Tuesday. I'm glad I know that. Sitting in the back you can sometimes see better what would clear up a misunderstanding because you're not being noticed. I hope my supervising teacher will give me some specific suggestions today. I know there are many things I could improve. I wouldn't feel hurt, and I certainly appreciate help. I'm sure the students would like the chance for better, fuller understanding, and that's what I want, too.
>
> Geometry went well for N. and me. Friday is their day for constructions. We presented them with the assignment and then walked around and helped them. At first, they were really shy about asking for help. Finally, some of the kids warmed up. The girls who needed help were even slower so I finally asked if they had any questions. That helped break the ice. Three guys sat in the back and didn't work after a while. One acts like he could be a troublemaker. So the big question—should I ask them to work on constructions and thus alienate them, maybe, or just let them be? I asked them. They didn't seem mad about it. I guess maybe Tuesday I'll find out if they will act hostile toward me.

This particular student teacher has taken an important step in professional development. He has begun to think about what he is doing and to consider alternative ways to working with his classes. One common topic of discussion, especially for beginning teachers, is the amount of student participation in class. Student participation is an important component of effective teaching (see

Research Highlight, "On Involvement"), and you may wish to take special note during observations of other classes about ways in which pupil participation is increased (see Chapter 12).

RESEARCH HIGHLIGHT | ON INVOLVEMENT

Over the years much research has explored various factors which might be related to the effectiveness of a teacher. Robitaille (1969) used a battery of tests which were found to classify teachers accurately as being effective or ineffective. Then he tested the hypothesis that the effective teacher of mathematics would encourage student participation significantly more often than would the minimally effective teacher. Using a Pupil Involvement Checklist, 50 lessons taught by the participating teachers were observed. The hypothesis was not rejected: student participation was significantly higher in the classes of effective teachers.

EVALUATING STUDENT TEACHING

Your college or university will have established procedures and criteria for evaluating your student teaching experience. We recommend that before you begin your student teaching you find out such things as the number of supervisory visits you can expect, who will make the visits, the expectations of the supervisors, and who should be included in the supervisory conferences (for example, will the department chairperson or other school faculty be involved?)

Figure 11-1 shows the form used by the University of Georgia for evaluating student teachers in mathematics. The form covers rather completely the various facets of student teaching and in many ways summarizes the main points of this book. We recommend that you examine the entire form in some detail and make your own assessment of your strengths and weaknesses as you enter student teaching. Such a form could also be useful in initiating interaction with your supervisor; you might ask for an evaluation, based on one or more parts of this form, at several stages in the course of your student teaching.

Critical Incidents Revisited

We believe that the critical incidents have been responded to in the comments of the student teachers quoted in this chapter. The feelings of Walter and Elaine are by no means unique to them. Walter has trouble seeing himself as a teacher,

University of Georgia

College of Education

STUDENT TEACHER EVALUATION
AND
STUDENT TEACHING RECORD

MATHEMATICS

Student Teacher_____

Supervising Teacher_____

School _____

Date _____

Figure 11-1 Evaluation form used by Department of Mathematics Education at the University of Georgia. (Reprinted by permission of Professional Experiences Laboratory, College of Education, University of Georgia.)

IMPORTANT INFORMATION

TO THE SUPERVISING TEACHER:

This evaluation guide and record should be discussed by the supervising teacher and student teacher early in the quarter. You will probably also want to discuss it with the college supervisor during his first visit. Items defining additional agreed upon goals which may be needed to fulfill the individual needs of a student teacher should be added under the appropriate **Part**. *Parts I, II and III should be completed by the supervising teacher together with the student teacher three times during the quarter;* approximately during the third and seventh weeks and at the end of the student teaching period.

The first two ratings are for the purpose of promoting student teacher growth. The final rating becomes a part of the student's record. Use the numeral *one* for the first rating, *two* for the second and *three* for the third. *Submit the entire guide and record to the college supervisor the last day of the student teaching period, after detaching your copy of the Student Teaching Record.*

Please use the following as a guide in determining the column in which to rate each item in Parts I, II and III. Parts IV, V and the Student Teaching Record are to be completed *at the end of the quarter.*

ABOVE AVERAGE - item evidenced to an extent that is *highly satisfactory* for a student teacher; can be considered *more* than adequate for a good beginning teacher

AVERAGE - item evidenced to an extent that is *satisfactory* for a student teacher; can be considered *adequate* for a good beginning teacher

BELOW AVERAGE - item evidenced to an extent that is *below the level regarded as satisfactory* for a student teacher; needs further attention before it can be regarded as *adequate* for a beginning teacher

NOTE: If at any time the student teacher's performance on an item is extremely low, the supervising teacher should immediately notify the college supervisor or the Coordinator of Student Teaching at the University and the person in the local school system who is responsible for placing student teachers so that remedial measures may be taken.

2

Figure 11-1   continued.

Part I - PERSONAL QUALITIES

All of these are to be related to a person who is yet a student, in the process of becoming a teacher

A. Demonstrates ability and understanding appropriate for a *student teacher* in:

	Above Average	Average	Below Average

1. INTERPERSONAL RELATIONSHIPS

Builds harmonious relationships with and among:

	Above Average	Average	Below Average
a. Pupils	. . ./	. . ./	. . ./
b. Staff members	. . ./	. . ./	. . ./
c. Parents and other adults	. . ./	. . ./	. . ./

Tactful, courteous, empathetic

2. PERSONAL ADEQUACY

	Above Average	Average	Below Average
a. Displays self-confidence	. . ./	. . ./	. . ./

Enthusiastic - creative

b. Shows maturity of judgement	. . ./	. . ./	. . ./

Identifies problems - offers realistic solutions to problems

c. Accepts constructive criticism without offense	. . ./	. . ./	. . ./
d. Wants and trys to improve	. . ./	. . ./	. . ./

Conscientious, sincere, dependable

3. PROFESSIONAL ATTRIBUTES

	Above Average	Average	Below Average
a. Uses voice effectively	. . ./	. . ./	. . ./
b. Grooms and dresses appropriately	. . ./	. . ./	. . ./
c. Reflects the vigor and stamina necessary for teaching	. . ./	. . ./	. . ./

4. ADDITIONAL INDIVIDUAL GOALS

Comments:

3

Figure 11-1   continued.

Part II - PROFESSIONAL PROCESSES

All of these are to be related to a person who is yet a student, in the process of becoming a teacher

A. Demonstrates ability and understanding appropriate for a *student teacher* in:

	Above Average	Average	Below Average

1. PLANNING PROCESSES

	Above Average	Average	Below Average
a. Plans teaching thoroughly *Sound objectives, sufficient detail, appropriate resources and evaluation*	. . ./	. . ./	. . ./
b. Uses plans effectively in teaching	. . ./	. . ./	. . ./

2. GROUP PROCESSES

	Above Average	Average	Below Average
a. Uses effective procedures in guiding individuals, and/or small groups, and entire class toward independent study and self-evaluation	. . ./	. . ./	. . ./
b. Communicates well with individuals and groups *Written and orally*	. . ./	. . ./	. . ./

3. MANAGEMENT PROCESSES

	Above Average	Average	Below Average
a. Maintains classroom physical environment suitable for effective teaching *Classroom comfort and organization*	. . ./	. . ./	. . ./
b. Handles routine matters efficiently	. . ./	. . ./	. . ./
c. Maintains classroom social environment suitable for effective teaching *Sound emotional climate, rapport with students*	. . ./	. . ./	. . ./
d. Views discipline as a matter of growth in self control *Helps individuals and class towards self-responsibility*	. . ./	. . ./	. . ./

4. LEARNING PROCESSES

	Above Average	Average	Below Average
a. Views students realistically *(1) Perceives students as differing individuals (2) Is acquiring accurate perceptions of individuals and of age groups involved*	. . ./	. . ./	. . ./
b. Analyzes teaching learning situations in terms of principles of learning *(1)Through observation of teachers and/or students, identifies specific learning principles in action (2) Through analysis, identifies alternatives to specific teacher actions (3) Discusses observations and teacher-student interaction in terms of educational principles*	. . ./	. . ./	. . ./

4

Figure 11–1   continued.

	Above Average	Average	Below Average
c. Makes effective use of learning principles in teaching _____ *Relates own teaching actions to learning principles*	. . ./	. . ./	. . ./

5. EVALUATION PROCESSES

a. Uses effective means for diagnosis of student behavior _____ *Counselor assistance, student records, and conferences, parent conferences, etc., as appropriate*	. . ./	. . ./	. . ./
b. Uses various effective devices to evaluate learning _____	. . ./	. . ./	. . ./

6. PROFESSIONAL DEVELOPMENT PROCESSES

a. Is developing a personal philosophy of education _____ *Verbalizes about own ideas of teaching*	. . ./	. . ./	. . ./
b. Is developing his/her own style of teaching _____ *Does not merely imitate supervising teacher*	. . ./	. . ./	. . ./
c. Evaluates his/her own teaching realistically _____ *Is acquiring accurate self-perceptions about his/her teaching strengths and weaknesses*	. . ./	. . ./	. . ./
d. Is becoming aware of the values of participation in professional organizations _____ *National, state, local, teaching field*	. . ./	. . ./	. . ./

7. ADDITIONAL INDIVIDUAL GOALS

Comments:

5

Figure 11–1   continued.

Part III - SPECIFIC MATHEMATICS TEACHING SKILLS

All of these are to be related to a person who is yet a student in the process of becoming a teacher

A. Demonstrates ability and understanding appropriate for a *student teacher* in:

	Above Average	Average	Below Average
1. Utilizing an adequate mathematical background in classroom situations *Including, in particular, appropriate review in anticipation of student questions and good judgement about responding to student questions*	. . ./	. . ./	. . ./
2. Designing lessons which would be classified under each of the headings below and which are appropriate to attain the objectives of mathematical topics	. . ./	. . ./	. . ./
(a) Developmental	. . ./	. . ./	. . ./
(b) Inductive	. . ./	. . ./	. . ./
(c) Discovery	. . ./	. . ./	. . ./
(d) Drill	. . ./	. . ./	. . ./
3. Emphasizing structure in mathematical systems	. . ./	. . ./	. . ./
4. Placing appropriate emphasis on the process of mathematics as opposed to acquisition of mathematical knowledge *In particular, the role of formal and information deduction*	. . ./	. . ./	. . ./
5. Using mathematical language that is both precise and appropriate for the particular classroom situation	. . ./	. . ./	. . ./
6. Assigning meaningful homework	. . ./	. . ./	. . ./
7. Presenting mathematical lessons in such a way that the students become involved as active participants	. . ./	. . ./	. . ./
8. Developing student abilities to form conjectures and to gain insight into the validity of these conjectures	. . ./	. . ./	. . ./
9. Stimulating student interest in mathematics	. . ./	. . ./	. . ./
10. Identifying potential resource material when confronted with a mathematics instructional problem	. . ./	. . ./	. . ./

6

Figure 11-1   continued.

	Above Average	Average	Below Average
11. Selecting appropriate manipulative materials for developing a given objective	. . ./	. . ./	. . ./
12. Facillitating the obtainment of a given objective by designing and producing duplicated materials, and constructing transparencies for an overhead projector	. . ./	. . ./	. . ./
13. Taking a position and defending it on relevent topics in mathematics education	. . ./	. . ./	. . ./
14. Being intentional in deciding the mathematical content, methodology, evaluation and classroom procedures	. . ./	. . ./	. . ./
15. Additional Individual Goals	. . ./	. . ./	. . ./

Comments:

7

Figure 11-1   continued.

**PARTS IV AND V
TO BE COMPLETED AT THE END OF THE QUARTER**

Part IV. - EXPERIENCE SUMMARY

A. Please indicate the amount of experience this student teacher has had in the areas noted below:

	Many	Some	Few	None
1. Observing other classroom situations	. . ./	. . ./	. . ./	. . ./
2. Working with the guidance counselor	. . ./	. . ./	. . ./	. . ./
3. Keeping a state register	. . ./	. . ./	. . ./	. . ./
4. Keeping a grade book	. . ./	. . ./	. . ./	. . ./
5. Attending professional meetings	. . ./	. . ./	. . ./	. . ./
6. Participating in extra-class activities	. . ./	. . ./	. . ./	. . ./
7. Conferring with parents	. . ./	. . ./	. . ./	. . ./
8. Completing report cards	. . ./	. . ./	. . ./	. . ./
9. Administering and/or scoring standardized tests	. . ./	. . ./	. . ./	. . ./
10. Utilizing results of standardized tests	. . ./	. . ./	. . ./	. . ./
11. Conferring with the principal	. . ./	. . ./	. . ./	. . ./
12. Requisitioning supplies and materials	. . ./	. . ./	. . ./	. . ./
13. Working with various audio-visual materials	. . ./	. . ./	. . ./	. . ./
14. Planning and utilizing the use of community resources	. . ./	. . ./	. . ./	. . ./

B. The student teacher assumed the *full range** of teaching responsibilities for .... weeks.

*Planning, teaching, evaluating, homeroom, etc. for all classes for which the supervising teacher is responsible

Part V - SUMMARY RATING AND COMMENTS

Basing your opinion upon evidences you have observed this quarter, please make a judgement of this student's capabilities	Above Average	Average	Below Average
Part I. Personal Qualities	. . ./	. . ./	. . ./
Part II. Professional Processes	. . ./	. . ./	. . ./
Part III. Specific Teaching Field Skills	. . ./	. . ./	. . ./

Comments:

8

Figure 11–1   continued.

THE UNIVERSITY OF GEORGIA

STUDENT TEACHING RECORD
(Revised – 11/20/74)

Student's Name _____ Degree & Major _____ Graduation Date _____

Subject Taught _____ Grade Taught _____

School Where Taught _____ When Taught _____

_____   _____
Supervising Teacher (Print)   College Supervisor (Print)

	Outstanding	Above Average	Average	Below Average	Unacceptable
Knowledge of Subject Matter					
Organizational Skills					
Understanding of Pupils					
Ability to Discipline					
Professional Attitude					
Personal Appearance					
Tact and Courtesy					
Enthusiasm					
Creativeness					
Demonstrated Ability to Teach					
Capacity for Development					

SUMMARY STATEMENT

_____
_____
_____
_____
_____
_____
_____
_____
_____
_____

    The ratings and comments on this form reflect my opinion of this student's ability and predicted success as a teacher. It may be incorporated in his permanent personnel file and sent to school officials. I understand that in accordance with the "Family Educational Rights and Privacy Act of 1974" the student will have access to the information contained herein.

Signature _____ Address _____

Official Title _____ Date _____

Figure 11-1 continued.

but one of our student teachers found that once she got up in front of her class and started teaching, she quickly adapted to her new role. Elaine is nervous about her first day of teaching. So were our student teachers. Maybe she can even interpret her nervousness as a good sign—that she is looking forward to teaching with a great deal of excitement, that she has worked hard preparing her lessons, and that she has set high goals for herself. If so, the chances are excellent that she will succeed—even beyond her expectations. Andrew had such an experience. Student teaching was fantastic! It can be the same for you.

Activities

1. Planning those first few minutes of teaching is extremely important. Work out in detail how the first five minutes of your first lesson might proceed. Consider possible introductory statements you might make and how you could present your examples. Anticipate student responses. Then consider possible directions your lesson might take in the light of student reactions. Summarize your commentary about the first five minutes of your first lesson in about 200 words.

2. Make arrangements to observe a secondary school mathematics lesson. Before your visit to the class, meet with the teacher and discuss his or her goals for the lesson. As you observe, identify strategies used to attain these goals. Make special note of individual students, how they respond to the teacher, their apparent interest in the lesson, and their success in doing the assigned work. After the lesson, discuss these with the teacher. Attempt to determine bases for the decision making which the teacher did during the lesson.

3. There is often much to be gained from observing classes in subjects other than mathematics. The woodworking shop, the band room, the physics laboratory, and the gymnasium all provide examples of various strategies employed to attain the instructors' goals. Perhaps less different from the mathematics class, but also useful to visit, will be the English, social science, and foreign language classes. Take opportunities to talk with the teachers and determine the various educational philosophies represented.

4. While doing your student teaching, make a point of talking to some of the noninstructional personnel at the school. To be sure, you will want to try to meet the school principal and a counselor; but do not underestimate the importance of talking with maintenance staff and secretaries. Attempt to determine views on education held by the various components which go to make up the school. Summarize briefly the views of each.

5. As part of the process of examining your new role as teacher, it may be

helpful to reexamine student roles in today's high school. One technique for doing this is the "shadow study." You choose one high school student and become his or her "shadow" for the day. That is, you follow one student through an entire day's sequence of classes, study halls, and so on. This technique is extremely helpful in visualizing school from the student's point of view. You might want to keep a record of the total number of minutes the student spends at a desk, the total amount of homework assigned, the number of times he or she receives personal recognition from a teacher, and so on.

6. In order to gain insight as to how today's secondary school students think (their views about teachers and student teachers, for example) you may be able to make arrangements to exchange letters with the students in some school. School students might find it valuable to write to you about life at college or to obtain help on projects they are engaged in. Obviously, to be successful, this activity requires extensive planning by your instructor and the school personnel.

7. Obtain a copy of the evaluation form that will be used by your supervisor or turn to the form in Figure 11–1. Summarize the major areas covered by the form. Assess your own strengths and weaknesses on the basis of the criteria of the form.

8. Role play a conference between you and your supervisor. Propose comments the supervisor might make concerning your strengths and weaknesses as a student teacher. Then prepare responses you might make to such comments.

9. As part of your overall evaluation of your own student teaching experience, you will find it extremely interesting and helpful to keep a log. A log is like a diary; it is a daily record of and commentry on your experiences. It may be simple, consisting of brief notes, or it may be more complex, including analyses of episodes which have occurred. (The commentaries of the student teachers in this chapter were taken from their log books. The student teachers were enthusiastic about keeping their logs, and we found them to be very helpful and informative to us as well.)

References

Elashoff, Janet D. and Snow, Richard E. *Pygmalion Reconsidered*. Worthington, Ohio: Charles A. Jones, 1971.

Fawcett, Harold P. *The Nature of Proof*. Thirteenth Yearbook of the National Council of Teachers of Mathematics. New York: Bureau of Publications, Teachers College, Columbia University, 1938.

Holt, John. *How Children Fail*. New York: Pitman, 1964.

Mendels, Glen E. and Flanders, James P. Teachers' Expectations and Pupil Performance. *American Educational Research Journal* 10: 203–212; Summer 1973.

Robitaille, David Ford. Selected Behaviors and Attributes of Effective Mathematics Teachers. (The Ohio State University, 1969.) *Dissertation Abstracts International* 30A: 1472–1473; October 1969.

Rosenthal, Robert and Jacobson, Lenore. *Pygmalion in the Classroom: Teacher Expectation and Pupils' Intellectual Development.* New York: Holt, Rinehart and Winston, 1968.

Silberman, Charles E. *Crisis in the Classroom.* New York: Random House, 1970.

Thorndike, R. L. Review of Pygmalion in the Classroom. *American Educational Research Journal* 5: 708–711; November 1968.

On Becoming Students
of Teaching

AFTER STUDYING CHAPTER 12, YOU WILL BE ABLE TO:

★ use videotape recordings to help you analyze your own teaching.
★ use a self-evaluation form to do a detailed analysis of your own teaching.
★ discuss Flanders' Interaction Analysis procedures with a supervisor.
★ list the merits and limitations of microteaching.
★ use pupil feedback to improve your own teaching.
★ identify several organizations whose aims are to promote the professional growth of mathematics teachers.
★ give the names of a dozen books and periodicals which could form a useful basic library for beginning mathematics teachers.

CRITICAL INCIDENTS

1. Philip has taught for four years now, and he's concerned. He has the feeling that he is not making progress. His classes don't give him trouble, but teaching has become a bore—turning pages, doing proofs, grading papers. Even though Philip has been teaching for four years, maybe he has had only one year's experience—four times over.

2. Barb is excited. She went to a workshop on materials for low achievers. She met a dozen other teachers who also teach ninth-grade general mathematics, and they all shared their experiences. Some of the teachers were doing really neat things that she wants to try with her class. And she received several handouts on games and activities that she thinks her students will really go for.
3. Now that he has taught for a few years, Larry finds that he wants to take time to improve himself professionally. He wonders how he should go about it.

EVALUATING YOU, THE TEACHER

Every occupation has its hazards. In some occupations, like coal mining or firefighting, the risks are obvious. In teaching, however, the dangers may not be as apparent. One very real danger is that teaching can demand so much of our time and energy that we become caught up in its tasks and find little opportunity to step back and think about what we are doing. As a result, like Philip in the first critical incident, we find that our professional growth is stunted.

Silberman (1970), in the Carnegie Study of United States schooling in the 1960s, pointed to the problem of the professional development of teachers:

> Despite the continuous contact with children . . . teaching is a lonely profession. Teachers rarely get a chance to discuss their problems or their successes with their colleagues, nor do they, as a rule, receive any meaningful help from their supervisors, not even in the first year of teaching. (p. 144)

VIDEOTAPE RECORDING

The development of easy-to-use, portable, and relatively inexpensive recording devices for both audio (sound only) and video (picture and sound) purposes has important applications in teacher education. These devices make it a simple matter to obtain direct feedback on your teaching. They allow us at least some measure of the capability the poet envisioned in his wish, "O wad some Power the giftie gie us / To see oursels as others see us!" (Robert Burns, "To a Louse," 1786).

The importance of feedback in learning skills was discussed in Chapter 3 (page 60). The videotape recording gives you the full story (see Research Highlight, "Videotapes—the Clear Picture"). Should you not have access to a videotape recorder, however, you will find audio cassette recorders (portable units are now available for as little as $30 to $50) extremely useful in improving your own teaching.

Your use of videotaping will not be as ambitious or extensive as that of the professor of the History of Engineering course, at least not initially. We suggest that you begin rather simply, increasing in complexity as indicated in the following discussion of levels of videotape usage.

RESEARCH HIGHLIGHT | VIDEOTAPES—THE CLEAR PICTURE

Perlberg and O'Bryant (1968) explored the use of playbacks of videotape recordings in helping instructors to improve their teaching. The course was History of Engineering for undergraduates at the University of Illinois; the class was divided into two groups of 20 students each because the professor stated he preferred class discussion to lecture and this could be done only in small groups.

Some preliminary taping of the History of Engineering course was done. When the professor saw the playbacks he found that, in contrast to his stated intentions, his style of teaching was lecture-centered. When there was student participation, it was mainly in response to lower-order questions involving simple interrogatives such as "who" and "when." It was found that much of what was in the lecture was in the textbook as well. This style had been the pattern for teaching this subject for many years and for many professors.

As a result of analyzing and discussing the videotapes, a course of action was proposed. The professor decided to concentrate on shifting from a lecture-centered approach to a style of teaching which emphasized teacher-student interaction. He studied and practiced a variety of questioning techniques and experimented with rearrangements of the classroom, such as putting the chairs into a semicircle that included the professor himself.

During the following semester, several changes in the style of teaching the class were observed. Deliberate attempts were made to involve all of the students. Emphasis was placed on the use of nonverbal communication techniques—gestures, facial expressions, movement around the classroom, and so on. The professor became so enthusiastic about the project that he used his newly acquired skills in his other classes as well!

LEVEL ONE: SELF-OBSERVATION AND EVALUATION

The simplest use of a videotape recorder (VTR) involves recording your own teaching of a lesson and then playing back the tape for your private observation and analysis. The process of simply seeing ourselves as others see us (or as we actually are) can have dramatic effects on the improvement of our teaching.

As you observe your teaching, you will find it helpful to have some checkpoints for evaluating what you see. One such list is given in the self-evaluation form of Figure 12–1. That list summarizes the five areas of mathematics teaching

Figure 12-1   Mathematics teaching: a self-evaluation form (adapted from Pingry and Kane, undated).

	Very Often	Several Times	A Few Times	Never
I. Teaching for Facts and Skills				
1. I use diagnostic tests to determine student weaknesses.				
2. I provide the students with practice, either orally or in writing, on the use of rules or procedures.				
3. I provide drill on facts and skills needed for success in subsequent learning.				
4. I use new words in classroom discussion so that students become familiar with vocabulary.				
II. Teaching for Understanding				
1. I use oral and written questions during the development of a concept to determine student learning.				
2. I provide periodic review of class work.				
3. I use teaching methods which assist students in discovering answers for themselves.				
4. I use chalkboard diagrams and examples different from those in the book to promote understanding rather than recall.				
5. I encourage class participation and allow sufficient time for students to think about their answers before responding.				
III. Teaching for Problem Solving				
1. I ask students to state problems in their own words.				
2. I encourage students to diagram or construct physical models of problem situations.				
3. I reinforce the solving of problems in a variety of ways.				
4. I attempt to provide problems at the student's own level of achievement.				
5. I encourage students to conjecture solutions to problems, then test their conjectures.				

	Very Often	Several Times	A Few Times	Never
IV. Teaching for Enjoyment and Enrichment				
1. I use variety in my presentations: colored chalk, charts, diagrams, models, etc.				
2. I point out rhythmic, symmetric and harmonic relationships which illustrate mathematics in nature, art, or music.				
3. I exhibit in class my own enjoyment of the beauty and power of mathematics.				
4. I call attention to applications of mathematics in industry, engineering, the sciences, and in the students' daily lives.				
V. Classroom Management				
1. My writing on the chalkboard is clear and flows logically.				
2. My instructional materials are arranged for easy access when needed.				
3. I give recognition to individual students by name.				
4. My students and I show signs of mutual respect.				
5. I provide opportunities for individual student participation and interaction.				

emphasized in this book. Although the list may be too long to deal with fully at any one time, it can be used in part as you concentrate individually on each of the five facets of teaching.

1. Teaching for facts and skills   (Chapter 3)
2. Teaching for understanding   (Chapter 4)
3. Teaching for problem solving   (Chapter 5)
4. Teaching for enjoyment and   (Chapters 1–6)
   enrichment (positive affect)
5. Classroom management   (Chapter 7)

This checklist is not exhaustive, however. As you observe playbacks of your teaching, do further reading, and observe other teachers, you will discover categories of teaching behavior especially important for your success. We encourage you to devise your own list and to practice those skills as well.

LEVEL TWO: SUPERVISED TEACHING

Self-observation has its limitations. Eventually the novelty of watching *even yourself* will wear off! You will profit from having an experienced, capable observer point out behaviors, cues, and activities which you have not noticed. You may find, for example, that you have been concentrating so much on what you have been doing as a teacher that you have neglected to pay careful attention to what the children are doing. How have they responded to your "motivating" examples? What children are receiving your attention? Are there "dead spots" in your classroom?— that is, are there regions of the classroom which you tend to ignore (see Research Highlight, "Teacher Wait Time")?

It will be a more informative and helpful supervisory experience if the super-

RESEARCH HIGHLIGHT | TEACHER WAIT TIME

Rowe (1973) was interested in the amount of time teachers provide for students to respond to questions. Her analysis of more than 800 tape recordings of science lessons revealed that the average teacher asks questions at the rate of two or three per minute. She further found that students must start replies on the average within one second after the question is asked. If the student does not, the question is rephrased, or someone else responds (p. 242).

The investigator, convinced that rapid question and reply was not a suitable teaching strategy for inquiry-oriented science lessons, conducted an in-service education project in training 50 teachers to provide increased wait times so that verbal interaction in the classroom would be encouraged.

When the . . . teachers returned to their classrooms and experimented with increased wait times, they reported that children who did not ordinarily contribute began to take a more active part in doing and talking about science.
. . . The teacher's own questioning behavior also varies with wait time.
As wait time increases, teachers begin to show much more variability in the kinds of questions they ask. Students receive more opportunity to respond to thought rather than straight memory questions. (pp. 255–257)

visor uses some guide or framework for classroom observation. Many such guides are now available. Descriptions of some of the better-known instruments, together with summaries of research involving the instruments, will be found in the *Handbook of Research on Teaching* (Gage, 1963) and the *Second Handbook of Research on Teaching* (Travers, 1973). We will acquaint you with three classroom observation instruments. The first centers upon the teacher, the second upon the student, and the third upon the interaction between teacher and student.

Teacher Behavior: Stanford Teacher Competence Appraisal Guide This observation form deals with four main components of instruction:

Goals (aims):   The clarity and appropriateness of the purposes of the lesson.

Planning: Organization of the lesson; selection of appropriate content; relation of materials to method of instruction.

Teaching: How the lesson was developed by the teacher; the extent and nature of pupil participation in the lesson.

Evaluating: The use of evaluative techniques to improve subsequent learning activities.

A copy of this form is shown in Figure 12–2. Since it was used extensively in the development of microteaching at Stanford University, considerable research data are available on this scale.

Pupil Feedback: Teacher-Image Questionnaire It is important to get feedback about your teaching from your pupils. The most valuable feedback you can get from them may be the unplanned-for, natural responses that occur in day-to-day teaching. Their smiles and frowns, handwavings and postures all help tell how you are doing. It takes experience to learn how to "read" a class. And each class will differ from the next.

Another way to get feedback from your students is simply to ask them to express either orally or in writing how they think you are doing as a teacher. The first few times you do this, you may find you have to muster up considerable courage, since students are pretty honest about such things. If you aren't prepared to accept what they say, you really shouldn't ask for their input. But we have found that students are usually very fair in their comments, and these often provide much insightful and helpful information.

A more structured way of getting pupil feedback is to use a student questionnaire such as the Teacher-Image Questionnaire developed at Western Michigan University, which appears in Figure 12–3.

Teacher-Student Interaction: Flanders' Interaction Analysis The heart of teaching is the interaction between teacher and student. One well-known procedure for studying this interaction is Flanders' Interaction Analysis (FIA). This method was developed by Flanders and his associates at the University of Minnesota between 1955 and 1960 and has been used in many research studies since then. A full discussion of how to use FIA and research findings using FIA appears in *Analyzing Teaching Behavior* (Flanders, 1970).

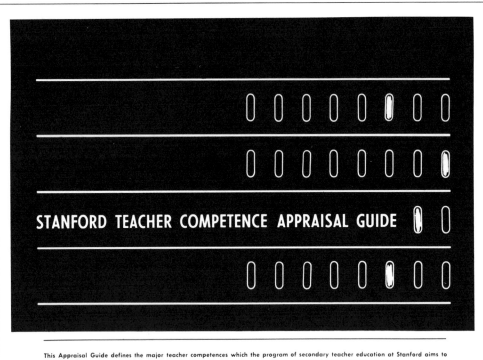

This Appraisal Guide defines the major teacher competences which the program of secondary teacher education at Stanford aims to develop. The total program of teacher education focuses on growth toward these standards as the common target.

To determine whether the program produces the desired growth, levels of competence must be appraised. Evidence for such appraisals may come from the trainee himself, from experienced teachers and administrators who supervise in the schools, from University teachers who instruct the trainees, and from the students taught.

This Appraisal Guide has been designed to assist in a cooperative effort to assess and to improve levels of competence in teaching. The basic sources of evidence are direct observations of the teacher followed by conferences and discussions related to observations. Secondary sources are communications with others who are in a position to observe and to know the teacher's work. The guide encourages the teacher (1) to accept with confidence his proper responsibility for continual self-improvement as a practicing professional in his specialty and (2) to contribute to the ongoing inquiry and the guiding body of theory by which he and his peers seek excellence in their area or specialty subject matter.

Purposely the guide avoids a rigid formula by defining 13 general practitioner competences, around which departmental specialists may build specific standards of expert practices appropriate to subject matter, grade levels, and groupings of students. The teacher being appraised is a most important one of these specialists and should be encouraged to participate in defining and improving standards for his specialty. Self-appraisals followed by observation and conferences with fellow teachers within a department will be useful to teachers as they accept increasing responsibility for self-improvement.

The conference following each observation stresses cooperative sharing of perceptions and ideas between professional teachers focused on the target of improving teaching, supervising, and learning. To facilitate this communication, the conference record is provided in duplicate so both participants may have copies for future use.

STANFORD UNIVERSITY SCHOOL OF EDUCATION

SECONDARY TEACHER EDUCATION PROGRAM

Figure 12–2 Stanford Teacher Competence Appraisal Guide (reprinted by permission of Stanford Center for Research and Development in Teaching, Stanford University, Robert N. Bush, Director). Form allows for assessment of two lessons for purposes of comparison.

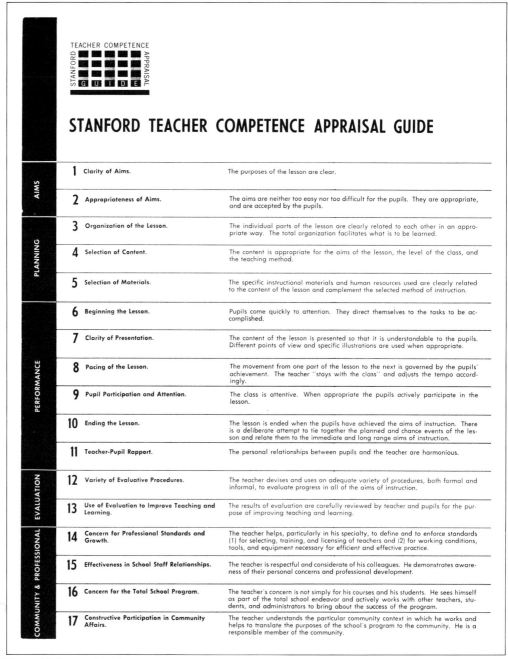

Figure 12–2   continued.

OBSERVATION NOTES

Name

Date

Class observed

Length of observation

Observer

0	30%	15%	15%	15%	15%	10%			
0 UNABLE TO OBSERVE	1 WEAK	2 BELOW AVERAGE	3 AVERAGE	4 STRONG	5 SUPERIOR	6 OUTSTANDING	7 TRULY EXCEPTIONAL		
0	1	2	3	4	5	6	7	1	AIMS
0	1	2	3	4	5	6	7	2	AIMS
0	1	2	3	4	5	6	7	3	PLANNING
0	1	2	3	4	5	6	7	4	PLANNING
0	1	2	3	4	5	6	7	5	PLANNING
0	1	2	3	4	5	6	7	6	PERFORMANCE
0	1	2	3	4	5	6	7	7	PERFORMANCE
0	1	2	3	4	5	6	7	8	PERFORMANCE
0	1	2	3	4	5	6	7	9	PERFORMANCE
0	1	2	3	4	5	6	7	10	PERFORMANCE
0	1	2	3	4	5	6	7	11	PERFORMANCE
0	1	2	3	4	5	6	7	12	EVALUATION
0	1	2	3	4	5	6	7	13	EVALUATION
0	1	2	3	4	5	6	7	14	COMMUNITY & PROFESSIONAL
0	1	2	3	4	5	6	7	15	COMMUNITY & PROFESSIONAL
0	1	2	3	4	5	6	7	16	COMMUNITY & PROFESSIONAL
0	1	2	3	4	5	6	7	17	COMMUNITY & PROFESSIONAL

GENERAL NOTES

Figure 12–2 continued.

CONFERENCE RECORD

I. Summarize notes relevant to the class setting:

II. Summarize interpretations of observation data: related to aims, planning, performance, and evaluation.

III. Summarize suggestions, possible resources and procedures for improvement of the teaching and learning. What should be retained, discarded, improved? How might this be done? What hypotheses for achieving more effective and efficient teaching and learning were discussed?

Date of Conference ————————————— Tentative plans for next observation ——————————————————

Figure 12–2   continued.

OBSERVATION NOTES

0	30%	15%	15%	15%	15%	10%			
0 UNABLE TO OBSERVE	1 WEAK	2 BELOW AVERAGE	3 AVERAGE	4 STRONG	5 SUPERIOR	6 OUTSTANDING	7 TRULY EXCEPTIONAL		
0	1	2	3	4	5	6	7	1	AIMS
0	1	2	3	4	5	6	7	2	
0	1	2	3	4	5	6	7	3	PLANNING
0	1	2	3	4	5	6	7	4	
0	1	2	3	4	5	6	7	5	
0	1	2	3	4	5	6	7	6	PERFORMANCE
0	1	2	3	4	5	6	7	7	
0	1	2	3	4	5	6	7	8	
0	1	2	3	4	5	6	7	9	
0	1	2	3	4	5	6	7	10	
0	1	2	3	4	5	6	7	11	
0	1	2	3	4	5	6	7	12	EVALUATION
0	1	2	3	4	5	6	7	13	
0	1	2	3	4	5	6	7	14	COMMUNITY & PROFESSIONAL
0	1	2	3	4	5	6	7	15	
0	1	2	3	4	5	6	7	16	
0	1	2	3	4	5	6	7	17	

Name

Date

Class observed

Length of observation

Observer

GENERAL NOTES

Figure 12–2 continued.

CONFERENCE RECORD

   I. Summarize notes relevant to the class setting:

   II. Summarize interpretations of observation data: related to aims, planning, performance, and evaluation.

   III. Summarize suggestions, possible resources and procedures for improvement of the teaching and learning. What should be retained, discarded, improved? How might this be done? What hypotheses for achieving more effective and efficient teaching and learning were discussed?

Date of Conference _____ Tentative plans for next observation _____

Figure 12–2   continued.

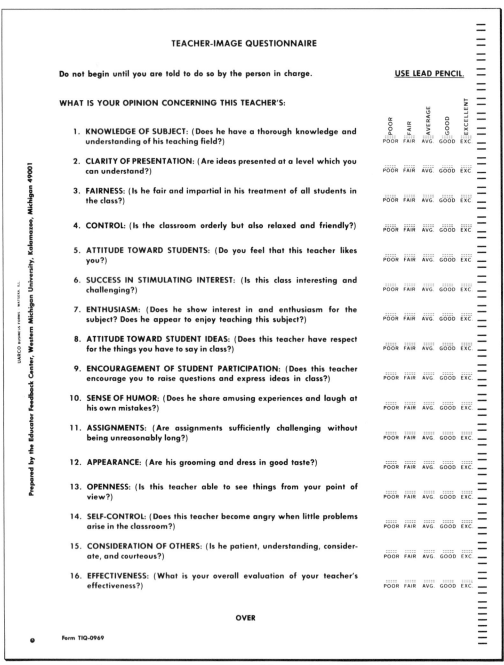

Figure 12–3 Teacher-Image Questionnaire (reprinted by permission of Research, Evaluation, Development, Experimentation Center, Western Michigan University, Kalamazoo, Michigan).

FIA classifies the verbal exchanges which take place in the classroom. Ten categories are used, as shown in Figure 12–4. Seven of the categories deal with teacher talk, two deal with student talk, and one is a silence or miscellaneous category. Data for FIA are obtained by a trained observer who sits in the classroom, or uses recordings made in the classroom, and classifies the communication taking place. Referring to a classification scheme such as that in Figure 12–4, the observer decides what category best describes the classroom activity at the moment and assigns a tally mark opposite the appropriate numeral, as shown in

Flanders' Interaction Analysis Categories* (FIAC)

Teacher Talk	Response	1. *Accepts feeling.* Accepts and clarifies an attitude or the feeling tone of a pupil in a nonthreatening manner. Feelings may be positive or negative. Predicting and recalling feelings are included. 2. *Praises or encourages.* Praises or encourages pupil action or behavior. Jokes that release tension, but not at the expense of another individual; nodding head, or saying "Um hm?" or "go on" are included. 3. *Accepts or uses ideas of pupils.* Clarifying, building, or developing ideas suggested by a pupil. Teacher extensions of pupil ideas are included but as the teacher brings more of his own ideas into play, shift to category five.
		4. *Asks questions.* Asking a question about content or procedure, based on teacher ideas, with the intent that a pupil will answer.
	Initiation	5. *Lecturing.* Giving facts or opinions about content or procedures; expressing *his own* ideas, giving *his own* explanation, or citing an authority other than a pupil. 6. *Giving directions.* Directions, commands, or orders to which a pupil is expected to comply. 7. *Criticizing or justifying authority.* Statements intended to change pupil behavior from nonacceptable to acceptable pattern; bawling someone out; stating why the teacher is doing what he is doing; extreme self-reference.
Pupil Talk	Response	8. *Pupil-talk—response.* Talk by pupils in response to teacher. Teacher initiates the contact or solicits pupil statement or structures the situation. Freedom to express own ideas is limited.
	Initiation	9. *Pupil-talk—initiation.* Talk by pupils which they initiate. Expressing own ideas; initiating a new topic; freedom to develop opinions and a line of thought, like asking thoughtful questions; going beyond the existing structure.
Silence		10. *Silence or confusion.* Pauses, short periods of silence and periods of confusion in which communication cannot be understood by the observer.

* There is *no* scale implied by these numbers. Each number is classificatory; it designates a particular kind of communication event. To write these numbers down during observation is to enumerate, not to judge a position on a scale.

Figure 12–4 Flanders' Interaction Analysis Categories (from Ned A. Flanders, *Analyzing Teaching Behavior*, 1970, with permission of the publisher, Addison-Wesley, Reading, Massachusetts).

Figure 12–5. Observations are recorded at the rate of about one every three seconds, with every effort made to keep the tempo as steady as possible.

Let us suppose that the data given in the tally sheet of Figure 12–5 were taken from a lesson which you had taught. What might these findings tell you? Notice that a large proportion of time was spent in teacher initiation of activities (categories 5, 6, 7) and a small proportion of time was spent in teacher response to students (categories 1, 2, 3). This may be a cue that you should become more responsive to your students' interests and to what they have to contribute to the class.

A more complex analytical procedure, which preserves some of the information about the sequence of behaviors in the classroom, involves the use of an interaction matrix. In this format, the observer writes down the numerals corresponding to the Flanders categories obtained. Let's say the observer made the following notations:

$$
\underset{\substack{\text{2nd pair}}}{\overset{\substack{\text{1st pair}}}{5 \qquad 5}} \quad \underset{\substack{\text{4th pair}}}{\overset{\substack{\text{3rd pair}}}{4 \qquad 8}} \quad \overset{\substack{\text{5th pair}}}{8 \qquad 10}
$$

This sequence is then transformed into five ordered pairs, and tally marks are placed in the appropriate cells of the 10×10 matrix as shown in Figure 12–6.

The resulting patterns of cells containing tally marks can be interpreted by experienced FIA analysts in order to describe the nature of the verbal interaction in a class. For example, heavy concentrations of tally marks in (4, 8) and (8, 4) cells indicate short-question, short-answer verbal interaction such as takes place in checking homework or reviewing simple facts. (Recall that category 4 is that of the teacher asking questions, and category 8 that of the pupil responding—

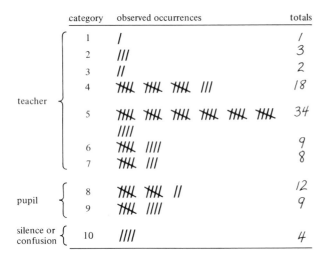

Figure 12–5 Hypothetical data resulting from Flanders' Interaction Analysis of a lession.

Observation Sequence (5, 5), (5, 4), (4, 8), (8, 8), (8, 10)

	Category	1	2	3	4	5	6	7	8	9	10	Total
Teacher	1											
Teacher	2											
Teacher	3											
Teacher	4								3rd pair			1
Teacher	5				2nd pair	1st pair						2
Teacher	6											
Teacher	7											
Pupil	8								4th pair		5th pair	2
Pupil	9											
Silence or Confusion	10											
	Total				1	1			2		1	5

Figure 12–6 Five tallies entered in an FIA matrix.

see Figure 12–4.) A more complex pattern of interaction describes creative inquiry in the classroom. Early phases of the lesson may be described by a preponderance of 4s and 8s as a problem is delineated for the class. However, as the teacher accepts or uses the ideas of the pupils (category 3) and praises or encourages (category 2), and as the students initiate talk by expressing their own ideas and introducing new topics (category 9), the result is increased proportions of tallies in such outlying cells as (3, 3), (3, 9), and (9, 9). Examples of such matrices and the interpretation of them appear in Flanders (1970, pp. 87–123).

As a result of considerable study and use of FIA, not only by himself but by many others (see Research Highlight, "Flanders' Interaction Analysis"), Flanders asserts:

> When classroom interaction shifts toward more consideration of pupil ideas, more pupil initiation, and more flexible behavior on the part of the teacher,

the present trend of research results would suggest that the pupils will have more positive attitudes toward the teacher and the schoolwork, and measures of subject-matter learning adjusted for initial ability will be higher. (p. 14)

RESEARCH HIGHLIGHT | FLANDERS' INTERACTION ANALYSIS

In a study of student teachers in English and Social Studies, Furst (1967) examined effects of Flanders' Interaction Analysis (FIA) training on teaching behaviors. She found that student teachers trained in FIA tended to be more accepting of their students and did more questioning (p. 326).

Soar (1968) analyzed data from 54 elementary school classrooms and found a curvilinear relationship between pupil academic growth and teacher *indirectness*. That is, as teachers became less direct with their classes (greater use of FIA categories 1, 2, 3, and 4) pupil achievement grew—up to a point. Then, for certain measures, achievement declined. Soar suggested that there exist optimal points of indirectness for various kinds of educational goals. For less complex goals (knowledge of facts or skills) the optimal point of indirectness is rather low. Beyond this point, as teachers become less direct, achievement declines. More complex goals (such as problem solving), however, appear to flourish under greater teacher indirectness. (See also Flanders, 1970, pp. 403–409.)

Flanders (1970) conducted an extensive five-state field study in an attempt to determine differences in classroom interaction which may have occurred as a result of using new mathematics curriculum materials. However, 97 percent of the classroom interaction remained about the same as before. Flanders concluded:

> The hoped for effects on pupil initiative in thinking did not appear during classroom interaction. Although there was more attention given to the structure of mathematics in the experimental classes, there was no corresponding change in the thinking of pupils. (p. 304)

LEVEL THREE: CLINICAL TRAINING (MICROTEACHING)

Microteaching is an advanced use of videotaping for the study and improvement of teaching. The technique was devised by Allen and his associates in the early 1960s at Stanford University. Microteaching was an important breakthrough in teacher education, since it made possible the identification and isolation of various components of the teaching act. These components may thus be studied and practiced individually for improved teaching by both pre-service and experienced teachers.

Microteaching scales down the teaching encounter to a manageable size for purposes of training teachers:

- Time: Rather than the usual period of 40–70 minutes, the lessons last 5–10 minutes.
- Scope: Topics are narrowed to specific subjects, such as learning how to factor an integer.
- Class size: Small groups of students (typically five or six) are used.

Microteaching also permits careful, detailed supervision and analysis of the lesson, including feedback from a video recorder and ratings from the students. (Figure 12–7 shows a student in a microteaching clinic.) Variations of microteaching have emerged in various parts of the United States and elsewhere.

The basic pattern of microteaching developed at Stanford University is organized along these lines: First, teaching skills to be learned are identified, and examples of each teaching skill are developed. Second, the teaching candidates are given a presentation on a particular skill. The presentation may consist of a

Figure 12–7   A student in a microteaching clinic. (Rogers, Monkmeyer)

handout sheet, a brief lecture, a sample videotaped or filmed lesson illustrating use of that skill, or some combination thereof. Then the candidate prepares and teaches a lesson in which this skill is to be practiced. After the lesson is taught, a supervisor, together with the candidate, reviews the playback of the lesson, considers feedback from the students, and discusses the strengths and weaknesses of the lesson. Finally, the candidate reteaches the lesson to a new class of five or six students, taking into account the information provided in the supervisory conference. A second supervisory conference typically ends the training sequence. It is this cycle of prepare–teach–receive feedback–reteach which makes microteaching a powerful teacher training tool (see Research Highlight, "Microteaching").

RESEARCH HIGHLIGHT | MICROTEACHING

Allen and Fortune (1965) analyzed data collected at microteaching institutes at Stanford University in 1963 and 1964. It was found that with this form of training, over an eight-week period involving less than ten hours per week, student teachers performed at a higher level of teaching competence than a comparable group of student teachers spending 20 to 25 hours per week in traditional programs. It was also found that teaching performance in microteaching was a statistically significant predictor of subsequent grades in student teaching.

In another study, Limbacher (1969) found that student teachers who had prior participation in microteaching were rated higher by their pupils than were those who had not done microteaching.

A fundamental (and natural) question to raise is "What skills should be taught in microteaching?" An obvious answer is "Those skills which are possessed by excellent teachers." But how does one decide what those skills are? The Stanford group solved this problem by identifying, on the basis of opinions of recognized experts or master teachers, those basic skills which are possessed by excellent teachers. To be sure, such a procedure will result in many lists of skills. However, the following skills are typically included in microteaching programs.

SKILL   COVERAGE IN THIS BOOK

establishing set—providing ways to motivate stu-   Chapter 7
   dents to study a unit of material

recognizing behavior—assisting the teacher to become more aware of the individual student and the student's interests, needs, and achievements   Chapter 8

providing reinforcement—offering methods of giving positive feedback to the learner   Chapter 3

questioning techniques—asking questions to promote desired instructional goals (see Research Highlight, "Improving Questioning Strategies")   Chapters 3–5

attaining closure—summarizing the important points covered in the lesson   Chapter 7

Detailed information about these skills and others may be found in Allen and Ryan (1969).

RESEARCH HIGHLIGHT | IMPROVING QUESTIONING STRATEGIES

Nisbet (1974) wanted to determine the effect of an instructional sequence which would help a group of secondary mathematics teachers to acquire more effective questioning practices. The nine teachers involved taught a lesson in a microteaching session. Then they were given instruction on effective questioning, particularly at the higher cognitive levels, with audiotape feedback on their microteaching and videotapes of a model teacher using higher-level questioning strategies. When a second microteaching session was taped, it was found that the percentage of higher-level questions significantly increased.

Suppose that you wished to use microteaching to learn a skill which you had identified as important for improving your own teaching. Let's say you want to get more student participation in your lessons and you have decided to attempt to do this by using an inductive teaching strategy. One such strategy might be to teach a concept, such as prime number, by using examples and nonexamples. Your planning of the microlesson might take the following form (you may wish to refer to the discussion of planning in Chapter 7).

MICROTEACHING LESSON PLAN

Prime Numbers

Length of Lesson: ten minutes
Number of Students: five

Goal:

Students shall be able to state a definition of "prime number" and to provide examples of numbers which are prime.

Instructional strategy:

The teacher provides examples of numbers which are prime and numbers which are not prime until the students can state the definition of "prime number."

Procedure:

Draw a vertical line on chalkboard to form two columns. Label one column "prime numbers" and the other column "nonprime numbers." State that whole numbers greater than one may be classified as either prime or nonprime. Give examples and nonexamples of prime numbers until the class determines the definition of "prime." When a student thinks he or she knows the definition, the student writes it down, then checks to see that it applies to each of the numbers classified on the chalkboard. Then the student raises a hand for the teacher to check the definition. (This approach will help prevent giving away the definition to others in the class. Once two or three students have the correct definition, they may be titled "experts" and allowed to assist the teacher in checking the written definitions of others in the class.) Begin by writing whole numbers in their correct columns, selecting some prime and some nonprime:

prime numbers	nonprime numbers
3	9
5	10
7	6
11	12

Summary:

Write some new examples on the board. Have students draw two columns on their papers and classify the numbers. Provide a few easy examples and a few harder ones. The discussion might lead to such questions as "How many prime numbers are there?" "Is there a formula for finding primes?" Some students might want to read about the sieve of Eratosthenes or Goldbach's conjecture.

Before actually teaching a lesson, you may find it a good idea to run through it mentally or to do a "dry run" on a friend. This will often point up problems of logical flow, appropriateness of examples, suitability of student assignments, and other practical aspects of lesson development. During the microlesson your supervisor may wish to use a form similar to the one in Figure 12–8. On this form comments may be made about key episodes in the lesson wherein the teacher made good (or not so good) use of student responses. Perhaps your facial expression or choice of words unintentionally tended to cut off student participation. Commentary might also be made on the use of the strategy itself. Were sufficient and appropriate examples used? Was adequate time given for students to formulate the definition? What proportion of the class appeared to have gained mastery of the definition? Commentary from students, either oral or written, might also be obtained.

After having reviewed the lesson with your supervisor and observed pertinent portions of the recorded lesson, you should revise your lesson plan and reteach. You may want to teach a new group of students, or you may need to modify your subject matter (while retaining the teaching skill you are practicing) so that you can teach the same group of students again.

PROMOTING PROFESSIONAL GROWTH

PROFESSIONAL ORGANIZATIONS

For many teachers, opportunities to interact with others about professional concerns come only rarely. Hence it is extremely important that you take advantage of what is available to you through professional associations. Glenadine Gibb, President of the National Council of Teachers of Mathematics in 1974–1976, has stated several reasons why a mathematics teacher ought to join a professional organization. These include: providing a personal means of keeping abreast of current issues and new insights into teaching, presenting opportunities for growing professionally with others, opening channels of communication to fellow educators at local, regional, and national levels, and providing leadership and service opportunities (Gibb, 1974, p. 1).

Table 12–1 lists organizations which are likely to be helpful to you as you seek to grow professionally as a mathematics teacher. As your interests develop, you may wish to explore other groups as well, but this list will be a good starting place. Some of these groups have regional organizations which will more directly address the needs and opportunities in the geographical location where you teach. By writing to the national offices or by reading their journals, you will be able to obtain information about such groups. And most professional organizations have student memberships at reduced rates.

We urge you to take the initiative in seeking out your local professional orga-

Teacher _____   Date _____

Observer (Supervisor) _____

Teaching skill being practiced _____

Topic of lesson _____

Observer's comments (Where appropriate, reference is given in left column to loca-
    tion on videotape of observed class episode.)

Location on Tape   *Comments*

Summary Comments

1. The primary objective of this lesson appears to be

_____

2. The level of effectiveness of the teaching skill being practiced was

_____

3. The lesson was a good one for the following reasons

_____

_____

_____

4. The following improvements in the lesson could be made

_____

_____

_____

Figure 12–8   Microteaching evaluation form.

nizations and participating fully in them. As a beginning teacher in a demanding
profession, you will be quickly introduced to leading educators who can be of
great assistance as you develop your own identity in the classroom. You will find
that these organizations are interested in bringing new teachers into their groups.

Table 12–1. Professional Organizations of Special Interest to Mathematics Teachers.

Organization	Level	Publications	Address
National Council of Teachers of Mathematics	Elementary school through community college and school of education	*Arithmetic Teacher* *Mathematics Teacher* *Journal for Research in Mathematics Education* *The Mathematics Student Newsletter* Yearbooks on selected topics	1906 Association Drive Reston, VA 22091
School Science and Mathematics Association	Elementary school through school of education	*School Science and Mathematics Newsletter*	P.O. Box 1614 Indiana University of Pennsylvania Indiana, PA 15701
Mathematical Association of America	College (junior and senior)	*American Mathematics Monthly* *Junior College Journal* *Mathematics Magazine* Carus Mathematics Monographs	1225 Connecticut Avenue, NW Washington, D.C. 20036
American Educational Research Association (Special Interest Group for Research in Mathematics Education)	Researchers and behavioral scientists in schools and colleges	*Educational Researcher* *AERA Journal* *Review of Educational Research* *Encyclopedia of Educational Research* (revised every 10 years)	1126 16th Street, NW Washington, D.C. 20036
Association of Teachers of Mathematics	Elementary school through school of education	*Mathematics Teaching*	Market Street Chambers Nelson, Lancashire England BB9 7LN

Volunteer to serve on committees (most committee chairpersons are eager for assistance), send a letter or an article to the newsletter of your local or state organization, and suggest topics for discussion at professional meetings. As a member of a professional organization, be prepared both to give and to receive help, encouragement, and ideas!

YOUR OWN PROFESSIONAL LIBRARY

Every teacher should possess a collection of books that are interesting and can serve as resources in daily and long-range planning. Here are our nominations for the 12 books "most likely to succeed" in being useful references in your professional library.

Aichele, Douglas and Reys, Robert E. *Readings in Secondary School Mathematics.* 2nd ed. Boston: Prindle, Weber and Schmidt, 1976.

Presents selected readings to provide a broader and deeper understanding of the current state of mathematics education in the secondary school.

Aleksandrov, A. D. et al., eds. *Mathematics, Its Content, Methods and Meaning.* Vols. 1–3. Cambridge, Massachusetts: M.I.T. Press, 1963.

Paperback translations of Russian articles which give a fairly comprehensive survey of major topics in mathematics. Written for the more advanced student in senior high school and in college.

Bell, Eric T. *Men of Mathematics.* New York: Simon and Schuster, 1937.

Well-written, easy-to-read account of the lives of some of the world's great mathematicians. Presents them as real people, providing considerable background and historical and anecdotal materials for making your mathematics classes more lively and interesting.

Berger, Emil J., ed. *Instructional Aids in Mathematics.* Thirty-fourth Yearbook of the National Council of Teachers of Mathematics. Reston, Virginia: The Council, 1973.

Contains a wealth of material on ways to enrich and enliven your teaching by means of instructional aids. Chapter titles include: "The Textbook as an Instructional Aid," Using Models as Instructional Aids," and "Mathematics Projects, Exhibits and Fairs, Games, Puzzles and Contests."

Biggs, Edith E. and MacLean, James. *Freedom to Learn: An Active Learning Approach, to Mathematics.* Toronto: Addison-Wesley, 1969.

Contains many ideas for presenting elementary and junior high school topics in an active way. Excellent source of suggestions for laboratory projects—and how to manage them.

Bergamini, David et al. *Mathematics.* Life Science Library. New York: Simon and Schuster, 1956.

Easy-to-read, profusely illustrated book which gives an excellent survey of mathematics for junior and senior high school students. Very good reference for independent reading or research projects.

Jacobs, Harold R. *Mathematics, a Human Endeavor: A Textbook for Those Who Think They Don't Like the Subject.* San Francisco: W. H. Freeman, 1970.

An engaging book which lives up to its title. Presents a good deal of mathematics in a manner which is captivating, humorous, and largely accessible, even to many junior high school students.

Kenna, L. A. *Understanding Mathematics with Visual Aids.* Paterson, New Jersey: Littlefield, Adams, 1962.

Rich with examples of how to present abstract mathematics in concrete ways. Especially suited for senior high school. Has chapters on graphing, curve stitching, wood models, paper folding and use of audio-visual aids.

Kline, Morris. *Mathematical Thought from Ancient to Modern Times.* New York: Oxford University Press, 1972.

A comprehensive survey of the history of mathematics. A monumental work, suitable for your own study and for advanced school students.

Newman, James R. *The World of Mathematics.* 4 vols. New York: Simon and Schuster, 1956.

An extensive collection of readings on major topics in mathematics by many well-known mathematicians. For advanced secondary school students.

Rosskopf, Myron F., ed. *The Teaching of Secondary School Mathematics.* Thirty-third

Yearbook of the National Council of Teachers of Mathematics. Washington: The Council, 1970.

> Has abundant material for beginning and veteran teachers alike. Contains chapters on such topics as "forces shaping today's mathematics program," "teaching for special outcomes," and "classroom applications."

Sobel, Max A. and Maletsky, Evan M. *Teaching Mathematics: A Sourcebook of Aids, Activities and Strategies*. Englewood Cliffs, New Jersey: Prentice-Hall, 1975.

> An excellent source of ideas for presenting familiar topics in new and interesting ways. Includes discussion of laboratory activities, use of audio-visual aids, and recreational mathematics. Although the authors have done extensive work with low achievers, this book will be useful as well for teachers of students who are average and high achievers.

INFORMAL SELF-IMPROVEMENT

Much professional improvement for teachers should take place on an informal basis. Opportunities for development are available to some extent in most school systems; you have the responsibility to seek out these opportunities and take advantage of them.

A good place to begin is with your own colleagues. Many of your fellow teachers will be happy to share ideas and materials with you. Other approaches and points of view are always worth considering. You might even be able to arrange with your administrators (department chairperson, principal) to visit other classes in your building and to visit other schools.

By "colleagues" we do not mean only other teachers of mathematics. Many important points of view, insights, and teaching strategies are used in other subject-matter areas that can be modified for use in mathematics with considerable success. For example, what approaches employed in the physics laboratory might be applied in a mathematics project? In a social science class, are there discussion techniques which might be effective in dealing with topics in the history of mathematics? In the shop, are there procedures in teaching the use of gauges or other precision instruments which might be used in teaching the use of protractors or the sextant (in preparation for a venture into the school grounds to do some applied trigonometry)?

The department chairperson is also a resource to be utilized; surprise him or her by asking for a visit to your class. Learn what resources are to be found in the audio-visual department—and how to use them. Ask about professional meetings in your geographic area. See what journals are subscribed to in your library. If your chairperson does not know of resources, maybe he or she will be motivated by you to find out!

CONTINUED FORMAL STUDY

In many states the master's degree is the accepted professional degree. That is, persons not having a master's degree are not considered by those states to be fully qualified. The trend toward increased requirements of formal education is expected to continue.

There are many ways in which you can pursue formal course work beyond the baccalaureate. There is probably a college or university within driving distance of your school that is interested in providing courses which teachers want and need. *Extension courses* for graduate credit are often available away from the college campus, taught by professors who commute by plane or automobile. *Correspondence courses* can usually be taken in their entirety by mail; books from the university library may be obtained through the mail as well. *Independent study courses* for graduate credit are another possibility. There you work on a topic of your choosing under the direction of a professor; you might read extensively, do library research, or devise and test curriculum materials. *Short courses, workshops, and institutes*, at times carrying graduate credit, are frequently available, and for some of these you may obtain released time from your school to attend. *Full-time study for an academic year* provides the greatest opportunity for intensive study and for interaction with other graduate students and with faculty. At times school districts are willing to provide leaves of absence, with some financial support, to assist teachers in their professional development.

Critical Incidents Revisited

Let's look again at the three teachers in this chapter's critical incidents. Philip is in a rut. He is surviving as a teacher, but he is not growing. Larry has the desire to improve but doesn't know how to go about it. How might some of the suggestions for professional development discussed in this chapter help Philip and Larry?

It's likely that Philip has no idea of how he appears to his students. A videotape of himself "turning pages, doing proofs" might startle him into looking around for help. Maybe Larry could begin by recording some of his lessons with a cassette recorder and evaluating them using the self-evaluation form in Figure 12–1.

Barb has discovered how valuable professional associations can be in providing encouragement and offering suggestions for doing a better job in the classroom. Maybe she will take advantage of other opportunities to grow, too, such as reading professional journals, starting her own professional library, and attending conferences or workshops. Barb is on her way to excellence and could well be in a position to offer leadership to fellow teachers in addition to providing first-rate teaching to her students.

Activities

1. Using either a video or audio recorder, teach a short lesson to some students or to your peers. Then analyze your lesson using selected sections of the self-evaluation form in Figure 12–1 such as teaching for

facts and skills or teaching for problem solving. You should repeat this activity every two or three weeks over a period of time so that changes in your teaching behavior can be noted.

2. Using a microteaching class or some other group of students with which you have been working, obtain feedback about your teaching by means of the Teacher-Image Questionnaire (Figure 12–3). What new information about your teaching does this feedback provide? You may want to discuss some of the responses with the students themselves in order to determine the reasons why they answered as they did.

3. Ask a fellow teacher or supervisor to evaluate your teaching using the Stanford Teacher Competence Appraisal Guide (Figure 12–2). Identify those areas of your teaching which need improvement and seek to improve them. Have another evaluation after several weeks and compare the ratings.

4. Determine how your teaching would be scored by Flanders' Interaction Analysis. Are your classes teacher-centered or student-centered according to this analysis? Are there aspects of your teaching which you would like to modify, on the basis of Flanders' Interaction Analysis?

5. Try the teaching skill discussed on pp. 415 ff, using a microteaching clinic if possible. Then identify another teaching skill which you would like to learn and prepare the lesson in a similar way to that of the sample lesson plan for teaching prime numbers. Teach the lesson and evaluate it. Then reteach the lesson on the basis of the evaluation and evaluate your teaching again.

6. The books suggested in this chapter for your professional library are but 12 of many possible titles which could be useful. Another list of titles was developed by Leake (1972), who obtained recommendations from 17 prominent mathematics educators. Compare his list, published in the *Mathematics Teacher*, with ours and select from his titles another three for inclusion in your prospective library.

7. Select one of the books from the list you developed in Activity 6, or from our list in this chapter, and prepare a report designed to convince a fellow student teacher, or teacher, that such a book is a good source of ideas for beginning teachers.

8. Many national professional organizations exist which, although not strictly devoted to mathematics teaching, are involved in matters of importance to mathematics education. Examples of such organizations are the Association for Computing Machinery, the American Psychological Association, and the Association for Supervision and Curriculum Development. Look for further information about these or other national groups, and determine which ones are of special interest to you. A good starting place for finding such information is their periodical publications, available in most college libraries.

9. Consult current issues of such journals as the *Mathematics Teacher* and *School Science and Mathematics* or such books as Simpson's *Teacher Self-Evaluation* (1966) for ideas on how you can develop a professional self-improvement program once you are in a full-time teaching position. Outline five major steps in such a program.
10. Obtain a program for a state, regional, or national meeting of mathematics teachers. These are sent to members and may also be available from the headquarters of the professional association. Identify five presentations on the program which appear to be of interest and help to you in your teaching.

References

Allen, Dwight W. and Fortune, J. C. An Analysis of Microteaching: A New Procedure in Teacher Education. Stanford University. Unpublished manuscript, 1965.

Allen, Dwight W. and Ryan, Kevin. *Microteaching*. Reading, Massachusetts: Addison-Wesley, 1969.

Dreikurs, Rudolph; Grunwald, Bernice B.; and Pepper, Floyd C. *Maintaining Sanity in the Classroom: Illustrated Teaching Techniques*. New York: Harper and Row, 1971.

Flanders, Ned A. *Analyzing Teaching Behavior*. Reading, Massachusetts: Addison-Wesley, 1970.

Furst, Norma. Effects of Training in Interaction Analysis on the Behavior of Student Teachers in Secondary Schools. In *Interaction Analysis: Theory, Research and Application* (edited by E. J. Amidon and J. B. Hough). Reading, Massachusetts: Addison-Wesley, 1967, pp. 315–328.

Gage, N. L., ed. *Handbook of Research on Teaching*. Chicago: Rand McNally, 1963.

Gibb, Glenadine. Professionally Speaking. *Newsletter of the National Council of Teachers of Mathematics* 11: 1; September 1974.

Leake, Lowell J. What Every Mathematics Teacher Ought to Read (Seventeen Opinions). *Mathematics Teacher* 65: 637–641; November 1972.

Limbacher, Phillip C. A Study of the Effects of Microteaching Experiences Upon Practice Teaching Classroom Behavior. (University of Illinois, Urbana, 1968.) *Dissertation Abstracts International* 30A: 189; July 1969.

Medley, Donald M. and Mitzel, Harold E. Measuring Classroom Behavior by Systematic Observation. In *Handbook of Research on Teaching* (edited by N. L. Gage). Chicago: Rand McNally, 1963, pp. 247–328.

Nisbet, Jean Ann. Instructional Sequence for Improving Teacher Question-Asking in Secondary Mathematics. (Arizona State University, 1974.) *Dissertation Abstracts International* 35A: 2841; November 1974.

Perlberg, Arye and O'Bryant, David C. The Use of Videotape Recording and Microteaching Techniques to Improve Instruction on the Higher Education Level. Urbana: University of Illinois, 1968. (Mimeo)

Pingry, Robert E. and Kane, Robert. A Form for Self-Evaluation of Mathematics Teaching. Urbana: University of Illinois. (Mimeo)

Rowe, Mary Budd. *Teaching Science as Continuous Inquiry*. New York: McGraw-Hill, 1973.

Silberman, Charles E. *Crisis in the Classroom: The Remaking of American Education*. New York: Random House, 1970.

Simon, Anita and Boyer, E. G., eds. *Mirrors for Behavior: An Anthology of Classroom Observation Instruments*. 14 vols. Philadelphia: Research for Better Schools, Inc., and the Center for the Study of Teaching, Temple University, 1967–1970.

Simpson, Ray H. *Teacher Self-Evaluation*. New York: Macmillan, 1966.

Soar, R. S. Optimum Teacher-Pupil Interaction for Pupil Growth. *Educational Leadership* 26: 275–280; December 1968.

Travers, R. M. W., ed. *Second Handbook of Research on Teaching*. Chicago: Rand McNally, 1973.

Then, Now, and Beyond

AFTER STUDYING CHAPTER 13, YOU WILL BE ABLE TO:

* identify events in history which have helped shape mathematics education.
* name several curriculum development projects playing major roles in the reform movement of the 1960s.
* cite characteristics of the "new mathematics" as it emerged in the 1960s.
* give examples of current curriculum trends in school mathematics.
* discuss probable future directions for school mathematics.

CRITICAL INCIDENTS

1. Miss Kelley is preparing for a conference with Suzy's parents concerning her work in algebra. Suzy is doing well, but her parents are unsure about the value of the "new math." They don't recognize all the symbols and expressions in the textbook, and they want assurance that Suzy is learning as much as they did in "the old math days."

2. Sometimes when Marge goes to a teachers' workshop she gets confused. There seem to be so many new ideas being tossed around that she's afraid she won't be able to keep up to date. She wants to do her best in the classroom and to see each student do as well as possible. But how can she decide what to select and what to ignore?

3. It seems to Tony that there are always new fads in education. He gets tired of people who are constantly pushing innovation and pressuring others to jump on the bandwagon. Tony's teaching works fine—and that's the way he intends to keep it.

4. Jason has had an argument with his roommate Tom. Jason is going into teaching and is excited about the challenge of working with people and ideas. Tom is going into marketing because "that's where the action really is." Tom argues, "All the latest technology is available to make marketing an exciting career. But mathematics teaching never changes. Two plus two will always be four, won't it?"

HISTORICAL ROOTS OF MATHEMATICS EDUCATION

Dusty books on dusty shelves: to many people, this is the first image that mention of the word *history* evokes (see Figure 13–1). Let's define it instead as "nostalgia" and trace a few of the highlights that we would remember had we lived through the years that saw the development of mathematics as a curricular area. Perhaps as we do this we will understand a little better how mathematics education came to be what it is today. (Sources for this historical review include Bell, 1937; Butts, 1955; Cajori, 1890; Johnson, 1904; Jones, 1970; Karpinski, 1925; and Smith and Ginsburg, 1934.)

FROM COLONIAL AMERICA TO 1900

MATHEMATICS   MATHEMATICS EDUCATION

1500

Juan Diez Freyle published first book in Western Hemisphere having mathematical content (published in Mexico City, in Spanish, 1556)

1600

Descartes's *Method* is published, introducing analytical geometry (1637)

Boston Latin Grammar School founded (1635)

I remember . . .
early colonial days, when schools existed only in towns and densely settled districts. They were established soon after

A

TREATISE

OF

ALGEBRA,

BOTH

𝕳𝖎𝖘𝖙𝖔𝖗𝖎𝖈𝖆𝖑 and 𝕻𝖗𝖆𝖈𝖙𝖎𝖈𝖆𝖑.

SHEWING,

The Original, Progreſs, and Advancement thereof, from time to time; and by what Steps it hath attained to the Heighth at which now it is.

With ſome Additional TREATISES,

I. Of the *Cono-Cuneus*; being a Body repreſenting in part a *Conus*, in part a *Cuneus*.

II. Of *Angular Sections*; and other things relating there-unto, and to *Trigonometry*.

III. Of the *Angle of Contact*; with other things appertaining to the *Compoſition of Magnitudes*, the *Inceptives of Magnitudes*, and the *Compoſition of Motions*, with the Reſults thereof.

IV. Of *Combinations, Alternations*, and *Aliquot Parts*.

By J O H N W A L L I S, *D. D. Profeſſor of* Geometry *in the* *Univerſity of* Oxford; *and a Member of the* Royal Society, London.

L O N D O N:
Printed by *John Playford*, for *Richard Davis*, Bookſeller, in the Univerſity of O X F O R D, M. DC. LXXXV.

Figure 13–1 Title page from John Wallis's *Treatise of Algebra*, published in 1685. (Courtesy Rare Book Room, University of Illinois, Urbana)

MATHEMATICS

MATHEMATICS EDUCATION

Newton and Leibniz create
the calculus (circa 1670)

1700

Euler publishes treatise
on the number *e* (1748)

Gauss formalizes modular
systems (1820)

Galois develops concept
of group (1831)

Harvard College founded
(1642)

Jesuit college for
secondary education
founded in Quebec (1655)

Isaac Greenwood's *Arith-
metic, Vulgar and Decimal*
is published (first English-
language book on mathe-
matics to be written
and published in North
America, 1729)

Yale requires arithmetic
for admission (1745)

Nicholas Pike's *A New
and Complete System of
Arithmetick* is published
(1788) (see Figure 13–2)

1800

Harvard requires arith-
metic for admission
(1807) (see Figure 13–3)

Warren Colburn's *An
Arithmetic on the Plan of
Pestalozzi* is published
(1821)

*settlement began, in Massachusetts in
1642 and in Connecticut in 1650. However, arithmetic was often not included in
the curriculum. The primary role of
the school was to teach reading; arithmetic
was offered either at a special fee or
from private tutors. Where it was included,
it consisted simply of learning to count
and to perform the fundamental operations
with whole numbers and "short excursions" into fractions. Such limited
arithmetic content was sufficient for the
needs of farmers, artisans, and those in
trade. Specialized knowledge for surveying,
navigation, and other vocations was
offered through vocational schooling.
Whether arithmetic was taught, and how
much of it was taught, depended on the
interest and knowledge of the teachers,
who frequently were clergy poorly
prepared for teaching arithmetic.*

*I remember . . .
    my arithmetic book. I copied it from my
teacher's book, which was written by
Nicholas Pike, one of the first American
textbook authors. I've heard it called "the
most influential arithmetic book published
in this country." Although it had
mostly arithmetic and a little algebra, it
was intended for use in academies
and colleges. Almost every page gave a
rule, followed by several examples.
    I've heard that arithmetic textbooks by
about 40 other authors were also available in America between 1705 and 1800.
Books by Nathan Daboll and Daniel
Adams, along with Pike's, were the
most popular. Fourteen geometry books
were published in America before 1820,*

Compound

Is extremely useful in finding the value of goods, &c.
And as in compound addition we carry from the lowest
denomination to the next higher so we begin and carry in
Compound Multiplication; one general rule being to multiply
the price by the quantity. *Case 1*

When the quantity does not exceed 12 yards, pounds, &c

Set down the price of 1, and place the quantity underneath the less

denomination, for the multiplier, and in multiplying by it,

observe the same rules for carrying from one denomination

to another, as in Compound Addition

Introductory Examples

Practical Questions

What will 9 yards of cloth at 5s 4d
per yard come to

2, 7 yards at [...] per yd

Abraham Dodge Jun
his Book

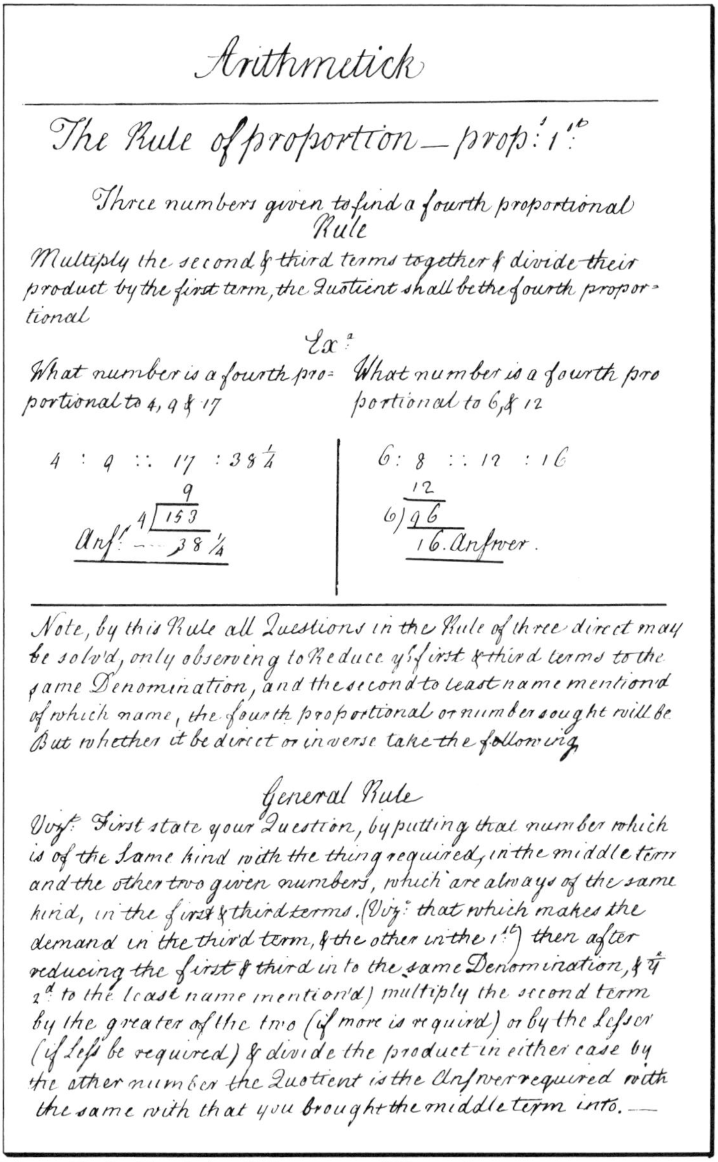

Arithmetick

The Rule of proportion — prop: 1st

Three numbers given to find a fourth proportional
Rule

Multiply the second & third terms together & divide their product by the first term, the Quotient shall be the fourth proportional

Ex:

What number is a fourth proportional to 4, 9 & 17

What number is a fourth pro
portional to 6, & 12

$4 : 9 :: 17 : 38\frac{1}{4}$

$6 : 8 :: 12 : 16$

$$\text{Ans}^r \quad 4\overline{)\begin{array}{c} 9 \\ 153 \end{array}}$$
$$38\frac{1}{4}$$

$$6\overline{)\begin{array}{c} 12 \\ 96 \end{array}}$$
$$16. \text{ Answer}.$$

Note, by this Rule all Questions in the Rule of three direct may be solv'd, only observing to Reduce ye first & third terms to the same Denomination, and the second to least name mention'd of which name, the fourth proportional or number sought will be But whether it be direct or inverse take the following

General Rule

Viz: First state your Question, by putting that number which is of the same kind with the thing required, in the middle term and the other two given numbers, which are always of the same kind, in the first & third terms. (Viz: that which makes the demand in the third term, & the other in the 1st) then after reducing the first & third into the same Denomination, & ye 2d to the least name mention'd) multiply the second term by the greater of the two (if more is required) or by the Lesser (if less be required) & divide the product in either case by the other number the Quotient is the Answer required with the same with that you brought the middle term into. —

Figure 13–3 The Rule of Three, part of Harvard's entrance requirement in arithmetic in 1807, as it appeared in a textbook of about 200 years ago. (The manuscript, from the rare books collection of the University of Illinois, Urbana, gives no indication of authorship, and bears on the title page only the inscription, "Mathematics and Its Applications, Date: About 1756.")

MATHEMATICS

MATHEMATICS EDUCATION

Babbage invents his "analytic engine," fore-runner of the computer (1830–1840)

Harvard drops study of arithmetic from freshman curriculum (1837)

almost all dealing with mensuration and surveying.

I remember . . .
    1789, when the teaching of arithmetic was made obligatory in Massachusetts and New Hampshire. Boston specified that it be studied by all those aged 11 years and older. Arithmetic, including the four operations and the "rule of three" (a problem involving use of ratio and proportion), was required for admission to Harvard in 1807. As late as 1850 only arithmetic and algebra through linear equations were required for admission to Dartmouth (see Figure 13–3).
    Academies were the college preparatory schools through the early 1800s, but a new type of secondary school emerged with the founding of a high school in Boston in 1821. Since public monies were being funneled into education, the people demanded a more practical kind of education than was available at the academies. The new school, designed for boys 12 years or older not planning to go to college, became very popular; in 1827 Massachusetts passed a law that such schools should be established in every town of 500 or more families. The new school curriculum emphasized English, social studies, and mathematics—all of a practical nature. By 1875 these schools were quite well established, at least in the northern states. Arithmetic remained in the curriculum of academies and high schools, not becoming an elementary school subject until the latter part of the nineteenth century. Algebra began to evolve as a mainstay of the high school curriculum when Harvard required it for admission in 1820 (Yale

1850

Charles Davies's *The Logic and Utility of Mathematics*, first book on teaching of mathematics in U. S. is published (1850)

Riemann completes his doctoral dissertation, "Foundations for a general theory of functions of a complex variable" (1851)

Lobachevsky's *Pangeometry* is published (1855)

Cayley originates matrices and matrix algebra (memoir of 1858)

Harvard requires geometry and logarithms for entrance (1870)

National Education Association founded (1870)

Cantor publishes revolutionary paper on set theory (1874)

MATHEMATICS   MATHEMATICS EDUCATION

and Princeton followed in 1847 and 1848, respectively). Geometry moved down from the college level just after the Civil War.

I remember . . .
   the first reform in arithmetic teaching in the United States. It began in 1821 with the publication of an arithmetic textbook by Warren Colburn, who had been greatly influenced by the ideas of the European educator Johann Pestalozzi. Pestalozzi believed that sense perceptions were extremely important in a child's mental development and devised series of lessons involving plants, animals, models, and other concrete objects for teaching concepts in language and number. Colburn was one of the first educators to adapt these methods to American schools, using inductive approaches to teaching and insisting that understanding must precede practice.

Edward Brooks's *The Philosophy of Arithmetic*, first book on the teaching of arithmetic to be published in U. S. (1880)

I remember . . .
   that, until the middle 1800s, one book often comprised the entire arithmetic program. During the late 1800s this number increased to include books on both mental and written arithmetic. In the majority of these, placement according to topic difficulty was unsystematic; the emphasis was almost entirely on mental discipline. Teaching became more and more formal as teaching to an entire class became the rule. It wasn't until after the turn of the century that separate textbooks for each grade were widely used.

1890

Committee of Ten appointed (1892)

American Mathematical Society founded (1894)

Hilbert publishes *Grundlagen der Geometrie* (1899)

I remember . . .
   one of the first books specifically on the

MATHEMATICS   MATHEMATICS EDUCATION

teaching of arithmetic, written by Edward Brooks in 1880. While he considered arithmetic basically a deductive procedure, he agreed that the inductive approach was possible. Many passages from this book appear to have been written about the curriculum of a hundred years later.

I remember . . .
    1892 the Committee of Ten on Secondary School Studies had been appointed by the National Education Association; the subcommittee on mathematics, composed of leading mathematicians, was the first national group to consider the goals and curriculum for mathematics education.

THE TWENTIETH CENTURY

MATHEMATICS   MATHEMATICS EDUCATION

1900

David E. Smith writes *The Teaching of Elementary Mathematics* (1900)

D. E. Smith joins faculty of Teachers College, Columbia University, as professor of mathematics (1901)

E. H. Moore addresses American Mathematical Society, urging mathematicians to become active in school mathematics (1902)

I remember . . .
    the beginnings of national professional organizations at the turn of the century and of committees to study and revise the school mathematics curriculum. There was a minor reform movement in mathematics education during the early 1900s. However, little recognition was given to the needs of the changing secondary school population, and courses remained largely college preparatory in nature.

MATHEMATICS	MATHEMATICS EDUCATION	
	Central Association of Science and Mathematics Teachers, later to become the School Science and Mathematics Association, is formed (1903)	*I remember . . .* *David Eugene Smith, the influential and unusually prolific writer of the early 1900s. He authored elementary-, secondary-, and college-level textbooks; books on the teaching of both elementary and secondary school mathematics; and standard works on the history of mathematics. He stressed mental discipline, mental arithmetic, usefulness, concrete problems, accuracy and speed, and understanding. He believed that the child should be led to feel that he or she is a discoverer. Professor Smith's work was so important that his appointment in 1901 as professor of mathematics at Teachers College, Columbia University, is regarded by some as the birth of mathematics education as a separate field of study. Also significant in mathematics education was the tenure of J. W. A. Young in a pedagogy of mathematics position in the mathematics department of the University of Chicago from 1892 until 1926.*
	Halsted's *Rational Geometry*, a high school textbook based on Hilbert's axioms, is published (1904)	
Einstein propounds theory of relativity (1905)		
	First doctoral theses in mathematics education at Teachers College are submitted (1906)	
	J. W. A. Young writes his *Teaching of Mathematics* (1907)	
	1910	
Russell and Whitehead publish *Principia Mathematica* (1910–1913)		
		I remember . . . *Professor E. H. Moore, a leading mathematician, who in his retiring presidential address to the American Mathematical Society made a plea for the improvement of mathematics education in the schools. During the same period, John Perry in England was urging that mathematics be made concrete by using approaches which we would now call "mathematics laboratories." Perry claimed that many*

MATHEMATICS   MATHEMATICS EDUCATION

	National Committee of Fifteen makes provisional report (1911)	*ideas usually postponed until later grades could be taught at lower grade levels.*
	College Entrance Examination Board established (1911)	*In 1911, when the Provisional Report of the National Committee of Fifteen on the Geometry Syllabus was published, it reflected some of the views of Moore and Perry in recommending more use of concrete examples and informal proofs, exclusion of topics on limits and incommensurables, and continued stress on logical structures.*
Mathematical Association of America founded (1915)		

1920

	National Council of Teachers of Mathematics founded (1920)	*I remember . . .*
Einstein delivers his lecture "Geometric und Erfahrung" (1921)	Publication of *Mathematics Teacher* taken over by National Council of Teachers of Mathematics (1921)	*the College Entrance Examination Board being established in 1911 to help articulate high school and college curriculums and to select high school graduates capable of succeeding in college. Forty-four years later (1955) the Board was to appoint a commission of experts to review the secondary school mathematics curriculum and make recommendations for its improvement. The report of the commission was to be extremely influential in the curriculum reform movement of the 1960s.*
	Report given by National Committee of Mathematical Requirements on reorganization of school mathematics (1923)	
Von Neumann's axiomatization of quantum mechanics appears (late 1920s)	Smith and Reeve publish *Teaching of Junior High Mathematics* (1927)	*I remember . . .*

1930

Goedel's proof announced (1931)		*when about 100 teachers, representing 20 states, met in Cleveland, Ohio, on February 24, 1920. After much debate, stimulated by proposals from the Men's Mathematics Club of Chicago, the National Council of Teachers of Mathematics was formed. I wonder how many of those attending would have thought that 50 years later the organization would be over 60,000 members strong!*

MATHEMATICS   MATHEMATICS EDUCATION

I remember . . .

when in 1923 the final report of the National Committee on Mathematical Requirements was published. It stressed such ideas as the reduction of elaborate manipulations in algebra and lessened emphasis on memorization of theorems and proofs in geometry. The importance and unifying nature of the function concept was clearly recognized. The report also advocated a general mathematics program for grades seven through nine (junior high schools were quite well established by this time).

Emil Artin develops abstract algebra and its application to number theory and geometry (1930s)

Nicolas Bourbaki, pseudonym for a group of French mathematicians, undertakes systematization of all modern mathematics (1939)

During the next 20 years, the recommendations of this committee were discussed and attempts were made to incorporate them into the curriculum. Integrated study of arithmetic, informal geometry, elementary algebra, statistics, and numerical trigonometry was promoted. Manipulation and memorization were reduced. Yet there was a failure to enact sweeping reforms in mathematics teaching, perhaps largely because of resistance to change on the part of teachers and administrators (as well as the scarcity of money to fund reform during the depression). The effectiveness of mathematics instruction continued to be questioned and enrollment in mathematics courses declined as school mathematics became an elective rather than a required subject.

1940

Topology recognized as separate branch of mathematics (1940s)

Joint Commission to Study the Place of Mathematics in Secondary Education gives report (1940)

I remember . . .

at the beginning of World War II the report of the Commission to study the Place of Mathematics in Secondary Education was published as the Fifteenth Yearbook of the National Council of

MATHEMATICS   MATHEMATICS EDUCATION

Committee on Function   *Teachers of Mathematics. According to*
of Mathematics in   *the Report's recommendations, mathe-*
General Education   *matics was to be required of all students*
reports (1940)   *through grade nine, with a two-track pro-*
                               gram in the ninth grade. Concurrently,
Birkhoff and Beatley's   *the Progressive Education Association*
high school geometry   *appointed a committee to study the role*
text *Basic Geometry*   *of mathematics in general education. This*
published (1941)   *committee examined the study and*
                               teaching of mathematics as it relates to the
National Council of   *whole process of general education and*
ENIAC, the first elec-   Teachers of Mathematics   *emphasized the importance of selected*
tronic computer, is   Commission on Post-War   *categories of mathematical behavior, such*
completed (1946)   Plans gives recom-   *as formulation and solution, function*
                          mendations (1944)   *and proof.*
                               The Commission on Post-War Plans
                               of the NCTM urged the need to reconsider
                               pre-war policies for mathematics educa-
                               tion. A by-product of wide-scale military
                               recruiting was the availability of data on
                               nationwide mathematics achievement.
                               Low overall proficiency appalled many
                               educators as well as policy-makers. The
                               pressure for reform began to build up
                               because of increased manpower needs,
                               concern for high-ability students, general
                               attacks on the profession, changed views of
                               mathematics and its usefulness, and
                               the continuing concern expressed by
                               mathematics educators and mathe-
                               maticians.

1950

National Science Founda-
tion created by act of
Congress (1950)

University of Illinois Committee on School Mathe-
matics appointed
(1951)

MATHEMATICS	MATHEMATICS EDUCATION	
	Commission on Mathematics established (1955)	
	U.S.S.R. launches Sputnik (1957)	*I remember . . .*
	School Mathematics Study Group formed (1958)	*the launching by the U.S.S.R. in 1957 of Sputnik, the first man-made satellite of the earth. The resulting shock waves shook loose the needed talents and*
	1960	*resources in America to set in motion*
Feit and Thompson prove that all finite groups of odd order are solvable (1963)	Cambridge Conference on School Mathematics held (1963)	*concerted curriculum reform efforts. Science and mathematics were first affected, but the spirit of reform was soon present in virtually every facet of the*
	National Longitudinal Study of Mathematical Abilities conducted (1962–1967)	*school curriculum.*
	Secondary School Mathematics Curriculum Improvement Study established (1966)	
	1970	
Matyasevich proves that there exists no algorithm for solving a Diophantine equation (early 1970s)	Conference Board of the Mathematical Sciences appoints National Advisory Committee on Mathematical Education (1974)	
Haken and Appel announce a solution to the Four Color Problem (1976)		
	1980	

This panorama may have given you some perspective on mathematics in the schools. Mathematics is not a fixed, static subject; it is alive, active, and growing. Mathematics for the schools did not end with Euclid or with Newton. Old theorems do not lose their truth values, but they do become obsolete as new and better ones are proved. Price (1961) has termed the twentieth century the "golden age of mathematics" since "more mathematics, and more profound mathematics has been created in this period than during all the rest of history" (p. 1). We should not be surprised to learn that the increased growth and importance of mathematics in our culture has had a profound impact on the schools, too (see Research Highlight, "Tracing the Role of Mathematics in the Curriculum").

RESEARCH HIGHLIGHT | TRACING THE ROLE OF MATHEMATICS IN THE CURRICULUM

Tracing the development of the content and the philosophy of teaching secondary school mathematics was given new impetus by the "new math" movement. Yasin (1962) traced reform movements since 1900. He noted stages which were defined by (1) concern for the learner, (2) attempts to reduce the number of mathematics courses, (3) inclusion of new topics, (4) weakening of content, and (5) emphasis on structure.

Hancock (1961) analyzed the instructional aims and recommendations of various groups from 1893 to 1960, noting that they reflected prevailing societal demands. He noted that little attention was given to methods of instruction. In another study of forces that had influenced change, Jamshaid (1969) noted that the first reform movement initiated around 1900 was child- and society-centered, while the reform movement which began in the 1950s was subject-centered.

Some researchers concentrated on specific areas of the curriculum. Geometry, for instance, has consistently received much attention: Quast (1968) analyzed recommendations and practices from 1890 to 1966, Hunte (1966) studied the role of demonstrative geometry from 1900 to 1965, and Pruitt (1969) analyzed the types of exercises in plane geometry textbooks from 1878 to 1966.

THE REVOLUTION IN MATHEMATICS EDUCATION

With the end of World War II and the publication of several prestigious reports on school reforms needed to meet the new demands of a post-war society, the stage was set for massive activity in curriculum research and development. In 1950 an act of Congress established the National Science Foundation to provide financial support for the promotion of basic research and education in the sciences (including mathematics). Private foundations were also to lend support to the efforts to improve the school curriculum.

UNIVERSITY OF ILLINOIS COMMITTEE ON
SCHOOL MATHEMATICS (UICSM)

In 1951, while Sputnik was little more than a gleam in the eye of Soviet scientists, a committee of mathematicians, engineers, and educators at the University of Illinois began investigating problems concerned with the teaching of secondary school mathematics and planning new directions in school mathematics programs. It was the first large-scale project to prepare materials for secondary school mathematics which expressed the contemporary view and role of mathematics. The chairman of the committee was the late Max Beberman (Figure 13–4).

The main developmental strategy of UICSM was to propose certain topics to be taught, to try them out with students, to revise the materials, and to try them again. Beberman's group also recognized the importance of having well-trained teachers for implementing the new materials, so for the first few years of its activity, UICSM would make available classroom sets of the textbooks only to teachers having training in the use of the materials.

The UICSM curriculum was characterized by precision of language and discovery of generalizations. Terms such as *pronumeral*** were introduced to clarify and make more precise the mathematics being presented. A variety of teaching strategies for encouraging students to make their own discoveries were developed. At one stage in the curriculum reform movement, "discovery teaching" and "UICSM mathematics" were virtually synonymous. As Beberman stated, "We believe that a student will come to understand mathematics when his textbook and teacher use unambiguous language and when he is enabled to discover generalizations by himself" (1958, p. 4).

The resulting curriculum for grades 9–12 was piloted in about a dozen schools in 1958. The materials dealt with most of the topics usually presented in the high school curriculum, but were couched in the distinctive UICSM approach. As the project evolved, the curriculum took the form of hardcover textbooks that were made commercially available. More recently, curriculum ventures were undertaken in the areas of mathematics for low-achieving junior high school students and applied mathematics for the ninth grade. At the time of Professor Beberman's death in 1971, UICSM had become heavily influenced by the work of such English mathematics educators as Edith Biggs and was on the verge of doing extensive work in reforming the curriculum of the elementary school. This concern for elementary school mathematics was continued when Robert B. Davis, Director of the Madison Project (see p. 445), came to the University

---

* In one UICSM text (Beberman et al., 1970), *pronumeral* is discussed in the teacher's commentary: "You may find it helpful to point out that real-number variables hold places for numerals in the same way that pronouns hold places for nouns. So real-number variables constitute a species of pronoun, and might reasonably be called *pronumerals*. This may help counteract the feeling that something which is called a variable should, somehow, vary" (p. TG 11).

Figure 13–4   Max Beberman. (Drawing by John Downs reprinted by permission of the *Chicago Daily News*.)

of Illinois in 1972. Davis was interested in the role of computers in mathematics education and undertook an extensive project in the development of instructional materials in elementary mathematics on PLATO, the University of Illinois teaching computer system.

SCHOOL MATHEMATICS STUDY GROUP (SMSG)

Two conferences of mathematicians in 1958 were sponsored by the National Science Foundation. Both recommended the appointment of a committee to restructure the school mathematics program. The American Mathematical Society, after conferring with the Mathematical Association of America and the National Council of Teachers of Mathematics, requested Edward G. Begle (Figure 13–5) of Yale University to direct such an effort. The result was the School Mathematics Study Group, which moved from Yale to Stanford University in the early 1960s.

The approach of the School Mathematics Study Group to curriculum development was unique in that it incorporated the ideas of many different groups— psychologists, evaluators, mathematicians, and mathematics educators. SMSG engaged large groups of mathematicians and mathematics teachers in summer writing sessions. The resulting materials were tried out in pilot classes around the country and then revised.

In 1959–1960, tryouts for materials for grades 7–12 were conducted in 45

Figure 13–5   Edward G. Begle.

states by more than 400 teachers and 42,000 pupils. During the tryouts experts from the project were available to the teachers for consultation and feedback purposes. Revisions were made on the basis of these field trials. The aim was to develop a rational sequence of courses in which algebra and geometry reinforced each other, while encouraging explorations by teachers and researchers of the hypotheses underlying mathematics education. SMSG ultimately produced sample textbooks for grades K–12, teachers' guides, enrichment materials, programmed materials for students at various ability levels, research monographs, and reports on studies of textbooks, students, and teachers as part of the National Longitudinal Study of Mathematical Abilities (see Research Highlight, "The National Longitudinal Study of Mathematical Abilities, 1962–1967").

RESEARCH HIGHLIGHT | THE NATIONAL LONGITUDINAL STUDY OF MATHEMATICAL ABILITIES, 1962–1967

This study, organized by the School Mathematics Study Group at Stanford University, was designed to determine long-term effects of various mathematics curricula on students. The project, funded by the National Science Foundation, had as its basis the assumption that

> a long-term study was needed to provide information for the further
> improvement of the school mathematics curriculum, to develop measures
> of mathematics achievement more sensitive to the wide range of outcomes
> expected from using various types of textbooks, to investigate the nature
> of mathematics achievement, to provide information for school personnel
> and to gain experience in operating a large-scale study in order to inform
> other investigators wishing to operate similar studies. (Wilson and Begle,
> 1972, p. vii)

The NLSMA Study was described by Cahen (1965 and Wilson and Begle, 1970). Detailed information concerning the findings of this extensive project is available in the form of NLSMA Reports, which may be obtained from A. C. Vroman, Incorporated, 2085 Foothill Boulevard, Pasadena, California 91109.

BOSTON COLLEGE MATHEMATICS INSTITUTE

Under the supervision of Stanley J. Bezuska at Boston College, secondary school mathematics materials were developed beginning in 1957. The emphasis was on the structure of mathematics approached from the historical point of view. Computer programming courses were added as a component of the project.

BALL STATE EXPERIMENTAL PROGRAM

The program, under the direction of Charles Brumfiel, developed courses in algebra and geometry, beginning in 1955. These emphasized, through a logical development, the axiomatic structure of mathematics and precision of language.

UNIVERSITY OF MARYLAND MATHEMATICS PROJECT (UMMaP)

In 1957, UMMaP began to prepare materials for junior high school students under the direction of John R. Mayor. These materials, designed to serve as a bridge between arithmetic and high school mathematics, stressed the language and structure of mathematics.

ELEMENTARY SCHOOL PROJECTS

Among the first projects were several whose focus was at the elementary school level. The University of Illinois Arithmetic Project, headed by David A. Page, and the Greater Cleveland Mathematics Program, initiated under the direction of B. H. Gundlach, provide two examples. A third, the Minnesota School Mathematics and Science Teaching Project, was also begun in the late 1950s, under the direction of Paul Rosenbloom. Students were taught fundamental mathematical ideas through a set of materials in which science and mathematics were coordinated. Each subject was used to support and reinforce the other where appropriate.

The Madison Project was originated in 1957, with Robert B. Davis as director. The intent was to develop a closer relationship between what was being taught in mathematics and mathematics education courses and what was going on in schools. A supplemental program was prepared which is designed to stimulate children to be creative in mathematics. Use of mathematics laboratory experiences with a discovery approach is a key aspect of the materials.

THE CAMBRIDGE CONFERENCE

In 1963 the Cambridge Conference brought together concerned persons, largely mathematicians and scientists, who outlined greatly accelerated programs as goals for attainment by the end of the century. Their report, *Goals for School Mathematics* (1963), proposed a curriculum which would in 13 years of schooling (grades K–12) offer training comparable to three years of study currently available in top-level colleges. Major points of agreement in the report included the parallel development of arithmetic and geometry in the elementary school, an introduction to probability and statistics early enough in the curriculum so that all students would study the topic (not only those electing advanced high school courses), and a "nodding acquaintance" with the calculus, which was labeled "one of the grandest edifices constructed by mankind" (p. 9).

COMPREHENSIVE SCHOOL MATHEMATICS PROGRAM (CSMP)

CSMP, under the direction of Burt Kaufman at Southern Illinois University, began its work in the late 1960s. The project seeks to develop a mathematics curriculum for all students aged 5–18 which is sound in content, enjoyable, and appropriate to the needs and abilities of the individual child. The involvement of mathematicians is stressed at all stages of the curriculum development process. At the secondary level, the *Elements of Mathematics Program* combines teacher-directed and independent study modes for the most highly capable students in grades 7–12. By means of a complete restructuring of the curriculum, this program makes it possible for the student to have encountered, by the time she or he graduates from high school, most of the topics typically taken by an undergraduate in mathematics. Subject matter covered in the program includes logic, sets, relations, linear algebra, probability, groups, rings, and fields. Several mathematicians from Europe have participated in the writing of the materials.

SECONDARY SCHOOL MATHEMATICS CURRICULUM IMPROVEMENT STUDY (SSMCIS)

This project, which grew out of conferences in Europe and the United States on needed reforms in secondary school mathematics, was developed by Howard F. Fehr (Figure 13–6) of Teachers College, Columbia University. The stated aim

Figure 13–6 Howard F. Fehr. (Photo courtesy Teachers College, Columbia University.)

of the project was to "bring forth a more unified, more efficient and more meaningful program in mathematical education for university-bound students" (Fehr, 1974, p. 25). The project has restructured the secondary school mathematics curriculum by doing away with the conventional organization of arithmetic, algebra, geometry, and analysis and replacing it with one program based upon such fundamental notions as sets, relations, mappings, groups, and fields. As Fehr states, "This structuring permits greater understanding and efficiency in learning and uncovers concepts and theories previously hidden by the traditional separation" (p. 26). In 1965 a group of mathematicians and mathematics educators from the United States and Europe met to establish the scope and sequence for such a program. Curriculum materials were then written for grades 7–12 in conjunction with classroom testing in pilot schools.

The impact of SSMCIS and of CSMP has not yet been determined, since their products have been in use for a relatively shorter time than the products of other projects.

TEACHER EDUCATION COMPONENTS

You may have wondered how the teacher, busy with a full complement of classes to teach, ever found the time to prepare for using the new materials which flooded the curricular market place. It was clear to the curriculum innovators that teacher education was an essential component of the reform package. Some curriculum projects, as has been noted of the University of Illinois Committee on School Mathematics, stressed the need for special teacher preparation prior to the use of their materials in the classroom. Largely through federal funding, notably from the National Science Foundation, in-service workshops flourished. Teachers went back to college in record numbers for short courses, for summer sessions, for evening programs, and for year-long institutes. As a long-term measure, undergraduate teacher education programs, stimulated by the recommendations of such groups as the Mathematical Association of America's Committee on the Undergraduate Program in Mathematics (CUPM), were revised to include considerably more formal coursework in mathematics.

MATHEMATICS EDUCATION FOR THE 1970s

Although by the mid-1970s the "revolution" of the 1960s had subsided, the school curriculum continued to change—in less dramatic ways. The new mathematics of the 1960s became the old mathematics of the 1970s, and what followed was a period of reassessment. Society's demands for higher levels of mathematical literacy were balanced against those components of the curriculum which seemed to preserve the best of what had been done in the past. A number of developments in the 1970s, however, appear to have considerable potential for determining future directions in school mathematics.

"EXTREMES" OF THE NEW MATHEMATICS ARE ATTACKED

Morris Kline, Professor of Mathematics at New York University, had been a vigorous and relentless critic of the curriculum movement of the 1960s, attacking what he termed the excessive formalism and symbolism of many of the new mathematics programs. In *Why Johnny Can't Add* (1973), he reasserted his concerns about what had been going on in the schools:

> The new mathematics as a whole is a presentation from the point of view of the shallow mathematician, who can appreciate only the petty deductive details and minor pedantic, sterile distinctions such as between number and numeral and who seeks to enhance trivia with imposing-sounding terminology and symbolism. . . . The formalism of this curriculum can lead only to an erosion of the vitality of mathematics and to authoritarianism in teaching, the rote learning of new routines far more useless than the traditional routines. (p. 102)

Then, in the final chapter entitled, "The Proper Direction for Reform," Kline concluded:

> What we should be fashioning and teaching, . . . beyond mathematics proper, are the relationships of mathematics to other human interests—in other words, a broad cultural mathematics curriculum which achieves an intimate communion with the main currents of thought and our cultural heritage. Some of these relationships can serve as motivation; others would be applications; and still others would supply interesting reading and discussion material that would vary and enliven the content of our mathematics courses. (pp. 145–146)

Concern over developments in curriculum arose in other sectors of society as well. In December 1973 the College Board reported persistent declines in average scores for high school students on both the verbal and mathematical portions of the Scholastic Aptitude Test, an admissions examination used by many institutions of higher education. On the mathematics portion of the examination, scores were reported by the Educational Testing Service as falling from an average of 502 in 1963 to an average of 481 in 1973. (The most selective colleges and universities typically require Scholastic Aptitude Test scores over 600 for admission.) An official of the College Board interpreted the results as suggesting that schools may not be preparing students in verbal and mathematical skills as well as they did in the past (*New York Times*, December 16, 1973, pp. 1, 26).

RELEVANCE IS SOUGHT

The activist climate of the 1960s, with its mass protests and demonstrations, promoted a suspicion of "merely academic" studies. Instead, "relevance" became the byword, and classroom study was to be related to the many critical problems confronting the world—food supply, ecology, and population control, among others.

A renewed interest in real-world applications of mathematics added impetus to a move to give probability and statistics a more important role in the secondary school curriculum. The Joint Committee on the Curriculum in Statistics and Probability, appointed by the American Statistical Association and the National Council of Teachers of Mathematics, prepared two publications suitable for high school use. *Statistics: A Guide to the Unknown* (Tanur et al., 1972) provides in easy-to-read, nontechnical form examples of applications of statistics in a variety of fields, from law to nutrition to meteorology to accounting. *Statistics by Example* (Mosteller et al., 1973) explains how statistics can be done, using a number of interesting, real-world examples.

"TRADITIONAL" SCHOOLING IS CALLED INTO QUESTION

More than just the mathematics curriculum became the object of concern and scrutiny in the 1970s. A flood of critiques appeared, beginning at the height of the curriculum reform movement in the mid-1960s, that included the writings of Holt (1964 and 1970), Kozol (1967), and Silberman (1970). Silberman's *Crisis in the Classroom* received the studied attention of educators and laymen alike. Alternatives to traditional schooling were explored, including the use of open

"Wilson, about your interest center on statistics and probability. . ."

© 1977 by Ford Button.

classrooms (developed at the elementary school level by British educators) and project or independent study approaches, such as the one by Seymour Papert of the Massachusetts Institute of Technology.

ACCOUNTABILITY OF EDUCATORS IS REQUIRED

Perhaps the demands for accountability arose as a natural reaction on the part of the taxpayer to the expensive, large-scale curriculum development projects and other educational activities of the 1960s. Perhaps they simply reflected an increased public awareness of education. Whatever the causes, accountability in education had been demanded of many teachers by 1970. A highway contractor delivers a road in return for pay, and an aircraft manufacturer, a supersonic jet; but what does the taxpayer get in return for his or her tax dollar? Often the product is difficult to describe precisely. Attempts were made to hold teachers responsible for producing student achievement in measurable terms. Teachers were requested to express instructional goals in the form of behavioral objectives and to show that they were indeed accountable by having predetermined numbers of their students attain specific levels of proficiency on those objectives. Thus it was expected that a teacher would meet a specification such as "75 percent of my class have passed 90 percent of the skill items for operations with rational numbers."

CALCULATORS AND COMPUTERS BECOME INCREASINGLY AVAILABLE

Sophisticated aids to calculating and data processing are becoming increasingly available in secondary schools (recall our discussion in Chapter 9). Calculators and computers are becoming faster, more compact, and less expensive. Calculators which can easily be held in the hand can perform a variety of mathematical operations—and yet can be purchased for as little as five dollars. At the other end of the spectrum, large-scale teaching computer systems, because they can service many customers on a time-sharing basis, are making access to powerful and sophisticated computers within the reach of many school systems, particularly in urban areas. With expert use of these tools, more complex mathematical topics can be made accessible to greater numbers of students.

THE ENVIRONMENT OF THE MATHEMATICS CLASSROOM IS ENRICHED

In many schools the mathematics classroom of the 1970s is becoming a very interesting, highly motivational place in which to be. The sterile rows of desks and seats fixed to the floor are giving way to rooms with flexible, functional furniture arrangements and even moveable walls. The staple textbook and chalkboard have been supplemented with a variety of instructional materials. One excellent reference for teachers wishing to find information on aids to instruction is the Thirty-fourth Yearbook of the NCTM (Berger, 1973). This attractive volume contains many practical suggestions for the use of overhead projectors, films, models, computers, and other aids. (See Appendix D for other references.)

MATHEMATICS EDUCATION EMERGES AS A DISCIPLINE

That branch of education devoted to mathematics teaching, learning, and curriculum development, has its roots in two traditions, the liberal arts and professional education. From the liberal arts comes the mathematical component, with well-developed theory, explicit rules of reasoning, and defined terms. The tradition of education brings components of educational psychology, philosophy, sociology, and instructional design. Kenneth B. Henderson (Figure 13–7), Professor of Mathematics Education at the University of Illinois from 1948 until 1974, did much to unite the two traditions of mathematics and of education into a single discipline. For example, he used concepts from set theory to denote teaching strategies and defined the teaching act as the ternary relation $T(X, Y, Z)$ where X teaches Y to Z (Henderson, 1963, p. 1007). Henderson urged that logical theory be used to clarify discussion in education, hypothesizing that "if semantic confusion is reduced, the theory will be more productive of questions worth asking . . . " (p. 1028).

THE NATIONAL ADVISORY COMMITTEE ON MATHEMATICAL EDUCATION

In 1974 the Conference Board of the Mathematical Sciences appointed a committee to prepare an overview and analysis of school mathematics in the United States. The committee invited input from all concerned individuals and profes-

Figure 13–7   Kenneth B. Henderson.

sional groups and prepared a highly informative report on the findings (National Advisory Committee on Mathematical Education, 1975). But perhaps the most important component of the report is its recommendations for the future course of mathematics in United States schools. For example, the Committee recommended that:

> beginning no later than the end of the eighth grade, a calculator should be available for each mathematics student during each mathematics class. Each student should be permitted to use the calculator during all of his or her mathematical work including tests. (p. 138)

And with respect to geometry, it made a plea for "new and imaginative approaches . . . in high school, junior high school and elementary school," and went on to say that "a rethinking of the role geometry should play in the objectives and goals of the mathematics curriculum and its relationship to the rest of the mathematics program would be timely and valuable" (p. 146).

This report made incisive comments on, and recommendations concerning, the major aspects of mathematics education in the 1970s. Its recommendations, it appears, will provide food for thought and goals for action for several years to come.

MATHEMATICS EDUCATION IN THE 1980s AND BEYOND

No one, of course, knows what the future holds for mathematics education. However, we can be confident that changes will continue to take place. The following are among those developments we expect to see.

INCREASING RESPONSIVENESS TO THE INDIVIDUAL NEEDS AND INTERESTS OF STUDENTS

We anticipate a greater variety of courses, in terms of length, content, and approach, which will be suited for students at different ability levels and with different learning styles. Minicourses (lasting a few weeks) will deal with a variety of topics in order to maintain the interests of students. We expect that many of these topics will be cross-disciplinary—relating mathematics to the social sciences, the arts, and industry. A problems-oriented course might take on the planning of a special edition of the class newspaper to deal with mathematics and ecology, or mathematics and energy supply. A general mathematics class might develop a project on computers and problem solving. As a result, student enrollments in elective mathematics courses can be expected to increase dramatically.

GREATER AND MORE CREATIVE USE OF INSTRUCTIONAL TECHNOLOGY

As society becomes increasingly technological, we expect that education will make better use of the technology available. Calculators and computers will become readily accessible. Many students will have access to computers from their homes—perhaps by using the telephone to transmit information to a central computer and then receiving output on the home television set. Papert and Solomon (1972) have asked:

> Why . . . should computers in schools be confined to computing the sum of the squares of the first twenty-odd numbers and similar so-called problem-solving uses? Why not use them to produce some action? There is no better reason than the intellectual timidity of the computers-in-education community, which seems remarkably reluctant to use the computers for any purpose that fails to look very much like something that has been taught in schools for the past centuries. (p. 9)

We expect that new and exciting uses will be made of computers in education and that many of the first steps will be taken in the mathematics classrooms of the 1980s.

Instructional television and movies will be more captivating and better suited to classroom needs. The concept of interactive television, a teaming of computers and television to permit interaction of the learner with the televised lesson, holds great promise for new directions in instruction. Textbooks will become individualized, making it possible for instructors—and even subgroups of their students—to specify topics for inclusion in a book.

MORE EMPHASIS ON APPLICATIONS OF SCHOOL MATHEMATICS TO REAL PROBLEMS

Courses will emphasize the integral role of mathematics in the sciences, with concern for environmental and economic factors. Cross-disciplinary teaching teams will be developed for certain courses. Effective teaming, with genuine communication between team members, can be mutually beneficial with respect to instructional goals and techniques in each subject matter area.

GREATER PROFESSIONALISM IN TEACHING

The mathematics teacher—at every level of the school—will become a highly regarded specialist. More preparation for these positions and greater responsibilities in the positions can be expected. We believe that more and better theory will be developed. Research will be more effective, providing results needed by classroom teachers. Teams of specialists—theorists, researchers, practicing teachers—will be formed to attack many of the problems facing education. As teachers better their training and promote improved education for their students, they will succeed in improving their professional lot as well. The need for adequate preparation time will be recognized. More classroom teachers will be given time to do research, to write, and to provide leadership in professional associations. We expect that "clinical professors of education," persons who hold joint appointments as professors of education and as practicing classroom teachers, will increase in number. Concurrently, professional organizations will grow and strengthen. These organizations will have a greater role in in-service education and they will be more successful in meeting the needs of classroom teachers.

CONCERTED ATTEMPTS TO RAISE LEVEL OF MATHEMATICAL LITERACY

The proverbial man on the street in the mid-1970s possesses only a meager grasp of mathematical knowledge. A report of the results of a survey of the National Assessment of Educational Progress (for details of this project, see Hazlett, 1975) stated that less than 50 percent of the 4200 adults sampled could correctly calculate the most economical package of rice and tuna, that only 55 percent could successfully read a federal income tax table, 16 percent could balance a checkbook, and 20 percent were able to compute a taxi fare. Roy H. Forbes, director of the National Assessment project, was quoted by Newhouse News Service as saying, "It is impossible to determine how many hundreds of dollars individual consumers waste each year because of their inability to use math skills to determine the best buy" (Fogg, 1975, p. 16). A second national assessment in mathematics is planned for the late 1970s.

We expect that during the 1980s various developments in mathematics education will result in a citizenry that is more literate mathematically—and therefore better equipped to function in tomorrow's world.

Critical Incidents Revisited

Miss Kelley, in our first critical incident, probably has some homework to do. Parents have questions about changes in curriculum, trends in education—a host of general issues—and they have a right to expect their child's teacher to be able to discuss these issues. Miss Kelley would do well to review the work of one or two of the curriculum projects of the past 10 or 15 years. She should be familiar with some of the controversy which has surrounded the curriculum reform movement. And she might familiarize herself with some research studies which have evaluated the newer curriculums.

Marge and Tony are reacting to change in education, but in different ways. It is difficult for Marge to make decisions about teaching because she lacks the experience. But, as we pointed out in Chapter 2, this book has been written to help in such decision-making. As she reads, plans, tries out her ideas, and then evaluates them, Marge will accumulate the knowledge and experience to help her become an outstanding teacher. Tony, on the other hand, does not see teaching as a dynamic process. Somehow, for him, education exists in isolation from the influences of a developing world. He needs to interact with other professionals—perhaps to attend meetings where he can identify persons he respects, from whom he will gain some broader perspectives of mathematics teaching. Maybe if he applied some self-evaluation techniques he would find that his teaching has more room for improvement than he thinks.

Jason's experience, in the fourth critical incident, provides us with an opportunity to conclude our book. We began by asserting that we teach mathematics to people. Mathematics continues to grow and develop; so do people. Hence, contrary to Tom's view, teaching is constantly changing, ever seeking more effective ways to provide the best education for others. Sorry, Tom, but you're wrong about teaching. As Jason will find out, teaching has its share of the action, too.

Activities

1. Read an account of mathematics education in the United States in the early 1900s. Summarize what you conclude to have been the prevailing view of mathematics instruction in those days. Two useful references are the Thirty-second Yearbook of the NCTM on the history of mathematics education (Jones, 1970) and the Golden Anniversary issue of the *American Mathematical Monthly* (1965).

2. Obtain a secondary school mathematics textbook published before 1940 and one published since 1960 but designed for the same course (freshman algebra, for example). Choose one topic, such as factoring algebraic expressions. How do the two presentations of the same topic differ? What new terminology has been introduced? How do the layouts of the texts compare?

3. During the curriculum reform movement, many controversial issues evolved concerning the teaching of mathematics. Examples are:
   a. the importance of distinguishing between a number and a numeral
   b. the question of whether to define an angle as an amount of rotation or as the union of two rays
   c. the value of teaching numeration systems other than base ten
   d. the efficacy of discovery teaching
   Study one of these issues or another that you can identify from your reading. State the various points of view concerning the issue, decide what position you would take, and defend your position in about 200 words.

4. Technology will have considerable impact on mathematics instruction in the next ten years. Choose some aspect of educational technology

(such as computers, videotape recording, information retrieval, or simulation) and investigate the new directions in education that are suggested through use of this technology. (A good reference: the periodical *Educational Technology*.)

5. Prepare a report on some topic in the history of mathematics which would add interest to your own mathematics class. Use the references at the end of this chapter as a starting place for your research.

6. Identify five major developments in the history of mathematics education. You will find Jones (1970) an invaluable resource.

7. Suppose you are in a leadership position in a mathematics department of a secondary school and a benefactor awards you $20,000 to improve mathematics education at that school. How would you spend the money? Why?

8. Look at the results of the most recent mathematics survey conducted by the National Assessment of Educational Progress. Summarize the important findings of the survey. What implications do these findings have for the improvement of mathematics education in the next ten years?

9. Do a detailed study of one of the curriculum development projects in school mathematics since 1960. What were the major purposes of the project? What did the project accomplish? To what extent was the project successful in attaining its goals?

10. What major development in mathematics education do you see as most essential for the 1980s? Why is this development important? What conditions are needed so that this development will take place?

11. Edwards et al. (1972) provide one expression of what it means to be mathematically literate in today's world. Read the article, summarize its main points, and then formulate your own definition of mathematical literacy.

References

Beberman, Max. *An Emerging Program of Secondary School Mathematics*. Inglis Lecture Series. Cambridge, Massachusetts: Harvard University Press, 1958.

Beberman, Max; Wolfe, Martin S.; and Zwoyer, Russell E. *Algebra 1: A Modern Course*. Lexington, Massachusetts: Heath, 1970.

Bell, E. T. *Men of Mathematics*. New York: Simon and Schuster, 1937.

Berger, Emil J., ed. *Instructional Aids in Mathematics*. Thirty-fourth Yearbook of the National Council of Teachers of Mathematics. Reston, Virginia: The Council, 1973.

Birkhoff, George David and Beatley, Ralph. *Basic Geometry*. Chicago: Scott, Foresman, 1941.

Brooks, Edward. *The Philosophy of Arithmetic*. Lancaster, Pennsylvania: Normal Publishing Company, 1880.

Butts, R. Freeman. *A Cultural History of Western Education*. New York: McGraw-Hill, 1955.

Cahen, Leonard S. An Interim Report on the National Longitudinal Study of Mathematical Abilities. *Mathematics Teacher* 58: 522–526; October 1965 and 58: 659; November 1965.

Cajori, Florian. *A History of Elementary Mathematics.* New York: Macmillan, 1890.

Colburn, Warren. *An Arithmetic on the Plan of Pestalozzi, with Some Improvements.* Boston, 1821.

Davies, Charles. *The Logic and Utility of Mathematics with the Best Methods Explained and Illustrated.* New York: A. S. Barnes and Co., 1850.

Edwards, E. L., Jr.; Nichols, Eugene D.; and Sharpe, Glyn H. Mathematical Competencies and Skills Essential for Enlightened Citizens. *Mathematics Teacher* 65: 671–677; November 1972.

Fehr, Howard F. The Secondary School Mathematics Curriculum Improvement Study: A Unified Mathematics Program. *Mathematics Teacher* 67: 25–33; January 1974.

Fogg, Susan. Lack of Basic Math Knowledge Hurts Consumers in Pocket Book. *The Daily Illini* 104: 16; July 25, 1975.

Greenwood, Isaac. *Arithmetick, Vulgar and Decimal.* Boston: S. Kneeland and T. Green, 1729.

Halsted, George B. *Rational Geometry; A Textbook for the Science of Space; Based on Hilbert's Foundations.* New York: Wiley, 1904.

Hancock, John David. The Evolution of the Secondary Mathematics Curriculum: A Critique. (Stanford University, 1961.) *Dissertation Abstracts* 22: 501–502; August 1961.

Hazlett, James A. A History of the National Assessment of Educational Progress, 1963–1973: A Look at Some Conflicting Ideas and Issues in Contemporary American Education. (University of Kansas, 1974.) *Dissertation Abstracts International* 35A: 5887; March 1975.

Henderson, Kenneth B. Research on Teaching Secondary School Mathematics. In *Handbook of Research on Teaching* (edited by N. L. Gage). Chicago: Rand McNally, 1963, pp. 1007–1030.

Holt, John. *How Children Fail.* New York: Pitman, 1964.

Holt, John. *What Do I Do Monday?* New York: Dutton, 1970.

Hunte, Beryl Eleanor. Demonstrative Geometry During the Twentieth Century: An Account of the Various Sequences Used in the Subject Matter of Demonstrative Geometry from 1900 to the Present Time. (New York University, 1965.) *Dissertation Abstracts* 26: 3979; January 1966.

Jamshaid, Mohammad. A Study of the Forces That Have Influenced Change in Secondary School Mathematics (Grades 7–12) in the United States Since World War II and the Possible Implications for Pakistan. (Indiana University, 1968.) *Dissertation Abstracts* 29A: 2890; March 1969.

Johnson, Clifton. *Old-Time Schools and Schoolbooks.* New York: Macmillan, 1904.

Jones, Phillip S., ed. *A History of Mathematics Education in the United States and Canada.* Thirty-second Yearbook of the National Council of Teachers of Mathematics. Washington: The Council, 1970.

Karpinski, Louis Charles. *The History of Arithmetic.* New York: Rand McNally, 1925.

Kline, Morris. *Why Johnny Can't Add: The Failure of the New Math.* New York: St. Martin's Press, 1973.

Kozol, Jonathan. *Death at an Early Age.* New York: Houghton Mifflin, 1967.

Lockard, J. David, ed. *Eighth Report of the International Clearinghouse on Science and Mathematics Curricular Developments 1972*. College Park, Maryland: Science Teaching Center, University of Maryland, 1972.

Lockard, J. David, ed. *Science and Mathematics Curricular Developments Internationally, 1956–1974: The Ninth Report of the International Clearinghouse on Science and Mathematics Curricular Developments*. College Park, Maryland: Science Teaching Center, University of Maryland, 1975.

Mosteller, Frederick et al., eds. *Statistics by Example*. 4 vols. Reading, Massachusetts: Addison-Wesley, 1973.

Nichols, Eugene D. The Many Forms of Revolution. In *The Continuing Revolution in Mathematics*. Washington: National Council of Teachers of Mathematics, 1968, pp. 16–37.

Papert, Seymour and Solomon, Cynthia. Twenty Things To Do with a Computer. *Educational Technology* 12: 9–18; April 1972.

Pike, Nicholas. *A New and Complete System of Arithmetick*. 7th ed. Boston: Thomas and Andrews, 1809.

Price, G. Baley. Progress in Mathematics and Its Implications for the Schools. In *The Revolution in School Mathematics: A Challenge for Administrators and Teachers*. Washington: National Council of Teachers of Mathematics, 1961.

Pruitt, Ralph Lewis. An Analysis of Types of Exercises in Plane Geometry Texts in the United States from 1878 to 1966. (The Ohio State University, 1969.) *Dissertation Abstracts International* 30A: 1414; October 1969.

Quast, William Garfield. Geometry in the High Schools of the United States: An Historical Analysis from 1890 to 1966. (Rutgers—The State University, 1968.) *Dissertation Abstracts* 28A: 4888; June 1968.

Silberman, Charles. *Crisis in the Classroom*. New York: Random House, 1970.

Smith, David E. *The Teaching of Elementary Mathematics*. New York: Macmillan, 1900.

Smith, David Eugene and Ginsburg, J. *A History of Mathematics in America Before 1900*. Chicago: Open Court Publishing Company, 1934.

Smith, David E. and Reeve, William D. *The Teaching of Junior High Mathematics*. Boston: Ginn, 1927.

Tanur, Judith et al., eds. *Statistics: A Guide to the Unknown*. San Francisco: Holden-Day, 1972.

Wilson, James W. and Begle, Edward G., "Evaluation of Mathematics Programs." In *Mathematics Education* (edited by Edward G. Begle) Sixty Ninth Yearbook of the Natl. Soc. for the Study of Educ., Part I. Chicago: University of Chicago Press, 1970.

Wilson, James W. and Begle, Edward G., eds. *Patterns of Mathematics Achievement in Grade 10, Z-Population*. NLSMA Reports, No. 16. Stanford, California: Stanford University, 1972.

Yasin, Said Taha. The Reform Movement in Secondary Mathematics—Its History and Present State. (Indiana University, 1961.) *Dissertation Abstracts* 22: 3084; March 1962.

Young, Jacob William Albert. *The Teaching of Mathematics in the Elementary and Secondary School*. New York: Longmans, Green, 1907.

Goals for School Mathematics, The Report of the Cambridge Conference on School Mathematics. Boston: Houghton Mifflin, 1963.

National Advisory Committee on Mathematical Education. *Overview and Analysis of School Mathematics, Grades K–12*. Washington: Conference Board of the Mathematical Sciences, 1975.

National Assessment of Educational Progress Newsletter. Adult Males on Plus-side in Math Basics. January–February 1975, pp. 1, 3.

New York Times. Students' Scores Again Show Drop. Sunday, December 16, 1973, pp. 1, 26.

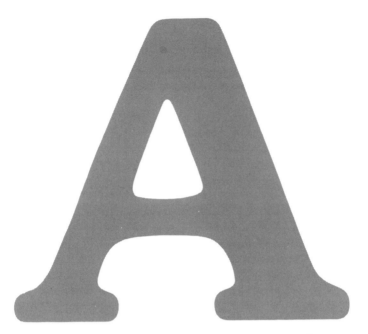

Current Research in Secondary School Mathematics

The research studies cited in this appendix were selected to illustrate the variety of types of research on secondary school mathematics.* The primary criterion was "Are the findings ones which a teacher might apply, either to instruction directly or to understanding how children learn?" An attempt was made to be evaluative, but the quality of the studies is nevertheless variable. Some relatively older reports are included so that you may see that certain themes recurred over the years. You will probably see that there are some findings you accept, some you question, and some you want to reject. The annotation is intended to convey the key finding or idea of the study; you will find that many other points are included in the research report itself. Some of the studies present materials as well as ideas which you can try out in your classroom.

* Many dissertations are included, since they "round out" the picture of research. Unfortunately, dissertations are not available in all libraries; however, the abstracts can usually be found, and they include more information than do the brief annotations provided here. Some instructors may be able to develop a library collection of dissertations on microfilm.

Aiken, Lewis R., Jr. Attitudes Toward Mathematics. *Review of Educational Research* 40: 551–596; October 1970.

> From a thorough review of research on attitudes toward mathematics and factors affecting those attitudes, it was noted that attitudes can be traced to childhood, with evidence that they are formed as early as the third grade. The results of a number of studies indicate that attitude toward mathematics becomes increasingly negative as the students go through school.

Albers, Dallas Frederick. An Investigation of the Effects of the Allocation of Class Time on Pupil Achievement and Scheduling Preferences. (University of Missouri-Columbia, 1972.) *Dissertation Abstracts International* 33A: 4700; March 1973.

> No significant differences in achievement were found between classes having geometry for 90 periods of 110 minutes or 180 periods of 55 minutes. Students preferred the double period.

Alexander, Vincent E. Seventh Graders' Ability to Solve Problems. *School Science and Mathematics* 60: 603–606; November 1960.

> For seventh graders, the ability to understand verbal concepts, mental age, reading comprehension and vocabulary, arithmetic concepts and computation, intelligence, ability to analyze problems, ability to interpret data, perception of relationships involving comparison of data, and recognition of limitations of given data appeared closely related to arithmetic reasoning. Going beyond the data and crude errors in interpreting data were associated with low achievement.

Anderson, Richard Mark. A Study of Cadet Teaching as a Method to Improve the Multiplication and Division Proficiency of a Selected Sample of Junior High School Students. (University of Iowa, 1970.) *Dissertation Abstracts International* 31A: 2782; December 1970.

> Tutoring itself did not increase proficiency, but the special instruction in multiplication and division skills needed by junior high school students for tutoring fifth graders resulted in significant gains in achievement.

Anttonen, Ralph G. A Longitudinal Study in Mathematics Attitude. *Journal of Educational Research* 62: 467–471; July/August 1969.

> Using a semantic differential technique, a significant positive correlation was found between the attitudes of students tested in grades 5 and 6 and retested in grades 11 and 12. A significant correlation was found between attitude and achievement.

Ashton, Sister Madeleine Rose. Heuristic Methods in Problem Solving in Ninth Grade Algebra. (Stanford University, 1962.) *Dissertation Abstracts* 22: 4289; June 1962.

> Teaching students to ask a set of questions about problems to be solved helped more than showing them how to solve a particular type of problem and then giving practice.

Austin, Joe Dan and Austin, Kathleen A. Homework Grading Procedures in Junior High Mathematics Classes. *School Science and Mathematics* 74: 269–272; April 1974.

> Grading a random half of the homework problems in each assignment seemed a valid alternative to grading every problem on each assignment.

Ayers, Jerry B.; Bashaw, W. L.; and Wash, James A. A Study of the Validity of the Sixteen Personality Factor Questionnaire in Predicting High School Academic Achievement. *Educational and Psychological Measurement* 29: 479–484; Summer 1969.

Correlations of personality factors with mathematics achievement in grade 10 were low. Good mathematics students tended to be withdrawn, conscientious, emotionally immature, and lacking in frustration tolerance.

Bachman, Alfred Morry. Factors Related to the Achievement of Junior High School Students in Mathematics. (University of Oregon, 1968.) *Dissertation Abstracts* 29A: 2139; January 1969.

Seventh graders in self-contained classes did not achieve significantly different from those in ability-grouped classes. A positive relationship was found between self-concept and achievement.

Bachman, Alfred Morry. The Relationship Between a Seventh-Grade Pupil's Academic Self-Concept and Achievement in Mathematics. *Journal for Research in Mathematics Education* 1: 173–179; May 1970.

General self-concept and self-concept in mathematics were each found to be significantly related to mathematical achievement, with mathematical self-concept related significantly more to such achievement than was general self-concept.

Baker, Betty Louise. A Study of the Effects of Student Choice of Learning Activities on Achievement in Ninth Grade Pre-Algebra Mathematics. (Northwestern University, 1971.) *Dissertation Abstracts International* 32A: 2895; December 1971.

See page 243.

Baley, John D. and Benesch, Mary P. *A System for Individualized Math Instruction in Secondary Schools*. July 1969. ERIC Document No. ED 050 555.

Multilevel team teaching with individualized instruction produced higher achievement in computational skills than did traditional methods.

Ballew, Julius Hunter. A Study of the Effect of Discovery Learning upon Critical Thinking Abilities of High School Algebra Students. (University of North Carolina at Chapel Hill, 1965.) *Dissertation Abstracts* 26: 3775; January 1966.

A major portion of the classwork in two algebra classes was devoted to the use of discovery exercises; another class was taught by an expository approach. Significant improvement on a test of critical thinking abilities was found in the discovery classes but not in the expository class, though no difference in mathematical achievement scores was found.

Barcaski, Peter Bartholomew. Cognitive and Motivational Factors, Methods of Teaching, and Their Effects on Achievement, Transfer, and Retention of Pre-Sequenced Concepts. (Columbia University, 1969.) *Dissertation Abstracts International* 31A: 639; August 1970.

The solved-examples-to-concepts method resulted in highest achievement, the concepts-and-solved-examples method was next, and the concepts-and-unsolved examples method was poorest, for ninth graders.

Bassler, Otto C.; Curry, Dick; Hall, Wayne; and Mealy, Ed. An Investigation of Two Instructional Variables in Learning Nonmetric Geometry. *School Science and Mathematics* 71: 441–450; May 1971.

No significant differences were found for groups of seventh graders receiving one, three, or five exercises per learning set for either an informal inductive or formal deductive mode.

Beaton, Mary Anne. A Study of Underachievers in Mathematics at the Tenth Grade Level in Three Calgary High Schools. (Northwestern University, 1966.) *Dissertation Abstracts* 27A: 3215–3216; April 1967.

Underachievers were found to have lower interest and attitude scores than achievers of comparable ability. A significantly greater number of the parents of tenth-grade underachievers indicated that they liked mathematics least of all school subjects, while parents of achievers considered mathematics to the twelfth-grade level important to students today.

Bechtold, Charles August. The Use of Extraneous Material in Developing Problem-Solving Ability of Pupils in Algebra I. (Columbia University, 1965.) *Dissertation Abstracts* 26: 3105; December 1965.

Instruction on solving problems with extraneous data resulted in better problem-solving scores than did instruction with problems having no extraneous data.

Becker, Jerry Page. An Attempt to Design Instructional Techniques in Mathematics to Accommodate Different Patterns of Mental Ability. (Stanford University, 1967.) *Dissertation Abstracts* 28A: 957–958; September 1967.

Ninth graders were assigned programs which used an expository or guided-discovery format, after determining their mathematical and verbal aptitudes. No significant interaction effects were found between either type of aptitude and the type of program.

Beckmann, Milton W. Ninth Grade Mathematical Competence—15 Years Ago and Now. *School Science and Mathematics* 69: 315–319; April 1969.

Students at the beginning of grade 9 in 1965 scored as well as those at the end of grade 9 in 1951.

Beckmann, Milton William. Eighth Grade Mathematical Competence—15 Years Ago and Now. *Arithmetic Teacher* 17: 334–335; April 1970.

Mean score on a 109-item test was 45.7 in 1951 and 54.9 in 1965.

Bell, Max S. Studies with Respect to the Uses of Mathematics in Secondary School Curricula. (The University of Michigan, 1969.) *Dissertation Abstracts International* 30A: 3813–3814; March 1970.

See page 7.

Bernstein, Allen. A Study of Remedial Arithmetic Conducted with Ninth Grade Students. *School Science and Mathematics* 56: 25–31; January 1956 and 429–437; June 1956.

Eighty percent of the errors with fundamentals made by ninth graders were in three categories: the use of zero in multiplication and division, borrowing in subtraction, and understanding of the decimal point in all four operations.

Beul, Bobbie Thatcher. An Evaluative Study of Teaching Seventh-Grade Mathematics Incorporating Team Teaching, Individualized Instruction and Team Supervision Utilizing the Strategy of Learning for Mastery. (Saint Louis University, 1973.) *Dissertation Abstracts International* 34A: 4685; February 1974.

A significant difference in achievement favored the group having the individualized program over the traditionally-taught group.

Bierden, James E. Behavioral Objectives and Flexible Grouping in Seventh Grade Mathematics. *Journal for Research in Mathematics Education* 1: 207–217; November 1970.

A combination of whole-class instruction and flexible intraclass grouping based on the achievement of behavioral objectives resulted in significant gains in computational skills, concept knowledge, and attitudes, as well as a reduction in anxiety.

Birr, Donald James. The Effects of Treatments by Parents and Teachers on the Self-

Concept of Ability Held by Underachieving Early Adolescent Pupils. (Michigan State University, 1969.) *Dissertation Abstracts International* 30A: 1354; October 1969.

> Self-concept and grade-point-average were not found to be related for seventh and eighth graders, but a significant correlation was found between the student's self-concept of ability and his parents' perception of his ability.

Bobier, Darold Thomas. The Effectiveness of the Independent Use of Programmed Textbooks in Aiding Students to Overcome Skill Weaknesses in English Mechanics and Arithmetic. (University of Denver, 1964.) *Dissertation Abstracts* 25: 3424–3425; December 1964.

> It was concluded that low-achieving students of limited ability were not sufficiently motivated to use programmed textbooks in basic skills on an independent study basis.

Boyd, Alvin Lyle. Computer Aided Mathematics Instruction for Low Achieving Students. (University of Illinois at Urbana-Champaign, 1972.) *Dissertation Abstracts International* 34A: 553; August 1973.

> No significant differences in achievement were found between groups of low-achieving students who used or did not use the computer to do calculations, solve problems, or execute simulations. The noncomputer group scored significantly higher on three of four attitude measures.

Bree, David Sidney. The Understanding Process as Seen in Geometry Theorems. (Carnegie-Mellon University, 1969.) *Dissertation Abstracts International* 30A: 1675; November 1969.

> Two "profiles" from the same student of the pattern followed while thinking aloud geometry proofs were more similar than two from different students. "Understanding" appeared to be composed of the simpler processes of consolidation, rephrasing, explaining, and predicting the steps of the solution.

Bright, George W. and Carry, L. Ray. The Influence of Professional Reference Groups on Decisions of Preservice Secondary School Mathematics Teachers. *Journal for Research in Mathematics Education* 5: 87–97; March 1974.

> See pages 344–346.

Brinke, Dirk Pieter Ten [sic]. Homework: An Experimental Evaluation of the Effect on Achievement in Mathematics in Grades Seven and Eight. (University of Minnesota, 1964.) *Dissertation Abstracts* 27A: 4176; June 1967.

> Seventh- and eighth-grade classes did not benefit more from homework than from supervised study. There was some indication that homework was more productive for upper-ability students while supervised study was better for low-ability students.

Brown, John Kenneth, Jr. Textbook Use by Teachers and Students of Geometry and Second-Year Algebra. (University of Illinois at Urbana-Champaign, 1973.) *Dissertation Abstracts International* 34A: 5795–5796; March 1974.

> See page 188.

Buchalter, Barbara Diane Elpern. The Validity of Mathematics Textbook Series in Grades 7–14 with Structure as an Objective. (University of Arizona, 1968.) *Dissertation Abstracts International* 30A: 198–199; July 1969.

> From an analysis of textbooks for grades 7–14, it was noted that the structure of mathematics was presented more often at the two lowest levels of cognitive learning—knowledge and comprehension—than at the four highest levels.

Buchman, Aaron L. Some Relationships Between Length of Courses in Elementary Algebra and Student Characteristics. (State University of New York at Albany, 1972.) *Dissertation Abstracts International* 33A: 2812; December 1972.

>   Studied were low achievers in ninth grade from schools providing only a two-semester algebra course, a three-semester course for low achievers and "slow workers," or a four-semester course. No differences in achievement were found, but some affective aspects were better in the lengthened courses.

Bull, Scott Spragg. A Comparison of the Achievement of Geometry Students Taught by Individualized Instruction and Traditional Instruction. (Arizona State University, 1971.) *Dissertation Abstracts International* 31A: 4616–4617; March 1971.

>   An individualized geometry program was studied in which the student paced his own learning, chose learning experiences to attain teacher-established objectives, and took tests when he felt prepared, with the teacher primarily helping individuals and small groups. The mean score of classes taught by the individualized method was significantly higher than that of classes taught by the traditional method.

Bunch, Martha Anne. A Study of the Effects on Retention and on the Problem-Solving Ability of Students When Geometry Is Used as an Aid in Teaching Factoring of Second-Degree Polynomials. (University of Missouri-Kansas City, 1972.) *Dissertation Abstracts International* 34A: 1057–1058; September 1973.

>   Insertion of geometrical interpretations into algebraic instruction did not significantly affect problem-solving performance in grade 8.

Bundrick, Charles Michael. A Comparison of Two Methods of Teaching Selected Topics in Plane Analytic Geometry. (Florida State University, 1968.) *Dissertation Abstracts International* 30A: 485–486; August 1969.

>   Students using a vector approach in Algebra II achieved significantly higher than those using a traditional approach.

Burbank, Irvin Kimball. Relationships Between Parental Attitude Toward Mathematics and Student Attitude Toward Mathematics, and Between Student Attitude Toward Mathematics and Student Achievement in Mathematics. (Utah State University, 1968.) *Dissertation Abstracts International* 30A: 3359–3360; February 1970.

>   Parental attitudes were significantly correlated with students' mathematics attitudes. Students' attitudes correlated with achievement in mathematical reasoning, concepts, and computation.

Burgess, Ernest Edward. A Study of the Effectiveness of the Planned Usage of Mathematical Games on the Learning of Skills and Concepts and on the Attitude Toward Mathematics and the Learning of Mathematics of Low Achieving Secondary Students. (Florida State University, 1969.) *Dissertation Abstracts International* 30A: 5333–5334; June 1970.

>   The regular use of mathematical games resulted in significantly different attitude scores, but no substantial relationships were found between attitude and achievement or ability.

Burke, Gerald Clayton. Case Studies of High School Students Using Physical Models to Study Mathematical Systems. (Michigan State University, 1973.) *Dissertation Abstracts International* 34A: 5574; March 1974.

>   Ten above-average students successfully used activity lessons for intermediate algebra.

Callahan, Walter J. Adolescent Attitudes Toward Mathematics. *Mathematics Teacher*
64: 751–755; December 1971.

> Twenty percent of the eighth graders surveyed felt that they disliked mathe-
> matics, 18 percent were neutral, and 62 percent liked it. The need for mathe-
> matics in life was named most frequently as the reason for liking it; not being
> good in mathematics was cited most often as the reason for disliking it.

Campbell, Azzie Leely. A Comparison of the Effectiveness of Two Methods of Class
Organization for the Teaching of Arithmetic in Junior High School. (Pennsylvania
State University, 1964.) *Dissertation Abstracts* 26: 813–814; August 1965.

> No significant difference in achievement was found between groups in grade 7
> having whole-class instruction or grouping within the class.

Cech, Joseph Phillip. *The Effect the Use of Desk Calculators Has on Attitude and Achievement in Ninth-Grade General Mathematics Classes*. June 1970. ERIC Document
No. ED 041 757.

> No significant differences in scores on tests of attitude and computational
> skills were found when desk calculators were used to check paper-and-pencil
> computation for seven weeks in grade 9.

Cohen, Martin Seymour. A Comparison of Effects of Laboratory and Conventional
Mathematics Teaching upon Underachieving Middle School Boys. (Temple University, 1970.) *Dissertation Abstracts International* 31A: 5026–5027; April 1971.

> Disadvantaged low achievers in grades 7 and 8 who used a conventional text-
> book/chalkboard/discussion approach had a significant increase in achieve-
> ment, when compared with a group taught using a laboratory approach with
> a variety of manipulative and multisensory materials.

Collins, Kenneth Michael. An Investigation of the Variables of Bloom's Mastery Learning
Model for Teaching Junior High School Mathematics. (Purdue University, 1971.)
Dissertation Abstracts International 32A: 3149; December 1971.

> Use of either a list of specific objectives or diagnostic-progress tests was
> sufficient for a significant increase in mastery of objectives by seventh graders.
> Eighth graders also profited from the use of alternative resources.

Computer System Helps Motivate High School Students. Application Note 145-7, Hewlett-Packard Corporation, March 1972.

> See page 289.

Cooney, Thomas J. and Henderson, Kenneth B. Ways Mathematics Teachers Help
Students Organize Knowledge. *Journal for Research in Mathematics Education*
3: 21–31; January 1972.

> This study was an attempt to identify methods of instruction which prove
> effective in helping students to structure their knowledge—that is, to organize
> in a meaningful way the concepts, facts, and principles they learn. From
> audiotapes of 44 instances of mathematics teaching by ten teachers in grades
> 7–12, they identified nine organizing relations: set membership, set inclusion,
> analysis, specifying, characterizing, explaining, implicating, generalizing, and
> abstracting.

Coppedge, Floyd L. and Hanna, Gerald S. Comparison of Teacher-Written and Empirically Derived Distractors to Multiple-Choice Test Questions. *Journal for Research
in Mathematics Education* 2: 299–303; November 1971.

> Students were administered a test in completion format; teachers were asked
> to generate three distractors to be used if the item were in a multiple-choice

format. There was much variability in teachers' ability to provide the most discriminating distractors, and to differentiate popular distractors from highly discriminating distractors. (See pages 361–363.)

Crosswhite, F. Joe. *Correlates of Attitudes Toward Mathematics.* NLSMA Report No. 20. Stanford, California: School Mathematics Study Group, 1972.

Many findings related to attitude are presented in this report on the analysis of data from NLSMA. The data indicate that attitudes tend to be less positive as students progress through secondary school.

Denmark, Ewell Thomas, Jr. A Comparative Study of Two Methods of Teaching Elementary Algebra Students to Use the Algebraic Technique to Solve Verbal Problems. (Florida State University, 1965.) *Dissertation Abstracts* 25: 5295–5296; March 1965.

Use of tables in organizing the information presented in a problem helped pupils to develop equations for solving algebraic verbal problems.

DeVenney, William S. *Final Report of an Experiment with Junior High School Very Low Achievers in Mathematics.* SMSG Reports, No. 7. Stanford, California: Stanford University, 1969. ERIC Document No. ED 042 630.

A program was developed with low-achieving seventh and eighth graders, incorporating daily worksheets, partially programmed lessons, and the use of tables to aid in computation. Students using these materials did significantly better on most SMSG tests and on attitude scales.

Dodson, Joseph Wesley. *Characteristics of Successful Insightful Problem Solvers.* NLSMA Report No. 31. Stanford, California: School Mathematics Study Group, 1972.

See page 151.

Drake, Richard. A Comparison of Two Methods of Teaching High School Algebra. *Journal of Educational Research* 29: 12–16; September 1935.

Students in grade 9 taught by a group method involving individual study achieved significantly higher scores than those doing individual work at their own rates.

Durall, Edwin Phillip. A Feasibility Study: Remediation by Computer Within a Computer-Managed Instruction Course in Junior High School Mathematics. (Florida State University, 1972.) *Dissertation Abstracts International* 33A: 2611–2612; December 1972.

Approximately 70 seventh-grade students worked in self-instructional booklets for 15 weeks; upon completion of each booklet, the student was evaluated by direct contact with a computer through teletype terminals. If the criterion of 80 percent was not attained, half of the students received remediation from an instructional sequence programmed into the computer, while the other half received remediation from their teacher. Both groups achieved comparably, but remediation from the teacher appeared to be more supportive for low-ability students.

Durrance, Victor Rodney. The Effect of the Rotary Calculator on Arithmetic Achievement in Grades Six, Seven, and Eight. (George Peabody College for Teachers, 1964.) *Dissertation Abstracts* 25: 6307; May 1965.

Used only in class for nine weeks in grades 6–8, calculators had no effect on achievement except for seventh-grade reasoning scores, nor did they affect correction of errors.

Dutton, Wilbur H. Attitudes of Junior High School Pupils Toward Arithmetic. *School Review* 64: 18–22; January 1956.

> From a study of attitudes in grades 7–9, it was reported that extreme dislike for mathematics was shown by the responses of a significant number of students (19 percent). Most students (87 percent) enjoyed problems when they knew how to work them, however, and the majority felt that arithmetic was as important as any other subject (83 percent). Reasons for liking mathematics included the practical aspects of the subject, the realization that it will be needed, and the enjoyment and challenge. Dislike centered on lack of understanding, difficulty in working problems, poor achievement, and its boring aspects.

Dutton, Wilbur H. Another Look at Attitudes of Junior High School Pupils Toward Arithmetic. *Elementary School Journal* 68: 265–268; February 1968.

> In a comparison of 1956 and 1966 attitudes of junior high school students, a slightly favorable change was found for the more recent group.

Dutton, Wilbur H. and Blum, Martha Perkins. The Measurement of Attitudes Toward Arithmetic with a Likert-Type Test. *Elementary School Journal* 68: 259–264; February 1968.

> About 30 percent of the students studied in grades 6–8 had very favorable attitudes toward modern mathematics, 53 percent were neutral, and 17 percent disliked the subject a great deal.

Ebert, Reuben S. Generalization Abilities in Mathematics. *Journal of Educational Research* 39: 671–681; May 1946.

> Students wrote mathematical patterns more easily from observed patterns than from sentence statements. Generalization achievement differed for different mental-ability and reading-ability eighth graders. The writing of general truths or facts in sentence statements was far more difficult than writing mathematical illustrations or relationships.

Edwards, Keith J. and DeVries, David L. *Learning Games and Student Teams: Their Effects on Student Attitudes and Achievement*. Baltimore: Johns Hopkins University, 1972. ERIC Document No. ED 072 391.

> See page 67.

Eldredge, Garth Melvin. Expository and Discovery Learning in Programed Instruction. (University of Utah, 1965.) *Dissertation Abstracts* 26: 5863; April 1966.

> A guided discovery program was more effective than an expository program. It was concluded that how materials are sequenced has an effect on learning.

Elkind, D. Quantity Concepts in Junior and Senior High School Students. *Child Development* 32: 551–560; 1961.

> Of the students tested in grades 7–12, 87 percent had abstract concepts of mass and weight; only 47 percent had an abstract concept of volume. More boys than girls had attained the volume concept.

Ellingson, James B. Evaluation of Attitudes of High School Students Toward Mathematics. (University of Oregon, 1962.) *Dissertation Abstracts* 23: 1604; November 1962.

> A significant positive correlation of attitudes toward mathematics was found, for students in grades 7–12, with teacher ratings of the students' attitudes and with achievement test scores.

Ellis, Dale Huband. An Analysis of Achievement Gains in Mathematics Classes Which Result from the Use of Student Tutors. (University of Utah, 1971.) *Dissertation Abstracts International* 32A: 1976–1977; October 1971.

> For classes in grades 9–11 in which above-median students tutored below-median students, greater achievement gains were made than in classes in which tutoring was not used.

Everett, Douglas Lavelle. The Effects of Tutoring on Achievement in and Attitude Toward Plane Geometry by Second Semester Tenth Grade Students. (University of Southern Mississippi, 1972.) *Dissertation Abstracts International* 33A: 1352; October 1972.

> No significant differences in achievement or attitude were found for students tutored or not tutored. Ability and achievement appeared to be positively related.

Farley, Sister Mary de Chantal. A Study of Mathematical Interests, Attitudes, and Achievements of Tenth and Eleventh Grade Students. (University of Michigan, 1968.) *Dissertation Abstracts* 29A: 3039–3040; March 1969.

> Boys' attitudes toward mathematics were more positive than girls' attitudes, with the difference more pronounced in grade 11 than in grade 10.

Fennema, Elizabeth H. Models and Mathematics. *Arithmetic Teacher* 19: 635–640; December 1972.

> From a review of research on the role of materials, it was concluded that research appears to indicate that the ratio of concrete to symbolic models used to convey mathematical ideas should reflect the developmental level of the learner. Thus, alternative models should be available so the learner can select the most meaningful one for him or her.

Fennema, Elizabeth. Mathematics Learning and the Sexes: A Review. *Journal for Research in Mathematics Education* 5: 126–139; May 1974.

> Thirty-eight studies concerned with sex differences in mathematics achievement were reviewed. No significant differences between boys' and girls' achievement were found before entry into elementary school or during the early elementary years. In upper elementary and early high school years significant differences were not always apparent. When significant differences did appear they were more apt to be in the boys' favor when higher-level cognitive tasks were being measured and in the girls' favor when lower-level cognitives were being measured. No conclusion was reached concerning high school learners.

Fey, James Taylor. *Patterns of Verbal Communication in Mathematics Classes.* New York: Teachers College Press, Columbia University, 1970.

> Tape-recorded lessons were analyzed to develop a profile of verbal activity in certain classes, with patterns described through the use of an instrument identifying interaction components. (See also page 97.)

Fitzgerald, William M. *Self-Selected Mathematics Learning Activities.* 1965. ERIC Document No. ED 003 348.

> In a study with students in grades 7 and 8, bright students (those with I.Q.s of 115 and over) did not learn as much in self-selection classes as in conventional classes. Those with I.Q.s below 115 learned equally well in both classes.

Flaherty, Eileen Gertrude. Cognitive Processes Used in Solving Mathematical Problems.

(Boston University School of Education, 1973.) *Dissertation Abstracts International* 34A: 1767; October 1973.

See page 107.

Foster, Thomas Edward. The Effects of Computer Programming Experiences on Student Problem Solving Behaviors in Eighth Grade Mathematics. (University of Wisconsin, 1972.) *Dissertation Abstracts International* 33A: 4239–4240; February 1973.

See page 306.

Fox, Lynn Hussey. Facilitating the Development of Mathematical Talent in Young Women. (The Johns Hopkins University, 1974.) *Dissertation Abstracts International* 35B: 3553; January 1975.

A special accelerated Algebra I program for seventh-grade girls was found to be more effective than two coeducational accelerated classes. At the mid-point in grade 8, those in the all-girls class were more knowledgeable about algebra than those in a traditional program.

Fredstrom, Paul Norman. An Evaluation of the Accelerated Mathematics Program in the Lincoln, Nebraska Public Schools. (University of Nebraska Teachers College, 1964.) *Dissertation Abstracts* 25: 5628–5629; April 1965.

Accelerated students in grade 12 achieved nearly as well as their older course-peers on tests, but had lower mathematics grade averages and less positive attitudes.

Friedman, Morton Lawrence. The Development and Use of a System to Analyze Geometry Teachers' Questions. (Columbia University, 1972.) *Dissertation Abstracts International* 33A: 4215–4216; February 1973.

See page 97.

Gadske, Richard Edward. A Comparison of Two Methods of Teaching First Year High School Algebra. *School Science and Mathematics* 33: 635–640; June 1933.

Students taught by an individualized procedure achieved significantly more than those taught as a group in grade 9.

Gaslin, William L. A Comparison of Achievement and Attitudes of Students Using Conventional or Calculator-Based Algorithms for Operations on Positive Rational Numbers in Ninth-Grade General Mathematics. *Journal for Research in Mathematics Education* 6: 95–108; March 1975.

Use of units in which fractional numbers were converted to decimals and examples and then solved on a calculator was found to be a "viable alternative" to use of conventional textbooks (including fractions) with or without a calculator, for low-ability or low-achieving students in grade 9.

Gawronski, Jane Donnelly. Inductive and Deductive Learning Styles in Junior High School Mathematics: An Exploratory Study. *Journal for Research in Mathematics Education* 3: 239–247; November 1972.

Eighth-grade students were classified as having inductive or deductive learning styles; each used programs developed inductively or deductively. No significant differences in achievement were found between groups, whether or not they had programs compatible with their learning style.

Gay, Lorraine R. Temporal Position of Reviews and Its Effect on the Retention of Mathematical Rules. *Journal of Educational Psychology* 64: 171–182; April 1973.

See page 64.

Goldberg, Miriam L. et al. *A Comparison of Mathematics Programs for Able Junior High*

School Students, Volume I—Final Report. May 1966. ERIC Document Nos. ED 010 056; ED 010 057.

> Acceleration resulted in greater achievement than did enrichment in grades 7–9.

Golledge, Margaret Ruth. The Development of Piaget-Type Formal and Concrete Reasoning. (University of Iowa, 1966.) *Dissertation Abstracts* 27A: 673–674; September 1966.

> Many students below age 16 had not mastered either formal or concrete reasoning, although improvement was evident with age. The formal reasoning scores progressed in a way consistent with Piaget's theory, but concrete reasoning items appeared to be more difficult than he described.

Gregory, John W. and Osborne, Alan R. Logical Reasoning Ability and Teacher Verbal Behavior Within the Mathematics Classroom. *Journal for Research in Mathematics Education* 6: 26–36; January 1975. (See also ERIC Document No. ED 064 178.)

> Twenty teachers and their seventh-grade classes were studied. One of each teacher's classes was audiotaped five times, and a reasoning test was administered to students at the beginning and end of the semester. The teachers were ranked on the basis of analysis of the frequency of their conditional moves: that is, how often they used "if-then" language in their teaching. Students of teachers who more frequently used such language outperformed students of teachers who made fewer such statements, on the reasoning tests.

Guiler, Walter Scribner. Difficulties in Fractions Encountered by Ninth-Grade Pupils. *Elementary School Journal* 46: 146–156; November 1945.

> Weaknesses in addition with fractions were demonstrated by 23 percent of the students, while approximately 40 percent had difficulty with each of the other operations with fractions. Faulty computation was a major source of error, as were changing fractions to a common denominator, lack of understanding of the process, use of the wrong process, "borrowing," and changing mixed numbers to improper fractions.

Guiler, Walter Scribner. Difficulties in Decimals Encountered by Ninth-Grade Pupils. *Elementary School Journal* 46: 384–393; March 1946.

> Almost 7 percent of the pupils had difficulty with multiplication with decimals; 33 percent, with addition and subtraction with decimals; 61 percent, with changing fractions to decimals and with division with decimals. Lack of understanding procedures and faulty computation were the chief problems.

Guiler, Walter Scribner. Difficulties in Percentage Encountered by Ninth-Grade Pupils. *Elementary School Journal* 46: 563–573; June 1946.

> Over 51 percent of the students had difficulty finding a percent of a number; over 47 percent, finding what percent one number is of another; 94 percent, finding a number when a percent of it is known; over 72 percent, finding the result of a percent increase or decrease; and over 88 percent, finding a percent of increase or decrease.

Guiler, W. S. and Edwards, Vernon. An Experimental Study of Methods of Instruction in Computational Arithmetic. *Elementary School Journal* 43: 353–360; February 1943.

> Diagnosis of difficulties and individualized group instruction for the needs of students in grades 7 and 8 resulted in greater gain than for pupils who did not have such help.

Haggard, Ernest A. Socialization, Personality, and Academic Achievement in Gifted Children. *School Review* 65: 388–414; December 1957.

> Specific characteristics of a group of high achievers in mathematics in grades 3–9 were identified. They tended to view their environment with curiosity, felt capable, had well-developed and healthy egos, could express feelings freely, were emotionally controlled and flexible, and showed independence of thought.

Hambleton, Ronald K. and Traub, Ross E. The Effects of Item Order on Test Performance and Stress. *Journal of Experimental Education* 43: 41–46; Fall 1974.

> Mathematics tests items arranged in easy-to-difficult order resulted in higher scores for the eleventh graders tested. Item order also had an effect on the stress generated during the test.

Hanna, Gerald S. A Summary of the Literature of Geometry Prediction with Emphasis upon Methodology and Theory. *School Science and Mathematics* 66: 723–728; November 1966.

> Results of studies on geometry prediction were summarized. It was noted that there are limits to the usefulness of such factors. In most studies, the nature of the geometry course was totally ignored.

Hanna, Gerald S. Testing Students' Ability To Do Geometric Proofs: A Comparison of Three Objective Item Types. *Journal for Research in Mathematics Education* 2: 213–217; May 1971.

> Multiple-choice items in which tenth-grade students selected (1) what was given and what was proved or (2) the "reason," were recommended over items which merely required the student to note whether a statement could be proved.

Hansen, Viggo Peter. Elementary Algebra Achievement as Related to Class Length and Teaching Method. (University of Minnesota, 1962.) *Dissertation Abstracts* 24: 198; July 1963.

> The effects of (1) lengthening the class period from 55 minutes to 110 minutes, but meeting on alternate school days only, and (2) using extended class discussions, a mathematics laboratory, library reading and research, class reports, and more instructional aids were studied. Achievement and attitudes were not different from those of students in the daily 55-minute classes, but results tended to favor the lengthened period.

Harris, Jasper William. An Analysis of the Effects of Using Quizzes and Modified Teaching Procedures to Increase the Unit Test Scores in Geometry, Algebra, and French Classes in an Inner City Senior High School. (University of Kansas, 1972.) *Dissertation Abstracts International* 33A: 2648; December 1972.

> For four geometry and algebra classes, prescribed content with set daily goals, feedback, and systematic reinforcement increased achievement in each course.

Hatfield, Larry L. and Kieren, Thomas E. Computer-Assisted Problem Solving in School Mathematics. *Journal for Research in Mathematics Education* 3: 99–112; March 1972.

> Use of computer programming as a problem-solving tool was especially helpful for average and above-average students in grade 7; in grade 11, it appeared best for average achievers.

Henderson, Kenneth B. and Rollins, James H. A Comparison of Three Stratagems for

Teaching Mathematical Concepts and Generalizations by Guided Discovery. *Arithmetic Teacher* 14: 503–508; November 1967.

Three inductive stratagems were found to be effective in teaching concepts and generalizations to eighth graders.

Hendrix, Gertrude. A New Clue to Transfer of Training. *Elementary School Journal* 48: 197–208; December 1947.

The effect of teaching for nonverbalized generalization was discussed, citing a set of studies with various groups of students.

Hernandez, Norma G. A Model of Classroom Discourse for Use in Conducting Aptitude-Treatment Interaction Studies. *Journal for Research in Mathematics Education* 4: 161–169; May 1973.

Great variability was noted among four eighth-grade teachers in the cognitive content of their statements. Memory was the most frequently coded cognitive process; little convergent and almost no divergent discourse was found. The semantic mode was the most frequently used.

Herriot, Sarah T. *The Slow Learner Project: The Secondary School "Slow Learner" in Mathematics.* SMSG Reports, No. 5. Stanford, California: Stanford University, 1967. ERIC Document No. ED 021 755.

Students in grades 7 and 9 who were classified as slow learners studied materials for two years. They achieved a greater gain than a high-ability group achieved in one year.

Hess, Robert D. and Tenezakis, Maria D. Selected Findings from "The Computer as a Socializing Agent: Some Socioaffective Outcomes of CAI." AV *Communication Review* 3: 311–325; Fall 1973.

Significant mean differences indicated that students who had used a remedial drill-and-practice program in basic arithmetic for one or two years regarded the computer in more positive terms than the teacher did. Non-CAI students also regarded the computer significantly more favorably: they had a less favorable image of the teacher than did the CAI group. For both CAI students and non-CAI students, the computer had a more favorable image than did the teacher, textbooks, and television news.

Hirsch, Christian Richard, Jr. An Experimental Study Comparing the Effects of Guided Discovery and Individualized Instruction on Initial Learning, Transfer, and Retention of Mathematical Concepts and Generalizations. (University of Iowa, 1972.) *Dissertation Abstracts International* 33B: 3194–3195; January 1973.

Significant differences in achievement and transfer were found for second-year algebra students, favoring the use of a guided-discovery treatment on complex numbers, when compared with individualized instruction packages using an expository or a branching format.

Hoffman, Nathan. Geometry in Mathematics: A Survey of Some Recent Proposals for the Content of Secondary School Geometry. (University of Montana, 1973.) *Dissertation Abstracts International* 34A: 3026; December 1973.

Recent proposals for the content of secondary school geometry were surveyed. It was concluded that geometry should be taught as an integrated course including vector, transformation, and coordinate methods. It should not be primarily a vehicle for teaching axiomatics. No justification was found for teaching geometry (and no other mathematics) in grade 10.

Hoffmann, Joseph Raymond, Jr. A Heuristic Study of Key Teaching Variables in Junior

High School Mathematics Classrooms. (University of Illinois at Urbana-Champaign, 1972.) *Dissertation Abstracts International* 34A: 665; August 1973.

> From recorded observations of 35 junior high school mathematics teachers, teacher-student transactions were identified.

Howitz, Thomas Allen. The Discovery Approach: A Study of Its Relative Effectiveness in Mathematics. (University of Minnesota, 1965.) *Dissertation Abstracts* 26: 7178–7179; June 1966.

> No significant differences were found between groups using expository or discovery-oriented textbooks on a standardized test, but the discovery-oriented group scored significantly higher on a nonstandardized test.

Hudson, James Alfred. A Pilot Study of the Influence of Homework in Seventh Grade Mathematics and Attitudes Toward Homework in the Fayetteville Public Schools. (University of Arkansas, 1965.) *Dissertation Abstracts* 26: 906; August 1965.

> The amount of homework assigned had no significant relationship to achievement on concepts, but may have influenced problem-solving scores.

Husén, Torsten, ed. *International Study of Achievement in Mathematics, Volumes I and II*. New York: Wiley, 1967.

> These volumes report on the intentions and background of the International Study. The hypotheses, the tests and scales, the findings, and the interpretations related to both achievement and attitudes in 12 countries are also included.

Jeffery, Jay M. Psychological Set in Relation to the Construction of Mathematics Tests. *Mathematics Teacher* 62: 636–638; December 1969.

> Students were found to develop a definite "set" toward problem solutions.

Johanson, Emma Jane Dixon. A Ninth Grade Piagetian Mathematics Curriculum. (University of Toledo, 1972.) *Dissertation Abstracts International* 33A: 223; July 1972.

> A nine-week curriculum for a ninth-grade class was developed, using apparatus and experiments which involved active manipulation, with game-playing, discussion, and children working in pairs or in small groups. The group taught with this curriculum scored higher on achievement and attitude measures than did a control group.

Johnson, Donovan A. A Study of the Relative Effectiveness of Group Instruction. *School Science and Mathematics* 56: 609–616; November 1956.

> No significant differences were found between twelfth-grade students taught in small groups by another student and regular teacher-led classes.

Johnson, Randall Erland. The Effect of Activity Oriented Lessons on the Achievement and Attitudes of Seventh Grade Students in Mathematics. (University of Minnesota, 1970.) *Dissertation Abstracts International* 32A: 305; July 1971.

> Use of activity-oriented lessons with seventh graders did not appear to be more effective than instruction with little or no emphasis on activities for units on number theory, geometry, measurement, and rational numbers, though activities did aid in the learning of some concepts by low- and middle-ability students.

Katz, Saul M. A Comparison of the Effects of Two Computer Augmented Methods of Instruction with Traditional Methods upon Achievement of Algebra Two Students in a Comprehensive High School. (Temple University, 1971.) *Dissertation Abstracts International* 32A: 1188–1189; September 1971.

Average-ability second-year algebra students who ran their own computer programs scored significantly lower than a group whose programs were run by aides or those who had regular instruction, on a full-year standardized test. On tests of only the topics that were related to computer-program-writing, there were no significant differences for any group.

Kennedy, George; Eliot, John; and Krulee, Gilbert. Error Patterns in Problem Solving Formulations. *Psychology in the Schools* 7: 93–99; January 1970.

See page 143.

Kennedy, Wallace A. and Walsh, John. A Factor Analysis of Mathematical Giftedness. *Psychological Reports* 17: 115–119; August 1965.

Mathematical ability appeared to be not a specific ability, but related to overall high ability.

Keough, John J. and Burke, Gerald W. *Utilizing an Electronic Calculator to Facilitate Instruction in Mathematics in the 11th and 12th Grades. Final Report.* July 1969. ERIC Document No. ED 037 345.

The group using calculators in grades 11 and 12 achieved significantly more than a group not using them.

Kester, Scott Woodrow. The Communication of Teacher Expectations and Their Effects on the Achievement and Attitudes of Secondary School Pupils. (University of Oklahoma, 1969.) *Dissertation Abstracts International* 30A: 1434–1435; October 1969.

See page 40.

Kilpatrick, Jeremy. Analyzing the Solution of Word Problems in Mathematics: An Exploratory Study. (Stanford University, 1967.) *Dissertation Abstracts* 28A: 4380; May 1968.

Students in grade 8 were asked to think aloud as they solved problems, and their answers were coded. Measures of quantitative ability, mathematics achievement, word fluency, general reasoning, and a reflective conceptual tempo were positively correlated with students' use of equations in solving word problems. Attitude toward mathematics was not correlated with the coded variables.

Kilpatrick, Jeremy and Wirszup, Izaak, eds. *Soviet Studies in the Psychology of Learning and Teaching Mathematics, Volumes I–XIV.* Stanford, California: School Mathematics Study Group, 1969, 1970, 1971, 1972, 1975.

These volumes contain translations of studies conducted in the Soviet Union. While many of them pertain to young children, implications for instruction may be pertinent to all teachers.

Kim, Sharon and Leton, Donald A. *Analysis of Mathematical Abilities Required for Success in Ninth-Grade Mathematics.* December 1966. ERIC Document No. ED 010 420.

From a survey of ninth-grade boys, it was concluded that mathematical ability is comprised of a number of aptitudes and is not simply a unitary trait.

King, Donald Thomas. An Instructional System for the Low-Achiever in Mathematics: A Formative Study. (University of Wisconsin, 1972.) *Dissertation Abstracts International* 32A: 6743; June 1972.

See page 306.

Kleckner, Lester Gerald. An Experimental Study of Discovery Type Teaching Strategies with Low Achievers in Basic Mathematics I. (Pennsylvania State University, 1968.) *Dissertation Abstracts International* 30A: 1075–1076; September 1969.

Nondiscovery classes of slow learners in grades 9 and 10 achieved significantly more than classes taught by discovery-type strategies in a mathematics laboratory setting.

Koopman, Norbert Earl. Evaluations by Superior High School Students of Their Problem Solving Performances. (University of Wisconsin, 1964.) *Dissertation Abstracts* 25: 3398; December 1964.

Boys made significantly more correct, as well as more incorrect, evaluations of their problem-solving accuracy than did girls, who were more unsure of their solutions. The sex difference on correct evaluations diminished and incorrect evaluations increased with age, though twelfth graders made more correct and fewer incorrect evaluations than did ninth graders.

Kort, Anthone Paul. Transformation vs. Non-Transformation Tenth-Grade Geometry: Effects on Retention of Geometry and on Transfer in Eleventh-Grade Mathematics. (Northwestern University, 1971.) *Dissertation Abstracts International* 32A: 3157–3158; December 1971.

For students in eleventh-grade mathematics classes, those who had studied tenth-grade geometry using a transformation approach showed some retention and transfer advantages over those students who had used a nontransformation approach in their study of tenth-grade geometry.

Kuhfittig, Peter K. The Relative Effectiveness of Concrete Aids in Discovery Learning. *School Science and Mathematics* 74: 104–108; February 1974.

Low-ability students who used concrete materials achieved better than low-ability students who did not use materials; no difference was found between high-ability students in grade 7. For intermediate-guidance groups, mean transfer-test scores for students using concrete aids were higher than scores of those not using aids; no difference was found for maximal-guidance students.

Kysilka, Marcella Louise. The Verbal Teaching Behaviors of Mathematics and Social Studies Teachers in Eighth and Eleventh Grades. (University of Texas at Austin, 1969.) *Dissertation Abstracts International* 30A: 2725; January 1970.

See page 97.

Lackner, Lois M. *The Teaching of Two Concepts in Beginning Calculus by Combinations of Inductive and Deductive Approaches. Final Report.* June 1968. ERIC Document No. ED 025 446.

A concrete inductive method (involving an example-to-rule method) resulted in better achievement with programmed calculus materials used for one semester with students in grades 11 and 12 than did an abstract deductive (rule-to-example) method.

Laing, Robert Andrew. Relative Effects of Massed and Distributed Scheduling of Topics on Homework Assignments of Eighth Grade Mathematics Students. (Ohio State University, 1970.) *Dissertation Abstracts International* 31A: 4625; March 1971.

See page 207.

Lankford, Francis G., Jr. *Some Computational Strategies of Seventh Grade Pupils. Final Report.* Charlottesville: The Center for Advanced Study, The University of Virginia, October 1972. ERIC Document No. ED 069 496.

The results of interviews with 176 students in grade 7 are presented. Frequency of right and wrong answers to examples for each operation with whole numbers and with fractions, strategies frequently used, the nature of wrong answers, and some characteristics of good and poor computers are presented.

Lash, Stark William Edward. A Comparison of Three Types of Homework Assistance for High School Geometry. (Temple University, 1971.) *Dissertation Abstracts International* 32A: 1984; October 1971.

Three modes of assistance used by students while doing geometry homework were studied. The group getting complete solutions achieved lower scores than groups getting hints or answers or no assistance.

Leake, Lowell, Jr. The Status of Three Concepts of Probability in Children of Seventh, Eighth and Ninth Grades. *Journal of Experimental Education* 34: 78–81; Fall 1965.

Students had a considerable knowledge of probability concepts such as the probability of a sample space, a simple event, or the union of two or more mutually exclusive events, before being taught them.

Leonard, Harold A. Difficulties Encountered by Elementary Algebra Students in Solving Equations in One Unknown—A Diagnosis of Errors and a Comparison After Forty Years. (Ohio State University, 1966.) *Dissertation Abstracts* 27A: 3778; May 1967.

Algebra students tested in 1966 obtained significantly better results in solving equations also attempted by a group in approximately 1926.

Lockwood, James Riley. An Analysis of Teacher-Questioning in Mathematics Classrooms. (University of Illinois at Urbana-Champaign, 1970.) *Dissertation Abstracts International* 31A: 6472–6473; June 1971.

Elements that would be helpful in explaining the question-asking behavior of teachers in the classroom were identified. Using audiotapes of 47 class sessions in grades 7–11 involving four carefully selected teachers, 16 cues (stimuli which act as signals to ask a question) and 17 factors (elements that have an influence on the question the teacher asks) were classified.

Loh, Elwood Lockert. The Effect of Behavioral Objectives on Measures of Learning and Forgetting on High School Algebra. (University of Maryland, 1972.) *Dissertation Abstracts International* 33A: 145; July 1972.

Use of behavioral objectives with two second-year algebra classes was investigated. Students who were informed of the behavioral objectives did not learn or retain better than students not informed of objectives.

Lorentz, Jerome Stephen. An Experimental Study of the Effects of Varying the Amount and Sequencing of Explanatory Materials in the Teaching of Formal Mathematical Definitions. (University of Georgia, 1970.) *Dissertation Abstracts International* 32A: 5853; November 1971.

For geometry content, materials in which explanation followed definition was not as effective as material developed in an explanation-definition-explanation pattern, explanation preceding definition, or definition alone.

Martin, Mavis Doughty. Reading Comprehension, Abstract Verbal Reasoning, and Computation as Factors in Arithmetic Problem Solving. (State University of Iowa, 1963.) *Dissertation Abstracts* 24: 4547–4548; May 1964.

Skills in reading, reasoning, process selection, and computation interacted and appeared crucial in the solution of problems in a verbal context.

Mastbaum, Sol. A Study of the Relative Effectiveness of Electric Calculators or Computational Skills Kits in the Teaching of Mathematics. (University of Minnesota, 1969.) *Dissertation Abstracts International* 30A: 2422–2423; December 1969.

Students in grades 7 and 8 learned to use the calculator to solve one-step computation problems, but this ability did not transfer to noncalculator situations. Neither achievement nor attitude was significantly improved.

Math Fundamentals: Selected Results from the First National Assessment of Mathematics. Mathematics Report No. 04-MA-01. Denver: National Assessment of Educational Progress, Education Commission of the States, January 1975.

See page 159.

Maxwell, Ann Alsobrook. An Exploratory Study of Secondary School Geometry Students: Problem Solving Related to Convergent-Divergent Productivity. (The University of Tennessee, 1974.) *Dissertation Abstracts International* 35A: 4987; February 1975.

Geometry students who scored high on divergent-type problems were found to make fewer generalizations and use trial-and-error methods more frequently. Trial-and-error played a major role initially in a problem-solving task, and a minor role as the solution was approached.

May, Daryle Cline. An Investigation of the Relationship Between Selected Personality Characteristics of Eighth-Grade Students and Their Achievement in Mathematics. (University of Florida, 1971.) *Dissertation Abstracts International* 33A: 555; August 1972.

Students were identified as "sensing" or "intuitive" personality type. The eighth-graders' scores on achievement and attitude measures were then compared. A significant difference in computation, concepts, and applications scores was found between sensing and intuitive types. No differences in attitudes toward mathematics were found. It was concluded that teachers should consider type of personality when planning instruction.

Maynard, Fred J. and Strickland, James F., Jr. *A Comparison of Three Methods of Teaching Selected Mathematical Content in Eighth and Ninth Grade General Mathematics Courses. Final Report.* Athens: University of Georgia, August 1969. ERIC Document No. ED 041 763.

No significant differences were found for boys in grades 8 and 9 who were taught by a nonverbalized discovery, a guided discovery, or an expository method for six weeks in general mathematics classes.

McCullouch, James Victor. The Effect of Using a Behavioral-Objectives Curriculum in Mathematics on the Achievement of Ninth-Grade Pupils in the Meridian Separate School District. (University of Alabama, 1970.) *Dissertation Abstracts International* 31A: 5114; April 1971.

The achievement of ninth-grade groups using a curriculum based on behavioral objectives and groups using a standard textbook program (which was also based on objectives, though presumably not behaviorally-stated objectives) was compared. No significant differences in achievement were found, though groups using the behavioral objectives curriculum made greater progress in arithmetic fundamentals and reasoning.

Meconi, L. J. Concept Learning and Retention in Mathematics. *Journal of Experimental Education* 36: 51–57; Fall 1967.

High-ability pupils in grades 8 and 9 learned and retained effectively the necessary concepts for problem-solving performance and retention regardless of whether they used rule-and-example, guided discovery, or rule-only programs.

Michael, R. E. The Relative Effectiveness of Two Methods of Teaching Certain Topics in Ninth Grade Algebra. *Mathematics Teacher* 42: 83–87; February 1949.

> Deductive procedures used in teaching a 45-day unit on integers in grade 9 produced significantly greater gains on a test of generalizations than did inductive procedures, while there were no significant differences on tests of computation and attitude. There was some evidence that those at higher I.Q. levels achieved more with the deductive procedure.

Miller, G. H. How Effective Is the Meaning Method? *Arithmetic Teacher* 4: 45–49; March 1957.

> The meaning method was found to be more effective for computation of fractions and for decimals and percentage; the rule method was superior for measurement. The meaning method seemed more effective for retention of computational processes, understanding of principles, and comprehension of complex analysis for seventh graders.

Moore, William J. and Cain, Ralph W. The New Mathematics and Logical Reasoning and Creative Thinking Abilities. *School Science and Mathematics* 68: 731–733; November 1968.

> Students using a "modern" program had significantly improved scores in logical reasoning, word fluency, and associational fluency.

Morrison, Roderick Ruel, Jr. A Study of the Effects of Departmental Organization on Academic Achievement in the Sixth and Seventh Grades. (University of Georgia, 1966.) *Dissertation Abstracts* 27A: 3270–3271; April 1967.

> Students in self-contained classes scored higher on reasoning and computation tests than those in departmentalized classes.

Neuhouser, David Lee. A Comparison of Three Methods of Teaching a Programed Unit on Exponents to Eighth Grade Students. (Florida State University, 1964.) *Dissertation Abstracts* 25: 5027; March 1965.

> See page 107.

Nibbelink, William Henry. The Use of an Anecdotal Style of Content Presentation as a Motivational and Instructional Device for Seventh Grade Under-Achievers in Mathematics. (Ohio State University, 1971.) *Dissertation Abstracts International* 32A: 3815; January 1972.

> The experimental booklets on counting and operations were found to be effective with both inner- and outer-city underachievers.

Nisbet, Jean Ann. Instructional Sequence for Improving Teacher Question-Asking in Secondary Mathematics. (Arizona State University, 1974.) *Dissertation Abstracts International* 35A: 2841; November 1974.

> See page 415.

Nix, George Carol. An Experimental Study of Individualized Instruction in General Mathematics. (Auburn University, 1969.) *Dissertation Abstracts International* 30A: 3367–3368; February 1970.

> Students in grade 8 with low I.Q.s, those with average mathematical ability, and boys achieved significantly more under individual instruction than under group-oriented instruction.

Norland, Charles R. Mathematics Achievement: Changes in Achievement Scores for Grades Six and Eight After Instruction in Modern Mathematics Programs for Four Years or More, 1969. (Northern Illinois University, 1971.) *Dissertation Abstracts International* 32A: 2363; November 1971.

Sixth- and eighth-grade groups who had five or more years of instruction using modern mathematics were compared with students who had instruction using primarily traditional materials. In general, students who had a traditional program scored significantly higher on computation tests, in six out of ten cases, than those who had a modern program. In only one case did significant differences favor the modern group. On tests of problem solving, the traditional groups were significantly higher in three of ten cases, and in only one case was the modern group significantly higher. In other cases, there were no significant differences.

Olley, Peter George. The Relative Efficacy of Four Experimental Protocols in the Use of Model Devices to Teach Selected Mathematical Constructs. (Washington State University, 1973.) *Dissertation Abstracts International* 34A: 4993; February 1974.

For transfer, use of concrete-to-abstract sequences were found to be preferable to pictorial-abstract or abstract sequences in teaching certain mathematical operations to seventh graders. No significant differences were found on retention tests.

Olson, Franklin Carl. The Effects of Pair Study on Student Attitude and Achievement in Plane Geometry. (University of Nebraska, 1971.) *Dissertation Abstracts International* 32A: 840–841; August 1971.

Achievement and attitude were not significantly different in geometry classes in which students studied in pairs or alone.

Ostheller, Karl Olney. The Feasibility of Using Computer-Assisted Instruction to Teach Mathematics in the Senior High School. (Washington State University, 1970.) *Dissertation Abstracts International* 31A: 4042; February 1971.

Seniors taught a unit on probability and statistics via a tutorial computer-assisted instruction program achieved as well as groups taught by programmed or regular textbooks. No significant differences were found in attitude, though students preferred student-teacher interaction to CAI.

Pack, Elbert Chandler. The Effect of Mode of Computer Operation on Learning a Computer Language and on Problem Solving Efficiency of College Bound High School Students. (University of California, Los Angeles, 1970.) *Dissertation Abstracts International* 31A: 6477; June 1971.

See page 303.

Paige, Donald D. Learning While Testing. *Journal of Educational Research* 59: 276–277; February 1966.

Immediate reinforcement after a testing situation resulted in significantly higher achievement scores for eighth graders.

Paige, Donald D. A Comparison of Team Versus Traditional Teaching of Junior High School Mathematics. *School Science and Mathematics* 67: 365–367; April 1967.

No significant differences in mathematical achievement, retention, or relearning ability were found for students taught by team teaching or by a single teacher.

Patterson, William Henry, Jr. The Development and Testing of a Discovery Strategy in Mathematics Involving the Field Axioms. (Florida State University, 1969.) *Dissertation Abstracts International* 30B: 5599; June 1970.

No significant differences were found between groups using a program on the field axioms with a discovery strategy in which the student received cues if she or he needed them and an expository strategy in which she or he was required to answer a set of questions.

Paulson, Casper F., Jr. *Slow Learners, Competition, and Programed Instruction*. August 31, 1964. ERIC Document No. ED 003 204.

Homogeneous groups scored higher than heterogeneous groups on algebra achievement measures. Heterogeneous groups having public display of scores achieved significantly higher than any of the other groups.

Pearl, Andrew Wilder. A Study of the Effects on Students' Achievement and Attitudes When They Work in Academic Teams of Three Members. (Cornell University, 1967.) *Dissertation Abstracts* 28A: 59–60; July 1967.

Students in grades 7 and 8 who worked in three-member teams achieved significantly better than those who worked individually.

Peterson, John Charles. Effect of Exploratory Homework Exercises upon Achievement in Eighth Grade Mathematics. (Ohio State University, 1969.) *Dissertation Abstracts International* 30A: 4339; April 1970.

See page 207.

Peterson, John Milo. A Comparison of Achievement in Traditional Mathematics Skills of Seventh Grade Students Using Three Different Types of Materials—Traditional, Transitional, and Modern. (Utah State University, 1966.) *Dissertation Abstracts* 27B: 2790–2791; February 1967.

Seventh graders using traditional materials achieved significantly lower in mechanical skills than did students using modern or transitional materials, while in application of skills those using transitional materials achieved the lowest scores.

Phillips, Robert Bass, Jr. Teacher Attitude as Related to Student Attitude and Achievement in Elementary School Mathematics. (University of Virginia, 1969.) *Dissertation Abstracts International* 30A: 4316–4317; April 1970.

See page 27.

Piatt, Robert George. An Investigation of the Effect the Training of Teachers in Defining, Writing and Implementing Educational Behavioral Objectives Has on Learner Outcomes for Students Enrolled in a Seventh Grade Mathematics Program in the Public Schools. (Lehigh University, 1969.) *Dissertation Abstracts International* 30A: 3352; February 1970.

See page 57.

Posamentier, Alfred S. Mathematical Achievement and Attitudinal Differences Among Students and Attitudinal Differences Among Teachers Under a Two-Semester and a Three-Semester Elementary Algebra Course. (Fordham University, 1973.) *Dissertation Abstracts International* 34A: 2279–2280; November 1973.

Achievement and attitude differences were found between low-achieving students in a two- or three-semester course in algebra, with some variability. The three-semester program was more effective in promoting achievement. Teacher preference was related to what they had taught.

Postlethwaite, T. N. International Association for the Evaluation of Educational Achievement (IEA)—The Mathematics Study. *Journal for Research in Mathematics Education* 2: 69–103; March 1971.

Procedures used in the IEA and data on tests and scales were presented. Among the many findings were: (1) age of entry into school was not an important variable in mathematics achievement, (2) type of school affected the achievement of 13-year-olds, and (3) correlations between achievement and attitude were small but positive. (Other articles in this *JRME* issue also review the IEA.)

Price, Jack. Discovery: Its Effects on Critical Thinking and Achievement in Mathematics. *Mathematics Teacher* 60: 874–876; December 1967.

    Tenth-grade general mathematics students had better reasoning and attitude scores when they used an inductive method requiring them to form generalizations rather than a deductive textbook-lecture method. Use of transfer materials resulted in a significant increase in critical thinking ability.

Prielipp, Robert Walter. The Effect of Textbook, Sex, and Setting of the Problem on the Ability of First Year Algebra Students to Recognize Three Properties of an Abelian Group. (University of Wisconsin, 1967.) *Dissertation Abstracts* 28A: 4545–4546; May 1968.

    In a study of the properties of an Abelian group with algebra students in grade 9, the commutative property was found to be the easiest of the properties studied, followed by the identity element and inverses.

Proctor, Charles McDavitt. An Experimental Study of the Relationship Between Certain Theoretically Postulated Elements in Classroom Learning and Student Achievement, Grade Distributions, and the Incidence of Certain Classroom Activities. (University of Maryland, 1967.) *Dissertation Abstracts* 28A: 4546; May 1968.

    A learning situation was studied in which (1) learning objectives were operationally clarified for students, (2) feedback was designed to provide teachers and students with knowledge as to the extent to which the student achieved those objectives, and (3) student achievement of the objectives was associated with marks assigned. While higher student achievement was associated with the use of operational objectives, classroom activities were not affected by the objectives.

Rappaport, David. Understanding Meanings in Arithmetic. *Arithmetic Teacher* 5: 96–99; March 1958.

    Students in grades 7 and 8 were not found to have an adequate understanding of meanings in arithmetic, assuming a score below 50 percent was inadequate. Computational skill was not an indication of the understanding of meanings of processes used in computation.

Ray, Willis E. Pupil Discovery vs. Direct Instruction. *Journal of Experimental Education* 29: 271–280; March 1961.

    Students in grade 9 taught by a discovery procedure retained and transferred more than those taught by an expository procedure. No significant interaction between ability and method was found.

Retzer, Kenneth A. and Henderson, Kenneth B. Effect of Teaching Concepts of Logic on Verbalization of Discovered Mathematical Generalizations. *Mathematics Teacher* 60: 707–710; November 1967.

    See page 107.

Reynolds, Philip Roger. Understanding Conditional Statements at the Tenth Grade Level. (University of Rochester, 1973.) *Dissertation Abstracts International* 34A: 125; July 1973.

    Students in Algebra II classes (before geometry) scored significantly higher on a test of conditional reasoning than did students in geometry classes (before Algebra II). For all students, scores were higher for test items written in the geometric rather than the algebraic content area.

Roberge, James J. A Study of Children's Abilities to Reason with Basic Principles of Deductive Reasoning. *American Educational Research Journal* 7: 583–596; November 1970.

Class reasoning was found to be significantly easier than conditional reasoning, though neither was consistently easier at all grade levels 4, 6, 8, and 10. Differences for content dimensions were significant: concrete-familiar was easiest, then suggestions, then abstract. Negation had a marked influence on the development of logical ability.

Robinson, G. Edith. An Investigation of Junior High School Students' Spontaneous Use of Proof to Justify Mathematical Generalizations. (University of Wisconsin, 1964.) *Dissertation Abstracts* 25: 2300; October 1964.

Three-fourths of the students in grades 7 and 9 gave at least one proof response. Most seventh graders could justify mathematical generalizations with a proof when the concepts were familiar to them.

Robitaille, David Ford. Selected Behaviors and Attributes of Effective Mathematics Teachers. (The Ohio State University, 1969.) *Dissertation Abstracts International* 30A: 1472–1473; October 1969.

See page 382.

Romberg, Thomas A. and Wilson, James W. The Development of Mathematics Achievement Tests for the National Longitudinal Study of Mathematical Abilities. *Mathematics Teacher* 61: 489–495; May 1968.

See page 161.

Schippert, Frederick Arthur. A Comparative Study of Two Methods of Arithmetic Instruction in an Inner-City Junior High School. (Wayne State University, 1964.) *Dissertation Abstracts* 25: 5162–5163; March 1965.

Students in an inner-city school had significantly higher achievement when they used a laboratory approach in which they manipulated actual models or representations of mathematical principles, than when they were taught with verbal or written descriptions of those principles.

Schultz, Margaret and Ohlsen, M. M. A Comparison of Traditional Teaching and Personalized Teaching in Ninth Grade Algebra. *Mathematics Teacher* 42: 91–96; February 1949.

Students in grade 9 taught with individual diagnosis and help achieved higher scores than those taught by expository procedures.

Scott, Allen Wayne. An Evaluation of Prescriptive Teaching of Seventh-Grade Arithmetic. (North Texas State University, 1969.) *Dissertation Abstracts International* 30A: 4696; May 1970.

Programmed materials on computation skills, selected to meet diagnosed needs, aided underachievers more than regular classroom practices.

Sederberg, Charles Herbert. A Comparison of Mathematics Teaching Methods for Average and Below-Average Ninth Grade Pupils. (University of Minnesota, 1964.) *Dissertation Abstracts* 25: 2384–2385; October 1964.

The general mathematics course resulted in higher computation and appeared to be better for low-ability ninth graders than a modified algebra course.

Sekyra, Francis, III. The Effects of Taped Instruction on Problem-Solving Skills of Seventh Grade Children. (University of Alabama, 1968.) *Dissertation Abstracts* 29A: 3473–3474; April 1969.

> Practice on problem solving using tape recordings resulted in improvement in the ability to extract and retrieve information, combine operations, and give correct responses.

Shaw, Judith A. *Reading Problems in Mathematics Texts.* August 1967. ERIC Document No. ED 016 587.

> Great internal variation was found in the reading level in textbooks for grades 1–8. In grade 7, high-ability texts had a fifth- to sixth-grade reading level, low-ability texts had a seventh-grade reading level, and middle-ability texts had a ninth- to tenth-grade reading level.

Sherrill, James M. The Effects of Different Presentations of Mathematical Word Problems upon the Achievement of Tenth Grade Students. *School Science and Mathematics* 73: 277–282; April 1973.

> Tenth graders having word problems with accurate pictorial representations scored significantly higher than students having no diagrams, who in turn scored significantly higher than those having distorted diagrams.

Shoecraft, Paul Joseph. The Effects of Provisions for Imagery Through Materials and Drawings on Translating Algebra Word Problems, Grades Seven and Nine. (University of Michigan, 1971.) *Dissertation Abstracts International* 32A: 3874–3875; January 1972.

> Except for low achievers, who seemed to derive particular benefit from representing problems with materials, students taught to translate selected types of algebra word problems directly performed comparably to those experiencing material referents and superior to those experiencing pictorial referents.

Short, Byrl G. and Szabo, Michael. Secondary School Teachers' Knowledge of and Attitudes Toward Educational Research. *Journal of Experimental Education* 43: 75–78; Fall 1974.

> See page 349.

Shumway, Richard J. Negative Instances in Mathematical Concept Acquisition: Transfer Effects Between the Concept of Commutativity and Associativity. *Journal for Research in Mathematics Education* 5: 197–211; November 1974.

> Significant differences in achievement favored the use of both positive and negative instances for associativity over the use of only positive instances for the ninth graders studied. Effects for negative instances transferred from commutativity to associativity.

Silbaugh, Charlotte Vance. A Study of the Effectiveness of a Multiple-Activities Laboratory in the Teaching of Seventh Grade Mathematics to Inner-City Students. (George Washington University, 1972.) *Dissertation Abstracts International* 33A: 205; July 1972.

> Twelve classes attended multiple-activities laboratories twice a week during the school year; 12 classes were housed in the same school but did not attend the laboratories; 12 classes were in schools with no laboratories. The students who attended the laboratories appeared to achieve significantly higher on a standardized test.

Simmons, Sadie Vee. A Study of Two Methods of Teaching Mathematics in Grades

Five, Six, and Seven. (University of Georgia, 1965.) *Dissertation Abstracts* 26: 6566–6567; May 1966.

>In a study with students in grades 5–7, students receiving instruction under a program of modern mathematics scored higher than those instructed under a traditional program, when achievement was measured by standardized tests designed to determine traditional achievement.

Skager, Rodney W. *Student Entry Skills and the Evaluation of Instructional Programs: A Case Study*. February 1969. ERIC Document No. ED 029 364.

>Teachers were found to select instructional objectives for low-achieving seventh graders that reflected skills already available to their students, and to gear instruction to skills already achieved by students at the time of their entry into the program.

Sligo, Joseph Richard. Comparison of Achievement in Selected High School Subjects. (State University of Iowa, 1955.) *Dissertation Abstracts* 15: 2136–2137; November 1955.

>A significant decline in algebra achievement test scores was found for students tested in 1954 when compared with students tested in 1934.

Smith, Lyle Ross. Aspects of Teacher Discourse and Student Achievement in Mathematics. (Texas A&M University, 1973.) *Dissertation Abstracts International* 34A: 3716; January 1974.

>See page 69.

Snyder, Henry Duane, Jr. A Comparative Study of Two Self-Selection–Pacing Approaches to Individualizing Instruction in Junior High School Mathematics. (University of Michigan, 1966.) *Dissertation Abstracts* 28A: 159–160; July 1967.

>No significant differences in achievement or in characteristics of students in grades 7 and 8 who selected either of two independent work approaches were found, though gains were greater than for students in regular classes.

Sobel, Max A. Concept Learning in Algebra. *Mathematics Teacher* 49: 425–430; October 1956.

>Students in grade 9 with high I.Q.s achieved and retained significantly better when taught with inductive rather than deductive procedures; no differences were found for the average I.Q. group.

Spickerman, William R. A Study of the Relationships Between Attitudes Toward Mathematics and Some Selected Pupil Characteristics in a Kentucky High School. (University of Kentucky, 1965.) *Dissertation Abstracts International* 30A: 2733; January 1970.

>A relationship was found between attitude toward mathematics and (1) mathematics course enrollment and (2) course mark aspiration. In grade 9, course marks, but not achievement scores, were related to attitude.

Steere, Bob F. *An Evaluation of a Nongraded Secondary School. Final Report*. December 1967. ERIC Document No. ED 018 003.

>Tenth graders in graded schools gained significantly more in mathematics reasoning than had students in nongraded schools.

Stilwell, Merle Eugene. The Development and Analysis of a Category System for Systematic Observation of Teacher-Pupil Interaction During Geometry Problem-Solving Activity. (Cornell University, 1967.) *Dissertation Abstracts* 28A: 3083; February 1968.

>See page 97.

Strickmeier, Henry Bernard, Jr. An Analysis of Verbal Teaching Behaviors in Seventh Grade Mathematics Classes Grouped by Ability. (University of Texas at Austin, 1970.) *Dissertation Abstracts International* 31A: 3428; January 1971.

> Patterns of teacher verbal behaviors in seventh-grade mathematics classes grouped by ability were described, and comparisons were made of teachers' perceptions of their verbal behaviors and expectations of students for classes of different ability levels. Although the ten teachers interviewed and observed had different perceptions and expectations for classes of different ability levels, such differences were not reflected by observable differences in the teachers' verbal behaviors.

Suydam, Marilyn N. *A Review of Research on Secondary School Mathematics.* Columbus, Ohio: ERIC Information Analysis Center for Science, Mathematics, and Environmental Education, March 1972.

> This is a review of research findings on a wide variety of mathematical topics, presented in the form of answers to questions which teachers might ask.

Swafford, Jane Oliver. A Study of the Relationship Between Personality and Achievement in Mathematics. (University of Georgia, 1969.) *Dissertation Abstracts International* 30A: 5353; June 1970.

> Eighth-grade students were classified into four groups on the basis of achievement, and then six personality factors were identified which distinguished among the groups.

Travers, Kenneth J. *Non-Intellective Correlates of Under- and Over-Achievement in Grades 4 and 6.* NLSMA Report No. 19. Stanford, California: School Mathematics Study Group, 1971.

> See pages 220–221.

Urwiller, Stanley LaVerne. A Comparative Study of Achievement, Retention, and Attitude Toward Mathematics Between Students Using Spiral Homework Assignments and Students Using Traditional Homework Assignments in Second Year Algebra. (University of Nebraska, 1971.) *Dissertation Abstracts International* 32A: 845; August 1971.

> See page 207.

Usiskin, Zalman Philip. The Effects of Teaching Euclidean Geometry Via Transformations on Student Achievement and Attitudes in Tenth-Grade Geometry. (University of Michigan, 1969.) *Dissertation Abstracts International* 31A: 688; August 1970.

> On a standardized test, scores of students using regular texts were significantly higher than scores of those using transformation-oriented texts.

Vance, James H. and Kieren, Thomas E. Laboratory Settings in Mathematics: What Does Research Say to the Teacher? *Arithmetic Teacher* 18: 585–589; December 1971.

> Recent research on mathematics laboratories was summarized by noting that (1) the research indicates that students can learn mathematical ideas in laboratory settings and (2) there is only limited evidence that laboratories promote better attitudes toward mathematics, though most students seem to prefer laboratory approaches to more class-oriented approaches.

Vance, James H. and Kieren, Thomas E. Mathematics Laboratories—More Than Fun? *School Science and Mathematics* 72: 617–623; October 1972.

> In this study with seventh and eighth graders, it was found that students who

spent one-fourth of the mathematics class time in informal exploration in a laboratory setting achieved as well as students in nonlaboratory programs. Students in the laboratory group and those in a discovery-oriented class scored higher than students in the regular program on cumulative achievement, transfer, and divergent-thinking tasks. The laboratory approach was strongly preferred.

Van Horn, Charles. *A Comparison Between Two Kinds of Secondary Mathematics Courses with Respect to Intellectual Changes.* October 1966. ERIC Document No. ED 011 059.

It appeared that the abilities which are most important in mathematics are those requiring cognition and convergent production.

Volchansky, Paul Robert. The Effects of Two Mathematical Instruction Approaches on Analytical Cognition. (University of New Mexico, 1968.) *Dissertation Abstracts* 29A: 4396; June 1969.

Eighth graders taught by a discovery method did significantly better in answering questions of an analytical nature than did those having an expository approach.

Weaver, J. Fred and Kilpatrick, Jeremy (editors). *The Place of Meaning in Mathematics Instruction: Selected Research Papers of William A. Brownell.* Volume 22, SMSG Studies in Mathematics. Stanford, California: School Mathematics Study Group, 1972.

Brownell played a definitive role in elementary school mathematics education, and his research had an impact on practice in mathematics classrooms at all levels. This volume contains a collection of reports on some of the research studies he conducted.

Werdelin, Ingvar. *The Value of External Direction and Individual Discovery in Learning Situations: 1. The Learning of a Mathematical Principle.* 1968. ERIC Document No. ED 029 541.

Students in grades 6 and 8 taught by instruction on a principle before application to examples tended to learn the mathematical idea best, but those taught by example only were superior on tests of retention and transfer.

White, Virginia Taffinder. An Evaluation Model to Test Teaching-Learning Units for Individualized Instruction in Mathematics. (University of Washington, 1972.) *Dissertation Abstracts International* 33A: 2247–2248; November 1972.

In a study with eight tenth-grade classes using geometry content, inquiry lessons used with individualized teaching-learning units significantly increased critical thinking, achievement, and retention scores for average- and high-ability students. Laboratory lessons significantly increased achievement and retention scores for low- and average-ability students. Students in the laboratory group made the greater gain in scores for attitude toward mathematics.

White, William F. and Aaron, Robert L. Teachers' Motivation Cues and Anxiety in Relation to Achievement Levels in Secondary School Mathematics. *Journal of Educational Research* 61: 6–9; September 1967.

Girls appeared to be more sensitive to the motive-arousing cues of mathematics teachers, and more in fear of failure. Significant differences in perception of achievement cues among achievement-level groups and between sexes were found for the eleventh and twelfth graders studied.

Willcutt, Robert E. Ability Grouping by Content Topics in Junior High School Mathematics. *Journal of Educational Research* 63: 152–156; December 1969.

> The achievement of seventh graders in self-contained classes was not significantly different from that in ability-grouped classes. Those in ability-grouped classes had a more positive attitude.

Wilson, Guy and Dalrymple, Charles O. Useful Fractions. *Journal of Educational Research* 30: 341–347; January 1937.

> See page 5.

Wilson, James William. Generality of Heuristics as an Instructional Variable. (Stanford University, 1967.) *Dissertation Abstracts* 28A: 2575; January 1968.

> For training tasks, problem-solving performance on functions and geometry tended to be independent of the level of generality of the heuristics. Students appeared to benefit on transfer tasks from having a wide range of heuristics available.

Wilson, James W.; Cahen, Leonard S.; and Begle, Edward G., eds. *NLSMA Reports.* Stanford, California: School Mathematics Study Group, 1968–1972.

> There are 32 volumes in the published series of reports on the National Longitudinal Study of Mathematical Abilities. Findings on the patterns of achievement at various grade levels are included in Reports 10–18; Report 26 is a summary of the correlates of achievement, while Report 30 presents a follow-up study for one portion of the population.

Wilson, James W. and Carry, L. Ray, eds. *Reviews of Recent Research in Mathematics Education.* Volume 19, SMSG Studies in Mathematics. Stanford, California: School Mathematics Study Group, 1969.

> Summaries of research on attitudes, teaching practices, computer use, and problem solving are included in this volume. (See also the *Review of Educational Research*, October 1969.)

Wolfe, Richard Edgar. Strategies of Justification Used in the Classroom by Teachers of Secondary School Mathematics. *School Science and Mathematics* 72: 334–338; April 1972.

> Eight strategies used by 11 mathematics teachers in algebra, general mathematics, and geometry classes were observed, in an investigation of the verbal activity of "justification" as it is carried out in the classroom. Criteria for identifying justification ventures and "moves" in such ventures were noted.

Wood, Nolan Earl, Jr. The Effect of an In-Service Training Program in Verbal Interaction Analysis on Teacher Behavior in the Classroom. (University of Houston, 1968.) *Dissertation Abstracts* 29A: 3788–3789; May 1969.

> The group having in-service instruction in interaction analysis became significantly more direct in their verbal behavior in the classroom, but did not significantly change attitudes.

Wright, R. E. Something Old, Something New. *School Science and Mathematics* 70: 707–712; November 1970.

> No significant differences were found in gains between eighth graders using a traditional or either of two modern programs in learning traditional concepts, but those in the modern programs achieved higher scores on a test of modern concepts.

Yasui, Roy Yoshio. An Analysis of Algebraic Achievement and Mathematical Attitude

Between the Modern and Traditional Mathematics Programs in the Senior High School: A Longitudinal Study. (University of Oregon, 1967.) *Dissertation Abstracts* 28A: 4967–4968; June 1968.

>   The achievement of students using modern programs was significantly higher than that of those using traditional programs on test items which both programs had in common.

Zahn, Karl George. The Optimum Ratio of Class Time To Be Allotted to Developmental Activities and to Individual Practice in Teaching Arithmetic. (University of Colorado, 1965.) *Dissertation Abstracts* 26: 6459; May 1966.

>   Eighth graders spent varying amounts of time for five months on developmental activities and on practice. Students who spent at least half of their time in developmental activities scored higher than those who spent the greater proportion of their time on practice.

CATEGORIZATION BY MATHEMATICS TOPIC

ORGANIZATIONAL PATTERNS

Bachman, 1969
Baley and Benesch, 1969
Beul, 1974
Buchman, 1972
Campbell, 1965
Goldberg et al., 1966
Morrison, 1967
Paige, 1967
Steere, 1967
Willcutt, 1969

TEACHING APPROACHES

Ashton, 1962
Ballew, 1966
Barcaski, 1970
Bassler et al., 1971
Becker, 1967
Bierden, 1970
Bull, 1971
Burgess, 1970
Cohen, 1971
Drake, 1935
Eldredge, 1966
Fitzgerald, 1965
Gawronski, 1972
Henderson and Rollins, 1967
Hirsch, 1973
Howitz, 1966

Johanson, 1972
Johnson, 1956
Johnson, 1971
Kleckner, 1969
Kuhfittig, 1974
Lackner, 1968
Lorentz, 1971
Maynard and Strickland, 1969
Meconi, 1967
Michael, 1949
Miller, 1957
Neuhouser, 1965
Nisbet, 1974
Patterson, 1970
Price, 1967
Rappaport, 1958
Ray, 1961
Retzer and Henderson, 1967
Robinson, 1964
Schippert, 1965
Schultz and Ohlsen, 1949
Sederberg, 1964
Silbaugh, 1972
Simmons, 1966
Skager, 1969
Snyder, 1967
Sobel, 1956
Vance and Kieren, 1971, 1972
Volchansky, 1969

Weaver and Kilpatrick, 1972
Werdelin, 1968
White, 1972
Wilson, 1968
Wilson and Carry, 1969
Wright, 1970
Zahn, 1966

CONTENT: SEQUENCING AND STRUCTURING

Bell, 1970
Bierden, 1970
Buchalter, 1969
Collins, 1971
Cooney and Henderson, 1972
Eldredge, 1966
Harris, 1972
Hoffman, 1973
Johanson, 1972
Leake, 1965
Loh, 1972
McCullouch, 1971
Piatt, 1970
Proctor, 1968
Reynolds, 1973
Skager, 1969
Usiskin, 1970
Wilson and Dalrymple, 1937

CONTENT: METHODS OF INSTRUCTION
General Mathematics

Cech, 1970
Gaslin, 1975
Howitz, 1966
King, 1972
Kleckner, 1969
Maynard and Strickland, 1969
Nix, 1970
Price, 1967
Sederberg, 1964

Algebra

Ashton, 1962
Ballew, 1966
Bechtold, 1965
Becker, 1967

Bell, 1970
Brown, 1974
Buchman, 1972
Bunch, 1973
Denmark, 1965
Drake, 1935
Flaherty, 1973
Fox, 1975
Gadske, 1933
Hansen, 1963
Harris, 1972
Hirsch, 1973
Katz, 1971
Kennedy et al., 1970
Leonard, 1967
Loh, 1972
Michael, 1949
Paulson, 1964
Posamentier, 1973
Prielipp, 1968
Proctor, 1968
Reynolds, 1973
Schultz and Ohlsen, 1949
Sederberg, 1964
Shoecraft, 1972
Sligo, 1955
Smith, 1974
Sobel, 1956
Urwiller, 1971
Yasui, 1968

Geometry

Albers, 1973
Bree, 1969
Brown, 1974
Bull, 1971
Bundrick, 1969
Coppedge and Hanna, 1971
Everett, 1972
Friedman, 1973
Hanna, 1966, 1971
Harris, 1972
Hoffman, 1973
Kort, 1971
Lash, 1971
Lorentz, 1971

Maxwell, 1975
Olson, 1971
Reynolds, 1973
Stilwell, 1968
Usiskin, 1970
White, 1972
Wilson, 1968

PROBLEM SOLVING

Alexander, 1960
Ashton, 1962
Bechtold, 1965
Bunch, 1973
Denmark, 1965
Dodson, 1972
Flaherty, 1973
Foster, 1973
Hatfield and Kieren, 1972
Kennedy et al., 1970
Kilpatrick, 1968
Koopman, 1964
Martin, 1964
Mastbaum, 1969
Maxwell, 1975
Meconi, 1967
Pack, 1971
Prielipp, 1968
Sekyra, 1969
Sherrill, 1973
Shoecraft, 1972
Stilwell, 1968
Wilson, 1968
Wilson and Carry, 1969

MATERIALS AND ACTIVITIES

Austin and Austin, 1974
Boyd, 1973
Brinke, 1967
Brown, 1974
Buchalter, 1969
Bunch, 1973
Burgess, 1970
Cech, 1970
Cohen, 1971
Collins, 1971
DeVenney, 1969
Durall, 1972

Durrance, 1965
Edwards and DeVries, 1972
Fennema, 1972
Foster, 1973
Gaslin, 1975
Hansen, 1963
Hatfield and Kieren, 1972
Hess and Tenezakis, 1973
Hewlett-Packard, 1972
Hudson, 1965
Johanson, 1972
Johnson, 1971
Katz, 1971
Keough and Burke, 1969
Kleckner, 1969
Kuhfittig, 1974
Laing, 1971
Lash, 1971
Lorentz, 1971
Mastbaum, 1969
Nibbelink, 1972
Olley, 1974
Ostheller, 1971
Pack, 1971
Peterson, 1967
Peterson, 1970
Prielipp, 1968
Reynolds, 1973
Roberge, 1970
Schippert, 1965
Sekyra, 1969
Shaw, 1967
Shoecraft, 1972
Silbaugh, 1972
Urwiller, 1971
Vance and Kieren, 1971, 1972
White, 1972
Wilson and Carry, 1969

LEARNING PACE
Low Achievers/Slow Learners

Anderson, 1970
Baker, 1971
Beaton, 1967
Bernstein, 1956
Bobier, 1964
Boyd, 1973

Brinke, 1967
Buchman, 1972
Burgess, 1970
Cech, 1970
Cohen, 1971
DeVenney, 1969
Durall, 1972
Ellis, 1971
Everett, 1972
Gaslin, 1975
Herriot, 1967
Hess and Tenezakis, 1973
Johnson, 1971
Kennedy et al., 1970
King, 1972
Kleckner, 1969
Kuhfittig, 1974
Mastbaum, 1969
Nibbelink, 1972
Posamentier, 1973
Scott, 1970
Sederberg, 1964
Skager, 1969

Faster Learners

Burke, 1974
Fitzgerald, 1965
Fox, 1975
Fredstrom, 1965
Goldberg et al., 1966
Haggard, 1957
Hatfield and Kieren, 1972
Kennedy and Walsh, 1965
Michael, 1949

Grouping

Bachman, 1969
Bierden, 1970
Bull, 1971
Campbell, 1965
Drake, 1935
Fitzgerald, 1965
Gadske, 1933
Guiler and Edwards, 1943
Hirsch, 1973
Johnson, 1956
Nix, 1970

Olson, 1971
Paulson, 1964
Pearl, 1967
Scott, 1970
Snyder, 1967
Strickmeier, 1971
Willcutt, 1969

Time

Albers, 1973
Buchman, 1972
Hansen, 1963
Herriot, 1967
Posamentier, 1973
Reynolds, 1973
Zahn, 1966

OTHER INDIVIDUAL DIFFERENCES
Sex Differences

Farley, 1969
Fennema, 1974
Fox, 1975
Koopman, 1964
White and Aaron, 1967

Personality

Ayers et al., 1969
Haggard, 1957
May, 1972
Swafford, 1970

Mathematical Ability

Bree, 1969
Kennedy and Walsh, 1965
Kim and Leton, 1966
Van Horn, 1966

ATTITUDES

Aiken, 1970
Anttonen, 1969
Bachman, 1969, 1970
Beaton, 1967
Birr, 1969
Burbank, 1970
Burgess, 1970
Callahan, 1971

Cech, 1970
Crosswhite, 1972
Dutton, 1956, 1968
Dutton and Blum, 1968
Ellingson, 1962
Farley, 1969
Fredstrom, 1965
Hess and Tenezakis, 1973
Hudson, 1965
Husén, 1967
Kester, 1969
Kilpatrick, 1968
Ostheller, 1971
Phillips, 1970
Posamentier, 1973
Postlethwaite, 1971
Spickerman, 1970
Vance and Kieren, 1971, 1972
White, 1972
White and Aaron, 1967
Wilson and Carry, 1969

EVALUATING PROGRESS
Achievement Evaluation

Bachman, 1969
Beckmann, 1969, 1970
Bernstein, 1956
Husén, 1967
Leonard, 1967
Moore and Cain, 1968
Norland, 1971
Peterson, 1967
Postlethwaite, 1971
Rappaport, 1958
Simmons, 1966
Sligo, 1955
Spickerman, 1970
Wilson et al., 1968–1972.
Wright, 1970
Yasui, 1968

Testing

Collins, 1971
Coppedge and Hanna, 1971
Hambleton and Traub, 1974
Hanna, 1971
Harris, 1972

Husén, 1967
Jeffery, 1969
NAEP, 1975
Paige, 1966
Postlethwaite, 1971
Romberg and Wilson, 1968

Parent and Teacher Effect

Beaton, 1967
Birr, 1969
Burbank, 1970
Kester, 1969
Phillips, 1970
Piatt, 1970
Robitaille, 1969
White and Aaron, 1967

Error Analysis

Bernstein, 1956
Durrance, 1965
Guiler, 1945, 1946a, 1946b
Guiler and Edwards, 1943
Kennedy et al., 1970
Koopman, 1964
Lankford, 1972
Leonard, 1967

LEARNING FACTORS

Ballew, 1966
Barcaski, 1970
Bassler et al., 1971
Bree, 1969
Bunch, 1973
Ebert, 1946
Edwards and DeVries, 1972
Elkind, 1961
Gawronski, 1972
Gay, 1973
Golledge, 1966
Gregory and Osborne, 1975
Harris, 1972
Henderson and Rollins, 1967
Hendrix, 1947
Hirsch, 1973
Jeffery, 1969
Johanson, 1972
Kilpatrick and Wirszup, 1969–1972

Kim and Leton, 1966
King, 1972
Kort, 1971
Loh, 1972
Meconi, 1967
Michael, 1949
Paige, 1966
Paulson, 1964
Price, 1967
Proctor, 1968
Ray, 1961
Retzer and Henderson, 1967
Reynolds, 1973
Roberge, 1970
Robinson, 1964
Shumway, 1974
Travers, 1971
Van Horn, 1966
Volchansky, 1969
Werdelin, 1968
White and Aaron, 1967

TEACHER BEHAVIOR

Bright and Carry, 1974
Cooney and Henderson, 1972
Fey, 1970
Friedman, 1973
Gregory and Osborne, 1975
Hernandez, 1973
Hoffman, 1973
Kester, 1969
Kysilka, 1970
Lockwood, 1971
Nisbet, 1974
Robitaille, 1969
Short and Szabo, 1974
Skager, 1969
Smith, 1974
Stilwell, 1968
Strickmeier, 1971
Wolfe, 1972
Wood, 1969

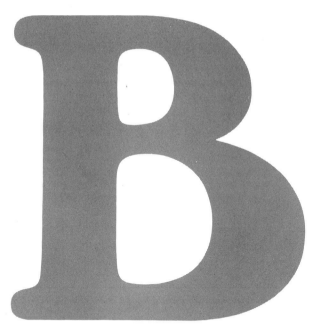

Mathematical Problems

PROBLEMS TO SOLVE

In this appendix, we present a variety of problems as examples of the diversity available. All of the problems are presented here for you to try and to think about. Most of them are too difficult for the average secondary school student, but some may be simplified. A special section of problems particularly suited to computer solutions is included at the end of the appendix.

1. Which regular polygons will tessellate (completely cover with no overlapping or gaps) a plane?
2. A farmer who lives on a square plot of land decides to retire. He retains one-fourth of his land for himself (as shown in Figure B–1) and gives the rest of the land to his four children. How may the land to be given away be divided so that the four children will each receive portions which are the same size and shape? (Kay, 1969, p. 21)
3. The device shown in Figure B–2 can be used to demonstrate binomial distributions. Such distributions approach normal distribution as we increase the number of little balls that are permitted to filter through the lattice. Assuming that the probability of a ball falling to the left or right as it hits the marked vertices of each of the hexagons is $\frac{1}{2}$, deter-

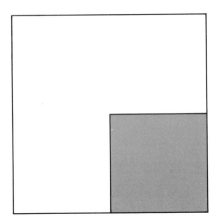

Figure B–1.

mine the probabilities for a ball falling in each compartment, one through eight.

4. Examine carefully the following six problems:
   a. Express $\frac{1}{2}$ as the sum of two unit fractions. (A unit fraction is of the form $1/n$, with n a positive integer.)
   b. Find all rectangles with integral sides whose area and perimeter are numerically equal.
   c. Which pairs of positive integers have 4 as their harmonic mean?
   d. The product of two integers is positive and equal to twice their sum. Find the possible pairs of integers.
   e. Given a point P, find all n such that the plane around P can be filled by nonoverlapping congruent regular n-gons.
   f. For which positive integers $n > 2$ does $n - 2$ divide $2n$?

   What do these problems have in common? On the surface, nothing! Actually, however, they have a very close relationship to each other because all of them can be solved using identical mathematics. Try solving them; if you get stuck, consult the article by Usiskin (1968, pp. 388–389) for a complete discussion of the problems.

5. Select any point on the base of an equilateral triangle. Draw the perpendiculars from this point to the other two sides as shown in Figure B–3. Prove that the sum of the lengths of these two perpendiculars is equal to the length of the altitude of the triangle. Extend this problem by selecting any point in the interior of an equilateral triangle and proving that the sum of the lengths of the perpendiculars from that point to the sides is equal to the length of the altitude of the triangle (see Figure B–4). (*The Mathematics Student*, February 1974) *These two problems can be quite helpful in solving the next problem.*

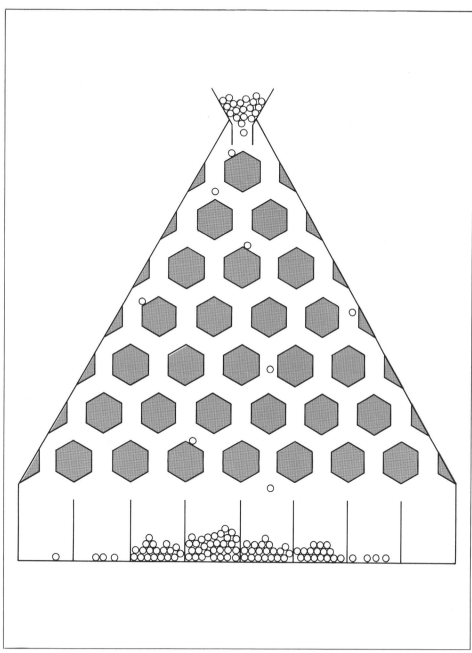

Figure B–2.

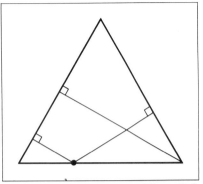

Figure B–3.

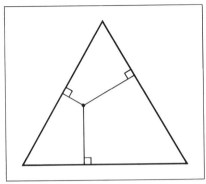

Figure B–4.

6. The Triangle Inequality states that the sum of the lengths of any two sides of a triangle is greater than the length of the third side. Suppose we are given a stick and told to break it into three parts; what is the probability that we will be able to form a triangle with the three pieces? (*The Mathematics Student*, February 1974)

7. Find the error in the following "proof":

$$\tfrac{1}{4} < \tfrac{1}{2}$$
$$(\tfrac{1}{2})^2 < \tfrac{1}{2}$$
$$\log (\tfrac{1}{2})^2 < \log (\tfrac{1}{2})$$
$$2 \log (\tfrac{1}{2}) < \log (\tfrac{1}{2})$$
$$2 < 1$$

8. A prisoner is given ten white marbles, ten black marbles, and two boxes. She is told that an executioner will draw one marble from one of the two boxes. If it is white, she will be set free; if it is black, she will be put to death. How should the prisoner distribute the marbles in the boxes so as to maximize her chances of survival?

9. In an effort to motivate a geometry student to do her homework, a mathematics teacher offered to pay the student 8¢ for each correct answer on her assignment and to fine her 5¢ for each incorrect answer. After the student had worked 26 problems, it was discovered that neither person owed money to the other. How many problems did the student solve correctly?

10. Suppose there were a steel band fitted tightly around the equator of the earth. Now suppose that you remove the band, cut it at one place, and splice in an additional piece 10 feet long, so that the new band is 10 feet longer than the original one. If you replace the band on the equator, it should fit more loosely than before. Determine how large a uniform

Figure B–5.

gap there would be between the band and the earth. Would it be large enough for
a. a man, 6 feet tall, to walk through?
b. a man to crawl through on hands and knees?
c. a piece of tissue paper just to slip through?
(See Figure B–5.)

11. There are many interesting problems related to the concept of the Golden Section. Before you consider the selection that follows, review the definition of the Golden Section in Chapter 1 (pp. 12–13).
    a. Prove that this construction procedure is correct (see Figure B–6):

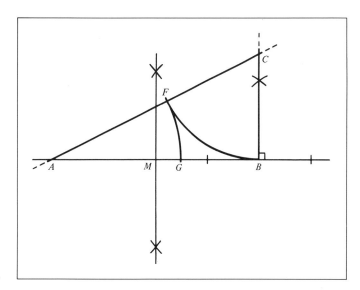

Figure B–6.

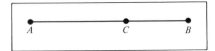

Figure B–7.

Procedure. Draw a line segment AB. Locate the midpoint of $\overline{AB}$ and call it M. At point B, construct $\overline{BC}$ perpendicular to $\overline{AB}$ and such that $BC = MB$. Draw $\overline{AC}$ and locate F on $\overline{AC}$ such that $CB = CF$. Then locate G on $\overline{AB}$ such that $AF = AG$. G is the required point of division.

b. Knowing that C forms the Golden Section in Figure B–7, we know by definition that $AB/AC = AC/CB$. Find the numerical value of these two ratios. (Hint: For computational convenience, let $AB = 1$ and let $AC = x$.) Note: The value AB/AC which you should have gotten occurs so frequently in literature related to the Golden Section that it is given a special name, τ (tau).

c. In Figure B–8, ABCDE is a regular pentagon. The inscribed star is called a pentagram and was reputedly the secret symbol of the ancient Pythagoreans. Using similar triangles, determine the ratio of a diagonal of the pentagon to one of its sides. For example,

$$EC/CB = ?$$

12. Find the sum of the measures of the five labeled angles in Figure B–9. (Weiss, 1972, p. 141)

Figure B–8.

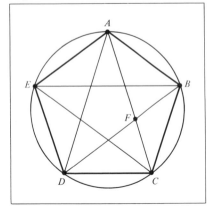

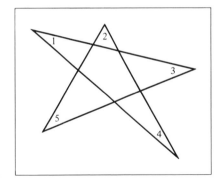

Figure B–9.

13. Knowing that the longer segments in Figure B–10 are twice the length of the shorter ones, can you form three congruent squares from these eight segments? (Weiss, 1972, p. 121)
14. Without taking your pencil off the paper, draw in Figure B–11 four line

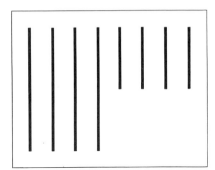

Figure B–10.

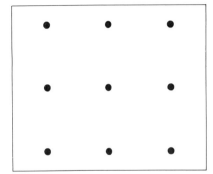

Figure B–11.

segments that pass through every point, and draw in Figure B–12 six line segments that pass through every point. (Weiss, 1972, p. 53)

15. *Fagnano's Problem.* In a given acute triangle, inscribe a triangle with the minimum perimeter. (Solution in Coxeter, 1967, p. 21.)

16. Refer to Figure B–13. Suppose that we were to cut pieces I, II, III, and IV from a square piece of cardboard whose dimensions are 8×8. We then rearrange the pieces to form the rectangle whose dimensions are 5×13. Does anything seem wrong? A moment's reflection reveals that there exists a discrepancy of one unit in the areas of the square and the rectangle. But how can this be? After all, the same pieces make up both figures! (Solution in Runion, 1972, pp. 75ff.)

17. Coplanar circles are constructed on each of the three sides of right triangle *ABC* in Figure B–14. The center of each circle is on the midpoint of a side of △*ABC*, and the length of the radius of a circle is equal to one-half the length of the side of △*ABC*. If the area of the triangular region *ABC* equals 12 square feet, what is the combined area of the smaller circular regions which are not intersected by the largest circular region?—that is, what is the area of the shaded region in Figure B–14? (*The Mathematics Student*, May 1969, p. 6)

18. Imagine a cube whose side length is 1 centimeter. Record the cube's volume and its surface area. Now slice the cube as shown in Figure B–15a and stack the four newly created rectangular parallelopipeds as indicated in Figure 5–15b. Record the volume and surface area of this shape. If, on the original cube, we perform 50 such transformations, determine the volume, height, and surface area of the final shape.

19. *The Problem of the Marked Foreheads.* A high school mathematics teacher wanted to find out which of her three students, Lucy, Tim, and Clayton, was the brightest. She seated the three of them around a circular table so that each could observe the other two. The teacher,

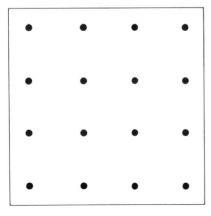

Figure B–12.

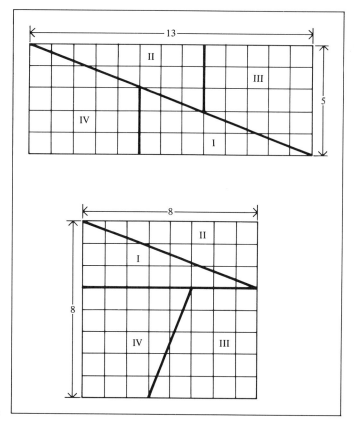

Figure B–13.

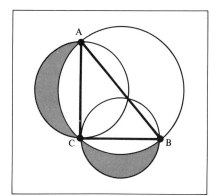

Figure B–14.

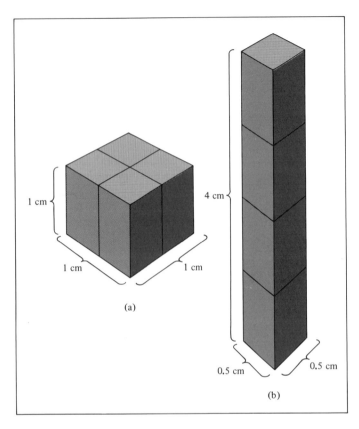

1 cm

1 cm   1 cm

(a)

4 cm

0.5 cm   0.5 cm

(b)

Figure B–15.

explaining that she planned to paste a sticker on the forehead of each, asked the three students to shut their eyes. Then she explained that the stickers would be marked with either a square or a star.

The rules were: (1) any student who saw a star should raise his or her hand; (2) as soon as a student knew what shape the sticker on her or his own forehead bore, he or she should fold his or her arms. When all eyes were closed, the teacher labeled each student's head with a star. Then she told them to open their eyes, whereupon each student immediately raised a hand. After a few moments, Lucy folded her arms. The teacher asked Lucy what kind of shape she had on her forehead. Lucy answered, "A star."

By what reasoning did she make this deduction?

20. In the house floor plan in Figure B–16, can you enter Door A and exit from Door B by passing through each room only once? (From the film, "Mr. Simplex Saves the Aspidistra.")

21. Two of the five Platonic solids mentioned in Problem (12), Chapter 5, are shown in Figure B–17 and Figure B–18.

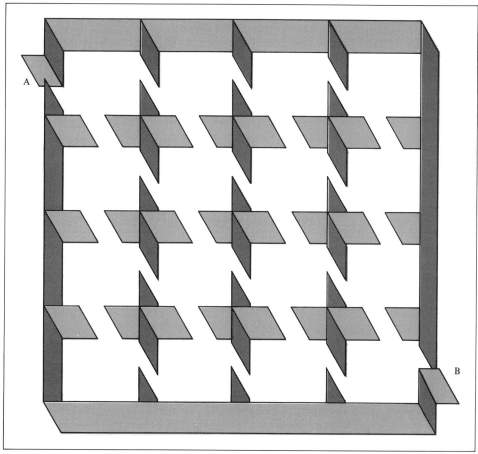

Figure B–16.

a. In the tetrahedron (Figure B–17), find the distance between any two of its opposite edges. Let the length of its edges be x.
b. Find the distance between any two opposite faces of the octahedron (Figure B–18). Let the length of any of its edges also be represented by x.

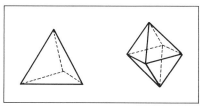

Figure B–17.   Figure B–18.

22. Construct a rhombus (using any instruments that you wish) that can be partitioned into four congruent isosceles triangles. After doing this, let the side length of the rhombus be 10, and find its area. This problem is a bit more challenging if you restrict yourself to a compass and straightedge only.

23. You have probably heard of the three famous construction problems of antiquity. They are the trisection of an angle, the duplication of a cube, and the squaring of a circle. These problems challenged mathematicians for years until it was proven that none of the problems could be solved using a compass and straightedge only. Below, however, is a procedure that could be used to trisect an angle. Prove that, using the procedure, angle AOB would indeed be trisected (see Figure B–19). State where the difficulty lies in attempting this procedure using only a compass and straightedge.

Let the angle AOB that is to be trisected be placed with its vertex at the center of a circle of any radius OA = OB. From O draw a radius OC ⊥ OB, and through A place a straight line AED in such a way that DE = OA. Finally through O draw OF parallel to AED. Then ⊀FOB is one-third the measure of angle AOB. (Boyer, 1968, pp. 285–286)

24. The arithmetic mean of a set of 50 numbers is 38. If two numbers of the set, namely 45 and 55, are discarded, what will be the arithmetic mean of the remaining set of numbers? (Salkind, 1966, p. 24)

25. Find the smallest positive integer x for which $1260x = n^3$, where n is an integer. (Salkind, 1966, p. 24)

26. Suppose you have a wooden cube as shown in Figure B–20a. You saw the cube into eight smaller cubes, as shown in Figure B–20b, and paint a dot on each exposed face.
   a. How many cubes have no dot on any of their faces?

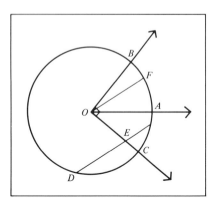

Figure B–19.

b. How many cubes have dots on exactly one face?

c. How many have dots on exactly two faces?

d. How many have dots on exactly three faces?

e. How many have dots on more than three faces?

Answer each of the above questions when the original cube is sawed into 27, 64, 125, . . . , 1,000, . . . , and n^3 smaller cubes.

27. Given the triangle shown in Figure B–21 with the segment lengths as shown, find the following:

a. the length of $\overline{DB}$

b. the lengths of the angle bisectors

c. the lengths of the three medians

d. the radius of the inscribed circle

e. the length of the three altitudes

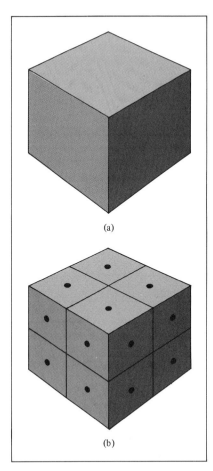

(a)

(b)

Figure B–20.

Ranucci, referring to this problem, claims that it can be solved by "relatively simple Euclidean methods untainted by excessively complex algebra." (Ranucci, 1970, p. 645)

COMPUTER PROGRAMMING PROBLEMS

The following problems have been selected because we found the experience of constructing computer programs to solve them to be challenging and educational.

28. Determine whether the roots of a given quadratic equation are imaginary, real unique, or real multiple.
29. Find a common fraction approximation, to specified accuracy, of the nth root of a given number. For example, what is $\sqrt{.75}$ expressed as a common fraction, to the nearest one-thousandth?
30. Given the BASIC functions SIN, COS, TAN, and ATN, write a program to calculate arcsines and arccosines.
31. Given the trigonometric functions of Question 30, find the corresponding hyperbolic functions.
32. Let us call the list of the measurements of the three sides and of the three angles of a triangle, "vital statistics." Given some of the "vital statistics" of a triangle,
    a. determine whether sufficient information has been given to attempt solution of the triangle
    b. determine whether any of the given information is self-contradictory
    c. determine whether the solution is unique
    d. solve the triangle
    (Suggestion: write the program so that the measurements of missing parts are entered as zeros.)
33. Find the mean and standard deviation of a set of numbers—such as students' scores on an examination.

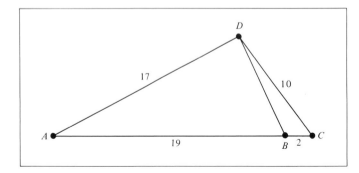

Figure B–21.

34. Find the Pearson Product Moment Coefficient of Correlation between two set of scores for the same group of students.
35. *The Stamp Problem*. Suppose the government wants to revise its postage system to contain seven (and only seven) different denominations of stamps. In order to automate the post office, it is also specified that a maximum of three stamps could be used on one envelope, although one stamp or two stamps may also be used. What must the seven denominations be to allow continuous integral combinations from one to seventy cents, using no more than three stamps at a time. It is not necessary that the stamps be of different denominations. (Adapted from Heimer and Langenbach, 1974. Solution available in that article.)
36. "Zeller's congruence" is an algorithm for calculating the day of the week for a given date. For example, on what day of the week was the Declaration of Independence signed?

    A date is expressed as:
    m = month number, with January and February taken as months 11 and 12 of the preceding year. March is then 01, April is 02, and so on. We take k as the day of the month, C for the century, and D for the year of the century.
    Example: for August 25, 1977, we have:

    $$m = 06, \ k = 25, \ C = 19, \ D = 77$$

    The congruence is

    $$f = \left\{ [2.6m - .2] + k + D + \left[\frac{D}{4}\right] + \left[\frac{C}{4}\right] - 2C \right\} \ mod \ 7$$

    where f is the number of the day of the week (Sunday is zero, Monday is one, and so on.) Square brackets [ ] denote "greatest integer in." Check the congruence using today's date. (Gruenberger and Jaffray, 1965, pp. 255–257)
37. Plot a trigonometric function of the form

    $a \sin x + b \sin^2 x + c \cos x + d \cos^2 x + e \tan x$
    $+ f \tan^2 x + g \cot x + h \cot^2 x$

    or any greater complexity you can think of, inputting the coefficients.
38. *The leaning bricks*. How many smooth, loose bricks, 8 inches long, are required to form a single pile with no part of the bottom brick under the top brick? (See Figure 7–10.) How far can such a pile lean? How many bricks are required to reach across a walk 24 inches (three bricks) wide? You can demonstrate this situation using a stack of computer cards. The problem involves use of the harmonic series. (From Gruenberger and Jaffray, 1965, pages 21–25.)
39. A *license plate problem*. Consider the right-hand two digits of auto-

1

There are two special rectangles. Their length and width measurements are integers. For each rectangle, the area and the perimeter equal the same number.

Find both rectangles!

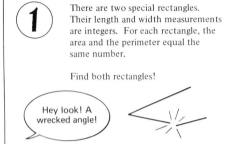

18

The square of **5** is **25**

And 5 is the last digit of 25.

The square of **90625** is **8,212,890,625**

And 90625 are the last five digits of 8,212,890,625.

Find some more numbers, less than 1000, which are the last or first digits of their squares.

24

Sol Lution, the mathematical sleuth, had become famous for his ability to find missing numbers. Another of his interesting cases was this one:

This number is divisible by 13.

When divided by 2, 3, 4, 5, 6, 7, 8, 9, 10, 11, or 12, there is always a remainder of 1.

What is the smallest number that fits these clues?

33

Aunt Maggie's father invested $1000 in a trust fund for her on the day she was born. The fund earned 8% interest compounded quarterly. Aunt Maggie forgot all about the trust fund until she was 88 years old. How much was the fund worth then?

MAKE A GUESS _____

Write a program that will show what the fund was worth when Maggie was 50, 55, 60, 65, 70, 75, 80, and 88 years old.

If Aunt Maggie takes all of the money out of the trust fund and lives to be 96 years old, how much could she spend each year?

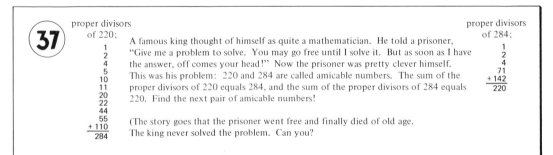

Figure B–22 Student cards (from Earl Orf and Diana Hestwood, *Computer Conversation (Speaking in BASIC)*, 1973, reprinted by permission the The Math Group, Inc., and the authors).

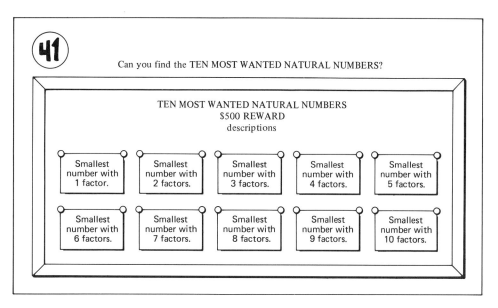

mobile license plates. There are 100 possibilities for digits in these two positions, 00 through 99. How many cars, at random, would it take to make an even-money bet that some two of them have the same last two digits on their plates? This problem generalizes to the familiar Birthday Problem. (Gruenberger, 1972, pages 20–24)

40. *The old woman and the eggs.* An old woman on her way to the market to sell her eggs is knocked down by a horseman by accident, causing all the eggs to be broken. He offers to pay the damages and wants to know how many eggs she had. She says she doesn't remember, but when she counted them two at a time there was one left over. Also there was one left over when they were taken 3, 4, 5, and 6 at a time, but when they were taken 7 at a time, it came out even. What are the possibilities for the number of eggs she had? This problem is related to the sieve of Eratosthenes. (Nievergelt et al., 1974, pp. 203–205)

41. Read an account of solving equations by iteration. Then apply an appropriate method to the solution of the transcendental equation

$$3x - \sqrt{1 + \sin x} = 0$$

(See, for example, Kovach, 1964, pp. 56–58. Kovach reports an estimate of the root as $x = 0.391847$.)

COMPUTER PROBLEMS

The problems that appear in Figure B–22 come from a set of student problem cards called *Computer Conversations* (Orf and Hestwood, 1973) and are reprinted with permission of The Math Group. The card format, the humorous approach, and the problem content are well suited for use in secondary school computer-oriented courses.

References

Barrodale, Ian; Roberts, Frank D.; and Ehle, Byron L. *Elementary Computer Applications in Science, Engineering and Business.* New York: Wiley, 1971.

Boyer, Carl B. *A History of Mathematics.* New York: Wiley, 1968.

Coxeter, H. S. M. and Greitzer, S. L. *Geometry Revisited.* Vol. 19, The New Mathematical Library. New York: Random House and L. W. Singer, 1967.

Gruenberger, Fred. *Computing with the BASIC Language.* San Francisco: Canfield Press, 1972.

Gruenberger, Fred and Jaffray, George. *Problems for Computer Solution.* New York: Wiley, 1965.

Heimer, Richard L. and Langenbach, Herbert. The Stamp Problem. *Journal of Recreational Mathematics.* 7: 235–250; Summer 1974.

Kay, David C. *College Geometry.* New York: Holt, Rinehart and Winston, 1969.

Kovach, Ladis D. *Computer-oriented Mathematics.* San Francisco: Holden-Day, 1964.

Nievergelt, Jurg; Farrar, J. Craig; and Reingold, Edward M. *Computer Approaches to Mathematical Problems.* Englewood Cliffs, New Jersey: Prentice-Hall, 1974.

Orf, Earl and Hestwood, Diana. *Computer Conversations (Speaking in BASIC): Student Cards.* Nos. 1, 18, 24, 33, 37, and 41. Minneapolis: The Math Group, 1973.

Ranucci, Ernest. Letter to the Editor. *Mathematics Teacher* 63: 645; December 1970.

Runion, Garth E. *The Golden Section and Related Curiosa.* Glenview, Illinois: Scott, Foresman, 1972.

Salkind, Charles T. *The MAA Problem Book II.* Vol. 17 of *The New Mathematical Library.* New York: Random House, 1966.

Usiskin, Zalman. Six Nontrivial Equivalent Problems. *Mathematics Teacher* 61: 388–389; April 1968.

Weiss, Sol. *Geometry: Content and Strategy for Teachers.* Belmont, California: Wadsworth, 1972.

"Mr. Simplex Saves the Aspidistra" (film). Available from Modern Film Rentals, 2323 New Hyde Park Road, New Hyde Park, New York, 11040.

The Mathematics Student, May 1969 and February 1974.

BASIC Programming

INTRODUCTION TO BASIC PROGRAMMING

Different computer installations have unique characteristics. For example, how you sign on or gain access to the computer will be devised to take into account local circumstances such as the amount of security needed, the computer time available, the source of funding for the computer, and so on. You will have to confer with personnel at the computer facility in order to obtain the necessary sign-on and authorization information.

Computer languages, too, may take on local characteristics, which might be called "dialects." Even BASIC, a fairly standard language, appears in dialects. For example, "#" may be used on some computer systems instead of "<>" to express the relation "not equal to", and "**" may be used instead of "↑" to mean "raising to a power."

In writing these programs, we have attempted to use the common features of BASIC which we expect to be available on most school computers. We suggest that you take one or both of the following approaches to determine the nonstandard features of BASIC which your computer employs, where by *nonstandard* we mean features of BASIC not used in this book. First, obtain a BASIC manual from your local computer facility. Often such a manual will point out features of the language peculiar to that particular facility. Second, you can use the programs as written in this book; then, as you implement the programs on your computer, error messages should help you spot nonstandard uses of BASIC (for example, that you used the wrong symbol for "not equal to"). These error messages, together with help that computer personnel are usually happy to provide, should take care of the majority of the problems you could encounter in successfully using the programs in this book.

STATEMENTS

A computer program consists of a sequence of statements. In BASIC the statements are ordered by line numbers, hence every BASIC statement must be numbered. When the program is executed by the computer, the line numbers determine the order in which the statements are to be taken, unless a command transfers control to some other line. In many computers, statements can be written in any order. The computer than places them in sequence, according to their line numbers. Line numbers are unique and are usually integral, from 1 to 99999. Often programers start a program with line number 100 and increment the numbers of succeeding lines by 10s. In this way, new statements can easily be inserted later, if desired, without renumbering.

CONSTANTS

A *numeric constant* is a number written with one or more digits and may be signed or unsigned. In many computers about 14-digit accuracy is maintained and scientific notation is employed for very large or very small numbers ($123.456E52$ means 123.456×10^{52}). Following are examples of valid numeric constants:

1	-1	1234.5678	$-$ .3	$-371.62E-14$
$-1.2E3$	5E52	.9876543210	$+6.0$	190000987

Note that commas are never used within a numeric constant.

A *string constant* is a series of letters, numerals, or other characters, except quotation marks, which is enclosed within quotation marks. Following are examples of string constants:

```
"TEACH MATHEMATICS"
"WHAT NUMBER DO YOU WANT?"
"TODAY'S DATE IS MAY 1, 1977"
```

VARIABLES

A variable represents a quantity which may change during execution of the program.

Numeric variables may be named by a single alphabetic character with or without a succeeding single digit. The variable's name may be thought of as an address in the computer's memory in which a number is stored. The following are examples of valid numeric variables in BASIC:

X1	Y3	A9
B	I	I0

String variables may be named by a single alphabetic character followed by a dollar sign. Examples are:

A$	B$	X$

Array variables (subscripted variables), including matrices, are allowed in most versions of BASIC (see p. 520).

OPERATORS

The following arithmetic operators and relational operators are available in BASIC:

OPERATOR	DEFINITION	BASIC EXAMPLES
↑	exponentiation	3↑2 , X↑A3
*	multiplication	3*X , 7*2*A3
/	division	18/Y, R3/2
+	addition	9+Y , 7E8+Z
−	subtraction	A3−A4, 7−2/.2
=	equal to	X5=9
<>	not equal to	A5<>(7+X2)
>	greater than	X>(7*X3)
<	less than	Y7<7+5
>=	greater than or equal to	A>=B
<=	less than or equal to	A3<=9

In the absence of parentheses the order of arithmetic operations is ↑, *, /, +, −. This is the same order as in conventional algebra. Note also that 2A is meaningless in BASIC; multiplication would be denoted by 2*A.

EXPRESSIONS

Arithmetic (or algebraic) expressions are used to compute numbers. Following are examples, using BASIC and using standard algebraic notation:

BASIC EXPRESSION	ALGEBRAIC EXPRESSION
A3+Z*R2↑3	$A_3 + Z(R_2)^3$
Z*(X1+X2)↑E	$Z(X_1 + X_2)^E$
−3.14*R↑2	$-3.14R^2$
.718E6*(S1−S2)↑A	$(.718 \times 10^6)(S_1 - S_2)^A$

NOTATION FOR DESCRIBING FUNCTIONS AND COMMANDS

In the following descriptions, these conventions are used:

NOTATION	EXAMPLE	MEANING
v,v1,v2...,vn	A6	variables are being named
e,e1,e2...,en	A3+Z*R5	arithmetic expressions are being named
n,n1,n2...,nn	140	program line numbers
i,i1,i2...,in	17	positive integers

BASIC expressions are written in upper case.

FUNCTIONS

The following functions are available in most versions of BASIC:

FUNCTION	EXAMPLE	MEANING
SIN(e)	SIN(3.14)	Sine of e, where e is the radian measure of an angle
COS(e)	COS(5E7)	Cosine of e, where e is the radian measure of an angle
TAN(e)	TAN(3*Y)	Tangent of e, where e is the radian measure of an angle
ATN(e)	ATN(5/7)	Archtangent of e in the principal value range of $-\pi/2$ to $+\pi/2$
EXP(e)	EXP(3.14)	Value of $e{\uparrow}e$, where e is the base of natural logarithm
LOG(e)	LOG(5*X)	Natural logarithm of e; $e > 0$
ABS(e)	ABS(Z3)	Absolute value of e
SQR(e)	SQR(Q1)	Square root of e; $e \geqslant 0$
INT(e)	INT(3.14*R)	Largest integer not greater than e Note: $(INT(-4.95) = -5)$
RND(1)	RND(1)	A uniformly distributed random number will be generated between 0 and 1, such as .41732168

COMMANDS

Following are BASIC commands, the form to use, examples and explanations:

COMMAND	FORM	EXAMPLE	EXPLANATION
LET	LET v=e	LET X=3*Y	The variable v is assigned the value of e.
PRINT	PRINT e1,e2,e3,e4	PRINT 3*A,Z,R+S	Print the value of the expressions.
		PRINT A$,Y	Print string variable and numeric variable (";" and ";"

			are used for print zone control).
END	END	END	Last statement in the program. Signals termination.
GOTO	GOTO n	GOTO 300	Transfer control to line n.
IF-THEN	IF e r e1 THEN n	IF A<B THEN 100	If the relation r between e and $e1$ is true, then transfer to line n; if not, continue to next statement.
READ	READ v1,v2,...vn	READ A1,X,Y	Assigns variables $v1$ through vn the sequence of values in the DATA statement.
DATA	DATA n1,n2,...,nn	DATA 5,9,3.4,126	Used with READ statements as a list of data.
INPUT	INPUT v1,v2,...,vn	INPUT R,S,T1	Causes a "?" to be printed, stops execution. The user enters values for $v1$ through vn, and returns control to computer.
REM	REM...	REM THIS PROGRAM	Permits program author to enter remarks in program listing. No action is taken by computer.
NODATA	NODATA n	NODATA 300	If all data have been used, control is transferred to line n; if not, next statement in regular sequence is executed.
RESTORE	RESTORE	RESTORE	Moves the data pointer to the beginning of DATA list.
FOR/TO	FOR v=e1 TO e2 STEP e3	FOR X=7 TO 50 STEP 5	Begins a loop. Variable v takes on value $e1$ and program continues until a NEXT v statement is encountered. Then v is incremented by $e3$ and the program continues from the FOR/TO statement. This process continues until v reaches or exceeds the value of $e2$. Note that $e3$ may be positive or negative. If STEP $e3$ is omitted, a STEP of 1 is assumed.

NEXT	NEXT v	NEXT X	Used with FOR as described above.
DEF	DEF FNv(v1)= f(x)	DEF FNA(X)= 3.14*X↑2	Provides a user-defined function $f(x)$ which is identified in the program by 3 letters and an argument (e.g., FNA(3) or FNB(2*Y1)).
GOSUB	GOSUB n	GOSUB 8000	Transfers control to line n. After a RETURN statement is encountered, control is transferred to the line following the GOSUB statement which initiated the subroutine.
ON	ON e GOTO n1,n2...nn	ON X3 GOTO 10,15,20	The expression e is evaluated and truncated to an integer, then control is transferred to line $n1$ if $e = 1$, to $n2$ if $e = 2$, . . . , to nn if $e = n$.
STOP	STOP	STOP	Equivalent to END, but STOP may appear at any point within the program. END must have the highest numbered line.
DIM	DIM v1(i1), v2(i2),...,vn(in)	DIM A(100), B1(2,50), C(10,30,20)	Used to dimension an array variable. In the example, A is a one-dimensional array variable with 100 entries, B1 is a two-dimensional array variable with 100 entries (2*50), and C is a 3-dimensional array variable with 6000 entries (10* 30*20). DIM must be used if any array variable has more than 10 entries in any dimension.

PROGRAMMING EXERCISES

These exercises provide practice in using BASIC programming skills. The exercises are grouped under topics in mathematics and are roughly in order of increasing difficulty.

1. WHOLE NUMBERS

a. Find the sum, product, difference, and quotient of pairs of numbers.
b. Find which members of a set of numbers are divisors of a given number, say 456.
c. Print the square root, square, and double of each of a given set of numbers.
d. Given two numbers, find which is the larger and print in ascending order.
e. Determine whether a given set of numbers is closed under a given mathematical operation.
f. Find whether addition modulo M is commutative and associative, for the integers 0, 1, 2, . . . , M − 1. Find an inverse for each integer. What is the identity element?
g. Find the union and intersection of two sets of numbers.
h. Find the union and intersection of two sets of letters.

2. NUMERATION

a. Change any number written in base 10 to its corresponding form in base 3.
b. Change any number written in base A to its corresponding form in base B.

3. NUMBER THEORY

a. Find all the factors of a given number.
b. Find the greatest common factor of two numbers.
c. Find the prime factorization of a number.
d. Generate the prime numbers through 200.
e. Determine if two numbers are relatively prime.
f. Write a program to test (partially) Goldbach's conjecture: Any even number greater than 4 can be written as the sum of two prime numbers.
g. Find the number of twin primes in a given interval.
h. Find the least common multiple of any three numbers.
i. Find triples which are related by $a^2 + b^2 = c^2$ (e.g., 3, 4, 5) for a, b, c each less than or equal to 100.

4. GEOMETRY

a. Find perimeter and area of a given set of rectangles.
b. Find the maximum area for a rectangle having a given perimeter.
c. Find the dimensions of a rectangle that will enclose the most area with 200 feet of fence where one side of the rectangle borders a river and therefore does not need fencing.
d. Find the circumference and area of a circle.
e. Find the volume and surface area of a given solid.
f. Find the minimum surface area which will enclose a given volume.
g. Find missing parts of a right triangle.

5. ALGEBRA

a. Find the slope of a line through two given points.
b. Find the equation of a line given the slope and y-intercept.
c. Write the equation of a line given two points on the line.
d. Print the slope and y-intercept of any equation of the form $ax + by = c$, where a and b are not both zero at the same time.
e. Find the sum of two polynomials.
f. Find the product of two polynomials.
g. Use synthetic substitution and/or division to find the value of a polynomial $P(x)$ for several values of the variable x.

6. MATRICES

Most BASIC compilers include commands for matrix operations such as matrix addition, matrix subtraction, matrix multiplication, scalar multiplication of matrices, transpose of a matrix, inverse of a matrix, and provision for a zero matrix, a constant matrix, and the identity matrix (see Kemeny and Kurtz, 1971, pp. 120–129). The following exercises may be completed without using the matrix commands.

a. Find the sum of two 3×3 matrices.
b. Find the product of 2×2 or 3×3 matrices using subscripted (array) variables.
c. Find the determinant of a 2×2 or 3×3 matrix.
d. Find the inverse of a 4×4 matrix.
e. Solve a set of four simultaneous linear equations in four variables, using Cramer's Rule.
f. Use the Gauss-Seidel (iterative) method for solving a system of simultaneous equations (see, for example, Barrodale et al., 1971, pp. 41–44).

SAMPLE PROGRAMS

REDUCE	A tutorial program to teach how to reduce common fractions to simplest form.
SCORES	Calculates summary statistics for a class of students over several tests. Finds means and standard deviations, converts raw scores to standard scores, computes weighted mean score for each student, and ranks students on basis of weighted mean scores.
PI	Six different ways of computing pi are demonstrated.
TEMP	Using a random walk technique, the tempera-

ture of a given point on a rectangular metal plate when temperatures at the edges are known, is calculated.

WORKsheet (developed by James Ulrich, Head of Mathematics Department, Arlington High School, Arlington Heights, Illinois, and used by permission)

Designed to aid the teacher in preparing student worksheets on topics in elementary algebra.

SIMEQ — Solving systems of simultaneous equations in two variables.

FACTOR — Factoring trinomials.

DISTR — Using the distributive property.

RAND — Generates a table of random digits.

REDUCE

Program

```
   5   LET A2=0
  10   RANDØMIZE
  20   PRINT
  30   LET N1=INT(RND(3)*5+2)
  40   LET D1=INT(RND(3)*5+2)
  50   LET X1=INT(RND(3)*22+2)
  60   IF N1<D1 THEN 110
  70   IF N1=D1 THEN 40
  80   LET N2=N1
  90   LET N1=D1
 100   LET D1=N2
 110   LET N1=N1*X1
 120   LET D1=D1*X1
 130   PRINT "REDUCE THIS FRACTIØN:";N1;"/";D1
 140   PRINT
 150   PRINT "TØ REDUCE THIS FRACTIØN YØU LØØK FØR ALL THE CØMMØN"
 160   PRINT "FACTØRS ØF THE NUMERATØR AND THE DENØMINATØR"
 170   PRINT "WHAT IS ØNE FACTØR THAT THE NUMERATØR AND"
 180   PRINT "DENØMINATØR HAVE IN CØMMØN?"
 190   INPUT A
 200   IF A>0 THEN 260
 210   IF A=0 THEN 240
 220   PRINT "USE ØNLY PØSITIVE NUMBERS. TRY ANØTHER NUMBER."
```

```
230    GØ TØ 170
240    PRINT "TUT TUT..NØ DIVIDING BY ZERØ..TRY AGAIN."
250    GØ TØ 170
260    IF A=INT(A) THEN 290
270    PRINT "USE ØNLY WHØLE NUMBERS, PLEASE !"
275    IF A2>0 THEN 400
280    GØ TØ 170
290    IF INT(N1/A)<>N1/A THEN 570
300    IF INT(D1/A)<>D1/A THEN 570
310    PRINT "CØRRECT ! ";A;" IS A FACTØR ØF BØTH THE"
320    PRINT "NUMERATØR AND THE DENØMINATØR."
330    PRINT "NØW YØU NEED TØ DIVIDE BØTH THE NUMERATØR AND"
340    PRINT "THE DENØMINATØR BY ";A;" WHAT IS YØUR"
350    PRINT "RESULT? (EXPRESS ANSWER N/D AS N,D)"
360    INPUT Y,Z
370    IF N1/A<>Y THEN 510
380    IF D1/A<>Z THEN 540
385    PRINT "VERY GØØD"
390    LET H=A
400    PRINT "CAN YØU FIND A GREATER CØMMØN FACTØR? (TYPE 1=YES, 2=NØ)"
410    INPUT N
420    IF N=1 THEN 440
430    IF N=2 THEN 610
440    PRINT "WHAT IS THE CØMMØN FACTØR?"
450    INPUT A2
460    IF A2<=H THEN 490
470    LET A=A2
480    GØ TØ 260
490    PRINT A2; "IS NØT GREATER THAN"; H; "."
500    GØ TØ 400
510    PRINT "YØU HAVE MADE A MISTAKE DIVIDING NUMERATØR BY"
520    PRINT "THAT FACTØR. TRY AGAIN."
530    GØ TØ 360
540    PRINT "YØU HAVE MADE A MISTAKE DIVIDING DENØMINATØR"
550    PRINT "BY THAT FACTØR. TRY AGAIN."
560    GØ TØ 360
570    PRINT "THAT NUMBER IS NØT A FACTØR ØF BØTH THE"
580    PRINT "NUMERATØR AND THE DENØMINATOR. SØ I ASK YOU AGAIN"
590    IF A=A2 THEN 400
600    GØ TØ 170
610    LET C=H
620    LET C=C+1
630    IF C>N1 THEN 680
```

```
640   IF C>D1 THEN 680
650   IF INT(N1/C)<>N1/C THEN 620
660   IF INT(D1/C)<>D1/C THEN 620
670   GØ TØ 730
680   PRINT "FANTASTIC! YØUR FRACTIØN IS REDUCED."
690   PRINT "WANT TØ TRY ANØTHER? (TYPE 1=YES, 2=NØ)"
700   INPUT S
710   IF S=1 THEN 20
720   GØ TØ 880
730   PRINT "BUT ISN'T"; C; "A CØMMØN FACTØR?"
740   PRINT "(TYPE 1=YES, 2=NØ)"
750   INPUT R
760   IF R=1 THEN 790
770   PRINT "CHECK AGAIN."
790   PRINT "NØW DØ THE DIVISIØN. WHAT IS THE RESULT?"
795   PRINT "(EXPRESS ANSWER N/D AS N,D)"
800   INPUT Y1, Z1
810   IF N1/C<>Y1 THEN 860
820   IF D1/C<>Z1 THEN 860
830   LET A=C
840   PRINT "NØW YØU ARE CØRRECT"
845   PRINT "THEREFØRE"; A; "IS A CØMMØN FACTØR"
850   GØ TØ 390
860   PRINT "YØU HAVE MADE A MISTAKE DIVIDING. TRY AGAIN."
870   GØ TØ 800
880   PRINT "GØØD-BYE!"
890   END
```

Sample Output

```
REDUCE THIS FRACTIØN: 32 / 64

TØ REDUCE THIS FRACTIØN YØU LØØK FØR ALL THE CØMMØN
FACTØRS ØF THE NUMERATØR AND THE DENØMINATØR
WHAT IS ØNE FACTØR THAT THE NUMERATØR AND
DENØMINATØR HAVE IN CØMMØN?
    ?3
THAT NUMBER IS NØT A FACTØR
ØF BØTH THE NUMERATØR AND THE DENØMINATØR. SØ I ASK YØU AGAIN
WHAT IS ØNE FACTØR THAT THE NUMERATØR AND
DENØMINATØR HAVE IN CØMMØN?
    ?2
```

CØRRECT ! 2 IS A FACTØR ØF BØTH THE
NUMERATØR AND THE DENØMINATØR.
NØW YØU NEED TØ DIVIDE BØTH THE NUMERATØR AND
THE DENØMINATØR BY 2 WHAT IS YØUR
RESULT? (EXPRESS ANSWER N/D AS N,D)
    ?14,19
YØU HAVE MADE A MISTAKE DIVIDING NUMERATØR BY
THAT FACTØR. TRY AGAIN.
    ?16,19
YØU HAVE MADE A MISTAKE DIVIDING DENØMINATØR
BY THAT FACTØR. TRY AGAIN.
    ?16,32
VERY GØØD
CAN YØU FIND A GREATER CØMMØN FACTØR? (TYPE 1=YES, 2=NØ)
    ?2
BUT ISN'T 4 A CØMMØN FACTØR?
(TYPE 1=YES, 2=NØ)
    ?2
CHECK AGAIN.
NØW DØ THE DIVISIØN. WHAT IS THE RESULT?
(EXPRESS ANSWER N/D AS N,D)
    ?7,16
YØU HAVE MADE A MISTAKE DIVIDING. TRY AGAIN.
    ?8,16
NØW YØU ARE CØRRECT
THEREFØRE 4 IS A CØMMØN FACTØR
CAN YØU FIND A GREATER CØMMØN FACTØR? (TYPE 1=YES, 2=NØ)
    ?1
WHAT IS THE CØMMØN FACTØR?
    ?8
CØRRECT ! 8 IS A FACTØR ØF BØTH THE
NUMERATØR AND THE DENØMINATØR.
NØW YØU NEED TØ DIVIDE BØTH THE NUMERATØR AND
THE DENØMINATØR BY 8 WHAT IS YØUR
RESULT? (EXPRESS ANSWER N/D AS N,D)
    ?4,8
VERY GØØD
CAN YØU FIND A GREATER CØMMØN FACTØR? (TYPE 1=YES, 2=NØ)
    ?1
WHAT IS THE CØMMØN FACTØR?
    ?2
2 IS NØT GREATER THAN 8 .
CAN YØU FIND A GREATER CØMMØN FACTØR? (TYPE 1=YES, 2=NØ)
    ?1

WHAT IS THE CØMMØN FACTØR?
    ?32
CØRRECT ! 32 IS A FACTØR ØF BØTH THE
NUMERATØR AND THE DENØMINATØR.
NØW YØU NEED TØ DIVIDE BØTH THE NUMERATØR AND
THE DENØMINATØR BY 32 WHAT IS YØUR
RESULT? (EXPRESS ANSWER N/D AS N,D)
    ?1,2
VERY GØØD
CAN YØU FIND A GREATER CØMMØN FACTØR? (TYPE 1=YES, 2=NØ)
    ?1
WHAT IS THE CØMMØN FACTØR?
    ?64
THAT NUMBER IS NØT A FACTØR
ØF BØTH THE NUMERATØR AND THE DENØMINATØR. SØ I ASK YØU AGAIN
CAN YØU FIND A GREATER CØMMØN FACTØR? (TYPE 1=YES, 2=NØ)
    ?2
FANTASTIC! YØUR FRACTIØN IS REDUCED.
WANT TØ TRY ANØTHER? (TYPE 1=YES, 2=NØ)
    ?2
GØØD-BYE!

SCORES*

Program

```
10    REM CLASS STATISTICS PRØGRAM
15    REM MØDIFIED FRØM CLASS PRØJECT
20    DIM V(50),N(50),R(50)
30    DIM A(15), Q(15), S(15)
40    DIM G(50,5),D(50,5)
50    REM NO IS NUMBER ØF STUDENTS IN CLASS (MAX 50)
60    REM N1 IS NUMBER ØF EXAMS (MAX 5)
70    REM G(I,J) IS SCØRE ØF STUDENT I ØN EXAM J
80    REM NEGATIVE SCØRE INDICATES STUDENT DID NØT TAKE EXAM J
90    REM D(I,J) IS STANDARD SCØRE ØF STUDENT I ØN EXAM J
100   REM N(I) IS IDENTIFYING NUMBER FØR STUDENT I
110   REM V(I) IS TØTAL WEIGHTED STANDARDIZED SCØRE FØR STUDENT I
120   REM A(J) IS MEAN SCØRE ØN EXAM J
130   REM S(J) IS STANDARD DEVIATIØN ØF SCØRES ØN EXAM J
140   REM W(J) IS WEIGHT ASSIGNED TØ EXAM J
150   REM Q(J) IS NUMBER ØF STUDENTS TAKING EXAM J
```

* Adapted from a class project in Secondary Education 356 at the University of Illinois, Urbana.

```
160   REM Q0 IS NUMBER ØF STUDENTS TAKING AT LEAST ØNE EXAM
170   REM READ DATA
180   GØSUB 300
190   REM—CØMPUTE MEANS AND STANDARD DEVIATIØNS
200   GØSUB 510
210   REM—CØMPUTE STANDARD SCØRES AND TØTAL SCØRES
220   REM—THESE STANDARD SCØRES HAVE A MEAN ØF 500 AND A STANDARD
230   REM—DEVIATIØN ØF 100
240   GØSUB 760
250   REM—PRINT RESULTS
260   GØSUB 970
270   REM—PRINT SUMMARY ØF TØTAL SCØRES
280   GØSUB 1420
290   GØTØ 1840
300   REM—SUBRØUTINE TØ READ DATA
310   READ C
320   PRINT "SCØRES FØR CØURSE NUMBER ";C
330   PRINT
340   READ N0
350   PRINT "NUMBER ØF STUDENTS IN CLASS=",N0
360   READ N1
370   PRINT "NUMBER ØF EXAMS=", N1
380   PRINT
390   FØR J=1 TØ N1
400   READ W(J)
410   PRINT "WEIGHT FØR EXAM", J, "=", W(J)
420   NEXT J
430   PRINT
440   FØR I=1 TØ N0
450   READ N(I)
460   FØR J=1 TØ N1
470   READ G(I,J)
480   NEXT J
490   NEXT I
500   RETURN
510   REM—SUBRØUTINE TØ CØMPUTE MEANS AND STANDARD DEVIATIØNS
520   FØR J=1 TØ N1
530   REM—CØMPUTE MEAN SCØRE
540   LET T=0
550   LET Q(J)=0
560   FØR I=1 TØ N0
570   IF G(I,J)<0 THEN 600
580   LET Q(J)=Q(J)+1
```

```
590   LET T=T+G(I,J)
600   NEXT I
610   LET A(J)=T/Q(J)
620   REM—CØMPUTE STANDARD DEVIATIØN ØF SCØRES
630   LET T=0
640   FØR I=1 TØ N0
650   IF G(I,J)<0 THEN 670
660   LET T=T+((G(I,J)−A(J))**2)
670   NEXT I
680   LET S(J)=(T/Q(J))**.5
690   REM—PRINT RESULTS
700   PRINT "EXAMINATIØN NUMBER", J
710   PRINT "MEAN=",A(J),"STD DEVIATIØN=",S(J)
720   PRINT Q(J), "STUDENTS TØØK THE EXAMINATIØN"
730   PRINT
740   NEXT J
750   RETURN
760   REM—SUBRØUTINE TØ CØMPUTE STANDARD SCØRES AND TØTAL SCØRES
770   LET Q0=0
780   FØR I=1 TØ N0
790   REM—CØMPUTE STANDARD SCØRES
800   LET T=0
810   LET T1=0
820   LET V(I)=9999
830   FØR J=1 TØ N1
840   IF G(I,J)>=0 THEN 870
850   LET D(I,J)=−1
860   GØ TØ 900
870   LET D(I,J)=INT(500+(100*(G(I,J)−A(J))/S(J))+.5)
880   LET T=T+(W(J)*D(I,J))
890   LET T1=T1+W(J)
900   NEXT J
910   REM—CØMPUTE TØTAL SCØRE
920   IF T=0 THEN 950
930   LET V(I)=T/T1
940   LET Q0=Q0+1
950   NEXT I
960   RETURN
970   REM—SUBRØUTINE TØ PRINT RESULTS
980   PRINT
990   FØR I=1 TØ N0
1000  PRINT "STUDENT NUMBER", N(I)
1010  PRINT "EXAM","RAW SC","STD SC","PCT ABØVE","PCT BELØW"
```

```
1020   REM—PRINT EXAM SCØRES
1030   FØR J=1 TØ N1
1040   IF G(I,J)>=0 THEN 1070
1050   PRINT J, "NØT TAKEN"
1060   GØ TØ 1210
1070   LET N3=0
1080   LET N4=0
1090   FØR K=1 TØ N0
1100   IF G(K,J)>G(I,J) THEN 1130
1110   IF G(K,J)<G(I,J) THEN 1150
1120   GØ TØ 1170
1130   LET N3=N3+1
1140   GØ TØ 1170
1150   IF G(K,J)<0 THEN 1170
1160   LET N4=N4+1
1170   NEXT K
1180   LET P1=INT((N3/Q(J))*100+.5)
1190   LET P2=INT(N4/Q(J))*100+.5)
1200   PRINT J,G(I,J),D(I,J),P1,P2
1210   NEXT J
1220   IF N1=1 THEN 1390
1230   REM—PRINT TØTAL SCØRES
1240   IF V(I)=9999 THEN 1390
1250   LET N3=0
1260   LET N4=0
1270   FØR K=1 TØ N0
1280   IF V(K)>V(I) THEN 1310
1290   IF V(K)<V(I) THEN 1340
1300   GØ TØ 1350
1310   IF V(K)=9999 THEN 1350
1320   LET N3=N3+1
1330   GØ TØ 1350
1340   LET N4=N4+1
1350   NEXT K
1360   LET P1=INT(((N3/Q0)*100)+.5)
1370   LET P2=INT(((N4/Q0)*100)+.5)
1380   PRINT "WEIGHTED SCØRE",,V(I),P1,P2
1390   PRINT
1400   NEXT I
1410   RETURN
1420   REM—SUBRØUTINE TØ PRINT SUMMARY ØF TØTAL SCØRES
1430   REM LØAD LIST R
1440   LET K=0
```

```
1450    FØR I=1 TØ N0
1460    IF V(I)=9999 THEN 1490
1470    LET K=K+1
1480    LET R(K)=I
1490    NEXT I
1500    REM—RANK STUDENTS
1510    LET N9=0
1520    FØR K=1 TØ Q0−1
1530    IF V(R(K))>=V(R(K+1)) THEN 1580
1540    LET R1=R(K)
1550    LET R(K)=R(K+1)
1560    LET R(K+1)=R1
1570    LET N9=1
1580    NEXT K
1590    IF N9=1 THEN 1510
1600    REM—PRINT SCØRES
1610    PRINT
1620    PRINT "STUDENTS RANKED BY WEIGHTED SCØRES"
1630    PRINT
1640    PRINT "STUDENT","WEIGHTED SCØRE"
1650    PRINT
1660    FØR K=1 TØ Q0
1670    PRINT N(R(K)),V(R(K))
1680    NEXT K
1690    RETURN
1710    REM—CØURSE NUMBER
1720    DATA 390
1730    REM—NUMBER ØF STUDENTS
1740    DATA 4
1750    REM—NUMBER ØF EXAMS GIVEN
1760    DATA 3
1770    REM—WEIGHTS ØF EXAMS 1, 2, 3 RESPECTIVELY
1780    DATA 1,1,2
1790    REM—SAMPLE STUDENT DATA
1795    REM— ID,SCØRE FRØM TEST 1,2,3
1800    DATA 1,80,66,88
1810    DATA 2,90,−1,95
1820    DATA 3,68,86,65
1830    DATA 4,75,96,67
1840    END
```

Sample Output

```
SCØRES FØR CØURSE NUMBER 390
NUMBER ØF STUDENTS IN CLASS=4
NUMBER ØF EXAMS=            3

WEIGHT FØR EXAM            1          =              1
WEIGHT FØR EXAM            2          =              1
WEIGHT FØR EXAM            3          =              2

EXAMINATIØN NUMBER         1
MEAN=        78.25     STD DEVIATIØN= 8.01171
   4        STUDENTS TØØK THE EXAMINATIØN

EXAMINATIØN NUMBER         2
MEAN=        82.6667    STD DEVIATIØN=   12.4722
   3        STUDENTS TØØK THE EXAMINATIØN

EXAMINATIØN NUMBER         3
MEAN=        78.75     STD DEVIATIØN=   13.0072
   3        STUDENTS TØØK THE EXAMINATIØN
```

STUDENT NUMBER 1

EXAM	RAW SC	STD SC	PCT ABØVE	PCT BELØW
1	80	522	25	50
2	66	366	67	0
3	88	571	25	50
WEIGHTED SCØRE		507.5	25	50

STUDENT NUMBER 2

EXAM	RAW SC	STD SC	PCT ABØVE	PCT BELØW
1	90	647	0	75
2	NØT TAKEN			
3	95	625	0	75
WEIGHTED SCØRE		632.333	0	75

STUDENT NUMBER 3

EXAM	RAW SC	STD SC	PCT ABØVE	PCT BELØW
1	68	372	75	0
2	86	527	33	33
3	65	394	75	0
WEIGHTED SCØRE		421.75	75	0

```
STUDENT NUMBER 4
EXAM          RAW SC        STD SC        PCT ABØVE     PCT BELØW
 1             75            459           50            25
 2             96            607            0            67
 3             67            410           50            25
WEIGHTED SCØRE              471.5          50            25


STUDENTS RANKED BY WEIGHTED SCØRES

STUDENT       WEIGHTED SCØRE

 2            632.333
 1            507.5
 4            471.5
 3            421.75
```

PI

Program

```
 10   REM PRØGRAMMED BY JØHN WILCØX
 20   REM STUDENT IN SEC ED 356 AT UNIV ØF ILL
 30   PRINT "SELECT THE NUMBER ØF THE METHØD YØU CHØØSE"
 40   PRINT "1 EARLY FØRMULAS (CHINESE AND INDIAN)"
 50   PRINT "2 WALLIS' METHØD"
 60   PRINT "3 LEIBNIZ' METHØD"
 70   PRINT "4 BUFFØN NEEDLE"
 80   PRINT "5 JAPANESE METHØD"
 90   PRINT "6 STRASSNITZKY'S METHØD"
100   PRINT "WHAT METHØD DØ YØU CHØØSE ";
110   INPUT Z
120   ØN Z GØTØ 130,270,530,870,1100,1330,1480
130   PRINT "AN EARLY CHINESE WØRKER (ABØUT 470 A.D.)"
140   PRINT "EXPRESSED PI AS THE RATIØNAL NUMBER 355/113"
150   PRINT
160   PRINT "ARYABHATA, AN INDIAN (ABØUT 510 A.D.)"
170   PRINT "GAVE PI AS THE RATIØNAL NUMBER 62,832/20,000"
180   PRINT "DØ YØU WANT THE CØMPUTER TØ FIND THE DECIMAL"
190   PRINT "EQUIVALENT FØR THESE ? 1=NØ; 2=YES";
200   INPUT C
210   ØN C GØTØ 240,220
220   PRINT "355/133 =";355/113;
230   PRINT "  62,832/20,000 =";62832/20000
```

```
240   PRINT "WHAT NØW ? 1=TRY ANØTHER METHØD; 2=STØP PRØGRAM"
250   INPUT C
260   ØN C GØTØ 100,1480
270   PRINT "JØHN WALLIS WAS A 17TH CENTURY MATHEMATICIAN WHØ WANTED"
280   PRINT "TØ CALCULATE THE AREA ØF A CIRCLE. UNFØRTUNATELY,"
290   PRINT "HE LIVED BEFØRE THE TIME ØF NEWTØN AND LEIBNIZ, SØ HE"
300   PRINT "WAS UNABLE TØ USE THE INTEGRAL CALCULUS. NØR CØULD HE"
310   PRINT "USE THE BINØMIAL EXPANSIØN. HE USED INSTEAD A SERIES ØF"
320   PRINT "INTERPØLATIØNS AND INDUCTIVE PRØCEDURES, ARRIVING AT"
330   PRINT "A VALUE FØR PI INVØLVING ØNLY RATIØNAL NUMBERS. HIS "
340   PRINT "EXPRESSIØN FØR CALCULATING PI WAS"
350   PRINT "        PI=4*(2/3)*(4/3)*(4/5)*(6/5)*(6/7)....."
360   PRINT "HØW MANY FACTØRS DØ YØU WANT IN THE PRØDUCT ?"
370   INPUT A
380   REM ØNLY EVEN NUMBER ØF FACTØRS CØMPUTED
390   REM SEE LEIBNIZ METHØD FØR GETTING INTEGRAL NUMBER
400   LET B=INT(A/2)
410   LET P=4
420   LET T=2
430   FØR I=1 TØ B
440   LET P=P*(T/(T+1))
450   LET P=P*((T+2)/(T+1))
460   LET T=T+2
470   NEXT I
480   PRINT "ESTIMATE FØR PI BY WALLIS' METHØD IS ";P
490   PRINT "WHAT NØW ? 1=ØBTAIN ANØTHER ESTIMATE; 2=TRY ANØTHER"
500   PRINT "METHØD; 3=STØP PRØGRAM"
510   INPUT C
520   ØN C GØTØ 360,100,1480
530   PRINT "LEIBNIZ, IN 1674, FØUND PI TØ BE THE LIMIT"
540   PRINT "ØF THE INFINITE SERIES BELØW"
550   PRINT "   PI=4*(1−1/3+1/5−1/7+1/9−1/11 .....)"
560   PRINT "HØW MANY TERMS DØ YØU WANT IN THE SERIES";
570   INPUT A
580   IF A=1 THEN 800
590   LET A1=A/2
600   IF INT(A1)=A1 THEN 710
610   LET P4=1
620   LET K=3
630   LET J=INT(A1)
640   FØR I=1 TØ J
650   LET C=−1/K
660   LET D=1/(K+2)
```

```
670   LET P4=P4+C+D
680   LET K=K+4
690   NEXT I
700   GØTØ 810
710   LET P4=0
720   LET K=1
730   FØR I=1 TØ A1
740   LET C=1/K
750   LET D=-1/(K+2)
760   LET P4=C+D+P4
770   LET K=K+4
780   NEXT I
790   GØTØ 810
800   LET P4=1
810   LET P=4*P4
820   PRINT "VALUE FØR PI IS ";P
830   PRINT "WHAT NØW ? 1=ØBTAIN ANØTHER ESTIMATE; 2=TRY ANØTHER METHØD"
840   PRINT "  3=STØP PRØGRAM"
850   INPUT C
860   ØN C GØTØ 560,100,1480
870   PRINT "BUFFØN'S NEEDLE METHØD USES A MØNTE CARLØ TECHNIQUE"
880   PRINT "TØ FIND PI. THE PRØGRAM GENERATES RANDØM PØINTS"
890   PRINT "REPRESENTING THE LØCATIØN AND ANGLE ØF THE FALL ØF"
900   PRINT "A NEEDLE ØN A PLANE RULED WITH EQUIDISTANT PARALLEL"
910   PRINT "LINES. PI IS CALCULATED FRØM AN ESTIMATE ØF THE"
920   PRINT "PROBABILITY THAT THE NEEDLE HITS ØNE ØF THE LINES."
930   PRINT "HØW MANY TIMES SHALL WE THRØW THE NEEDLE";
940   INPUT N
950   LET T=0
960   FØR I=1 TØ N
970   REM GENERATE RANDØM X,Y SUCH THAT"
980   REM 0<=X<=PI/2 AND 0<=Y<=1
990   LET Y=RND(1)
1000  LET X=RND(1)*3.141596/2
1010  LET S=SIN(X)
1020  IF Y>S THEN 1040
1030  LET T=T+1
1040  NEXT I
1050  PRINT "PI IS ESTIMATED AS ";2/(T/N)
1060  PRINT "WHAT NØW ? 1=GET ANØTHER ESTIMATE 2=TRY ØTHER"
1070  PRINT "METHØD 3=STØP";
1080  INPUT C
1090  ØN C GØTØ 930,100,1480
```

```
1100    PRINT "JAPANESE METHØD—DEVELØPED BY MATHEMATICIAN"
1110    PRINT "TAKEBE IN 1722. IT TAKES THE FIRST QUADRANT"
1120    PRINT "ØF A CIRCLE CENTERED AT ØRIGIN WITH RADIUS 1"
1130    PRINT "AND DIVIDES THE X-AXIS FRØM 0 TØ 1 INTØ N"
1140    PRINT "INTERVALS. IF LINES THRØUGH THESE N PØINTS"
1150    PRINT "ARE DRAWN TØ THE CIRCLE THE AREAS ØF THE"
1160    PRINT "RESULTING RECTANGLES ARE CØMPUTED AND SUMMED,"
1170    PRINT "GIVING AN ESTIMATE ØF PI/4."
1180    PRINT "HØW MANY INTERVALS N DØ YØU WANT ?";
1190    INPUT D1
1200    LET B=1
1210    LET A=0
1220    FØR I=1 TØ D1
1230    LET C=SQR(1−(I/D1)↑2)
1240    LET A=A+.5*(1/D1)*(B+C)
1250    LET B=C
1260    NEXT I
1270    LET A=4*A
1280    PRINT "ESTIMATE FØR PI IS ";A
1290    PRINT "WHAT NØW ? 1=GET ANØTHER ESTIMATE; 2=TRY ANØTHER METHØD"
1300    PRINT "3=STØP ";
1310    INPUT C
1320    ØN C GØTØ 1180,100,1480
1330    PRINT "THE VIENNESE MATHEMATICIAN STRASSNITZKY GAVE JØHANN"
1340    PRINT "MARTIN ZACHARIAS DAZE THIS FØRMULA FØR PI"
1350    PRINT "  PI/4=ARCTAN(1/2)+ARCTAN(1/5)+ARCTAN(1/8)."
1360    PRINT "THE AMAZING MR. DAZE WAS ABLE TØ USE THIS FØRMULA"
1370    PRINT "TØ MENTALLY CØMPUTE PI TØ 205 DECIMAL PLACES,"
1380    PRINT "ALL ØF WHICH THE LAST 5 TURNED ØUT TØ BE CORRECT!"
1390    PRINT "DØ YØU WANT TØ SEE WHAT VALUE FØR PI THE CØMPUTER"
1400    PRINT "GETS USING THIS FØRMULA ? 1=NØ 2=YES"
1410    INPUT C
1420    ØN C GØTØ 1450,1430
1430    LET P=4*(ATN(.5)+ATN(.2)+ATN(.125))
1440    PRINT "CØMPUTER GETS THIS VALUE FØR PI",P
1450    PRINT "WHAT NØW ? 1=TRY ANØTHER METHØD; 2=STØP PRØGRAM"
1460    INPUT C
1470    ØN C GØTØ 100,1480
1480    END
```

Sample Output

SELECT THE NUMBER ØF THE METHØD YØU CHØØSE
1 EARLY FØRMULAS (CHINESE AND INDIAN)
2 WALLIS' METHØD
3 LEIBNIZ' METHØD
4 BUFFØN NEEDLE
5 JAPANESE METHØD
6 STRASSNITZKY'S METHØD
WHAT METHØD DØ YØU CHØØSE ?1
AN EARLY CHINESE WØRKER (ABØUT 470 A.D.)
EXPRESSED PI AS THE RATIØNAL NUMBER 355/113

ARYABHATA, AN INDIAN (ABØUT 510 A.D.),
GAVE PI AS THE RATIØNAL NUMBER 62,832/20,000.
DØ YØU WANT THE CØMPUTER TØ FIND THE DECIMAL
EQUIVALENT FØR THESE ? 1=NØ; 2=YES ?2
355/113=3.14159 62,832/20,000=3.1416
WHAT NØW ? 1=TRY ANØTHER METHØD; 2=STØP PRØGRAM
     ?1
WHAT METHØD DØ YØU CHØØSE ?2

JØHN WALLIS WAS A 17TH CENTURY MATHEMATICIAN WHØ WANTED
TØ CALCULATE THE AREA ØF A CIRCLE. UNFØRTUNATELY,
HE LIVED BEFØRE THE TIME ØF NEWTØN AND LEIBNIZ, SØ HE
WAS UNABLE TØ USE THE INTEGRAL CALCULUS. NØR CØULD HE
USE THE BINØMIAL EXPANSIØN. HE USED INSTEAD A SERIES ØF
INTERPØLATIØNS AND INDUCTIVE PRØCEDURES, ARRIVING AT
A VALUE FØR PI INVØLVING ØNLY RATIØNAL NUMBERS. HIS
EXPRESSIØN FØR CALCULATING PI WAS
        PI=4*(2/3)*(4/3)*(4/5)*(6/5)*(6/7).....
HØW MANY FACTØRS DØ YØU WANT IN THE PRØDUCT ?
   ?40
ESTIMATE FØR PI BY WALLIS' METHØD IS 3.17921
WHAT NØW ? 1=ØBTAIN ANØTHER ESTIMATE; 2=TRY ANØTHER METHØD
    3=STØP PRØGRAM
    ?2
WHAT METHØD DØ YØU CHØØSE ?3
LEIBNIZ, IN 1674, FØUND PI TØ BE THE LIMIT
ØF THE INFINITE SERIES BELØW
    PI=4*(1−1/3+1/5−1/7+1/9−1/11 .....)
HØW MANY TERMS DØ YØU WANT IN THE SERIES ?50
VALUE FØR PI IS 3.12159

WHAT NØW ? 1=ØBTAIN ANØTHER ESTIMATE; 2=TRY ANØTHER METHØD
   3=STØP PRØGRAM
   ?2
WHAT METHØD DØ YØU CHØØSE ?4
BUFFØN'S NEEDLE METHØD USES A MØNTE CARLØ TECHNIQUE
TØ FIND PI. THE PRØGRAM GENERATES RANDØM PØINTS
REPRESENTING THE LØCATIØN AND ANGLE ØF THE FALL ØF
A NEEDLE ØN A PLANE RULED WITH EQUIDISTANT PARALLEL
LINES. PI IS CALCULATED FRØM AN ESTIMATE ØF THE
PRØBABILITY THAT THE NEEDLE HITS ØNE ØF THE LINES.
HØW MANY TIMES SHALL WE THRØW THE NEEDLE ?800
PI IS ESTIMATED AS 3.07692
WHAT NØW ? 1=GET ANØTHER ESTIMATE 2=TRY ØTHER
METHØD 3=STØP ?1
HØW MANY TIMES SHALL WE THRØW THE NEEDLE ?500
PI IS ESTIMATED AS 3.34448
WHAT NØW ? 1=GET ANØTHER ESTIMATE 2=TRY ØTHER
METHØD 3=STØP ? 1
HØW MANY TIMES SHALL WE THRØW THE NEEDLE ?500
PI IS ESTIMATED AS 3.07692
WHAT NØW ? 1=GET ANØTHER ESTIMATE 2=TRY ØTHER
METHØD 3=STØP ?2
WHAT METHØD DØ YØU CHØØSE ?5
JAPANESE METHØD—DEVELØPED BY MATHEMATICIAN
TAKEBE IN 1722. IT TAKES THE FIRST QUADRANT
ØF A CIRCLE CENTERED AT ØRIGIN WITH RADIUS 1
AND DIVIDES THE X-AXIS FRØM 0 TØ 1 INTØ N
INTERVALS. IF LINES THRØUGH THESE N PØINTS
ARE DRAWN TØ THE CIRCLE THE AREAS ØF THE
RESULTING RECTANGLES ARE CØMPUTED AND SUMMED,
GIVING AN ESTIMATE ØF PI/4.
HØW MANY INTERVALS N DØ YØU WANT ? ?100
ESTIMATE FØR PI IS 3.14042
WHAT NØW ? 1=GET ANØTHER ESTIMATE; 2=TRY ANØTHER METHØD
3=STØP ?2
WHAT METHØD DØ YØU CHØØSE ?6
THE VIENNESE MATHEMATICIAN STRASSNITZKY GAVE JØHANN
MARTIN ZACHARIAS DAZE THIS FØRMULA FØR PI
   PI/4=ARCTAN(1/2)+ARCTAN(1/5)+ARCTAN(1/8).
THE AMAZING MR. DAZE WAS ABLE TØ USE THIS FØRMULA
TØ MENTALLY CØMPUTE PI TØ 205 DECIMAL PLACES,
ALL ØF WHICH EXCEPT THE LAST 5 TURNED ØUT TØ BE CØRRECT!
DØ YØU WANT TØ SEE WHAT VALUE FØR PI THE CØMPUTER

```
          GETS USING THIS FØRMULA ? 1=NØ 2=YES
              ?2
          CØMPUTER GETS THIS VALUE FØR PI              3.14159
          WHAT NØW ? 1=TRY ANØTHER METHØD; 2=STØP PRØGRAM
              ?2
```

TEMP*

Program

```
 10   DIM T(4)
 20   FØR C=1 TØ 4
 30   REM READ IN TEMP AT EACH EDGE
 40   READ T(C)
 50   NEXT C
 60   REM WE WANT TØ FIND TEMP AT 5 DIFFERENT PØINTS
 70   FØR D=1 TØ 5
 80   LET S=0
 90   REM READ IN CØØRDINATES ØF EACH PØINT
100   READ I1,J1
110   REM TAKE RANDØM WALK UNTIL EDGE ØF PLATE IS REACHED
120   REM PLATE IS 10 UNITS BY 5 UNITS
130   FØR X=1 TØ 500
140   LET I=I1
150   LET J=J1
160   REM CHØØSE RANDØM DIRECTIØN FØR 1 UNIT WALK
170   REM CAN GØ EITHER EAST,WEST,NØRTH,SØUTH
180   LET Y=INT(RND(1)*4+1)
190   ØN Y GØTØ 200,240,280,320
200   LET I=I+1
210   REM HAVE WE REACHED AN EDGE ?
220   IF I=10 THEN 370
230   GØTØ 180
240   LET I=I−1
250   REM HAVE WE REACHED AN EDGE ?
260   IF I=0 THEN 370
270   GØTØ 180
280   LET J=J+1
290   REM HAVE WE REACHED AN EDGE ?
300   IF J=5 THEN 370
```

* Programmed by Tom Nall, senior at Buffalo Grove High School, Arlington Heights, Illinois, in May 1975.

```
310   GØTØ 180
320   LET J=J−1
330   REM HAVE WE REACHED AN EDGE ?
340   IF J=0 THEN 370
350   GØTØ 180
360   REM ADD TEMP AT EDGE ØF PLATE
370   LET S=S+T(Y)
380   NEXT X
390   PRINT "ESTIMATED TEMP AT PØINT (";I1;",";J1;")=";S/500;" DEG C"
400   PRINT
410   NEXT D
420   REM TEMPS AT EDGE ØF PLATE
430   DATA 30,3,15,40
440   REM PØINTS (X,Y) FØR WHICH TEMP WANTED
450   DATA 4,2,1,3,5,2.5,2,4,3,1
460   END
```

Sample Output

```
ESTIMATED TEMP AT PØINT ( 4 , 2 ) = 28.432 DEG C
ESTIMATED TEMP AT PØINT ( 1 , 3 ) = 10.158 DEG C
ESTIMATED TEMP AT PØINT ( 5 , 2.5 ) = 17.202 DEG C
ESTIMATED TEMP AT PØINT ( 2 , 4 ) = 14.73 DEG C
ESTIMATED TEMP AT PØINT ( 3 , 1 ) = 32.106 DEG C
```

SIMEQ

Program

```
10   REM THIS PRØGRAM GENERATES 8 SYSTEMS ØF EQUATIØNS, EACH SYSTEM
20   REM CØNSISTING ØF TWØ EQUATIØNS IN TWØ VARIABLES. ALL (X,Y)
30   REM PAIRS ØF ANSWERS ARE SUCH THAT EACH ØF X AND Y RANGES ØVER
40   REM THE INTEGERS FRØM −10 TØ +10. THE CØEFFICIENTS ØF X AND Y
50   REM IN EACH EQUATIØN HAVE THE SAME RANGE. THE (X,Y) SØLUTIØNS
60   REM AND THE CØEFFICIENTS ARE GENERATED AT RANDØM WITHIN THE LIMITS
70   REM STATED ABØVE
80   REM IF THE USER WISHES TØ RUN THIS PRØGRAM WITH A DITTØ MASTER, HE
90   REM SHØULD:
100  REM        A.  KNØW THAT HE WILL GET EIGHT PRØBLEMS — ØTHER-
110  REM            WISE, CHANGE P (LINE 270) ACCØRDINGLY;
120  REM        B.  LET THE ANSWERS PRINT ØN THE MASTER. WHEN RUNNING
```

```
130 REM                 CØPIES HE CAN THEN RUN A FEW WITH ANSWERS AND MASK
140 REM                 THEM ØUT BEFØRE RUNNING ØTHER (STUDENT) CØPIES.
150 DIM A(6)
160 PRINT
170 PRINT
180 PRINT "ALGEBRA          NAME_ _ _ _ _ _ _ _ _ _ _ _ _ _ _ _ _ _ _ _ _ _ _ _ _ _ _"
190 PRINT
200 PRINT "SØLVE EACH ØF THE FØLLØWING SYSTEMS ØF EQUATIØNS. FØLLØW"
210 PRINT "THE DIRECTIØNS GIVEN BY YØUR TEACHER. WRITE YØUR ANSWERS"
220 PRINT "IN THE SPACES PRØVIDED AT THE RIGHT ØF EACH PRØBLEM."
230 PRINT
240 PRINT "PRØBLEM"
250 REM CHANGE P IF YØU WISH TØ GENERATE A DIFFERENT NUMBER ØF EXERCISES
270 LET P=8
280 FØR N=1 TØ P
290 REM LINES 310 TØ 380 GENERATE THE + AND − SIGNS FØR X AND Y AND
300 REM FØR THE CØEFFICIENTS.
310 FØR K=1 TØ 6
320 LET Z=INT(RND(8)*10+1)
330 IF Z<5 THEN 360
340 LET S=−1
350 GØ TØ 370
360 LET S=1
370 LET A(K)=S
380 NEXT K
390 REM LINES 400 TØ 490 FØRM X, Y, THE CØEFFICIENTS, AND THE CØNSTANTS
400 LET X=INT(RND(8)*10+1)
410 LET X1=A(5)*X
420 LET Y=INT(RND(8)*10+1)
430 LET Y1=A(6)*Y
440 LET A1=INT(RND(8)*10+1)
450 LET B1=INT(RND(8)*10+1)
460 LET A2=INT(RND(8)*10+1)
470 LET B2=INT(RND(8)*10+1)
480 LET C1=A(1)*A1*X1 + A(2)*B1*Y1
490 LET C2=A(3)*A2*X1 + A(4)*B2*Y1
500 PRINT " ";N;"."
505 PRINT " ";A(1)*A1;"X + ";A(2)*B1;"Y = ";C1;TAB(40);"X=";X1
510 PRINT " ";A(3)*A2;"X + ";A(4)*B2;"Y = ";C2;TAB(40);"Y=";Y1
550 PRINT
560 PRINT
570 NEXT N
580 END
```

Sample Output

ALGEBRA   NAME_ _

SØLVE EACH ØF THE FØLLØWING SYSTEMS ØF EQUATIØNS. FØLLØW
THE DIRECTIØNS GIVEN BY YØUR TEACHER. WRITE YØUR ANSWERS
IN THE SPACES PRØVIDED AT THE RIGHT ØF EACH PRØBLEM.

PRØBLEM

1 .
$2 X + -1 Y = -16$ $X = -9$
$8 X + -7 Y = -58$ $Y = -2$

2 .
$-9 X + -7 Y = 119$ $X = -7$
$-5 X + -8 Y = 99$ $Y = -8$

3 .
$-6 X + 9 Y = -72$ $X = 9$
$5 X + -5 Y = 55$ $Y = -2$

4 .
$9 X + -10 Y = -2$ $X = 2$
$-1 X + -3 Y = -8$ $Y = 2$

5 .
$10 X + 3 Y = -44$ $X = -5$
$6 X + 2 Y = -26$ $Y = 2$

6 .
$1 X + -1 Y = 7$ $X = 9$
$-10 X + 5 Y = -80$ $Y = 2$

7 .
$3 X + -10 Y = 72$ $X = -6$
$2 X + -3 Y = 15$ $Y = -9$

8 .
$7 X + -7 Y = 77$ $X = 3$
$-9 X + -9 Y = 45$ $Y = -8$

FACTOR

Program

```
 10   RANDØMIZE
 20   FØR J=1 TØ 5
 30   PRINT
 40   NEXT J
 50   PRINT "        SUPPLEMENTARY EXERCISES IN FACTØRING FØR TEACHERS"
 60   PRINT "        _____"
 70   PRINT "        FACTØRS ØF TRINØMIALS ØF THE FØRM:"
 80   PRINT "                              2"
 90   PRINT "                   AX    +    BX    +    C"
100   PRINT
110   PRINT "        A,B,C ARE RANDØMLY PØSITIVE ØR NEGATIVE"
130   PRINT
140   PRINT "        EXERCISES INCREASE GRADUALLY IN LEVEL ØF DIFFICULTY."
150   PRINT "        SØME EXERCISES MAY DUPLICATE ØTHERS DUE TØ RANDØM"
160   PRINT "        GENERATIØN ØF THE CØEFFICIENTS. THE FACTØRED FØRMS"
170   PRINT "        PRØDUCED MAY CØNTAIN CØMMØN MØNØMIAL FACTØRS."
180   PRINT
190   LET P=1
200   LET N=1
210   LET M=5
220   LET M1=6
230   R1=INT(RND(8)*M+1)
240   R2=INT(RND(8)*M+1)
250   S1=INT(RND(8)*M1+1)
260   IF R1=S1 THEN 250
270   IF R1/2<>INT(R1/2) THEN 290
280   IF S1/2=INT(S1/2) THEN 250
290   IF R1/3<>INT(R1/3) THEN 310
300   IF S1/3=INT(S1/3) THEN 250
310   S2=INT(RND(8)*M1+1)
320   IF R2=S2 THEN 310
330   FØR K=2 TØ 3
340   IF R2/K<>INT(R2/K) THEN 360
350   IF S2/K=INT(S2/K) THEN 310
360   NEXT K
370   IF N>=8 THEN 390
380   IF R1*R2>9 THEN 230
390   IF N<=16 THEN 410
400   IF R1*R2<10 THEN 230
```

```
410   IF N<20 THEN 440
420   GØSUB 770
430   LET R1=R1*T
440   GØSUB 770
450   LET S1=S1*T
460   GØSUB 770
470   LET S2=S2*T
480   LET A=R1*R2
490   LET B=R1*S2+R2*S1
500   LET C=S1*S2
510   PRINT
520   PRINT N;".";"          2"
530   PRINT "          ";A;"X    + ";B;"X    + ";C;"=";
535   PRINT "(";R1;"X + ";S1;")(";R2;"X +";S2;")"
540   IF N=6 THEN 570
550   PRINT
560   GØ TØ 590
570   LET M=M+1
580   LET M1=M1+1
590   LET N=N+1
600   IF N=11 THEN 640
610   IF N=24 THEN 640
620   IF N=37 THEN 640
630   GØ TØ 230
640   PRINT
650   PRINT "                        PAGE"; P
660   LET P=P+1
670   FØR K=1 TØ 10
680   PRINT
690   NEXT K
700   IF N=37 THEN 840
710   IF N=12 THEN 730
720   GØ TØ 230
730   LET M=M+2
740   LET M1=M1+3
750   GØ TØ 230
760   STØP
770   LET Z7=INT(RND(8)*9+1)
780   IF Z7>5 THEN 810
790   LET Z7=-1
800   GØ TØ 820
810   LET Z7=1
820   LET T=Z7
830   RETURN
840   END
```

Sample Output

```
                    SUPPLEMENTARY EXERCISES IN FACTØRING FØR TEACHERS
            - - - - - - - - - - - - - - - - - - - - - - - - - - - - -
        FACTØRS ØF TRINØMIALS ØF THE FØRM:
                              2
                          AX   +   BX   +   C

            A,B,C ARE RANDØMLY PØSITIVE ØR NEGATIVE

        EXERCISES INCREASE GRADUALLY IN LEVEL ØF DIFFICULTY.
        SØME EXERCISES MAY DUPLICATE ØTHERS DUE TØ RANDØM
        GENERATIØN ØF THE CØEFFICIENTS. THE FACTØRED FØRMS
        PRØDUCED MAY CØNTAIN CØMMØN MØNØMIAL FACTØRS.


    1 .         2
            3 X   + 23 X + 30 =( 1 X + 6 )( 3 X + 5 )


    2 .         2
            6 X   + 11 X + 3 =( 2 X + 3 )( 3 X + 1 )


    3 .         2
            4 X   + 0 X + −25 =( 2 X + 5 )( 2 X +−5 )


    4 .         2
            3 X   + −11 X + 10 =( 3 X + 5 )( 1 X + 2 )


    5 .         2
            6 X   + −5 X + −4 =( 2 X + 1 )( 3 X +−4 )


    6 .         2
            6 X   + −7 X + 2 =( 3 X + −2 )( 2 X +−1 )


    7 .         2
            6 X   + 11 X + −2 =( 6 X + −1 )( 1 X + 2 )


    8 .         2
            4 X   + −29 X + 30 =( 4 X + −5 )( 1 X +−6 )


    9 .         2
           12 X   + −8 X + 1 =( 2 X + −1 )( 6 X +−1 )


   10 .         2
            6 X   + −25 X + 4 =( 1 X + −4 )( 6 X +−1 )


                            PAGE 1
```

DISTR

Program

```
  25    RANDØMIZE
  50    REM THIS PRØGRAM GENERATES 30 EXERCISES ØF THE TYPE
 100    REM          A(BX+C)=
 150    REM WHERE THE STUDENT IS EXPECTED TØ PRØDUCE THE
 200    REM PRØDUCT:     ABX+AC
 300    REM ANSWERS ARE NØT INCLUDED IN THE PRØGRAM! !
 400    REM A DITTØ MASTER MAY BE MADE IN THE USUAL WAY PRØVIDED        .
 450    REM THE PAPER IS SET PRØPERLY (EXPERIMENT A BIT)
 500    LET N=1
 550    DIM A(6)
 600    FØR L=1 TØ 5
 650    PRINT
 700    NEXT L
 750    PRINT "ALGEBRA                      NAME_ _ _ _ _ _ _ _ _ _ _ _ _ _ _ _ _ _ _ _"
 800    PRINT
 850    PRINT "USE THE DISTRIBUTIVE PRØPERTY TØ CØMPLETE THE FØLLØWING:"
 950    PRINT
1000    PRINT
1050    FØR K=1 TØ 6
1100    LET S=INT(RND(8)*9+1)
1150    IF S>5 THEN 1300
1200    LET Z=1
1250    GØ TØ 1350
1300    LET Z=−1
1350    LET A(K)=Z
1400    NEXT K
1450    LET A1=INT(RND(8)*9+1)
1500    LET B1=INT(RND(8)*9+1)
1550    LET C1=INT(RND(8)*9+1)
1600    LET A2=INT(RND(8)*9+1)
1650    LET B2=INT(RND(8)*9+1)
1700    LET C2=INT(RND(8)*9+1)
1750    IF N>=4 THEN 1800
1760    IF A1<=4 THEN 1800
1770    IF A(1)>=0 THEN 1800
1780    GØ TØ 1050
1800    IF N<=7 THEN 1850
1810    IF A1>=5 THEN 1850
1820    GØTØ 1450
```

```
1850  LET A1=A(1)*A1
1900  LET B1=A(2)*B1
1950  LET C1=A(3)*C1
2000  LET A2=A(4)*A2
2050  LET B2=A(5)*B2
2100  LET C2=A(6)*C2
2150  PRINT N;".   ";A1;"(";B1;"X + ";C1;") =";
2160  PRINT N+15;".   ";A2;"(";B2;"X +";C2;") ="
2350  PRINT
2400  PRINT
2450  IF N=15 THEN 2600
2500  LET N=N+1
2550  GØ TØ 1050
2600  FØR J=1 TØ 4
2625  PRINT
2650  NEXT J
2800  END
```

Sample Output

ALGEBRA   NAME_ _
USE THE DISTRIBUTIVE PRØPERTY TØ CØMPLETE THE FØLLØWING:

1 .	7 (−9 X + −1) =	10 .	8 (−4 X + 6) =
2 .	7 (−3 X + 7) =	11 .	9 (−6 X + 1) =
3 .	5 (1 X + 4) =	12 .	−7 (−7 X + −4) =
4 .	−6 (1 X + −8) =	13 .	6 (2 X + 6) =
5 .	−5 (3 X + −4) =	14 .	−8 (6 X + 3) =
6 .	8 (3 X + 2) =	15 .	−5 (−3 X + 3) =
7 .	4 (9 X + 3) =	16 .	−1 (5 X + 2) =
8 .	6 (−6 X + −3) =	17 .	8 (9 X + 3) =
9 .	−6 (−2 X + 8) =	18 .	7 (−6 X + 8) =

19 .   5 (−7 X + −6) =   25 .   −9 (8 X + −6) =

20 .   7 (1 X + −6) =   26 .   9 (2 X + −9) =

21 .   −5 (8 X +   3) =   27 .   −7 (−7 X + −3) =

22 .   3 (−8 X +   8) =   28 .   −5 (4 X +   3) =

23 .   2 (−8 X +   3) =   29 .   −1 (7 X +   8) =

24 .   1 (−1 X + −1) =   30 .   6 (−5 X + −1) =

RAND

Program

```
10   READ B,N,Z
20   DATA 5,4,25
30   FOR K=1 TØ Z
40   FØR J=1 TØ N
50   FØR I=1 TØ B
60   PRINT INT(10*RND(1));
70   NEXT I
80   PRINT "      ";
90   NEXT J
100  IF K/5<>INT(K/5) GØTØ 120
110  PRINT
120  NEXT K
130  END
```

Sample Output (500 Random Digits)

```
2 6 2 9 4     7 8 1 1 0     7 6 9 8 9     8 9 6 6 7
8 6 4 7 5     2 1 6 2 7     8 1 5 8 4     4 2 7 6 9
0 2 1 1 8     9 0 2 0 0     2 0 9 0 4     1 9 2 5 1
0 7 7 1 3     3 8 1 0 0     9 4 3 8 1     5 4 8 5 8
2 9 1 2 1     5 9 5 1 6     2 7 6 6 8     8 3 9 5 4

9 8 3 7 8     3 8 7 6 7     6 6 0 6 3     3 8 3 9 4
5 6 3 2 8     6 7 8 5 9     0 6 1 7 8     1 1 7 4 3
9 5 6 5 6     0 9 4 2 7     5 0 3 7 9     3 0 2 0 6
7 1 3 7 9     2 0 5 1 5     8 4 3 6 3     6 9 4 8 9
6 8 4 9 1     6 4 5 9 1     4 6 6 6 9     6 7 0 6 2
```

```
3 0 2 0 4   6 5 3 4 9   5 8 1 4 9   2 0 9 7 3
5 8 0 4 5   6 4 9 3 6   8 5 0 0 8   8 6 9 3 9
6 7 2 5 5   1 1 2 3 4   7 8 5 5 9   1 9 3 7 6
5 2 9 5 8   0 5 0 1 9   1 9 1 5 2   0 1 6 1 2
6 9 2 3 3   1 4 9 2 3   8 3 3 8 8   5 7 6 1 9

5 5 7 6 2   6 6 1 7 3   2 3 4 9 4   1 7 2 8 5
7 6 9 3 0   4 4 0 6 3   8 1 0 8 9   7 4 7 0 8
3 6 7 3 1   6 6 7 7 4   9 8 4 0 6   5 8 0 5 0
9 6 1 9 9   2 1 8 6 3   0 7 8 4 8   4 2 5 6 3
6 4 1 5 2   1 4 2 6 2   9 4 7 6 1   0 9 6 6 8

6 8 7 8 2   4 5 4 1 1   3 7 5 6 8   8 6 0 5 3
3 8 4 0 4   4 8 4 2 9   9 8 7 7 3   1 6 7 1 8
4 0 2 5 1   2 1 0 2 5   7 1 4 7 7   3 0 4 8 8
7 1 2 8 1   9 1 5 2 2   9 6 2 3 5   9 4 3 2 1
8 9 1 6 2   1 3 5 1 3   0 7 2 6 5   8 6 7 3 1
```

References

Barrodale, Ian; Roberts, Frank D. K.; and Ehle, Byron L. *Elementary Computer Applications in Science, Engineering and Business.* New York: Wiley, 1971.
Kemeny, John G. and Kurtz, Thomas E. *BASIC Programming.* New York: Wiley, 1971.

Teaching Aids and Resources

This list includes a variety of books and publications which teachers have found to be helpful. It is intended as a supplement to those references cited at the end of each chapter. Certain items will, of course, be more suitable for some teachers and classes than they will be for others.

To aid you in selecting appropriate materials, we have indicated certain categories by a letter in front of each item:

- B Basic arithmetic
- A Algebra
- G Geometry
- C Computers
- P Problems
- M Games, activities
- H History
- R Reference
- O Other

548

BOOKS

H Aaboe, Asger. *Episodes from the Early History of Mathematics*. New York: Random House, 1964.
> Historical account of some important mathematical concepts.

G Abbott, Edwin A. *Flatland*. New York: Barnes and Noble, 1963.
> Imaginative account of one- and two-dimensional worlds.

G Adler, C. F. *Modern Geometry*, 2nd ed. New York: McGraw-Hill, 1967.
> Introduction to several geometries.

B Adler, Irving. *Magic House of Numbers*. New York: John Day, 1957.
> Mathematical curiosities and riddles. Introduction to the basis of our number system.

B Adler, Irving. *The Giant Golden Book of Mathematics*. Wayne, New Jersey: Golden Press, 1960.
> Introduction to history of standards and measures, arithmetic, geometry, trigonometry, algebra, and calculus.

O Adler, Irving. *Probability and Statistics for Everyman*. New York: John Day, 1963.
> Guide to fundamentals of probability and statistics. Concepts of sample space, probability, random variable, expectations, standard deviations, and correlations explained.

O Adler, Irving. *Logic for Beginners*. New York: John Day, 1964.
> Entertaining introduction to logic.

O Adler, Irving, ed. *Readings in Mathematics*, Books 1 and 2. Lexington, Massachusetts: Ginn, 1972.
> Book 1—Collections of mathematical readings on numbers.
> Book 2—Collections of mathematical readings on probability, growth, surfaces, volumes, natural phenomena, rhythms, and making your own discoveries.

O Adler, Irving. *The Changing Tools of Science: from Yardstick to Synchrotron*. New York: John Day, 1973.
> Revised edition of *The Tools of Science* (1958). Focus on measurement.

C Ahl, David H., ed. *101 BASIC Computer Games*. Maynard, Massachusetts: Digital Equipment Corporation, 1973.
> Contains name of game, author, brief description, mention of any computer limitations, program listing, sample run, and, where appropriate, some diagrams.

C Ahl, David H. *Getting Started in Classroom Computing*. Maynard, Massachusetts: Digital Equipment Corporation, 1974.
> Written for secondary students. Provides introduction to several computer-related concepts through a set of six classroom games.

O Ahrendt, Myrl H. *The Mathematics of Space Exploration*. New York: Holt, Rinehart and Winston, 1965.
> Discusses the mathematics of motion and gravity in relation to space exploration. Exercises and solutions.

O Albarn, Keith et al. *The Language of Pattern*. New York: Harper & Row, 1974.
> Discusses many varied types of patterns, with diagrams as well as explanations.

C Albrecht, Robert L.; Lindberg, Eric; and Mann, Walter. *Computer Methods in Mathematics*. Reading, Massachusetts: Addison-Wesley, 1968.
>Eight chapters on BASIC language. Includes section on functions, sequences and series, and arrays. Last three chapters on FORTRAN language.

C Allison, Ronald et al. *Problems for Computer Mathematics*. Maynard, Massachusetts: Digital Equipment Corporation, 1971.
>Contains 80 problems for computer solution, ranging from simple to complex.

O Anderson, Raymond W. *Romping Through Mathematics*. New York: Knopf, 1947.
>Mathematics from basic numbers to differential calculus.

C Andree, Richard B. *Computer Programming and Related Mathematics*. New York: Wiley, 1967.
>Introduces problem-solving processes and mathematics related to programming and analysis.

B Asimov, Isaac. *Realm of Numbers*. Boston: Houghton Mifflin, 1959.
>Topics include finger counting, square roots, logarithms, rational and irrational numbers, and infinity.

A Asimov, Isaac. *Realm of Algebra*. Boston: Houghton Mifflin, 1961.
>Includes consideration of quadratic and cubic equations, simultaneous equations, and those involving imaginary and transcendental numbers. More theory than drill.

B Asimov, Isaac. *Quick and Easy Math*. Boston: Houghton Mifflin, 1964.
>Guide to rapid calculations.

O Asimov, Isaac. *Science, Numbers and I*. Garden City, New York: Doubleday, 1968.
>Enrichment for advanced high school students with mathematics and science background.

M Bakst, Aaron. *Mathematical Puzzles and Pastimes*. New York: Van Nostrand, 1954.
>Set of mathematical recreations. Examples from pure number theory, algebra, geometry, and trigonometry.

C Ball, Marion J. *What Is a Computer?* Boston: Houghton Mifflin, 1972.
>Introduction to computers. Includes history of computer development, operation, and fundamentals of flowcharting.

M Ball, W. W. Rouse. *Mathematical Recreation and Essays*. Riverside, New Jersey: Macmillan, 1939.
>Classic book on recreational mathematics.

M Bernard, Douglas. *Adventures in Mathematics*. New York: Hawthorne Books, 1965.
>Recreational mathematics for the general reader.

H Barnard, Douglas. *It's All Done by Numbers*. New York: Hawthorne Books, 1968.
>Introduction to the history of mathematics.

O Barnett, I. A. *Some Ideas about Number Theory*. Washington: National Council of Teachers of Mathematics, 1961.
>Informal account of some of the more elementary results of number theory. Includes classroom applications.

G Beard, Col. R. S. *Patterns in Space*. Palo Alto: Creative Publications, 1973.
>Includes topics on polygons, the Golden Section, Fibonacci numbers, curves, conics, spirals and triangles.

R Bell, Eric T. *Men of Mathematics*. New York: Simon and Schuster, 1937.
> Biographies of 30 great mathematicians with dominant ideas of modern mathematics.

R Bendick, Jeanne and Livin, Marcia. *Mathematics Illustrated Dictionary*. New York: McGraw-Hill, 1965.
> Guide to important mathematical facts and people. Information on the Greek, Roman, and Arabic number systems.

O Bergamini, David and the Editors of Life. *Mathematics*. Morristown, New Jersey: Silver Burdett, 1963.
> Account of some of the highlights of the history and development of mathematics. Includes considerations of the uses of mathematics in the arts, professions, etc.

H Bidwell, James K. and Clason, Robert G., eds. *Reading in the History of Mathematics Education*. Washington: National Council of Teachers of Mathematics, 1970.
> Complement to 32nd NCTM Yearbook. Contains committee reports and other primary source documents.

A Boyle, Patrick. *Graph Gallery*. Palo Alto: Creative Publications, 1971.
> Twenty graphs with equations to give specific pictures. Equations and graphs of result. Gives practice in plotting Cartesian coordinates and curves.

A Boyle, Patrick. *Palatable Plotting*. Palo Alto: Creative Publications, 1972.
> Twenty graphs with equations necessary to graph them. Equations and graphs of result.

A Boyle, Patrick and Juarez, William. *Accent on Algebra*. Palo Alto: Creative Publications, 1972.
> Emphasizes some of the vocabulary, concepts, and skills of elementary algebra in novel ways.

A Brant, V. and Keedy, M. L. *Introduction to Functions: Relations, Functions, and Graphs*. Boston: Holt, Rinehart and Winston.
> Treatment of relations and functions emphasizing examples of many relations. Includes absolute values, inequalities, conjunctions, and disjunctions.

B Brooke, Maxey. *150 Puzzles in Crypt Arithmetic*. New York: Dover, 1963.
> Collection of arithmetical puzzles with letters substituted for numbers. Solved by examination of the number relations between groups of symbols.

G Burger, Dionys. *Sphereland*. New York: Crowell, 1965.
> Fantasy about the expansion of a two-dimensional world.

G Carroll, Lewis. *Euclid and His Modern Rivals*. New York: Dover, 1973.
> Republished. Written as a play, with dialogue dealing with problems, pairs of lines, symmetry, incommensurables, angles, and proof. Indicates that no one has yet found the perfect geometry.

B Chandler, A. M., ed. *Experiences in Mathematical Ideas*, Volumes 1 and 2. Washington: National Council of Teachers of Mathematics, 1970.
> Thirteen units designed to provide meaningful mathematical experiences for low achievers in grades 5–8.

P Charosh, Mannis, compiler. *Mathematical Challenges: Selected Problems from The*

Mathematics Student Journal. Washington: National Council of Teachers of Mathematics, 1965.

    Includes a variety of challenging problems not usually met in the classroom.

C Corlett, P. N.; Tinsley, J. D.; and Court, R. A. *Practical Programming*, 2nd ed. New York: Cambridge University Press, 1972.

    Computing course. Some mathematics background assumed. Series of practical computer problems. Summary of ALGOL and FORTRAN programming.

G Coxeter, H. S. M. *Regular Polytopes*, 3rd ed. New York: Dover, 1973.

    Polygons, polyhedra, Euler's formula, rotation groups, star-polyhedra, truncation, kaleidoscopes, Petrie polygons, star-polytopes included, with historical summaries. Numerous diagrams, examples.

C Crawford, Rudd A., Jr. and Copp, David H. *Introduction to Computer Programming*. Boston: Houghton Mifflin, 1969.

    Workbook with answers. Does not require access to an actual computer.

C Crowley, Thomas. *Understanding Computers*. New York: McGraw-Hill, 1967.

    Provides explanation of how computers work.

G Cundy, H. Martyn and Rollet, A. P. *Mathematical Models*, 2nd ed. New York: Oxford University Press, 1961.

    Detailed instructions for making wide variety of models; includes dissections, paperfolding, curve stitching, plane tessellations, polyhedra, models for logic, and mechanical models.

M Dalton, LeRoy C. and Snyder, Henry D., eds. *Topics for Mathematics Clubs*. Reston, Virginia: National Council of Teachers of Mathematics, 1973.

    Collection of exciting topics not usually discussed in the classroom.

B Davis, Philip J. *The Lore of Large Numbers*, No. 6, New Mathematical Library. New York: Random House, 1961.

    Deals with aspects of very large numbers. Includes pi taken to 4000 places, exponents, and rapidity of growth of sequences.

A Del Grande, J. J.; Egsgard, J. C.; and Mulligan, H. A. *An Introduction to the Nature of Proof*. Toronto: W. J. Gage Ltd., 1967.

    Introduction to mathematical proof and logic. Includes inductive and deductive reasoning, negation and conjunction, disjunction, and quantifiers, with exercises at end of each section.

G Diggins, Julia. *String, Straightedge, and Shadow: The Story of Geometry*. New York: Viking, 1965.

    Chronicle of how man first noticed the beauty of geometric forms in nature, then put them to practical use.

C Dorn, William S. and Greenberg, Herbert J. *Mathematics and Computing, with FORTRAN Programming*. New York: Wiley, 1965.

    Reference for teachers. Problems and research material. Emphasizes flowcharts. Contains sections on linear equations, iteration procedures, logic, probability, and number theory.

G Elliott, H. A.; MacLean, James R.; and Jorden, Janet M. *Geometry in the Classroom— New Concepts and Methods*. Toronto: Holt, Rinehart and Winston of Canada, 1968.

    Many ideas for teachers on geometric models and geometric recreations.

M Emmet, E. R. *Puzzles for Pleasure*. New York: Emerson Books, 1972.
    Problems and puzzles that could interest many students.

H Eves, Howard. *An Introduction to the History of Mathematics*, rev. ed. New York: Holt, Rinehart and Winston, 1964.
    History of mathematics for reference or self-study.

R Eves, H. W. *Mathematical Circles Revisited: a Second Collection of Mathematical Stories and Anecdotes*. Boston: Prindle, Weber & Schmidt, 1971.
    Contains 300 anecdotes about mathematics and mathematicians.

R Eves, H. W. *The Other Side of the Equation*. Boston: Prindle, Weber & Schmidt, 1972.
    A collection of facts and fables.

B Eves, Howard and Newsom, Carroll V. *An Introduction to the Foundations and Fundamental Concepts of Mathematics*. New York: Holt, Rinehart and Winston, 1965.
    Book with numerous exercises.

R Fadiman, Clifton, ed. *Fantasia Mathematica*. New York: Simon and Schuster, 1958.
    Collection of entertaining narratives on topics from Archimedes to the Moebius strip.

O Feldzaman, A. N. and Henle, Faye. *The Calculator Handbook*. New York: Berkley Publishing Corporation, 1973.
    How to use the calculator for various real-life applications.

C Feng, Chuan C. *Computer Related Mathematics and Science Curriculum Materials*. Boulder, Colorado: Boulder Valley School District, 1967.
    Collection of 22 teacher-written projects dealing with classroom application of computers. Uses BASIC.

R Feravolo, Rocco. *Wonders of Mathematics*. New York: Dodd, Mead, 1963.
    Development of mathematics with simple activities and problems to demonstrate what mathematics can do.

G Fouke, George R. *A First Book of Space Form Making*. San Francisco: GeoBooks, 1974.
    Handbook with 46 tasks.

R Freedman, Miriam K. and Perl, Teri. *A Sourcebook for Substitutes and Other Teachers*. Reading, Massachusetts: Addison-Wesley, 1974.
    Collection of ideas not only for substitute teachers but for all teachers.

O Freund, John E. *Modern Elementary Statistics*, 3rd ed. Englewood Cliffs, New Jersey: Prentice-Hall, 1967.
    Acquaints beginning students with fundamentals of modern statistics.

B Friend, J. Newton. *Numbers: Fun & Facts*. New York: Scribner, 1972.
    Study of numbers, their origin, peculiarities and traditions, legends and superstitions that have collected about those numbers.

M Fujii, John N. *Puzzles and Graphs*. Washington: National Council of Teachers of Mathematics, 1966.
    Collection of geometric puzzles.

M Fults, John Lee. *Magic Squares*. LaSalle, Illinois: Open Court Publishing Co., 1974.
    Comprehensive discussion of many kinds of magic squares and the methods of their construction.

M Gardner, Martin. *Mathematics, Magic, and Mystery*. New York: Dover, 1955.
    One hundred fifteen diversions, magical tricks arising from mathematical principles. Topological and geometric vanishing tricks, cards, dice.

O Gardner, Martin. *Logic Machines and Diagrams*. New York: McGraw-Hill, 1958.
　　　　Careful explanations of logic instances.

M Gardner, Martin. *Mathematical Puzzles and Diversions*. New York: Simon and Schuster, 1959.
　　　　Collection of columns from *Scientific American*. Includes paradoxes and paperfolding, Moebius variations and mnemonics, brain teasers, and other recreations with mathematical commentaries.

M Gardner, Martin. *Mathematical Puzzles of Sam Loyd*. New York: Dover, 1959.
　　　　Collection of puzzles.

G Gardner, Martin. *The Ambidextrous Universe*. New York: Basic Books, 1964.
　　　　Left-right symmetry and asymmetry in nature and science.

M Gardner, Martin. *The Unexpected Hanging and Other Mathematical Diversions*. New York: Simon and Schuster, 1969.
　　　　Fourth in a collection of puzzles from the author's column in *Scientific American*.

M Gardner, Martin. *Perplexing Puzzles and Tantalizing Teasers*. New York: Simon and Schuster, 1969.
　　　　Collection of both old and new puzzles with solutions.

C Gear, C. William. *Computer Organization and Programming*. New York: McGraw-Hill, 1969.
　　　　Second course in programming. Students presumed to have completed a basic course in a procedure-oriented language. Discusses machine language, design of computers, software systems, programming philosophy, compilers, and many other computer-related topics.

A Gelfand, I. M., general ed. *The Method of Coordinates* and *Functions and Graphs*. "Library of School Mathematics" series. Cambridge, Massachusetts: M. I. T. Press.
　　　　Two books which can be used as supplements.

M Gellis, Marilyn J. *Thinkisthenics: Exercises for the Brain*. Palm Springs, California: Thinkisthenics, 1971.
　　　　Contains games, designs, and puzzles.

M Gellis, Marilyn J. *Centuous Math*. Palm Springs, California: Thinkisthenics, 1973.
　　　　Handbook of mathematical ticklers, teasers, tricks, and twisters.

R Glaser, Anton. *Neater by the Meter*. Southampton, Pennsylvania: Anton Glaser, 1974.
　　　　Guide to the metric system.

B Gold, Marvin and Carlberg, Robert E. *Modern Applied Mathematics*. Boston: Houghton Mifflin, 1971.
　　　　Emphasis on applications, with illustrations and activities.

M Golomb, Solomon. *Polyominoes*. New York: Scribner, 1965.
　　　　Mathematical recreation and a way to learn more about mathematics. Includes mathematical proofs for many problems and discusses theorems about counting.

B Goodwin, A. Wilson and O'Neil, David R. *Lost in Space*. Des Moines: Central Iowa Low Achievers Mathematics Project, 1969.
　　　　Has goal of teaching rational number system operations via ordered pairs of whole numbers.

P   Graham, L. A. *Ingenious Mathematical Problems and Methods*. New York: Dover, 1959.
> One hundred problems, with solutions. Twenty-eight mathematical nursery rhymes.

M   Greenblatt, M. H. *Mathematical Entertainments*. New York: Crowell, 1965.
> Puzzle book.

C   Gruenberger, Fred and Jaffray, George. *Problems for Computer Solution*. New York: Wiley, 1965.
> Presents interesting problems with a wide range of difficulty. Problems deal with polynomials, percentages, interest, probability, number theory, Monte Carlo methods, and other topics.

R   Gruver, Howell L. *School Mathematics Contests: a Report*. Washington: National Council of Teachers of Mathematics, 1968.
> Includes samples of rules, procedures, and contest questions.

R   Hafner, Lawrence E. *Improving Reading in Middle and Secondary Schools: Selected Readings*, 2nd ed. New York: Macmillan, 1974.
> Two sections deal with reading in mathematics.

R   Hardgrove, Clarence Ethel and Miller, Herbert F. *Mathematics Library—Elementary and Junior High School*. Reston, Virginia: National Council of Teachers of Mathematics, 1973.
> Annotated bibliography of enrichment books, classified by level.

C   Hawkes, Nigel. *The Computer Revolution*. New York: Dutton, 1972.
> Historical material, how a computer works, and applications of computing in areas such as business, science, the arts, and artificial intelligence.

A   Heimer, Ralph T.; Kocher, Frank; and Lottes, John J. *A Program in Contemporary Algebra*. Boston: Holt, Rinehart and Winston, 1963.
> Sequence of five programmed booklets.

B   Henderson, George L. and Glunn, Lowell D. *Let's Play Games in Mathematics*. Skokie, Illinois: National Textbook Co., 1972.
> One hundred seventy-seven objective-associated games and puzzles.

M   Henderson, George L. and Glunn, Lowell D. *Let's Play Games in Metrics*. Skokie, Illinois: National Textbook Co., 1974.
> One hundred seventy-seven objective-associated games and activities with history of measurement and basic units of metric measure.

R   Henderson, Kenneth B. *Teaching Secondary School Mathematics*. Washington: National Education Association, 1969.
> Discusses pertinent research related to curriculum and instruction.

A   Herber, E. A. *Number Sets. Mathematical Systems. Linear Equations & Systems. Functions*. Cupertino, California: Hewlett-Packard, 1972, 1973.
> Supplementary lab books for students who have completed at least one year of algebra. Emphasis on program writing.

B   Herrick, Marian Cliffe; Zartman, Jane; and Conrow, Thomas R., Jr. *Modern Mathematics for Achievement, Courses I & II*. Boston: Houghton Mifflin, 1972.
> Two one-year general mathematics texts for low achievers. Eight paperback booklets for each course. Inductive approach with minimum of reading material.

G Higgins, A. *Geometry Problems*. Portland, Maine: J. Weston Walch, 1971.
Collection of geometry problems to supplement a regular course.

G Hilbert, David and Cohn-Vossen, Stephen. *Geometry and the Imagination*. Bronx, New York: Chelsea, 1952.
Discussions of some aspects of two- and three-dimensional geometry.

P Hill, Thomas, ed. *Mathematical Challenges II—Plus Six*. Reston, Virginia: National Council of Teachers of Mathematics, 1974.
Includes 100 problems selected from *The Mathematics Student Journal* since 1965, plus six articles.

H Hogben, Lancelot. *Mathematics for the Million*. New York: Norton, 1951.
An account of the development of mathematics.

O Hogben, Lancelot Thomas. *The Wonderful World of Mathematics*. Garden City, New York: Doubleday, 1955.
Mathematical concepts presented with graphs and pictures.

G Holden, A. *Shapes, Space, and Symmetry*. New York: Columbia University Press, 1971.
Illustrates numerous relations between solids.

B Holt, M. and Dienes, A. *Let's Play Math*. New York: Walker, 1973.
Games for teaching children at all levels to learn, think, and do mathematics.

O Holt, M. and Marjoram, D. T. E. *Mathematics in a Changing World*. New York: Walker, 1973.
Applications of mathematical models.

O Honsberger, Ross. *Mathematical Gems*. The Dolciani Mathematical Expositions, No. 1. Washington: The Mathematical Association of America, 1973.
Includes 13 different simple themes ranging from number theory, equilateral triangles, combinatorics, and recursion. Includes historical facts. Useful as enrichment material for the mathematically interested student.

B Hooten, J. R., Jr. and Mahaffey, Michael L. *Elementary Mathematics Laboratory Experiences*. Columbus, Ohio: Merrill, 1973.
Includes 11 laboratory experiences which might be used in middle school or junior high school classrooms.

R Hopkins, Robert A. *The International (SI) Metric System and How It Works*, 2nd ed. Tarzana, California: Polymetric Services, 1974.
Includes information on development of metric system, current status, and the system itself.

O Huff, Darrell and Geis, Irving. *How to Take a Chance*. New York: Norton, 1959.
Discussions of various aspects of chance, probability, and error applied to everyday life.

B Hunter, J. *Math Brain Teasers*. New York: Bantam.
Collection of math puzzles.

O Hunter, William L. *Getting the Most out of Your Electronic Calculator*. Blue Ridge Summit, Pennsylvania: Tab Books, 1974.
How to use the calculator for homework and for various real-life applications.

G Huntley, H. E. *The Divine Proportion*. New York: Dover, 1970.
Analysis of relationship between geometry and aesthetics. Poetry, patterns in philosophy, psychology, music, and certain mathematical figures used to show the divine proportion.

R Iadvurian, H. M. *How to Study—How to Solve*. Reading, Massachusetts: Addison-Wesley, 1957.
>   Covers arithmetic, algebra, trigonometry, and brief presentation of differentiation. Examples and illustrative problems with methods of solutions.

G Jayne, Caroline Furness. *String Figures and How to Make Them*. New York: Dover, 1962.
>   Describes how to make over 100 figures by maneuvering a loop of string. Illustrated. Some games require four hands.

C Johnson, David C. et al. *Computer Assisted Mathematics Program (CAMP)*. Glenview, Illinois: Scott, Foresman, 1968–1970.
>   Set of six supplementary student booklets for grades 7–12, containing many exercises. Students write computer programs in BASIC to study concepts and solve problems.

M Johnson, Donovan A. *Excursions in Outdoor Measurement*. Portland, Maine: J. Weston Walch, 1974.
>   Ideas to interest students in measurement.

O Johnson, Donovan A.; Glenn, W. H.; and Norton, M. Scott. "Exploring Mathematics on Your Own." Manchester, Missouri: McGraw-Hill, 1960, 1961, 1963.
>   Mathematics enrichment booklets. Many forms of modern mathematics and new applications of traditional mathematics:
>> *Adventures in Graphing*.
>> *Basic Concepts of Vectors*.
>> *Curves in Space*.
>> *Computing Devices*.
>> *Finite Math Systems*.
>> *Fun with Mathematics*.
>> *Geometric Constructions*.
>> *Invitation to Mathematics*.
>> *Logic and Reasoning*.
>> *Numeration System*.
>> *Probability and Chance*.
>> *Pythagorean Theorem*.
>> *Sets*.
>> *Shortcuts in Computing*.
>> *Topology*.
>> *World of Measurement*.
>> *World of Statistics*.

R Johnson, Donovan A. and Olander, C. E. *How to Use Your Bulletin Board*. Washington: National Council of Teachers of Mathematics, 1953.
>   Booklet on bulletin board suggestions.

R Johnson, Donovan A. and Rising, Gerald R. *Guidelines for Teaching Mathematics*, 2nd ed. Belmont, California: Wadsworth, 1972.
>   Contains a list of suggested references and materials.

O Judd, Wallace. *Games, Tricks, and Puzzles for a Hand Calculator*. Menlo Park, California: Symax, 1974.
>   Various recreational activities for the calculator.

G Kelly, George W. *Trisection of the 120 Degree Angle*. New York: Vantage Press, 1973.
>   Includes step-by-step proofs. Challenging reading.

H Kline, Morris. *Mathematical Thought from Ancient to Modern Times*. New York: Oxford University Press, 1972.
> The history of mathematics, organized around the central ideas of mathematical thought.

C Koetke, Walter. *Computers in the Classroom: Teacher's Resource Manual for Algebra*. Maynard, Massachusetts: Digital Equipment Corporation, 1971.
> Contains demonstrations programs, possible assignments for students (with solutions), and remedial drill programs.

M Kordemsky, B. A. *The Moscow Puzzles: 359 Mathematical Recreations*. New York: Scribner, 1972.
> Challenging problems and puzzles.

C Kovach, Ladis D. *Computer-Oriented Mathematics*. San Francisco: Holden-Day, 1964.
> "Mainstream" ideas in computer mathematics: iteration, interpolation, Monte Carlo methods included.

R Krulik, Stephen. *A Handbook of Aids for Teaching Junior-Senior High School Mathematics*. Philadelphia: Saunders, 1971.
> Includes teaching aids for pre-algebra, algebra, geometry, trigonometry, and additional topics such as the Moebius strip and probability curves aids.

M Krulik, Stephen. *A Mathematics Laboratory Handbook for Secondary Schools*. Philadelphia: Saunders, 1972.
> Activities for a mathematics laboratory. Includes sections on pre-algebra, algebra, geometry, topology, and probability and statistics.

R Krulik, Stephen and Kaufman, Irwin. *How to Use the Overhead Projector in Mathematics Education*. Washington: National Council of Teachers of Mathematics, 1966.
> Presents specific suggestions for making transparencies and presenting mathematics lessons using the overhead projector.

C LaFave, L. J.; Milbrandt, G. D.; and Garth, D. W. *Problem Solving: The Computer Approach*. New York: McGraw-Hill, 1972.
> Presents an arbitrary machine language and the FORTRAN language in a problem-solving context.

G Laycock, Mary. *Dual Discovery Through Straw Polyhedra*. Palo Alto: Creative Publications, 1970.
> How to make regular polyhedra out of straws; stellate the polyhedra; show relationships between polyhedra through stellations; get reference material on polyhedra.

R Lenchner, George. *The Overhead Projector in the Mathematics Classroom*. Reston, Virginia: National Council of Teachers of Mathematics, 1974.
> Emphasizes applications to teaching mathematics and describes techniques for making more extensive use of the aid.

G Levi, Howard. *Foundations of Geometry and Trigonometry*. Englewood Cliffs, New Jersey: Prentice-Hall, 1960.
> Beginning geometry without appeal to the concept of distance. Introduces assumptions as they are needed.

M Levine, Sidney R. *Verses to Reckon with . . . the Fun-amentals of Mathematics*. Skokie, Illinois: Magna Publications, 1974.
> Collection of 101 verses, each of which contains a problem to solve.

O Lichtenberg, Betty K. and Troutman, Andria P., eds. *Fostering Creativity Through Mathematics*. Tampa: Florida Council of Teachers of Mathematics, 1974.
>   Contains 26 activities, most appropriate for grades 6–9. Described in detail; some include sample worksheets or other illustrations.

O Linderholm, C. E. *Mathematics Made Difficult: A Handbook for the Perplexed*. New York: World Publishing, 1972.
>   Answers to questions that might be asked.

G Lindgren, H. *Recreational Problems in Geometric Dissections and How to Solve Them*, 2nd ed. New York: Dover, 1972.
>   Challenging problems involving geometric dissections.

R Lockard, J. David, ed. *Science and Mathematics Curricular Developments Internationally, 1956–1974: The Ninth Report of the International Clearinghouse on Science and Mathematics Curricular Developments*. College Park, Maryland: Science Teaching Center, University of Maryland, 1975.
>   Biennial publication on curriculum development projects throughout the world. Indicates purpose, status, materials developed, and other pertinent information.

G Lockwood, E. H. *A Book of Curves*. New York: Cambridge University Press, 1967.
>   Special curves and ways of finding new curves. Includes chapters on cardioids, the limacon, the deltoid, the lemniscate of Bernoulli, conchoids, strophoids, spirals, glissettes, and many other curves.

G Loomis, Elisha Scott. *The Pythagorean Proposition*. Washington: National Council of Teachers of Mathematics, 1968.
>   Historical review, presenting 370 demonstrations of the Pythagorean Theorem.

G Mandell, Alan. *The Language of Science*. Washington: National Science Teachers Association, 1974.
>   Correlates science and mathematics.

R Marks, Robert W. *The New Mathematics Dictionary and Handbook*. New York: Bantam.
>   Answers to questions about mathematics presented in nontechnical language.

B Menninger, Karl. *Number Words and Number Symbols—a Cultural History of Numbers*. Translated by Paul Broneer from the revised German edition. Cambridge, Massachusetts: M. I. T. Press, 1969.
>   Our number system—its characteristics, peculiarities, and history. Comparisons of other cultures' number systems and numerals.

O Merrill, Arthur A. *How Do You Use a Slide Rule?* New York: Dover, 1961.
>   Step-by-step explanation of slide rule.

B Meyer, Jerome S. *Arithmetricks*. New York: Scholastic Book Services, 1972.
>   Stunts and tricks with numbers.

B Meyer, Jerome S. *Getting a Line on Mathematics*. New Rochelle, New York: Cuisenaire, 1965.
>   Supplement or workbook. Accompanied by 13 charts, instructions, and examples.

P Moses, Stanley. *The Art of Problem-Solving*. London: Transworld Publishers, 1974.
>   Presents strategies on how to solve problems.

O Mosteller, F. et al. *Statistics by Example*. Reading, Massachusetts: Addison-Wesley, 1973.
>   Series offering real-life examples of a variety of statistical situations: "Exploring data," "Detecting Patterns," "Finding Models," "Weighing Chances."

O Mullish, Henry. *How to Get the Most out of Your Pocket Calculator*. New York: Collier Books, 1973.
>   How to use the calculator for various real-life applications.

G Munari, Bruno. *Discovery of the Square*. New York: George Wittenborn, 1970.
>   History of the square, with illustrations.

G Neely, Henry M. *Triangles*. New York: Crowell, 1962.
>   Uses of triangles and why triangles are important outside mathematics.

C Newey, A. J. *One Hundred Computer Programming Problems*. New York: Pitman Publishing, 1973.
>   Compendium of traditional computer-programming problems, most appropriate at the Algebra II level.

R Newman, James R. *The World of Mathematics* (4 volumes). New York: Simon and Schuster, 1956.
>   A collection of selections from mathematical literature.

B Nibbelink, William. *How It Might Have Been*. Des Moines: Central Iowa Low Achiever Mathematics Project, 1969.
>   Narrative with summary and workbook about a cave people who developed a base eight counting system.

G Nibbelink, William and Van Buren, Nelda. *Area Measurement*. Des Moines: Central Iowa Low Achiever Mathematics Project, 1969.
>   Treatment of the problems involved in measuring the area of such figures as the triangle, square, rectangle, parallelogram, and circle.

C Nievergelt, Jurg; Farrar, J. Craig; and Reingold, Edward M. *Compute Approaches to Mathematical Problems*. Englewood Cliffs, New Jersey: Prentice-Hall, 1974.
>   Presents wide variety of problems that require both mathematical theory and computer methods for their solution.

G Niman, John and Postman, Robert. *Mathematics on the Geoboard*. New Rochelle, New York: Cuisenaire, 1974.
>   Demonstrates uses of the geoboard.

G O'Daffer, Phares G. and Clemens, Stanley R. *Geometry—An Investigative Approach*. Reading, Massachusetts: Addison-Wesley, 1976.
>   Guides the learning of geometry through explorations and investigations.

P Ogilvy, C. S. *Tomorrow's Math: Unsolved Problems for the Amateur*, 2nd ed. New York: Oxford University Press, 1972.
>   Challenging problems, particularly appropriate for advanced students.

M Olson, Alton T. *Mathematics Through Paper Folding*. Reston, Virginia: National Council of Teachers of Mathematics, 1975.
>   Revision of Donovan Johnson's *Paper Folding for the Mathematics Class* (1967). Directions for folding basic and more advanced constructions; includes theorems that can be demonstrated by paper folding.

C O'Neil, David R. *An Introduction to Flow Charting*. Des Moines: Central Iowa Low Achievers Mathematics Project, 1969.
>   Guide for teachers.

R Osen, Lynn M. *Women in Mathematics*. Cambridge, Massachusetts: M. I. T. Press, 1974.
   Biographies of eight women mathematicians.

M Peck, Lyman C. *Secret Codes, Remainder Arithmetic, and Matrices*. Washington: National Council of Teachers of Mathematics, 1961.
   Presents codes to introduce mathematical ideas.

G Pedoe, Daniel. *Circles*. Long Island City, New York: Pergamon, 1957.
   Exposition of some topics from pure mathematics which are related to the circle.

M Phillips, Hubert (Caliban). *My Best Puzzles in Logic and Reasoning*. New York: Dover, 1961.
   One hundred puzzles, with solutions, for both beginners and experts.

M Phillips, Hubert (Caliban). *Caliban's Problem Book—Mathematical, Inferential and Cryptographic Puzzles*. New York: Dover, 1961.
   One hundred five problems, with solutions. Mostly problems of intermediate or advanced level.

R Polya, George. *Mathematics and Plausible Reasoning*. Princeton, New Jersey: Princeton University Press, 1954.
   Guide to the practical art of plausible reasoning. First volume: induction and analogy in mathematics. Second volume: patterns of plausible inference.

P Polya, George. *How to Solve It*. Garden City, New York: Doubleday, 1957.
   Effort to teach the knack of solving mathematical problems.

P Polya, George and Kilpatrick, Jeremy. *The Stanford Mathematics Problem Book with Hints and Solutions*. New York: Teachers College Press, 1974.
   Contains all of the problems used from 1946 to 1965 on the Comprehensive Examination for High School Seniors, a yearly test given by Stanford University. None of the problems is trivial; none requires high-powered mathematics.

G Posamentier, Alfred S. and Salkind, Charles T. *Challenging Problems in Geometry*. New York: Macmillan, 1970.
   Numerous geometry problems. Hints given for solutions.

G Posamentier, A. S. and Wernick, W. *Geometric Constructions*. Portland, Maine: J. Weston Walch, 1973.
   Contains illustrations of a variety of constructions.

C Post, Dudley L., ed. *The Use of Computers in Secondary School Mathematics*. Newburyport, Massachusetts: Entelek, 1970.
   Types of installations, languages, student uses, costs, and results of experimental computer curriculums in some schools discussed.

R Raab, Joseph A. *Audiovisual Materials in Mathematics*, rev. ed. Reston, Virginia: National Council of Mathematics, 1976.
   Films, filmstrips, loops, transparencies, and videotapes listed according to subject matter and mode of presentation.

P Rademacher, Hans and Toeplitz, Otto. *The Enjoyment of Mathematics—Selections from Mathematics for the Amateur*. Princeton, New Jersey: Princeton University Press, 1966.

Requires only background in plane geometry and elementary algebra. Sampler of mathematical problems.

B Razzell, Arthur G. and Watts, K. G. *This Is 4*. Garden City, New York: Doubleday, 1967.
    Elementary explanation of the number four.

G Read, Ronald C. *Tangrams—330 Puzzles*. New York: Dover, 1965.
    Puzzles with solutions for use with both 7-piece and 15-piece tangrams.

B Reichmann, W. J. *The Fascination of Numbers*. New York: Oxford University Press, 1957.
    Discussion of methods and results of elementary number theory.

O Reichmann, W. J. *Use and Abuse of Statistics*. Baltimore: Penguin Books, 1971.
    Offers many statistical examples and counter examples.

M Rice, Trevor. *Mathematical Games and Puzzles*. New York: St. Martin's Press, 1973.
    Over 40 games and puzzles, on dissections, puzzle models, binary notation, mathematical magic, and mathematics games. Indicates that all such games and puzzles have mathematical explanations.

B Ringenberg, Laurence A. *A Portrait of 2, a Brief Survey of the Elementary Number Systems*. Washington: National Council of Teachers of Mathematics, 1964.
    Presents concepts of the number 2 in modern number theory.

B Roberts, Edward M. *Fingertip Math*. Dallas: Texas Instruments, 1974.
    Explains how to use a hand-held calculator effectively.

R Ross, Frank, Jr. *The Metric System—Measures for All Mankind*. New York: S. G. Phillips, 1974.
    Comprehensive history of measurement from primitive man to the present.

G Runion, Garth E. *The Golden Section and Related Curiosa*. Glenview, Illinois: Scott, Foresman, 1972.
    Eleven sections with exercises and answers. Includes definition and constructions of the Golden Section with relationships of the Golden Section to pentagons, Fibonacci numbers, and other topics. Applications and examples in art and nature.

A Russell, Donald. *Algebra Problems*. New York: Barnes and Noble, 1968.
    Self-study review of algebra.

C Sage, Edwin R. *Problem-Solving with the Computer*. Newburyport, Massachusetts: Entelek, 1969.
    Examples and sample programs using BASIC. Exercises in algebra, geometry, advanced algebra, and analysis. Stresses use of flow diagrams.

P Salkind, C. T. and Earl, J. M., compilers. *The MAA Problem Book III*. New York: Random House, 1973.
    Collection of challenging problems from MAA contests.

O Sawyer, W. W. *Mathematician's Delight*. Baltimore Penguin, 1943.
    Reading for those who have a dread of mathematics.

M Schaaf, William L. *A Bibliography of Recreational Mathematics*, Volumes 1, 2, and 3. Reston, Virginia: National Council of Teachers of Mathematics, 1970, 1971, 1973.
    Includes classroom games, recreational activities, and a glossary.

R Schaaf, William L. *The High School Mathematics Library*, 5th ed. Reston, Virginia: National Council of Teachers of Mathematics, 1973.

Contains more than 1000 entries.

C Serisky, Melvin. *Computer Math Experiences*. Hartsdale, New York: Olcott Forward, 1970.

Programs and exercises. Reference of various levels. Includes algebra, geometry, number theory, business mathematics, calculus, theory of equations, and recreational mathematics problems and exercises.

C Sessions, Peter L. and Doerr, Christine, eds. *Calculus*. Cupertino, California: Hewlett-Packard, 1973.

Adapted from computer curriculum materials developed under Project SOLO in Pittsburgh. Primarily a problem book, but does contain explanatory material and BASIC executions.

G Seymour, Dale G. and Chandler, Reuben A. *Creative Constructions*. Palo Alto: Creative Publications, 1968.

Basic constructions with illustrations of many possible constructions.

M Seymour, Dale et al. *Aftermath*. Palo Alto: Creative Publications, 1971.

Set of four books with numerous puzzles, games, and activities.

G Sharpton, Robert E. *Designing Pictures with String*. New York: Emerson Books, 1974.

Presents the basic principles relating straight lines to the generation of curves. Includes helpful drawings and photographs with explanations for designing pictures.

P Shklarsky, D. O.; Chentzov, N. N.; and Yaglom, I. M. *The USSR Olympiad Problem Book*. San Francisco: Freeman, 1972.

Three hundred twenty unconventional problems in algebra, arithmetic, elementary number theory, and trigonometry, with solutions.

B Shoemaker, Richard W. *Perfect Numbers*. Reston, Virginia: National Council of Teachers of Mathematics, 1973.

Acquaints students with the history of perfect numbers and points out ties with secondary school mathematics.

C Sippl, Charles J. and Sippl, Charles P. *Computer Dictionary*, 2nd ed. Indianapolis: Howard Sams & Co., 1974.

Comprehensive reference designed to aid in identifying, classifying, and interpreting terms and concepts of electronic data processing, information technology, and computer science.

R Smart, James R. *Metric Math: The Modernized Metric System*. Monterey, California: Brooks/Cole, 1974.

Resource book which includes description of the International System with suggestions for laboratory activities.

B Smith, Harold and Keiffer, Mildred. *Pathways in Mathematics*. Boulder, Colorado: Pawnee Publishing Co., 1973.

For use with low-achieving ninth graders; variety of topics designed to motivate and stimulate interest.

B Sobel, Max. *Teaching General Mathematics*. Englewood Cliffs, New Jersey: Prentice-Hall, 1967.

Sourcebook of ideas for teaching slow learners.

M Sobel, Max A. and Maletsky, Evan M. *Teaching Mathematics: a Sourcebook of Aids, Activities, and Strategies*. Englewood Cliffs, New Jersey: Prentice-Hall, 1975.

Presents a wide variety of specific suggestions and materials.

C Spencer, Donald D. *Computer Dictionary*. Ormond Beach, Florida: Abacus Computer Corporation, 1973.
>   Contains approximately 1200 entries. Includes many acronyms.

C Spencer, Donald D. A *Guide to Teaching About Computers in Secondary Schools*. Ormond Beach, Florida: Abacus Computer Corporation, 1973.
>   Guide to planning, content, methodology. Annotated listings of available resources and materials.

C Spencer, Donald D. *Computers in Action*. Rochelle Park, New Jersey: Hayden, 1974.
>   Describes development, structure, and operation of the computer. Discusses what a computer program is, computer languages, and BASIC.

C Spencer, Donald D. *Computers in Society: The Wheres, Whys, and Hows of Computer Use*. Rochelle Park, New Jersey: Hayden, 1974.
>   Explains the impact of computers on the individual and society and discusses future roles of computers.

M Steinhaus, H. *Mathematical Snapshots*, 3rd ed. New York: Oxford University Press, 1969.
>   Mathematics as it relates to real world. Problems and tricks.

G Stepelman, Jay. *Milestones in Geometry*. New York: Macmillan, 1970.
>   Constructions, history and approximations of pi, various proofs of the Pythagorean theorem, Euler's formula, and many other geometry-related topics.

O Summers, G. J. *Test Your Logic*. New York: Dover, 1972.
>   Problems to challenge students.

R Swain, Henry. *How to Study Mathematics: a Handbook for High School Students*. Washington: National Council of Teachers of Mathematics, 1970.
>   Suggestions about doing homework, making the most of class periods, and taking tests.

O Tanur, J. M.; Mosteller, F. et al., eds. *Statistics: a Guide to the Unknown*. San Francisco: Holden-Day, 1972.
>   Contains 44 essays which explain how statistics and probability were used to solve real-life problems in a variety of fields.

B Thomason, Mary E. *Modern Math Games, Activities and Puzzles*. Belmont, California: Lear Siegler, 1970.
>   Games, activities, and puzzles. Designed to strengthen mathematical skills and broaden concepts.

P Tietz, Heinrich. *Famous Problems of Mathematics*. Baltimore, Maryland: Graylock Press, 1965.
>   Translated from an earlier German edition. Lists many problems both solved and unsolved.

B Travers, Kenneth J.; Runion, Garth; and LeDuc, John. *Teaching Resources for Low Achieving Mathematics Class*. Columbus, Ohio: ERIC/SMEAC, 1971. ERIC Document No. ED 053 980.
>   Bibliography of materials for use with low achievers.

M Trigg, Charles. *Mathematical Quickies*. Manchester, Missouri: McGraw-Hill, 1967.
>   Recreational mathematics book.

R Turner, Nura D., ed. *Mathematics and My Career*. Washington: National Council of Teachers of Mathematics, 1971.

Discussions of the usefulness of mathematics in their careers by seven former mathematics contest winners.

G Valens, Evans G. *The Number of Things: Pythagoras, Geometry, and Humming Strings*. New York: Dutton, 1964.

Story of the Greek philosopher Pythagoras of Samos and related items.

G Wenninger, Magnus J. *Polyhedron Models for the Classroom*. Washington: National Council of Teachers of Mathematics, 1966.

Directions for constructing various models, with notes on the history and mathematics of polyhedra.

G Wenninger, M. J. *Polyhedron Models*. New York: Cambridge University Press, 1971.

Definitive set of instructions for making models of the 75 known uniform polyhedra plus some stellated forms.

P Wickelgren, Wayne A. *How to Solve Problems: Elements of a Theory of Problems and Problem Solving*. San Francisco: Freeman, 1974.

Presents problems and leads reader toward solutions.

G Winter, John. *String Sculpture*. Palo Alto: Creative Publications, 1972.

Starts with two-dimensional designs. Progresses to more complex designs.

R Wittgenstein, Ludwig. *Remarks on the Foundations of Mathematics*. Translated by G. E. M. Anscombe. Cambridge, Massachusetts: M.I.T. Press, 1967.

Remarks on the philosophy of mathematics and logic.

G Yates, R. C. *The Trisection Problem*. Washington: National Council of Teachers of Mathematics, 1971.

An examination of various attempts, successful and otherwise, to construct an angle trisector.

G Yates, Robert C. *Curves and Their Properties*. Reston, Virginia: National Council of Teachers of Mathematics, 1974.

Teaching supplement on a variety of unusual plane curves.

R Youse, Bevan K. *Mathematics: a World of Ideas*. Boston: Allyn & Bacon, 1974.

Portrait of a mathematician and his biography introduces each chapter. A variety of topics is considered.

R Zelenik, Mary Ella. *An Annotated Bibliography of Math Materials*. Los Angeles: Instructional Materials Center for Special Education, University of Southern California. ERIC Document No. ED 085 950.

Includes information on approximately 500 mathematics materials for students from preschool through high school. Emphasizes manipulative aids, kits, and prepackaged programs.

B Zimmerman, Joseph. *E.S.P. Booklet*. Des Moines: Central Iowa Low Achiever Mathematics Project, 1969.

Enrichment Student Projects booklet. Description and instructions for using such things as a Tower of Hanoi, ten-men-in-a-boat puzzle, and numerous other puzzles.

PUBLICATIONS

C *BASIC Application Programs*. Maynard, Massachusetts: Digital Equipment Corporation, 1971.

Programs designed to demonstrate how the computer can be applied to the problems of many disciplines. Simple.

C *BASIC Simulation Programs: Volumes III, Mathematics.* Maynard, Massachusetts: Digital Equipment Corporation, 1971.

Twenty programs developed as part of the Huntington Computer Project.

R *Careers in Statistics.* Committee of Presidents of Statistical Societies. Washington: American Statistical Association, 1974.

Answers questions about the work of a statistician.

B Central Iowa Low Achievers Mathematics Project. Des Moines: CILAMP, 1969.

Several of the booklets from this project have already been included in this listing. Other booklets include:

First Probability Program. Most basic concepts of probability covered. May be used as overview or review.

Gimmick. Booklet of drill exercises, puzzles and games, and new concepts.

Math in Sports. Collection of worksheets. Seven sports used to help motivate.

Measurement. Make-believe system of measurement developed, to teach conversion within and between measurement systems.

B *Computers and Computations.* San Francisco: Freeman, 1971.

Readings from *Scientific American.* Topics range from games and music to computers and uses of the computer.

C *Computer Mathematics in Secondary Schools.* White Plains, New York: IBM.

Course administration and curriculum guide. Explains flow diagrams, FORTRAN, and has section with sample problems.

O "Contemporary School Mathematics Series." Boston: Houghton Mifflin.

Paperbacks which include many examples, diagrams, exercises, and solutions.

Computers—Books 1 and *2.*
An Introduction to Probability and Statistics.
Matrices—Books 1 and *2.*
Sets and Logic—Books 1 and *2.*
Shape, Size, and Place.

B *Experiences in Mathematical Discovery.* Washington: National Council of Teachers Mathematics, 1966, 1967, 1970, 1971.

A general mathematics series using the discovery approach for the non-college-bound.

(1) *Formulas, Graphs, and Patterns.*
(2) *Properties of Operations with Numbers.*
(3) *Mathematical Sentences.*
(4) *Geometry.*
(5) *Arrangements for Selections.*
Answers for Units 1–5.
(6) *Mathematical Thinking.*
(7) *Rational Numbers.*
(8) *Positive and Negative Numbers.*

R *Goals for School Mathematics.* Education Development Center, Inc. Boston: Houghton Mifflin, 1963.

Report of the Cambridge Conference on School Mathematics. Gives views on the nature of a good mathematics curriculum.

C *How the Computer Gets the Answer*. Life Education Reprint. New York: Time, Inc., 1967.

Reprint of an article giving background on computers.

R *Index to Instructional Media Catalogs: A Multi-Indexed Directory of Materials and Equipment for Use in Instructional Programs*. New York: R. R. Bowker Co., 1974.

Three-part index, a guide to the suppliers of instructional materials, organized by subject, vendors, and companies.

C *Information*. Scientific American editors. San Francisco: Freeman, 1966.

Collection of articles about computers.

P *Mathematics Contest Problems*. Saint Mary's College. Palo Alto: Creative Publications, 1972.

Collection of elementary and advanced problems with solutions. Junior and senior high school level.

O "Mathematics Enrichment Series." Boston: Houghton Mifflin.

Booklets. Exercises with answers, bibliographies, illustrations, and suggestions for projects. Designed to add depth, scope, and flexibility.

The Conics—a Geometric Approach.

Fibonacci and Lucas Numbers.

Four by Four.

Four-Dimensional Geometry.

Graphs, Groups, and Games.

Induction in Mathematics.

Introduction to the Theory of Numbers.

Legislative Apportionment.

Mosaics.

Sequences.

Stereograms.

Topics from Inversive Geometry.

O "Modern Math Booklets." Greenfield, Massachusetts: Channing L. Bete Co.

About Mathematical Sets.

Scriptographic guide to mathematics sets.

About the "New Math."

Covers the major subject areas of the "new mathematics," explaining the importance of each. Defines and illustrates terminology involved.

C *More About Computers*. Armonk, New York: IBM, 1969.

Discusses how computers work. Includes glossary of computer terms.

O "New Mathematics Library." New York: Random House.

Series of short expository monographs on various mathematical subjects.

The Contest Problem Book I.

The Contest Problem Book II.

Continued Fractions.

Elementary Cryptanalysis—a Mathematical Approach.

Episodes from the Early History of Mathematics.

First Concepts of Topology.

Geometric Inequalities.
Geometric Transformations I.
Geometric Transformations II.
Geometry Revisited.
Graphs and Their Uses.
Groups and Their Graphs.
Hungarian Problem Book I.
Hungarian Problem Book II.
Ingenuity in Mathematics.
Introduction to Inequalities.
Invitation to Number Theory.
The Lore of Large Numbers.
Mathematics of Choice.
Numbers: Rational and Irrational.
From Pythagoras to Einstein.
Uses of Infinity.
What is Calculus About?

R *Pathways to Probability: History of the Mathematics of Certainty and Chance.* New York: Holt, Rinehart and Winston.

Survey of contributions of great mathematicians of past and present to our understanding of such subjects as odds, combinations, and permutations and annuities.

B "Programmed Units in Modern Mathematics" Series. Boston: Houghton Mifflin.

Programmed instruction booklets which may be used for enrichment.

Arithmetic of Directed Numbers.
Equations and Inequalities.
Introduction to Coordinate Geometry.
Introduction to Exponents.
Introduction to Sets.

O "School Mathematics Study Group Reprint Series." Pasadena, California: A. C. Vroman.

Pamphlets containing articles on special topics. Articles taken from periodic publications.

Computation of Pi.
Finite Geometry.
Geometric Constructions.
Geometry, Measurement and Experience.
The Golden Measure.
Infinity.
Mascheroni Constructions.
Mathematics and Music.
Memorable Personalities in Mathematics: Nineteenth Century.
Memorable Personalities in Mathematics: Twentieth Century.
Nature and History of Pi.
Prime Numbers and Perfect Numbers.
Space, Institution and Geometry.
The Structure of Algebra.
What is Contemporary Mathematics?

O "School Mathematics Study Group Supplementary and Enrichment Series." Pasadena, California: A. C. Vroman.

Pamphlets designed to help promote enjoyment of the study of mathematics. Generally cover topics usually taught in a regular mathematics course, with different presentation. Teacher's commentaries available for many of the pamphlets.

Absolute Value.
Algebraic Structures.
Circular Functions.
The Complex Number System.
Factors and Primes.
Functions.
Inequalities.
Mathematical Systems.
Mathematical Theory of the Struggle for Life.
The Mathematics of Trees and Other Graphs.
Numeration.
Non-Metric Geometry.
1 + 1 = ?
Order and the Real Numbers: a Guided Tour.
Plane Coordinate Geometry.
Radioactive Decay.
The System of Vectors.
Systems of First Degree Equations in Three Variables.

O "Thinking with Mathematics Series." Lexington, Massachusetts: Heath.

Paperback series. Supplemental program for more able secondary mathematics students.

The Complex Numbers.
The Concept of a Function.
Congruence and Motion in Geometry.
Finite Mathematical Structures.
Graphing Relations and Functions.
The Integers.
An Introduction to Linear Programming.
An Introduction to Sets and the Structure of Algebra.
An Introduction to Transfinite Mathematics.
Mathematics Projects Handbook.
The Natural Numbers.
The Rational Numbers.
The Real Numbers.

O "Topics in Mathematics" (translations from the Russian). Lexington, Massachusetts: Heath.

Paperbound booklets. Range of topics beyond scope of high school mathematics. Enrichment materials to supplement high school mathematics courses.

Algorithms and Automatic Computation Machines.
Areas and Logarithms.
Computation of Areas of Oriented Figures.

Configuration Theorems.
Convex Figures and Polyhedra.
Eight Lectures in Mathematical Analysis.
Equivalent and Equidecomposable Figures.
The Fibonacci Numbers.
Geometric Constructions in the Plane.
Geometry of the Straightedge and Geometry of the Compass.
How to Construct Graphs and Simplest Maxima and Minima Problems.
Hyperbolic Functions.
Induction in Geometry.
Infinite Sets.
An Introduction to the Theory of Games.
Isoperimetry.
The Method of Mathematical Induction.
Mistakes in Geometric Proofs.
Multicolor Problems.
Probability and Information.
Problems in the Theory of Numbers.
Proof in Geometry.
Random Walks.
Summation of Infinitely Small Quantities.
What is Linear Programming?

R   Publications of the National Council of Teachers of Mathematics. Reston, Virginia. NCTM publishes three journals for teachers (the *Arithmetic Teacher,* the *Mathematics Teacher,* and the *Journal for Research in Mathematics Education*), plus *The Mathematics Student,* for secondary school use. In addition to various supplementary books (which have been included in this listing), the NCTM makes available up-to-date "information resources" on such topics as "Free Materials for the Teaching of Mathematics" and "Sources of Publications Available on Careers in Mathematics." A list of these information resources and single copies of each are available on request from the NCTM (1906 Association Drive, Reston, Virginia 22091), as is an up-to-date list of current publications. The NCTM also publishes yearbooks, a list of which follows. Several of these are especially appropriate for the elementary school level but contain some useful ideas for secondary school teachers.

Austin, Charles M. et al., eds. *A General Survey of Progress in the Last Twenty-Five Years,* 1st Yearbook, 1926.

Reeve, W. D. et al., eds. *Curriculum Problems in Teaching Mathematics,* 2nd Yearbook, 1928.

Clark, J. R. and Reeve, W. D., eds. *Selected Topics in the Teaching of Mathematics,* 3rd Yearbook, 1928.

Reeve, W. D., ed. *Significant Changes and Trends in the Teaching of Mathematics Throughout the World Since 1910,* 4th Yearbook, 1929.

Reeve, W. D. ed. *The Teaching of Geometry,* 5th Yearbook, 1930; reprint 1966.

Reeve, W. D., ed. *Mathematics in Modern Life*, 6th Yearbook, 1931.

Reeve, W. D., ed. *The Teaching of Algebra*, 7th Yearbook, 1932.

Reeve, W. D., ed. *The Teaching of Mathematics in the Secondary School*, 8th Yearbook, 1933.

Hamley, Herbert Russell. *Rational and Functional Thinking in Mathematics*, 9th Yearbook, 1934.

Reeve, W. D., ed. *The Teaching of Arithmetic*, 10th Yearbook, 1935.

Reeve, W. D., ed. *The Place of Mathematics in Modern Education*, 11th Yearbook, 1936.

Bakst, Aaron. *Approximate Computation*, 12th Yearbook, 1937.

Fawcett, Harold P. *The Nature of Proof*, 13th Yearbook, 1938.

Turner, Ivan Stewart. *The Training of Mathematics Teachers*, 14th Yearbook, 1939.

Reeve, W. D., ed. *The Place of Mathematics in Secondary Education*, 15th Yearbook, 1940.

Reeve, W. D., ed. *Arithmetic in General Education*, 16th Yearbook, 1941.

Olds, Edwin G. et al., compilers. *A Source of Mathematical Applications*, 17th Yearbook, 1942.

Reeve, W. D., ed. *Multi-Sensory Aids in the Teaching of Mathematics*, 18th Yearbook, 1945.

Kiely, Edmond R. *Surveying Instruments: Their History and Classroom Use*, 19th Yearbook, 1947.

Johnson, J. T. et al., compilers. *The Metric System of Weight and Measures*, 20th Yearbook, 1948.

Fehr, Howard F., ed. *The Learning of Mathematics: Its Theory and Practice*. 21st Yearbook, 1953.

Clark, John R., ed. *Emerging Practices in Mathematics Education*, 22nd Yearbook, 1954.

Wren, F. Lynwood, ed. *Insights into Modern Mathematics*, 23rd Yearbook, 1957.

Jones, Philip S., ed. *The Growth of Mathematical Ideas, Grades K–12*, 24th Yearbook, 1959.

Grossnickle, Foster E., ed. *Instruction in Arithmetic*, 25th Yearbook, 1960.

Johnson, Donovan A., ed. *Evaluation in Mathematics*, 26th Yearbook, 1961.

Hlavaty, Julius H., ed. *Enrichment Mathematics for the Grades*, 27th Yearbook, 1963.

Hlavaty, Julius H., ed. *Enrichment Mathematics for High School*, 28th Yearbook, 1963.

John, Lenore, ed. *Topics in Mathematics for Elementary School Teachers*, 29th Yearbook, 1964.

Beckenbach, Edwin F., ed. *More Topics in Mathematics for Elementary School Teachers*, 30th Yearbook, 1969.

Hallerberg, Arthur E., ed. *Historical Topics for the Mathematics Classroom*, 31st Tearbook, 1969.

Jones, Philip S., ed. A *History of Mathematics Education in the United States and Canada*, 32nd Yearbook, 1970.

Rosskopf, Myron F., ed. *The Teaching of Secondary School Mathematics*, 33rd Yearbook, 1971.

Berger, Emil J., ed. *Instructional Aids in Mathematics*, 34th Yearbook, 1973.

Lowry, William C., ed. *The Slow Learner in Mathematics*, 35th Yearbook, 1972.

Henderson, Kenneth B., ed. *Geometry in the Mathematics Curriculum*, 36th Yearbook, 1973.

Payne, Joseph N., ed. *Mathematics Learning in Early Childhood*, 37th Yearbook, 1975.

Nelson, L. Doyal, ed. *Measurement in School Mathematics*, Yearbook, 1976.

PUBLISHERS

Abacus Computer Corporation, Suite 222, 110 East Granada Avenue, Ormond Beach, Florida 32074

Academic Press, Inc., 111 Fifth Avenue, New York, New York 10003

Addison-Wesley Publishing Company, Inc., 508 South Street, Reading, Massachusetts 01867

Aero Publishers, Inc., 329 Aviation Road, Fallbrook, California 92028

Allyn & Bacon, Inc., 470 Atlantic Avenue, Boston, Massachusetts 02210

American Book Company, 450 West 33rd Street, New York, New York 10001; 300 Pike Street, Cincinnati, Ohio 45202

American Statistical Association, 806 Fifteenth Street, N. W., Washington, D. C. 20005

Association Press, 291 Broadway, New York, New York 10007

Bantam Books, Inc., 666 Fifth Avenue, New York, New York 10019

Barnes and Noble, Inc., 10 East 53rd Street, New York, New York 10022

Basic Books, Inc., Publishers, 10 East 53rd Street, New York, New York 10022

Berkley Publishing Corporation, 200 Madison Avenue, New York, New York 10016

Channing L. Bete Company, Inc., 45 Federal Street, Greenfield, Massachusetts 01301

Blaisdell Publishing Company, 275 Wyman Street, Waltham, Massachusetts 02154

Bobbs-Merrill Co., Inc., 4300 West 62nd Street, Indianapolis, Indiana 46268

R. R. Bowker Co., National Information Center for Educational Media, 1180 Avenue of the Americas, New York, New York 10036

Brooks/Cole Publishing Co., Belmont, California 94002

Cambridge University Press, 32 East 57th Street, New York, New York 10022

Chelsea Publishing Company, Inc., 159 East Tremont Avenue, Bronx, New York 10453

Collier Books, 866 Third Avenue, New York, New York 10022

Columbia University Press, 562 West 113th Street, New York, New York 10025

Creative Publications, P. O. Box 328, Palo Alto, California 94303

Thomas Y. Crowell Company, 666 Fifth Avenue, New York, New York 10019

Cuisenaire Company of America, Inc., 12 Church Street, New Rochelle, New York 10805

The John Day Company, Inc., 257 Park Avenue South, New York, New York 10010

Dickenson Publishing Co., Inc., Encino, California 91316

Digital Equipment Corporation, 146 Main Street, Maynard, Massachusetts 01754

Dodd, Mead & Company, Inc., 79 Madison Avenue, New York, New York 10016

Doubleday & Company, Inc., Garden City, New York 11530; 277 Park Avenue, New York, New York 10017

Dover Publications, 180 Varick Street, New York, New York 10014

E. P. Dutton and Company, Inc., 201 Park Avenue South, New York, New York 10003

Educators Publishing Service, Cambridge, Massachusetts 02139

Emerson Books, Inc., 251 West 19th Street, New York, New York 10011

Encyclopaedia Britannica, Inc., 425 North Michigan Avenue, Chicago, Illinois 60611

Entelek, Inc., 42 Pleasant Street, Newbury, Massachusetts 01950

Fearon Publishers, Education Division of Lear Siegler, Inc., 6 Davis Drive, Belmont, California 94002

W. H. Freeman and Company, Publishers, 660 Market Street, San Francisco, California 94104

Ginn and Company, 191 Spring Street, Lexington, Massachusetts 02173

Golden Press, 150 Parish Drive, Wayne, New Jersey 07470

Graylock Press, 428 East Preston Street, Baltimore, Maryland 21202

Grolier Educational Corporation, 845 Third Avenue, New York, New York 10022

Harcourt Brace Jovanovich, Inc., 757 Third Avenue, New York, New York 10017

Harper & Row, Publishers, Inc., 10 East 53rd Street, New York, New York 10022

Hawthorne Books, Inc., 70 Fifth Avenue, New York, New York 10011

Hayden Book Co., 116 West 14th Street, New York, New York 10011

D. C. Heath and Company, 125 Spring Street, Lexington, Massachusetts 02173

Herder and Herder, Inc., 232 Madison Avenue, New York, New York 10016

Hewlett-Packard Company, 10900 Wolfe Road, Cupertino, California 95014

Holden-Day, Inc., 500 Sansome Street, San Francisco, California 94111

Holt, Rinehart and Winston, Inc., 383 Madison Avenue, New York, New York 10017

Houghton Mifflin Company, 2 Park Street, Boston, Massachusetts 02107

International Business Machines Corporation (IBM), Armonk, New York 10504

Alfred A. Knopf, Inc., 201 East 50th Street, New York, New York 10022

Laidlaw Brothers, Thatcher and Madison Streets, River Forest, Illinois 60505

Learning Research Associates, 1501 Broadway, New York, New York 10036

J. B. Lippincott Company, East Washington Square, Philadelphia, Pennsylvania 19105

Litton Educational Publishing, Inc., 450 West 33rd Street, New York, New York 10010

Lyons and Carnahan, 407 East 25th Street, Chicago, Illinois 60616

The Macmillan Company, 866 Third Avenue, New York, New York 10022; Front and Brown Streets, Riverside, New Jersey 08075

McCormick-Mathers Publishing Company, Inc., 450 West 33rd Street, New York, New York 10001

McGraw-Hill Book Company, 330 West 42nd Street, New York, New York 10036

Charles E. Merrill Publishing Company, 1300 Alum Creek Drive, Columbus, Ohio 43216

National Council of Teachers of Mathematics, 1906 Association Drive, Reston, Virginia 22091

National Education Association, 1201 Sixteenth Street, N.W., Washington, D.C. 20036

National Textbook Company, 8259 Niles Center Road, Skokie, Illinois 60076

New American Library, Inc., 1301 Avenue of the Americas, New York, New York 10019

W. W. Norton and Company, Inc., 55 Fifth Avenue, New York, New York 10003
Open Court Publishing Company, 1307 Seventh Street, LaSalle, Illinois 61301
Oxford University Press, 200 Madison Avenue, New York, New York 10016
Penguin Books, 3300 Clipper Mill Road, Baltimore, Maryland 21211
Pergamon Press, Inc., Maxwell House, Fairview Park, Elmsford, New York 10523
Prentice-Hall, Inc., Englewood Cliffs, New Jersey 07632
Princeton University Press, Princeton, New Jersey 08541
Prindle, Weber & Schmidt, Inc., 20 Newbury Street, Boston, Massachusetts 02116
Rand McNally and Company, Box 7600, Chicago, Illinois 60680
Random House, Inc., 201 East 50th Street, New York, New York 10022
W. B. Saunders, West Washington Square, Philadelphia, Pennsylvania 19105
Scholastic Book Services, 50 West 44th Street, New York, New York 10036
Science Research Associates, Inc. (SRA), 259 East Erie Street, Chicago, Illinois 60611
Scott, Foresman and Company, 1900 East Lake Avenue, Glenview, Illinois 60025
Charles Scribner's Sons, 597 Fifth Avenue, New York, New York 10017
Silver-Burdett Company, 250 James Street, Morristown, New Jersey 07960
L. W. Singer Company, Inc., 249 West Erie Boulevard, Syracuse, New York 13202
St. Martin's Press, Inc., 175 Fifth Avenue, New York, New York 10017
Teachers Publishing Company, 23 Leroy Avenue, Darien, Connecticut 06820
Frederick Ungar Publishing Company, Inc., 250 Park Avenue South, New York, New York 10003
Van Nostrand Reinhold Company, 450 West 33rd Street, New York, New York 10010
Viking Press, Inc., 625 Madison Avenue, New York, New York 10022
A. C. Vroman, 2085 East Foothill Boulevard, Pasadena, California 91109
Wadsworth Publishing Company, Inc., Belmont, California 94002
J. Weston Walch, Box 1075, Portland, Maine 04104
Walker and Company, 720 Fifth Avenue, New York, New York 10019
Webster Division, McGraw-Hill Book Company, Manchester Road, Manchester, Missouri 63011
John Wiley and Sons, Inc., 605 Third Avenue, New York, New York 10016
The World Publishing Company, 250 West 57th Street, New York, New York 10019

PERIODICALS

PROFESSIONAL

These are some of the journals which provide information on mathematics content and on mathematics methods and activities.

American Mathematical Monthly, Mathematical Association of America, 1225 Connecticut Avenue, N. W., Washington, D. C. 20036

Concerned with mathematics at the college undergraduate and graduate levels. Most articles of a mathematical nature, with one section on mathematics education.

Arithmetic Teacher, National Council of Teachers of Mathematics, 1906 Association Drive, Reston, Virginia 22091

Articles of interest to teachers in grades K–8, with many activities suggested.

Creative Computing, Box 1036, Concord, Massachusetts 01742
> Suggests a variety of computer problems and activities.

Fibonacci Quarterly, Department of Mathematics, St. Mary's College, Maraga, California 94575
> Concerned with mathematics related to the Fibonacci numbers, including implications and applications.

Investigations in Mathematics Education, 244 Arps Hall, The Ohio State University, Columbus, Ohio 43210
> Contains abstracts and critiques of reports on mathematics education research.

Journal for Research in Mathematics Education, National Council of Teachers of Mathematics (address above)
> Concerned with research related to the teaching and learning of mathematics at all levels.

Mathematics Magazine, Mathematical Association of America (address above)
> Contains articles dealing with mathematical content, primarily at the undergraduate level. Includes many problems and solutions.

Mathematics Teacher, National Council of Teachers of Mathematics (address above)
> Articles of interest to teachers in grades 7–12 and junior college. Many activities suggested, mathematical topics discussed, and materials reviewed.

Mathematics Teaching, Association of Teachers of Mathematics, Market Street Chambers, Nelson, Lancashire BB9 7LN, England
> Concerned with the teaching of mathematics at both elementary and secondary levels. Contains many suggestions for classroom activities.

Mathematics in the School, Longman Group Ltd., 33 Montgomery Street, Edinburgh EH7 5JX, Scotland
> Provides a variety of articles for both elementary and secondary teachers, with many activities suggested.

The Pentagon, Kappa Mu Epsilon, Central Michigan University, Mount Pleasant, Michigan 48858
> Concerned primarily with mathematics rather than the teaching of mathematics.

School Science and Mathematics, Lewis House, P. O. Box 1614, Indiana University of Pennsylvania, Indiana, Pennsylvania 15701
> Publishes articles on both science and mathematics, with some on content and some on methodology.

Scientific American, 415 Madison Avenue, New York, New York 10017
> Contains articles related to scientific developments; section on mathematical games in each issue.

Scripta Mathematica, Yeshiva University, Amsterdam Avenue and 186th Street, New York, New York 10003
> Focuses on mathematical content.

FOR STUDENTS

The following journals or newsletters contain articles (usually brief) and a variety of games, problems, and other activities for mathematics students at the secondary school and college levels.

The Mathematical Log, Mu Alpha Theta, Department of Mathematics, 601 Elm, Room 423, University of Oklahoma, Norman, Oklahoma 73069

Mathematical Pie, Alpha House, The Lane, Rowington, Warwickshire, England

The Mathematics Student, National Council of Teachers of Mathematics (address above)

Mathematical Spectrum, Oxford University Press, Education Department, Walton Street, Oxford, England

O. U. Mathematics Letter, O. U. Mathematics Service Committee, University of Oklahoma, Norman, Oklahoma 73069

Pi Mu Epsilon Journal, 1000 Asp Avenue, Room 215, University of Oklahoma, Norman, Oklahoma 73069

Pythagoras, Fanfare Educational Publishing Co., Fanfare House, 174 Chingford Mount Road, London, England

TEACHING AIDS, GAMES, AND MODELS

This list includes some of the instructional aids useful for introducing a topic or for providing practice. A variety of interesting materials, selected to meet specific goals, helps to motivate students as well. The letters preceding each item indicate that you might find it most useful in teaching basic arithmetic (B), algebra (A), geometry (G), or other topics (O).

Item	Supplier
B Abacus	Ideal
O Acquire	3M
A Amusements in Mathematics	Dover
G Angle Mirror	SEE
B Attribute Games	SEE; McGraw-Hill
O Basis	Wff'N Proof
G Bee Line	SEE
O Chrominoes	Creative Publications
G Circlometer (circumference and diameter ratio demonstrator)	Science Materials Center
O Circular Slide Rule	Math-Master
B Classroom Construction Kit	Ideal
G Configurations	Wff'N Proof
G Conics	Lano
A Coordinate Grids (rectangular and polar)	Lano
B Cross-Number Puzzles (Whole Numbers and Fractions)	SRA
G Curve Stitching	Invicta Plastics
A Data Guides (charts)	Math-Master
O Digi-Comp	Childcraft
A Equations	Wff'N Proof
G Folio Tool Kit (charts, compass, protractor, T-square, slide rule, etc.)	Math-Master

A	FOO (Fundamental Order of Operations game)	Cuisenaire
B	Fraction Dominoes	SEE
G	Geoboard	Cuisenaire; Concept
G	Geo-O-Stix	Childcraft; Cuisenaire
G	Geometry Equal Volume Solids	Math-Master
G	Geo-Strips	Invicta Plastics
B	Heads Up	Creative Publications; E. S. Lowe
B	Here to There	Scott, Foresman
O	Hexapawn	IBM
O	Hexstat	Lano
O	Jumpin	3M
O	Kalah	Creative Publications
B	Krypto	Creative Publications
A	Laws and Symbols Charts	Teachers Publishing
B	Make One	Garrard Press
B	Mathfacts Games: Addition and Subtraction, Multiplication and Division	Scott, Foresman
B	Math Match	Creative Publications
G	Miniature Geometric Shapes	Edmund Scientific
G	Mira Math	Cuisenaire
G	Moby Linx Construction Kit	Kindrey
A	Modern Logic	Wff'N Proof
B	Money Matters	Creative Publications
O	Monopoly	Parker Brothers
B	Multifacto	Scott, Foresman
B	Multo	Scott, Foresman
B	Numble (Number Scrabble)	Creative Publications
B	Numo (Bingo)	Midwest Publications
A	On-Set	Wff'N Proof
B	Operations Bingo	Creative Publications
G	Optical Illusions in Geometry (set of 18 posters)	J. Weston Walch
B	Orbit the Earth	Scott, Foresman
G	Pattern Block Activity Cards	SEE
G	Pattern Block Stickers	Creative Publications
G	Plane Figures Set	Lano
G	Polyhedra	Scott, Foresman
O	Polyhedral Dice	Creative Publications
B	Prime Drag	Creative Publications
O	Probability and Statistics Lab Unit	Math-Masters
B	Producto	Scott, Foresman
A	Propaganda	Wff'N Proof
G	Psyche-Paths (puzzle)	Cuisenaire
B	Quations	Math Shop
A	Queries 'N Theories	Wff'N Proof

G	Qwik/Sane (topological puzzle)	WffN Proof
B	Ranko	Midwest Publications
A	Real Numbers	WffN Proof
G	Regular Solids	Math-Master
O	Respond! (mathematical baseball)	ME-Press
B	Rook	Parker Brothers
G	Sage Kit	LaPine Scientific
O	Slide Rule	Cuisenaire
G	Solid Shapes Lab	Science Materials Center
G	Soma Puzzles	Cuisenaire; Edmund Scientific; Creative Publications
G	Space Spider (curve stitching)	Childcraft
O	Stocks and Bonds	3M
G	Strip Puzzle	Concept
B	Sumup	3M
G	SuperStructures	Cuisenaire
G	Symmetry Cominoes	SEE
A	Tac-Tickle	WffN Proof
G	Tangrams	Creative Publications
A	Trigtracker (unit circle demonstrator)	Lano
A	Tri-Nim	WffN Proof
G	Vectors	Cuisenaire; SEE
G	Yuke, The Game of Euclid	Midwest Publications

SUPPLIERS

Are-Jay Game Company, Inc., 7509 Denison Avenue, Cleveland, Ohio 44102

Berger Scientific, 37 William Street, Boston, Massachusetts 02119

Central Scientific Company, 2600 Kostner Avenue, Chicago, Illinois 60623

Childcraft Equipment Company, Inc., 155 East 23rd Street, New York, New York 10010

Concept Company, Inc., P. O. Box 273, Belmont, Massachusetts 02178

Creative Playthings, Edinburgh Road, Cranbury, New Jersey 08512

Creative Publications, P. O. Box 328, Palo Alto, California 94303

Denoyer-Geppert Company, 5235 Ravenswood Avenue, Chicago, Illinois 60640

Early Stages, James Falt Company, Ltd., Cheadle, Cheshire, England

Edmund Scientific Company, Edscorp Building, Barrington, New Jersey 08007

Educational Aids Service, 3034 Thayer Street, Evanston, Illinois 60201

Gamco Industries, Inc., Box 310, Big Spring, Texas 79720

General Learning Corporation, 250 James Street, Morristown, New Jersey 07960

Geyer Instructional Aids Company, Inc., 1229 Maxine Drive, Fort Wayne, Indiana 46806

Hubbard Scientific Company, P. O. Box 195, Northbrook, Illinois 60062

Ideal School Supply Company, Oak Lawn, Illinois 60453

Imout Arithmetic Drill Games, 706 Williamson Building, Cleveland, Ohio 44114

The Instructo Corporation, Paoli, Pennsylvania 19301

Invicta Plastics, Educational Aids Division, Oadby, Leicester, England

Kindrey Manufacturing, P. O. Box 11606, Palo Alto, California 94306

Kohner Brothers, Inc., Tyrne Game Division, P. O. Box 294, East Patterson, New Jersey 07407

Lano Company, 4741 West Liberty Street, Ann Arbor, Michigan 48103

La Pine Scientific Company, 6001 South Knox Avenue, Chicago, Illinois 60629

E. S. Lowe Company, Inc., 27 West 25th Street, New York, New York 10010

Madison Project, Math Media Division, H & M Associates, Danbury, Connecticut 06810

Math-Master Labs, Division of Gamco Industries, Inc., Box 1911, Big Spring, Texas 79720

Math Media, Inc. P. O. Box 345, Danbury, Connecticut 06810

Math Shop, 5 Bridge Street, Watertown, Massachusetts 02172

Math Shortcuts, 4043 N. E. 28th Avenue, Portland, Oregon 97212

Math-Um-Matic, Inc., 3017 North Stiles, Oklahoma City, Oklahoma 73105

Midwest Publications Company, Inc., P. O. Box 129, Troy, Michigan 48084

Milton Bradley Company, 74 Park Street, Springfield, Massachusetts 01101

3M—Minnesota Mining and Manufacturing Company, 3M Center, St. Paul, Minnesota 55101

Modern Learning Aids, 16 Spear Street, San Francisco, California 94105

Nasco Science and Mathematics Materials, 901 Janesville Avenue, Fort Atkinson, Wisconsin 53538

Frederick Post Company, 333 Sibley Street, St. Paul, Minnesota 55101

Sargent-Welch Scientific Company, 7300 North Linder Avenue, Skokie, Illinois 60076

Science Related Materials, P. O. Box 1009, Evanston, Illinois 60204

Scott Scientific Inc., Box 2121, Fort Collins, Colorado 80521

SEE—Selective Educational Equipment, Inc., 3 Bridge Street, Newton, Massachusetts 02195

Sigma Scientific, Inc., P. O. Box 1302, Gainsville, Florida 32601

TUF—Avalon Hill Company, 4517 Harford Road, Baltimore, Maryland 21214

Wabash Instruments and Specialties, Box 194, Wabash, Indiana 46992

Wff'N Proof Publishers, P. O. Box 71, New Haven, Connecticut 06501

World Wide Games, Box 450, Delaware, Ohio 43015

Yoder Instruments, East Palestine, Ohio 44413

CALCULATORS AND COMPUTERS

Some of the manufacturers of various types of calculators and computers are listed. Many also provide some educational materials and services.

Adler Business Machines, 1600 Route 22, Union, New Jersey 07083

Canon U. S. Inc., Minicalculator Division, 10 Nevada Drive, Lake Success, New York 11040

Casio, Inc., Suite 14011, 1 World Trade Center, New York, New York 10049

Clary Corporation, 408 Juniper Street, San Gabriel, California 91776

Commodore Business Machines, Inc., 390 Reed Street, Santa Clara, California 95050

Computer Control Company, Old Connecticut Path, Framingham, Massachusetts 13017

Control Data Corporation, 8100 South 34th Avenue, Minneapolis, Minnesota 55420

Frieden Inc., 2350 Washington Avenue, San Leandro, California 94577

Hewlett Packard, 10900 Wolfe Road, Cupertino, California 95014

Honeywell, 2701 South 4th Avenue, Minneapolis, Minnesota 55408

IBM, Data Processing Division, 112 East Post Road, White Plains, New York 10601

Melcor Electronics Corporation, 1750 New Highway, Farmingdale, New York 11735

Monroe, The Calculator Company, 550 Central Avenue, Orange, New Jersey 07051
National Cash Register, Main and K Streets, Dayton, Ohio 45409
Olivetti Corporation of America, 500 Park Avenue, New York, New York 10022
Panasonic Calculators, 200 Park Avenue, New York, New York 10017
Sharp Electronics Corporation, 10 Keystone Place, Paramus, New Jersey 07652
Sperry Rand Corporation, UNIVAC Division, 1290 Avenue of the Americas, New York, New York 10019
Summit International Corporation, 180 West 2950 South, Salt Lake City, Utah 84115
Texas Instruments, Inc., P. O. Box 22013, Dallas, Texas 75221
Unicom Systems, Rockwell International Corporation, 500 Fifth Avenue, Suite 1010, New York, New York 10036
Unisonic Corporation, 16 West 25th Street, New York, New York 10010
Victor Comptometer Corporation, 3900 North Rockwell Street, Chicago, Illinois 60618
Wang Laboratories, 836 North Street, Tewksbury, Massachusetts 01876
Westinghouse Electric Corporation, Westinghouse Building, Gateway Center, Pittsburgh, Pennsylvania 15222

FILMS AND FILMSTRIPS

Films and filmstrips for use in mathematics classes should be selected to meet the goals for the class. Therefore, filmed materials should always be previewed by the teacher to ensure the appropriateness of their use. The letters preceding each item indicate that you might find it most useful in teaching basic arithmetic (B), algebra (A), geometry (G), about computers (C), or other topics (O).

Item	Distributor
A Algebra: Relations, Function, and Variation (bw/11 min)	Coronet
A Algebra: A Way of Thinking about Numbers (bw/14 min)	Coronet
G Angles and Their Measurement (bw/11 min)	Coronet
O Arrangements and Combinations (bw/30 min)	Associated Films
C The Computer Revolution (c/22 min)	Film Associates of California
C Computers (c/11 min)	BFA Educational Media
G Dance Squared (c/5 min)	International Film Bureau
G Donald in Mathmagic Land (c/26 min)	Walt Disney Productions
G Flatland (c/15 min)	McGraw-Hill
A Graphs of Linear Equations (c/9 min)	Film Associates of California
A The Language of Graphs (bw/15 min)	Coronet
G Look At It This Way	Holt, Rinehart & Winston
B Lost in the Mish-Mosh: Area Measure (c/14 min)	Xerox Films
G Mathematics of the Honeycomb (c/13 min)	Moody Institute of Science
B Mr. Simplex Saves the Aspidistra (c/33 min)	Modern Learning Aids

A Modern Algebra Film Series (bw)   Modern Learning Aids
    Addition and Subtraction of Rational Num-
      ers
    Algebra of Points and Lines
    Algebraic and Complex Fractions
    More Solutions of Linear Equations
    Multiplication of Rational Numbers
    Natural Numbers, Integers and Rational Num-
      bers
    Quadratic Equations
    Solving Equations of Fractional Form
    Solving Simultaneous Linear Equations
    Special Products and Factoring
A Modern Algebra Filmstrips   Society for Visual Education
    The Closure, Commutative and Associative
      Properties
    The Distributive Property
    Equivalent Open Sentences
    Identity and Inverse Properties
    Introduction to the Number Line
    The Language of Sets
    Negative Numbers
    Open Sentences
    Order Properties
    Standards and Measurements
    Measurement Systems and Theory
    Subtraction and Division
G Modern Geometry Filmstrips   Society for Visual Education
    Angles
    Arcs and Areas of Circles
    Areas of Polygonal Regions
    Circles and Spheres
    Congruence
    Coordinate Geometry
    Distance and Betweenness
    Parallelism
    Points, Lines and Planes
    Polygons
    Proof
    Volumes of Solids
G Newton's Equal Areas (c/8 min)   International Film Bureau
G Notes on a Triangle (c/5 min)   International Film Bureau
O Prime Time (c/9 min)   Xerox Films
O Probableman (c/13 min)   Xerox Films
B Sets, Crows, and Infinity (c/12 min)   Film Associates of California
O Expanding Math Skills with the   Encyclopaedia Britannica
    Minicalculator (c/16 min)

R Teaching High School Mathematics   Modern Learning Aids
      First (series of 50 films developed by UICSM
      for algebra teachers) (bw/17–47 min)
G Trio for Three Angles (c/7 min)   International Film Bureau
B Using Modern Mathematics Filmstrips (ap- Society for Visual Education
      proximately 25 filmstrips presenting basic
      arithmetic and geometry content)
G Volume of Cubes, Prisms, Cylinders (c, bw) Cenco
B The Weird Number (c/13 min)   Xerox Films

DISTRIBUTORS

Associated Films, Inc., 600 Madison Avenue, New York, New York 10022
Bailey Films, Inc., 6509 DeLongpre Avenue, Los Angeles, California 90028
BFA Educational Media, 221 Michigan Avenue, Santa Monica, California 90404
Cenco Educational Films, 2600 South Kostner Avenue, Chicago, Illinois 60623
Colonial Films, Inc., 752 Spring Street, N. W., Atlanta, Georgia 30308
Coronet Films, 65 East South Water Street, Chicago, Illinois 60601
Davidson Films, 3701 Buchanan Street, San Francisco, California 94123
Denoyer-Geppert Audio Visuals, 5235 Ravenswood Avenue, Chicago, Illinois 60640
Walt Disney Productions, Educational Film Division, 350 South Buena Vista Avenue,
      Burbank, California 91503
Educational Service Inc., 47 Galen Street, Watertown, Massachusetts 02100
Encyclopaedia Britannica Educational Corporation, 425 North Michigan Avenue,
      Chicago, Illinois 60611
Eye Gate House, 146–01 Archer Avenue, Jamaica, New York 11435
Film Associates of California, 11559 Santa Monica Boulevard, Los Angeles, California
      90025
Gateway Productions Inc., 1859 Powell Street, San Francisco, California 94111
General Electric Educational Films, 60 Washington Avenue, Schenectady, New York
      12305
General Motors Corporation, 3044 West Grand Boulevard, Detroit, Michigan 48202
International Film Bureau, 332 South Michigan Avenue, Chicago, Illinois 60604
Jam-Handy Organization, 2781 East Grand Boulevard, Detroit, Michigan 48211
McGraw-Hill Text Films, 330 West 42nd Street, New York, New York 10036
Modern Film Rentals, 2323 New Hyde Park Road, New Hyde Park, New York 11040
Modern Learning Aids, 16 Spear Street, San Francisco, California 94105
Moody Institute of Science, 1200 East Washington Boulevard, Whittier, California 90606
Moreland-Latchford Productions, Ltd., 299 Queen Street West, Toronto, Ontario,
      Canada
Oxford Films, 1136 North Las Palmas Avenue, Los Angeles, California 90038
Sigma Educational Films, Inc., Box 1235, Studio City, California 91604
Society for Visual Education, 1345 Diversey Parkway, Chicago, Illinois 60614
United Transparencies, Box 688, Binghamton, New York 13902
Xerox Films, Xerox Education Group, 1200 High Ridge Road, Stamford, Connecticut
      06905

Many films and filmstrips are available for rental from various university audio-visual service departments, such as those at the University of Illinois, The Ohio State University, and The Pennsylvania State University.

Index